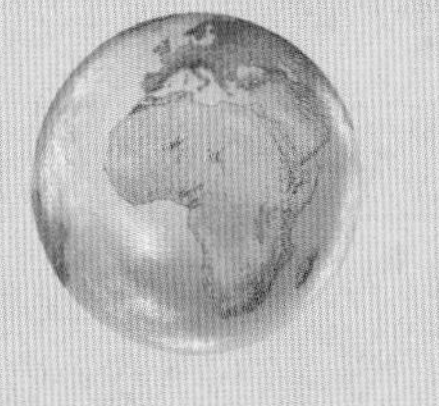

# OXFORD
## Student
# ATLAS

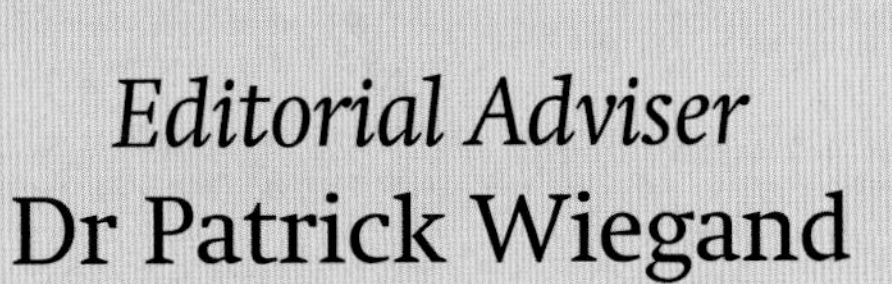

*Editorial Adviser*
Dr Patrick Wiegand

# OXFORD
## UNIVERSITY PRESS

Great Clarendon Street, Oxford OX2 6DP

Oxford University Press is a department of the University of Oxford.
It furthers the University's objective of excellence in research, scholarship,
and education by publishing worldwide in

Oxford   New York

Auckland   Cape Town   Dar es Salaam   Hong Kong   Karachi
Kuala Lumpur   Madrid   Melbourne   Mexico City   Nairobi
New Delhi   Shanghai   Taipei   Toronto

With offices in

Argentina   Austria   Brazil   Chile   Czech Republic   France   Greece
Guatemala   Hungary   Italy   Japan   Poland   Portugal   Singapore
South Korea   Switzerland   Thailand   Turkey   Ukraine   Vietnam

Oxford is a registered trade mark of Oxford University Press
in the UK and in certain other countries

ISBN 0 19 832164 3 (hardback)

ISBN 0 19 832163 5 (paperback)

3 5 7 9 10 8 6 4

Printed in Singapore

## Acknowledgements

The publishers would like to thank the following for permission to reproduce photographs:

Corbis UK Ltd, p.15; FAO-UN, p.76; NASA, p.76, 84, 97, 106, 113, 127; Oxford Scientific Films, p.114, 115; spaceimaging.com, p.87;
Science Photo Library, p.108, 113, 114; US Geological Survey, p.126; Visual Insights, p.131.

Cover image: Visible Earth / Rich Irish, Landsat 7 Team, NASA GSFC; data provided by EROS Data Center. Globes: Chapman Bounford.

The page design is by Adrian Smith.

The publishers are grateful to the following colleagues in geography education for their helpful comments and advice during
the development stages of this atlas: Pam Boardman, Graham Butt, Kathryn Clayton, Alan Cottle, Ruth Crossley, Rachel Dean,
Bob Digby, Ian Douglass, Tony Field, Martyn Gill, Joel Griffiths, Matthew Gunn, Gareth Huws, Kathryn Jones, Irfon Morris Jones,
David Langham, Patrick Lewis, Bob Newman, Andrew Parkinson, Liz Roodhouse, John Sadler, Toni Schiavone, Natasha Sirin,
Andrea Wade, Patrick Wherity, Steve Wilkes, and Malcolm Yates.

The publishers would also like to thank the many individuals, companies societies, and institutions
who gave assistance in the gathering of data.

# 2 Contents

*page*

**Maps and satellite images** ........... 4
**Map projections and graphics** ........... 5

### British Isles   Topographic maps

British Isles ........... 6
Orkney and Shetland Islands ........... 7
Scottish Highlands and Islands ........... 8–9
Northern Ireland  Southern Scotland ........... 10–11
Northern England ........... 12–13
Wales  West and East Midlands ........... 14–15
South East England ........... 16–17
South West England ........... 18–19
Ireland ........... 20–21

topographic maps of the British Isles

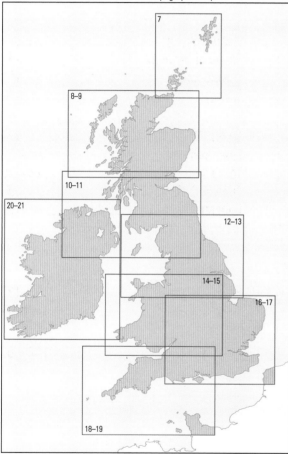

### British Isles   Thematic maps

| | | | |
|---|---|---|---|
| Political | 22–23 | Employment | 33 |
| Physical | 24 | Population | 34–35 |
| Geology | 25 | Sport  Tourism | 36 |
| Climate | 26–27 | Quality of life | 37 |
| Land use | 28 | Conservation | 38 |
| Farming | 29 | Pollution | 39 |
| Water  Minerals | 30 | Transport | 40 |
| Energy | 31 | Trade | 41 |
| Manufacturing industry | 32 | | |

### Europe   Thematic maps

Physical ........... 42
Climate  Ecosystems ........... 43
Political ........... 44
Population  Wealth ........... 45
Economy ........... 46
Tourism ........... 47

### Europe   Topographic maps

Benelux and the Ruhr ........... 48
Scandinavia and Iceland ........... 49
France ........... 50
Spain and Portugal ........... 51
Middle Europe ........... 52–53
Italy and the Balkans ........... 54–55
East Europe and Turkey ........... 56–57

topographic maps of Europe

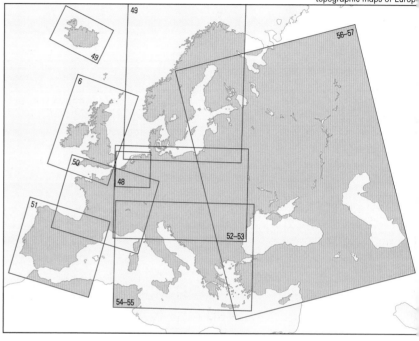

### Asia   Thematic maps

Physical ........... 58
Political ........... 59
Climate  Ecosystems ........... 60
Economy  Population ........... 61

### Asia   Topographic maps

Northern Eurasia ........... 62–63
Middle East ........... 64–65
South Asia ........... 66–67
China ........... 68–69
Japan ........... 70
South East Asia ........... 71

topographic maps of Asia

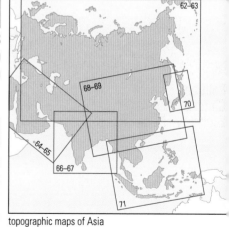

### Oceania   Topographic map

Australia and New Zealand ........... 72–73

### Oceania   Thematic maps

Climate  Ecosystems Population  Economy ........... 74

topographic map of Oceania

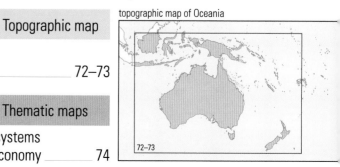

# Contents  3

**Africa**  Thematic maps

Physical  Political _____ 75
Satellite images _____ 76
Climate  Ecosystems _____ 77
Economy  Population _____ 78
Tourism _____ 79

**Africa**  Topographic maps

Northern Africa _____ 80–81
Southern Africa _____ 82
East Africa _____ 83

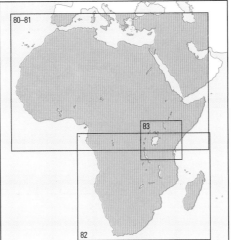

topographic maps of Africa

**North America**  Thematic maps

Physical  Political _____ 84
Climate  Ecosystems _____ 85
Economy  Population _____ 86
Tourism _____ 87

**North America**  Topographic maps

Canada _____ 88–89
United States of America _____ 90–91
Great Lakes _____ 92–93
Mexico, Central America,
and the Caribbean _____ 94–95

**South America**  Topographic map

South America _____ 96–97

**South America**  Thematic maps

Political _____ 97
Climate  Ecosystems _____ 98
Economy  Population _____ 99

topographic maps of North America

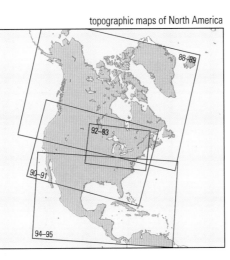

topographic map of South America

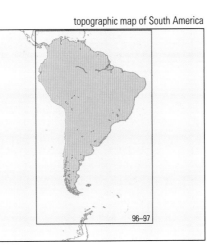

**World**  Thematic maps                 *page*

Political _____ 102–103
Physical _____ 104–105
Earthquakes and volcanoes _____ 106–107
Temperature and ocean currents _____ 108

Pressure, winds, and precipitation _____ 109
Climate regions _____ 110–111
Climate data _____ 112
Climate hazards _____ 113

Ecosystems _____ 114–115
Population _____ 116–117
Population trends _____ 118
Population migration _____ 119

Quality of life _____ 120–121
Economic activity _____ 122
Development _____ 123
Energy _____ 124

Trade  Transport _____ 125
Environmental issues _____ 126–127
Tourism _____ 128–129
Time and distance _____ 130
Cyberspace _____ 131

Flags _____ 132–133
Datasets _____ 134–141

**Index** _____ 142–160

**Oceans and Poles**  Topographic maps

Pacific Ocean _____ 100
Arctic Ocean  Antarctica _____ 101

topographic map of the Poles

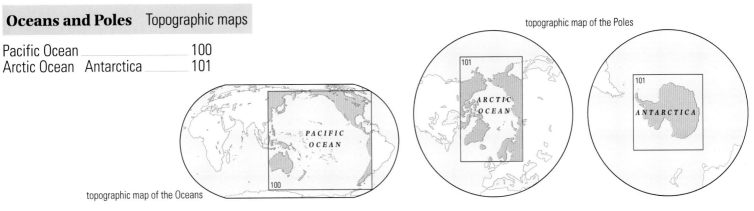

topographic map of the Oceans

© Oxford University Press

# 4 Maps and satellite images

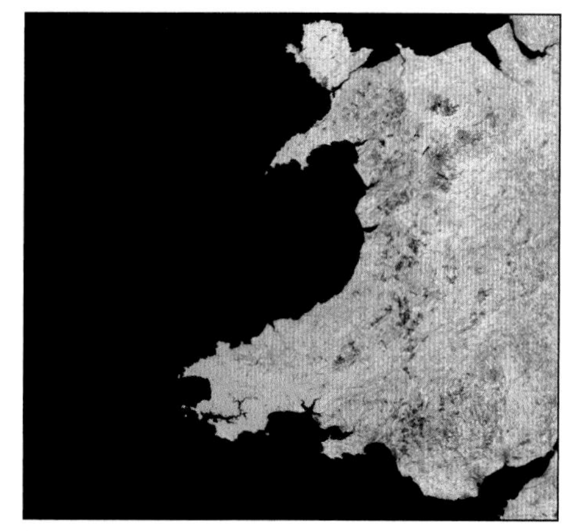

Satellite scanners 'read' the Earth's radiation. The data can be organised by computer to form a visual image. In this image the colours are not real but have been arranged to show how the land is used.

Orange: rough pasture
Red: forest and woodland
Green: improved pasture
Dark blue: urban areas

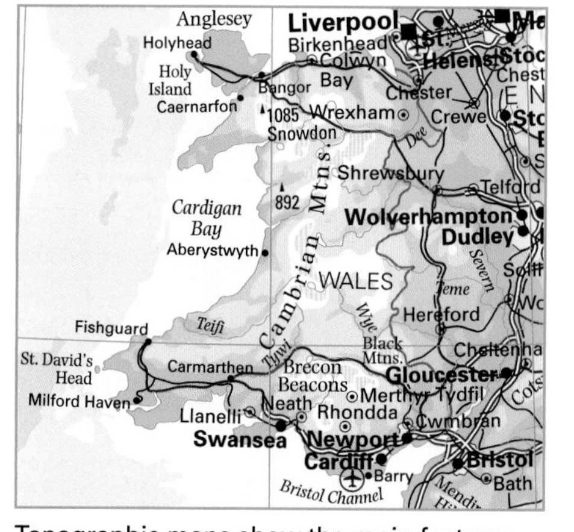

Topographic maps show the main features of the physical landscape as well as settlements, communications, and boundaries. Background colours show the height of the land.

Greens: low land
Browns: high land

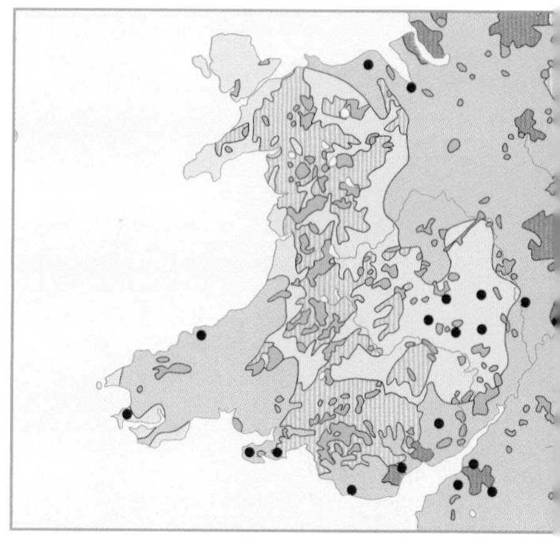

Thematic maps show information about special topics such as agriculture, industry, population, the environment, and quality of life. This map shows land use.

Dark green: forest and woodland
Purple: built-up area

## Symbols on thematic maps

### Point symbols

**Dot map**
Each black dot represents 100 000 sheep.
*From p29*

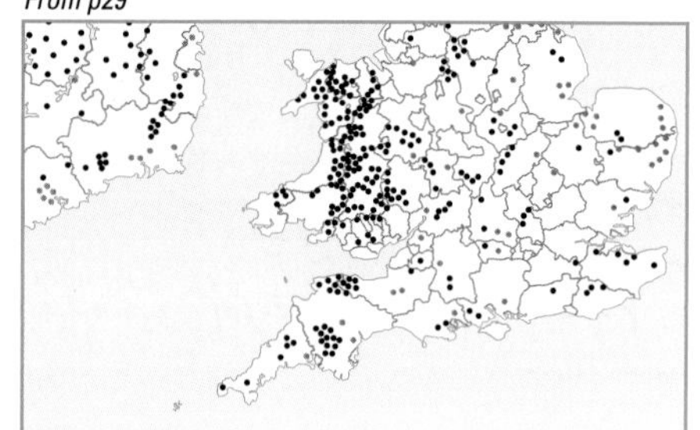

**Economic map**
Blue squares represent a main centre of the motor vehicle industry.
*From p32*

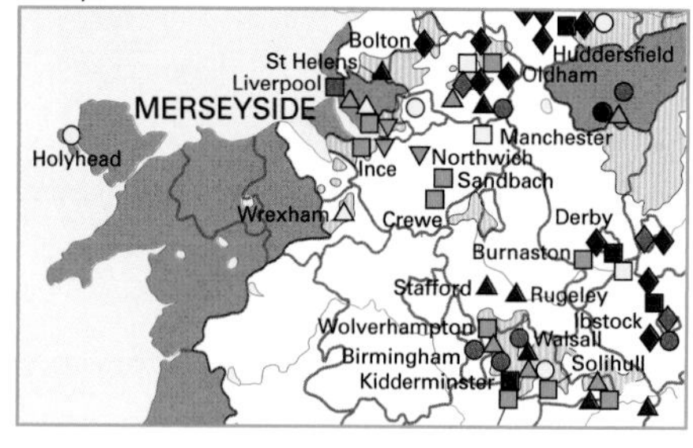

**Proportional symbols**
The size of the circles is proportional to the amount of greenhouse gas emission.
*From p127*

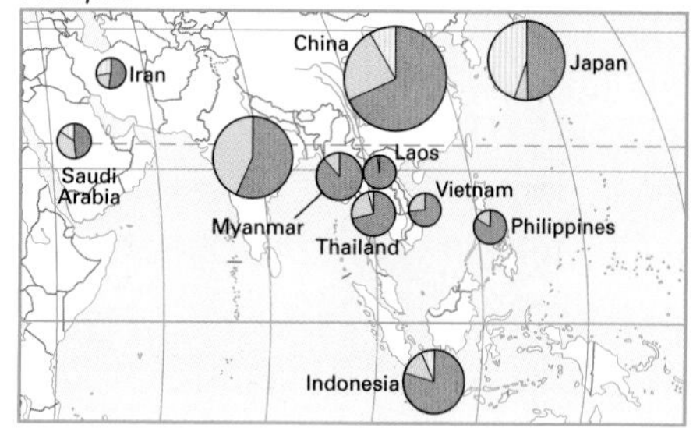

### Line symbols

**Isopleth map**
Lines join places with equal amounts of sunshine.
*From p27*

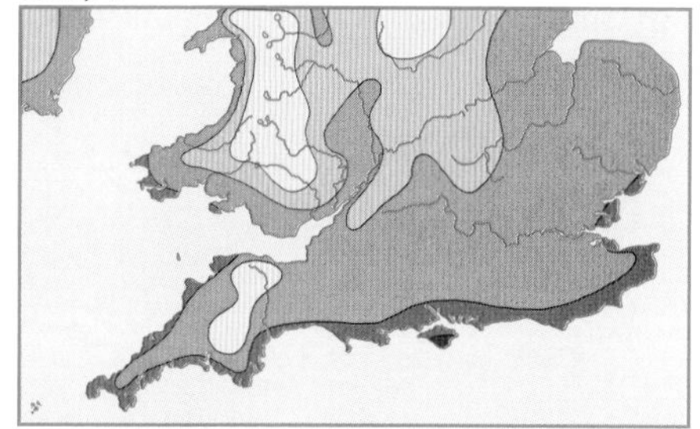

**Isotherm map**
Some isopleths have special names. Isotherms join places with equal temperature.
*From p26*

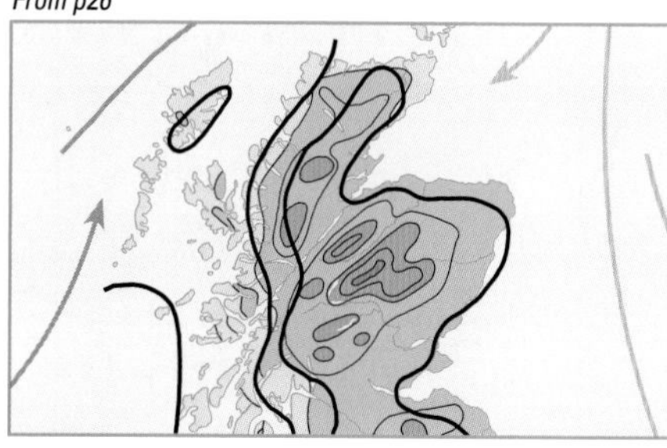

**Flowline map**
The thickness of the line is proportional to the amount of internet traffic.
*From p131*

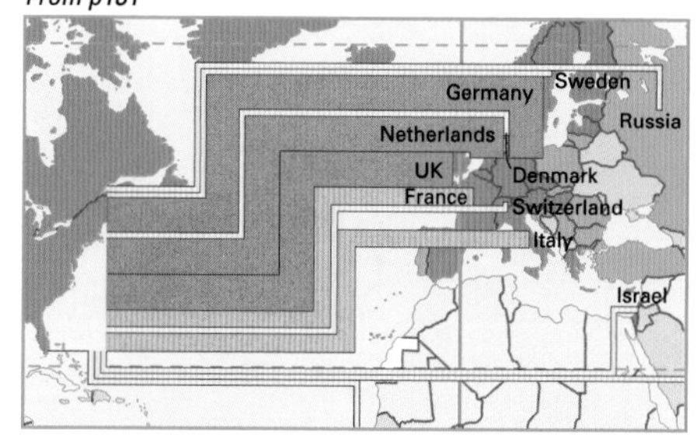

### Area symbols

**Choropleth map**
Darker colours show areas with a higher percentage of land used for growing potatoes.
*From p29*

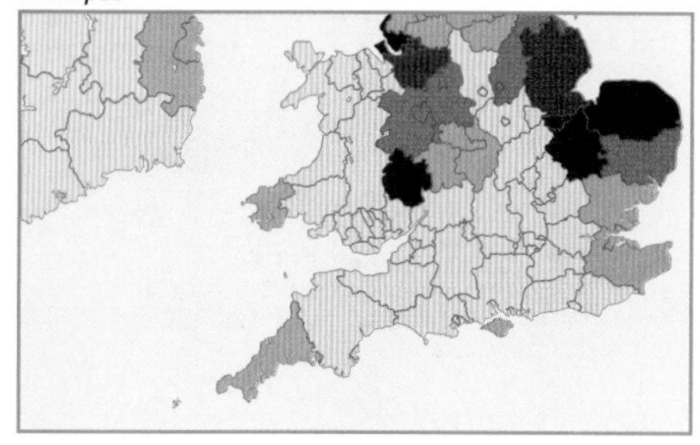

**Environmental map**
Each colour represents an ecosystem. Purple stands for mountains.
*From p115*

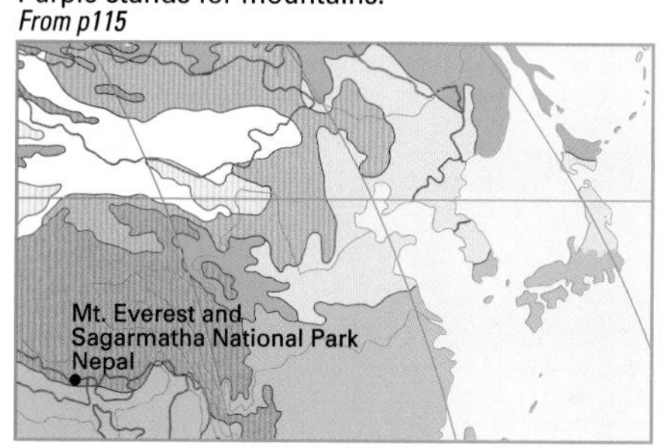

**Political map**
Colours have no meaning but are simply used to show where one country ends and another begins.
*From p102*

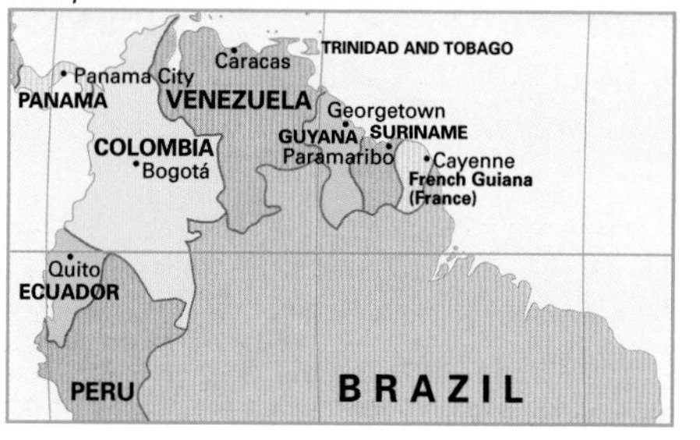

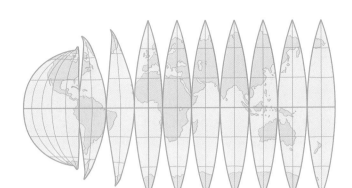

This map has been made by unpeeling strips from the Earth's surface. It would be difficult to use because gaps are left in the land and sea.

Parallels of latitude and meridians of longitude form a grid. Different grid patterns, called projections, can be used to turn the spherical surface of the Earth into a flat world map. It is impossible to make a world map in which both the sizes and shapes of the Earth's land masses are shown accurately. All world maps are distorted in some way.

There are many map projections. It is important that the projection used for a world map is suitable for the information shown on it.

**Polar**
Most world maps do not show Antarctica or the Arctic Ocean accurately. Polar projections give a better view of the poles.

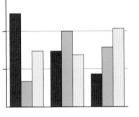

**Eckert IV**
Equal area. The land masses are the correct size in relation to each other but there is some distortion in shape.

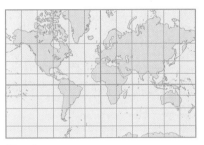

**Mercator**
Conformal. The shape of the land masses is correct but their size becomes larger further away from the equator.

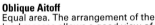

**Oblique Aitoff**
Equal area. The arrangement of the land masses allows a good view of the northern hemisphere.

## Graphical representation of data

**Clustered column**
Compares values across categories
*See example p35*

**Line**
Shows trend over time across categories
*See example p33*

**Pie**
Shows the contribution of each value to a total
*See example p124*

**Stacked column**
Compares the contribution of each value to a total across categories
*See example p32*

**Triangular**
Compares trios of values
*See example p123*

**Simple bar**
Length of bar is proportional to each value
*See example p128*

**Scatter**
Compares pairs of values
*See example p61*

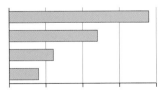

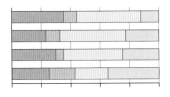

**100% stacked bar**
Compares the percentage each value contributes to a total across categories
*See example p116*

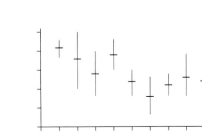

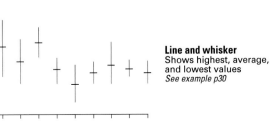

**Line and whisker**
Shows highest, average, and lowest values
*See example p30*

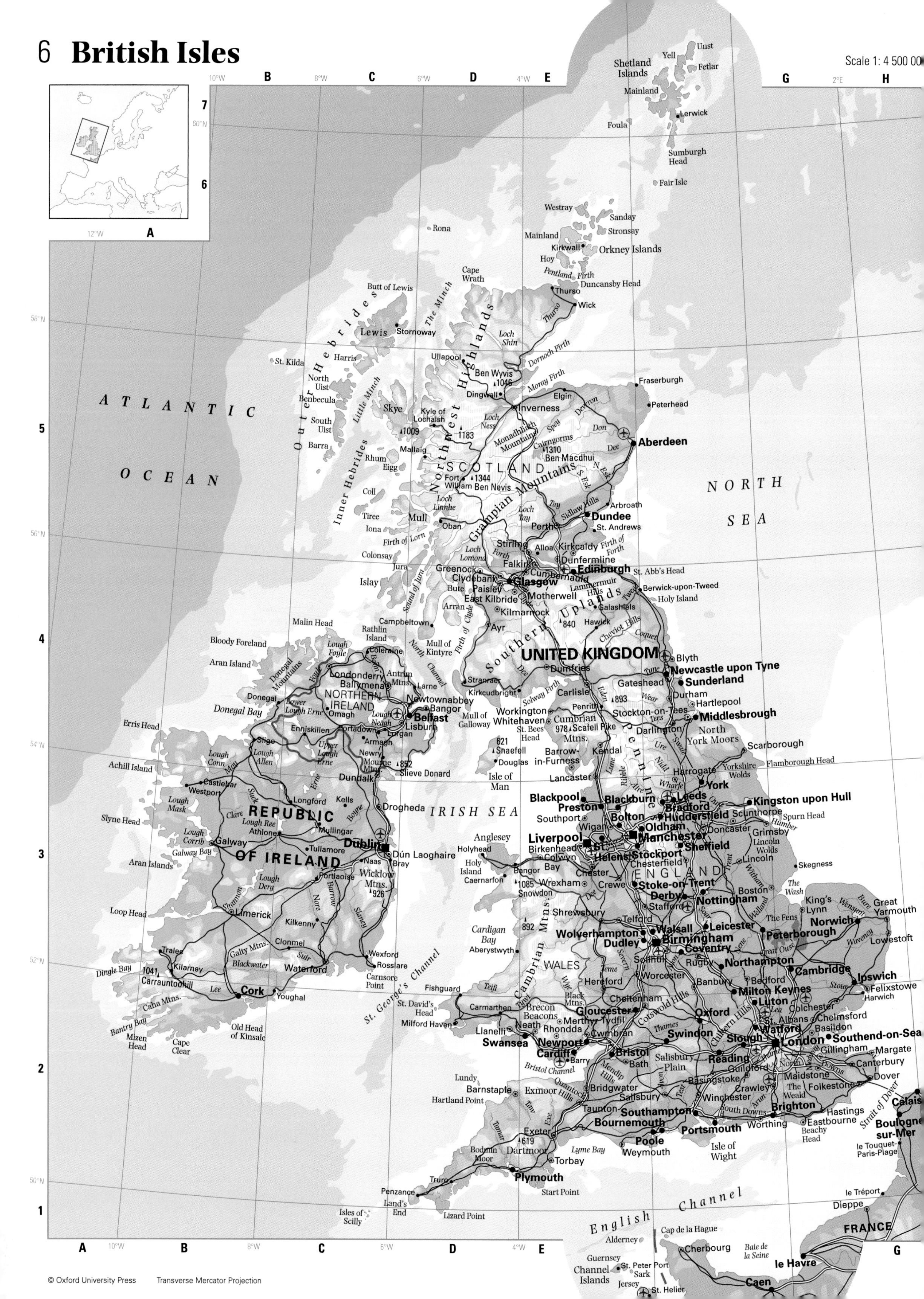

6 British Isles
Scale 1: 4 500 000
10°W
8°W
6°W
4°W
2°E
60°N
58°N
56°N
54°N
52°N
50°N
12°W
ATLANTIC OCEAN
NORTH SEA
IRISH SEA
English Channel
St. George's Channel
Bristol Channel
Shetland Islands
Yell
Unst
Fetlar
Mainland
Lerwick
Foula
Sumburgh Head
Fair Isle
Westray
Sanday
Stronsay
Mainland
Kirkwall
Orkney Islands
Hoy
Pentland Firth
Duncansby Head
Thurso
Wick
Cape Wrath
Butt of Lewis
The Minch
Outer Hebrides
Lewis
Stornoway
Harris
St. Kilda
North Uist
Benbecula
South Uist
Barra
Little Minch
Skye
Kyle of Lochalsh
Mallaig
Rhum
Eigg
Coll
Tiree
Iona
Mull
Oban
Inner Hebrides
Firth of Lorn
Colonsay
Jura
Islay
Sound of Jura
Rona
Loch Shin
Dornoch Firth
Moray Firth
Ullapool
Ben Wyvis
1046
Dingwall
Northwest Highlands
Inverness
Loch Ness
Monadhliath Mountains
Spey
Cairngorms
1310
Ben Macdhui
Elgin
Deveron
Don
Dee
Fraserburgh
Peterhead
Aberdeen
1009
1183
SCOTLAND
Fort William
1344
Ben Nevis
Grampian Mountains
Loch Linnhe
Loch Tay
Tay
Sidlaw Hills
N. Esk
S. Esk
Arbroath
Dundee
St. Andrews
Perth
Stirling
Forth
Alloa
Kirkcaldy
Firth of Forth
Dunfermline
Loch Lomond
Falkirk
Edinburgh
St. Abb's Head
Greenock
Clydebank
Glasgow
Cumbernauld
Bute
Paisley
East Kilbride
Motherwell
Lammermuir Hills
Berwick-upon-Tweed
Holy Island
Tweed
Arran
Kilmarnock
Southern Uplands
Galashiels
Firth of Clyde
Ayr
840
Hawick
Cheviot Hills
Coquet
Campbeltown
Mull of Kintyre
Malin Head
Rathlin Island
North Channel
Bloody Foreland
Aran Island
Lough Foyle
Coleraine
Bann
Donegal Mountains
Foyle
Londonderry
Ballymena
Antrim Mts.
Larne
NORTHERN IRELAND
Newtownabbey
Bangor
Belfast
Lough Neagh
Donegal
Donegal Bay
Lower Lough Erne
Omagh
Enniskillen
Portadown
Lisburn
Lurgan
Armagh
Upper Lough Erne
Newry
Mourne Mts.
852
Slieve Donard
Dundalk
Erris Head
Sligo
Lough Allen
Lough Conn
Achill Island
Castlebar
Westport
Moy
Erne
Suck
Longford
Kells
Lough Mask
Clare
REPUBLIC OF IRELAND
Lough Ree
Athlone
Mullingar
Boyne
Drogheda
Slyne Head
Lough Corrib
Galway
Galway Bay
Tullamore
Dublin
Dún Laoghaire
Naas
Bray
Aran Islands
Wicklow Mts.
926
Lough Derg
Portlaoise
Nore
Barrow
Slaney
Loop Head
Shannon
Limerick
Kilkenny
Tralee
Galty Mts.
Clonmel
Suir
Wexford
Rosslare
Carnsore Point
Dingle Bay
1041
Carrauntoohill
Kilarney
Blackwater
Waterford
Lee
Cork
Youghal
Caha Mts.
Bantry Bay
Mizen Head
Cape Clear
Old Head of Kinsale
UNITED KINGDOM
Stranraer
Kirkcudbright
Dumfries
Dee
Solway Firth
Tyne
Blyth
Newcastle upon Tyne
Sunderland
Gateshead
Carlisle
Eden
893
Wear
Durham
Hartlepool
Mull of Galloway
Workington
Whitehaven
St. Bees Head
Cumbrian Mts.
978
Scafell Pike
Penrith
Stockton-on-Tees
Tees
Middlesbrough
Darlington
North York Moors
Scarborough
Swale
Ure
Pennines
621
Snaefell
Douglas
Isle of Man
Barrow-in-Furness
Kendal
Lancaster
Lune
Nidd
Harrogate
Wharfe
Yorkshire Wolds
Flamborough Head
York
Ribble
Aire
Blackpool
Preston
Blackburn
Leeds
Bradford
Ouse
Kingston upon Hull
Huddersfield
Scunthorpe
Spurn Head
Humber
Grimsby
Southport
Bolton
Wigan
Oldham
Manchester
Doncaster
Lincoln Wolds
Anglesey
Holyhead
Liverpool
Birkenhead
St. Helens
Stockport
Sheffield
Colwyn Bay
Chesterfield
Lincoln
Witham
Skegness
Holy Island
Bangor
Caernarfon
Chester
ENGLAND
Trent
1085
Snowdon
Wrexham
Crewe
Stoke-on-Trent
Boston
The Wash
Derby
Nottingham
Stafford
King's Lynn
Wensum
Bure
Great Yarmouth
Shrewsbury
Soar
Welland
The Fens
Norwich
Telford
Leicester
Peterborough
Waveney
Lowestoft
Cardigan Bay
892
Cambrian Mtns.
Wolverhampton
Walsall
Birmingham
Nene
Great Ouse
Dudley
Coventry
Aberystwyth
Solihull
Rugby
Northampton
Severn
Avon
WALES
Cambridge
Teme
Ipswich
Wye
Hereford
Worcester
Banbury
Bedford
Stour
Felixstowe
Harwich
Milton Keynes
Fishguard
Teifi
Tywi
Black Mtns.
Cheltenham
Luton
Lea
Colchester
St. David's Head
Carmarthen
Brecon Beacons
Gloucester
Oxford
Chelmsford
Cotswold Hills
Chiltern Hills
Milford Haven
Merthyr Tydfil
St. Albans
Llanelli
Neath
Rhondda
Cwmbran
Thames
Swindon
Watford
Basildon
Swansea
Newport
Southend-on-Sea
Slough
London
Gillingham
Margate
Cardiff
Barry
Bristol
Bath
Salisbury Plain
Reading
Canterbury
Medway
North Downs
Mendip Hills
Guildford
Maidstone
Dover
Lundy
Basingstoke
Folkestone
Barnstaple
Exmoor
Quantock Hills
Bridgwater
Crawley
The Weald
Strait of Dover
Hartland Point
Salisbury
Winchester
Test
Arun
Brighton
Hastings
Calais
Taunton
South Downs
Southampton
Worthing
Eastbourne
Beachy Head
Boulogne-sur-Mer
Taw
Exe
Bournemouth
Portsmouth
Tamar
Exeter
619
Dartmoor
Poole
Isle of Wight
le Touquet-Paris-Plage
Bodmin Moor
Lyme Bay
Weymouth
Torbay
Plymouth
Truro
Start Point
Penzance
Land's End
le Tréport
Dieppe
Isles of Scilly
Lizard Point
Cap de la Hague
Alderney
Cherbourg
Baie de la Seine
FRANCE
Guernsey
St. Peter Port
Sark
Channel Islands
le Havre
Jersey
St. Helier
Caen
A
B
C
D
E
G
H
1
2
3
4
5
6
7
© Oxford University Press
Transverse Mercator Projection

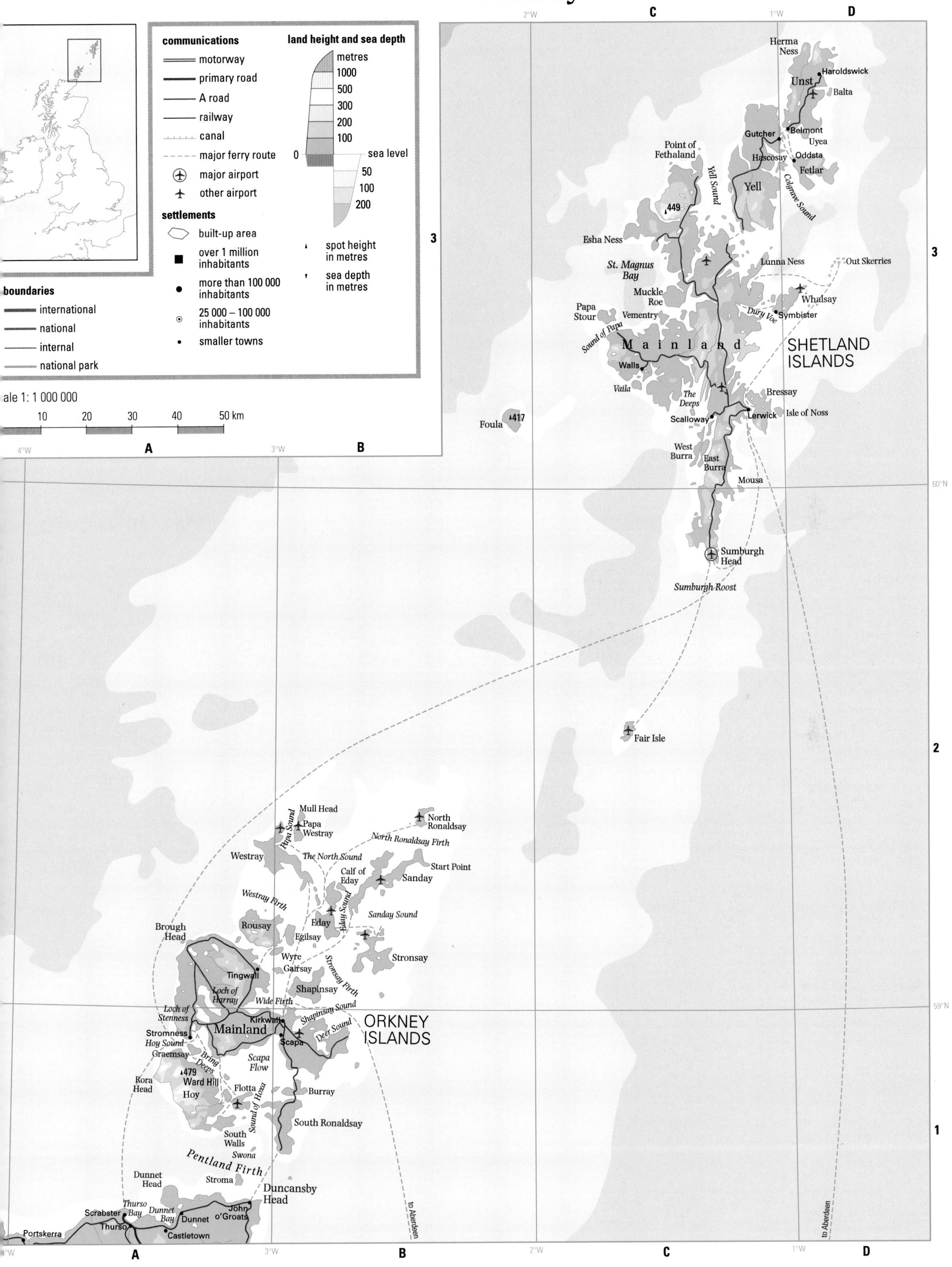

**communications**
═══ motorway
━━━ primary road
─── A road
─── railway
┈┈┈ canal
╌╌╌ major ferry route
✈ major airport
✈ other airport

**settlements**
⬡ built-up area
■ over 1 million inhabitants
● more than 100 000 inhabitants
◉ 25 000 – 100 000 inhabitants
• smaller towns

**land height and sea depth**

metres
1000
500
300
200
100
0 — sea level
50
100
200

▲ spot height in metres
▼ sea depth in metres

**boundaries**
─── international
─── national
─── internal
─── national park

Scale 1: 1 000 000

0  10  20  30  40  50 km

**SHETLAND ISLANDS**

Herma Ness
Haroldswick
Unst
Balta
Point of Fethaland
Gutcher · Belmont
Uyea
Hascosay · Oddsta
Fetlar
▲449
Yell Sound
Yell
Colgrave Sound
Esha Ness
Lunna Ness
Out Skerries
St. Magnus Bay
Muckle Roe
Whalsay
Papa Stour
Vementry
Dury Voe · Symbister
Mainland
Walls
Bressay
Vaila
The Deeps
Isle of Noss
Scalloway
Lerwick
Foula ▲417
West Burra
East Burra
Mousa
60°N
Sumburgh Head
Sumburgh Roost
to Aberdeen

Fair Isle

**ORKNEY ISLANDS**

Mull Head
Papa Westray
North Ronaldsay
Westray
Papa Sound
The North Sound
North Ronaldsay Firth
Calf of Eday
Start Point
Westray Firth
Sanday
Rousay
Eday
Eday Sound
Sanday Sound
Brough Head
Egilsay
Stronsay
Wyre
Gairsay
Tingwall
Shapinsay
Stronsay Firth
Loch of Harray
Wide Firth
Shapinsay Sound
Loch of Stenness
Kirkwall
Deer Sound
Stromness
Mainland
Scapa
Hoy Sound
Graemsay
Bring Deeps
Scapa Flow
59°N
Rora Head
▲479 Ward Hill
Flotta
Burray
Hoy
Sound of Hoxa
South Ronaldsay
South Walls
Swona
Pentland Firth
Dunnet Head
Stroma
Duncansby Head
Portskerra
Scrabster
Thurso Bay
Dunnet Bay
Dunnet
John o'Groats
to Aberdeen
Thurso
Castletown

Transverse Mercator Projection  © Oxford University Press

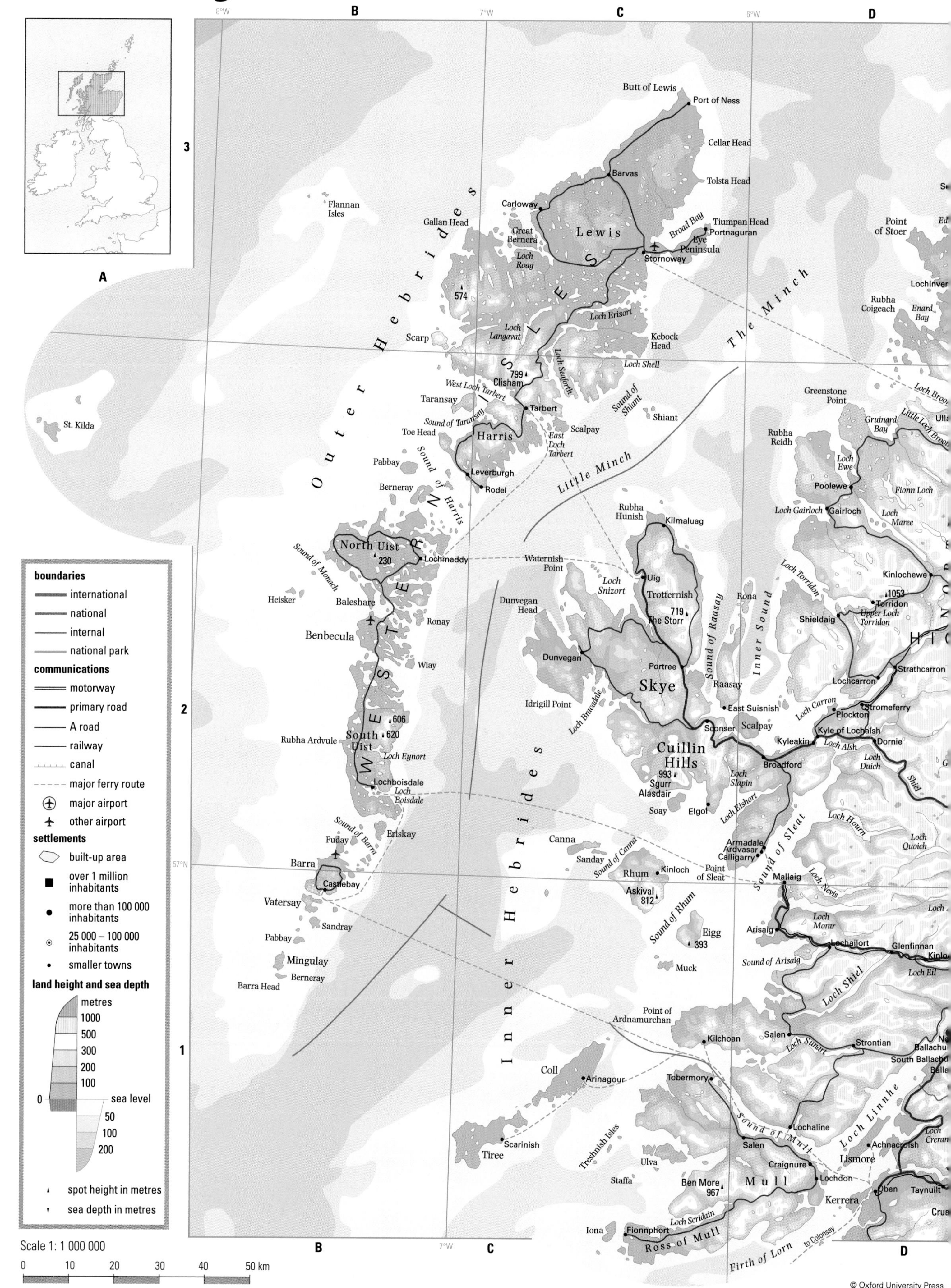

boundaries
— international
— national
— internal
— national park

communications
— motorway
— primary road
— A road
— railway
⊢⊣ canal
--- major ferry route
⊕ major airport
✈ other airport

settlements
⬠ built-up area
■ over 1 million inhabitants
● more than 100 000 inhabitants
⊙ 25 000 – 100 000 inhabitants
• smaller towns

land height and sea depth

metres
1000
500
300
200
100
0 — sea level
50
100
200

▲ spot height in metres
▼ sea depth in metres

Scale 1: 1 000 000

0  10  20  30  40  50 km

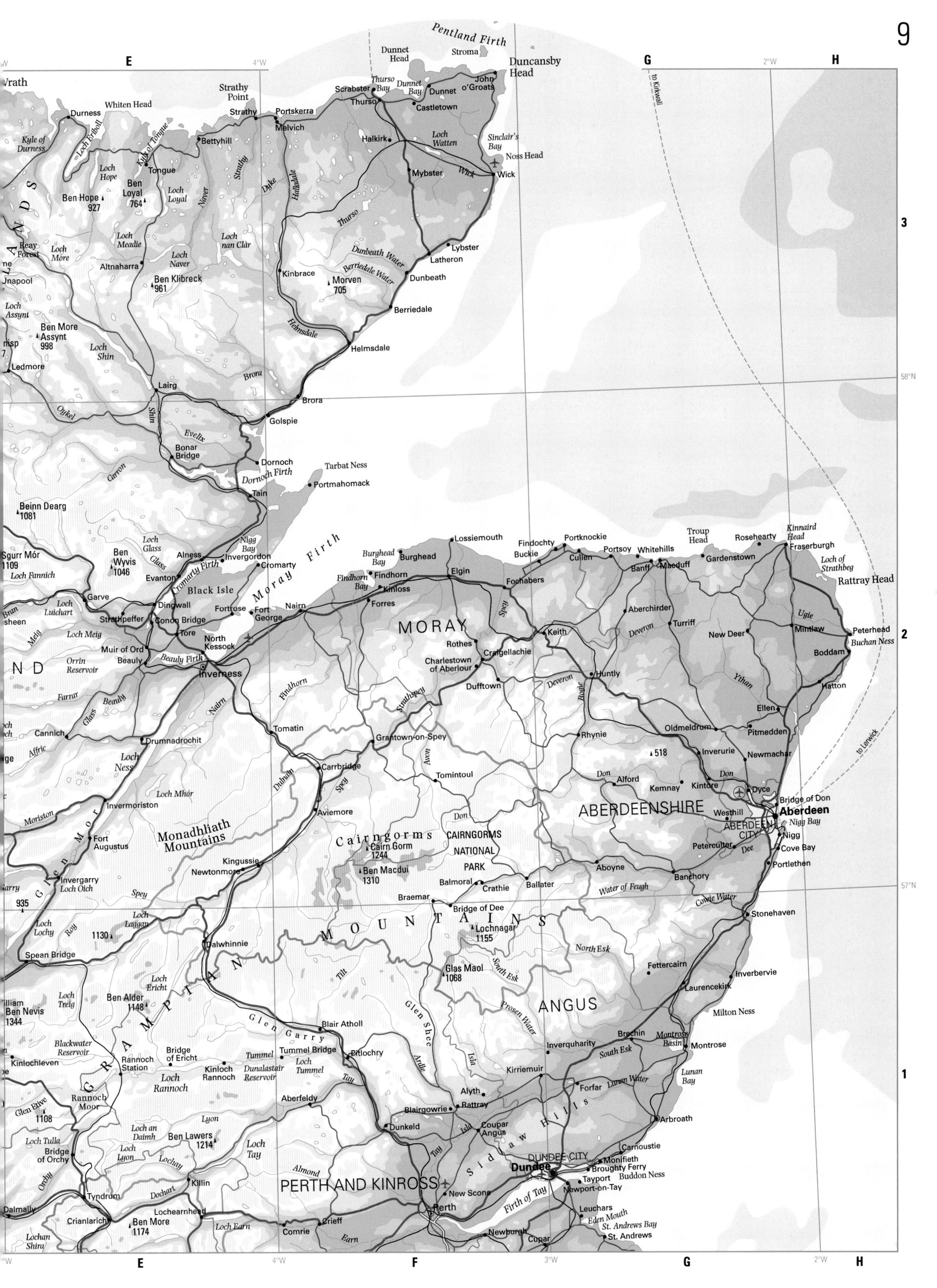
Pentland Firth
Dunnet Head
Stroma
Duncansby Head
John o'Groats
Dunnet
Dunnet Bay
Thurso Bay
Scrabster
Thurso
Castletown
Strathy Point
Strathy
Portskerra
Melvich
Whiten Head
Durness
Kyle of Durness
Loch Eriboll
Kyle of Tongue
Bettyhill
Tongue
Loch Hope
Ben Hope 927
Ben Loyal 764
Loch Loyal
Naver
Strathy
Halladale
Halkirk
Loch Watten
Sinclair's Bay
Noss Head
Wick
Mybster
Thurso
Lybster
Latheron
Dunbeath Water
Berriedale Water
Dunbeath
Kinbrace
Morven 705
Berriedale
Helmsdale
Loch Meadie
Loch nan Clàr
Loch Naver
Reay Forest
Loch More
Altnaharra
Ben Klibreck 961
Loch Assynt
Ben More Assynt 998
Loch Shin
Ledmore
Lairg
Brora
Golspie
Oykel
Shin
Evelix
Bonar Bridge
Dornoch
Dornoch Firth
Tarbat Ness
Portmahomack
Tain
Carron
Beinn Dearg 1081
Sgurr Mór 1109
Loch Fannich
Loch Glass
Ben Wyvis 1046
Alness
Nigg Bay
Invergordon
Cromarty
Cromarty Firth
Evanton
Moray Firth
Black Isle
Garve
Loch Luichart
Strathpeffer
Dingwall
Conon Bridge
Fortrose
Fort George
Nairn
Tore
North Kessock
Loch Meig
Muir of Ord
Beauly
Beauly Firth
Inverness
Orrin Reservoir
Farrar
Beauly
Glass
Cannich
Drumnadrochit
Affric
Loch Ness
Loch Mhór
Invermoriston
Moriston
Glen Mor
Fort Augustus
Monadhliath Mountains
Invergarry
Loch Oich
935
Loch Lochy
Roy
1130
Spean Bridge
Loch Laggan
Spey
Kingussie
Newtonmore
Dalwhinnie
Loch Ericht
Loch Treig
Ben Alder 1148
Ben Nevis 1344
Blackwater Reservoir
Kinlochleven
Rannoch Station
Bridge of Ericht
Kinloch Rannoch
Loch Rannoch
Rannoch Moor
Glen Etive 1108
Loch Tulla
Bridge of Orchy
Orchy
Tyndrum
Dalmally
Crianlarich
Ben More 1174
Lochan Shira
Loch an Daimh
Lyon
Ben Lawers 1214
Loch Lyon
Lochay
Killin
Dochart
Lochearnhead
Loch Earn
Comrie
Crieff
Earn
GRAMPIAN MOUNTAINS
Glen Garry
Blair Atholl
Tummel Bridge
Tummel
Dunalastair Reservoir
Loch Tummel
Pitlochry
Tay
Aberfeldy
Loch Tay
Almond
PERTH AND KINROSS
Dunkeld
Blairgowrie
Rattray
Alyth
Coupar Angus
Isla
Perth
New Scone
Firth of Tay
Newburgh
Cupar
Sidlaw Hills
Dundee
DUNDEE CITY
Monifieth
Broughty Ferry
Buddon Ness
Tayport
Newport-on-Tay
Leuchars
Eden Mouth
St. Andrews Bay
St. Andrews
Carnoustie
Arbroath
Forfar
Lunan Water
Lunan Bay
Kirriemuir
Inverquharity
South Esk
Brechin
Montrose Basin
Montrose
Milton Ness
ANGUS
Prosen Water
Glen Shee
Ardle
Isla
Glas Maol 1068
South Esk
North Esk
Lochnagar 1155
Fettercairn
Laurencekirk
Inverbervie
Stonehaven
Cowie Water
Braemar
Bridge of Dee
Balmoral
Crathie
Ballater
Water of Feugh
Aboyne
Banchory
Portlethen
Cove Bay
Nigg
Nigg Bay
Peterculter
Dee
ABERDEEN CITY
Aberdeen
Bridge of Don
Westhill
Dyce
Kintore
Kemnay
Alford
Don
ABERDEENSHIRE
518
Inverurie
Newmachar
Oldmeldrum
Pitmedden
Ellen
Ythan
Hatton
Boddam
Buchan Ness
Peterhead
Mintlaw
Ugie
New Deer
Turriff
Deveron
Aberchirder
Rattray Head
Loch of Strathbeg
Fraserburgh
Kinnaird Head
Rosehearty
Troup Head
Gardenstown
Macduff
Banff
Whitehills
Portsoy
Cullen
Portknockie
Findochty
Buckie
Fochabers
Spey
Keith
Huntly
Bogie
Deveron
Rhynie
Lossiemouth
Burghead Bay
Burghead
Findhorn Bay
Findhorn
Kinloss
Forres
Elgin
MORAY
Rothes
Craigellachie
Charlestown of Aberlour
Dufftown
Strathspey
Findhorn
Nairn
Tomatin
Grantown-on-Spey
Avon
Tomintoul
Carrbridge
Dulnain
Spey
Aviemore
Don
Cairngorms
Cairn Gorm 1244
Ben Macdui 1310
CAIRNGORMS NATIONAL PARK
to Kirkwall
to Lerwick
58°N
57°N
4°W
3°W
2°W
E
F
G
H
1
2
3

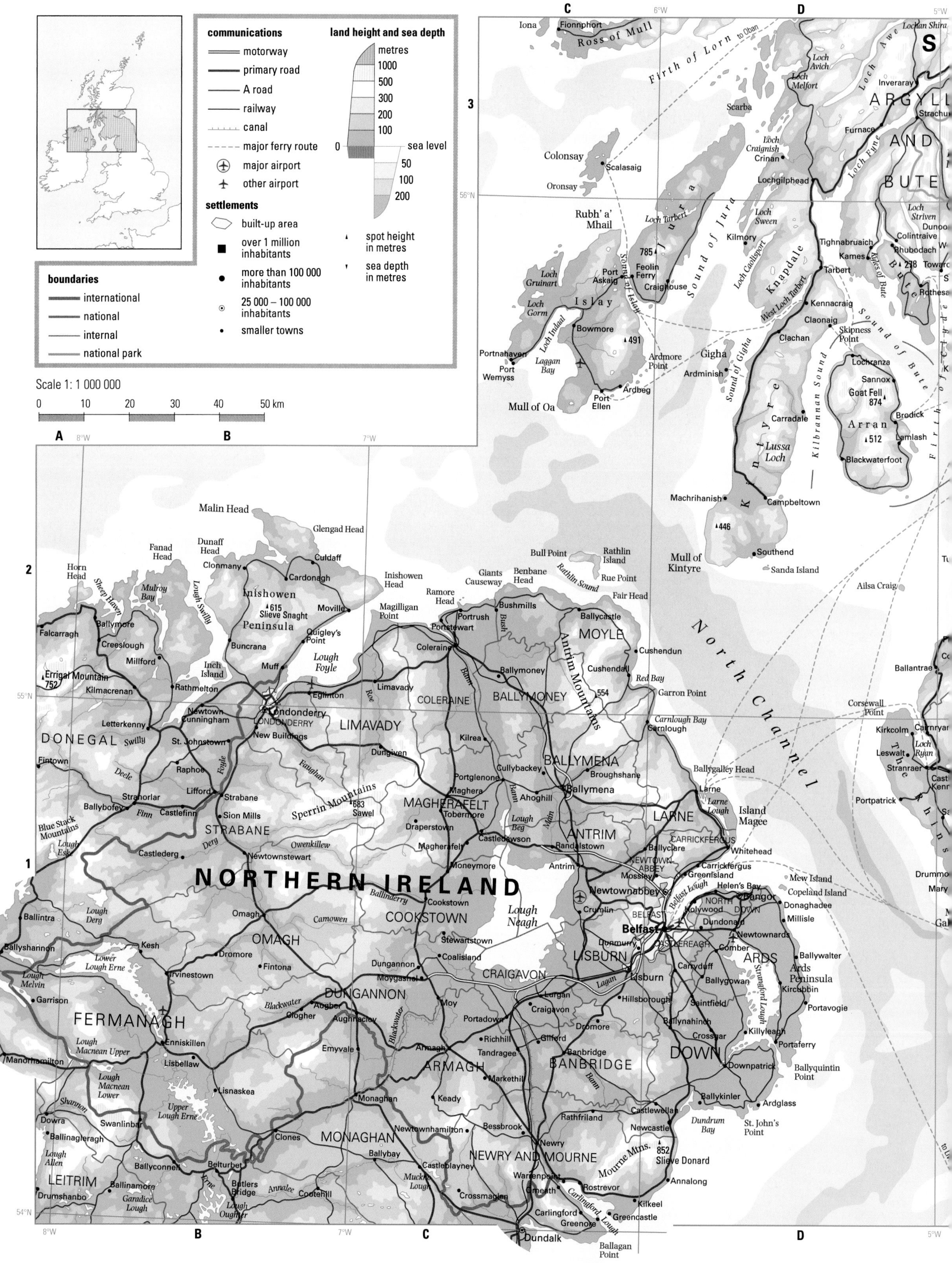

**communications**
- motorway
- primary road
- A road
- railway
- canal
- major ferry route
- ✈ major airport
- ✈ other airport

**settlements**
- built-up area
- ■ over 1 million inhabitants
- ● more than 100 000 inhabitants
- ◉ 25 000 – 100 000 inhabitants
- • smaller towns

**land height and sea depth**
metres
1000
500
300
200
100
0 sea level
50
100
200

▲ spot height in metres
▼ sea depth in metres

**boundaries**
- international
- national
- internal
- national park

Scale 1: 1 000 000

0   10   20   30   40   50 km

© Oxford University Press

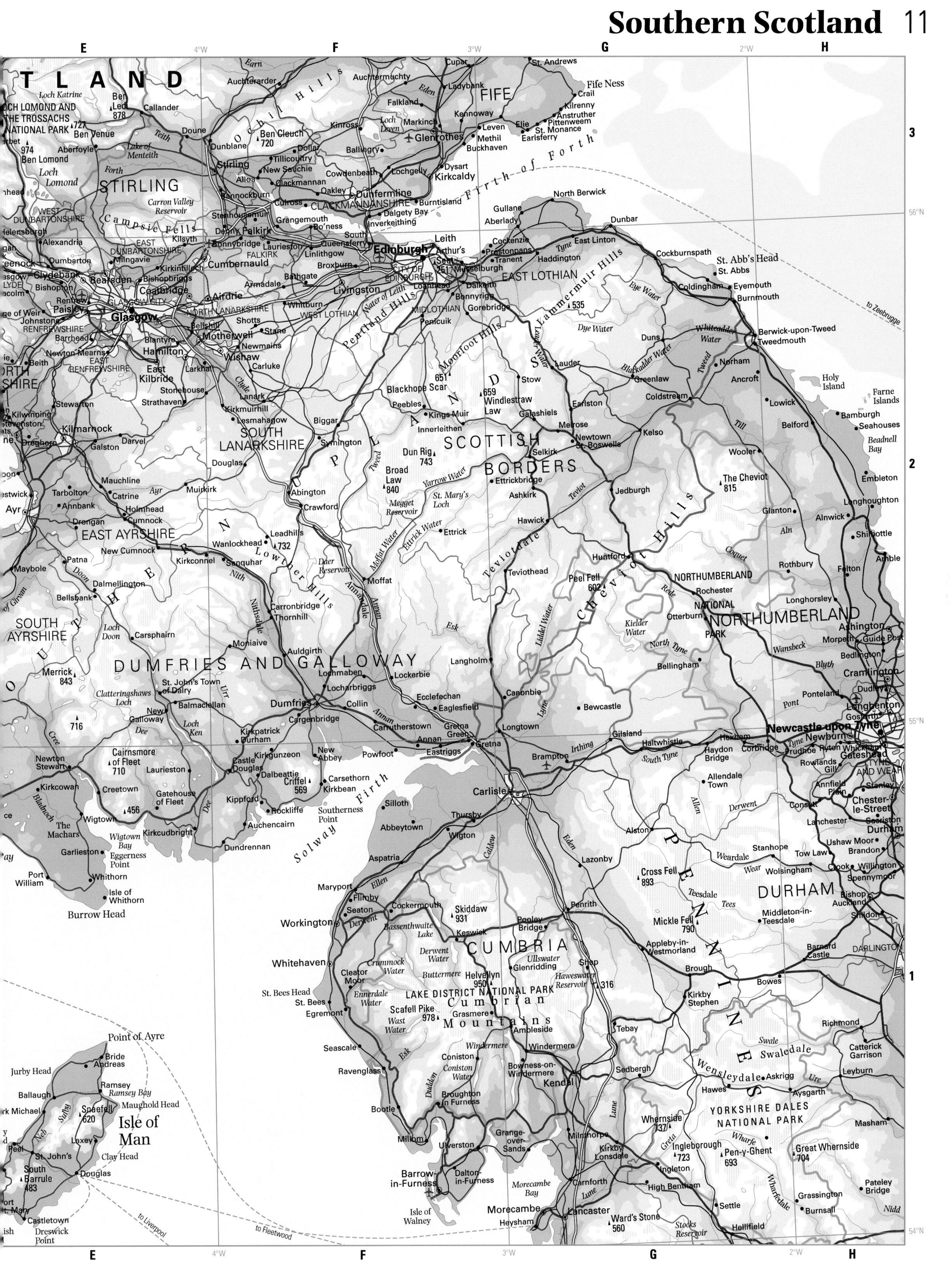

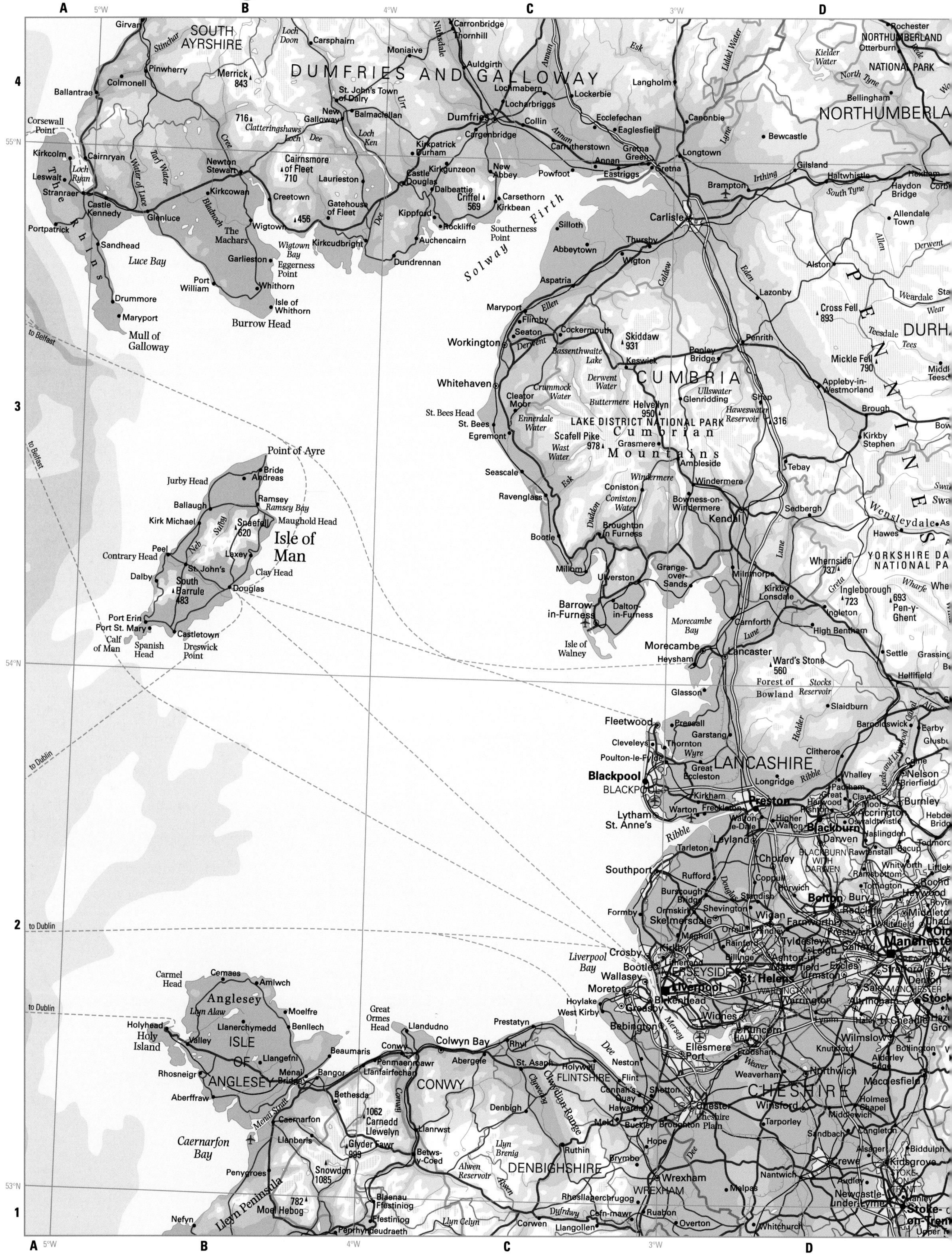

Transverse Mercator Projection    © Oxford University Press

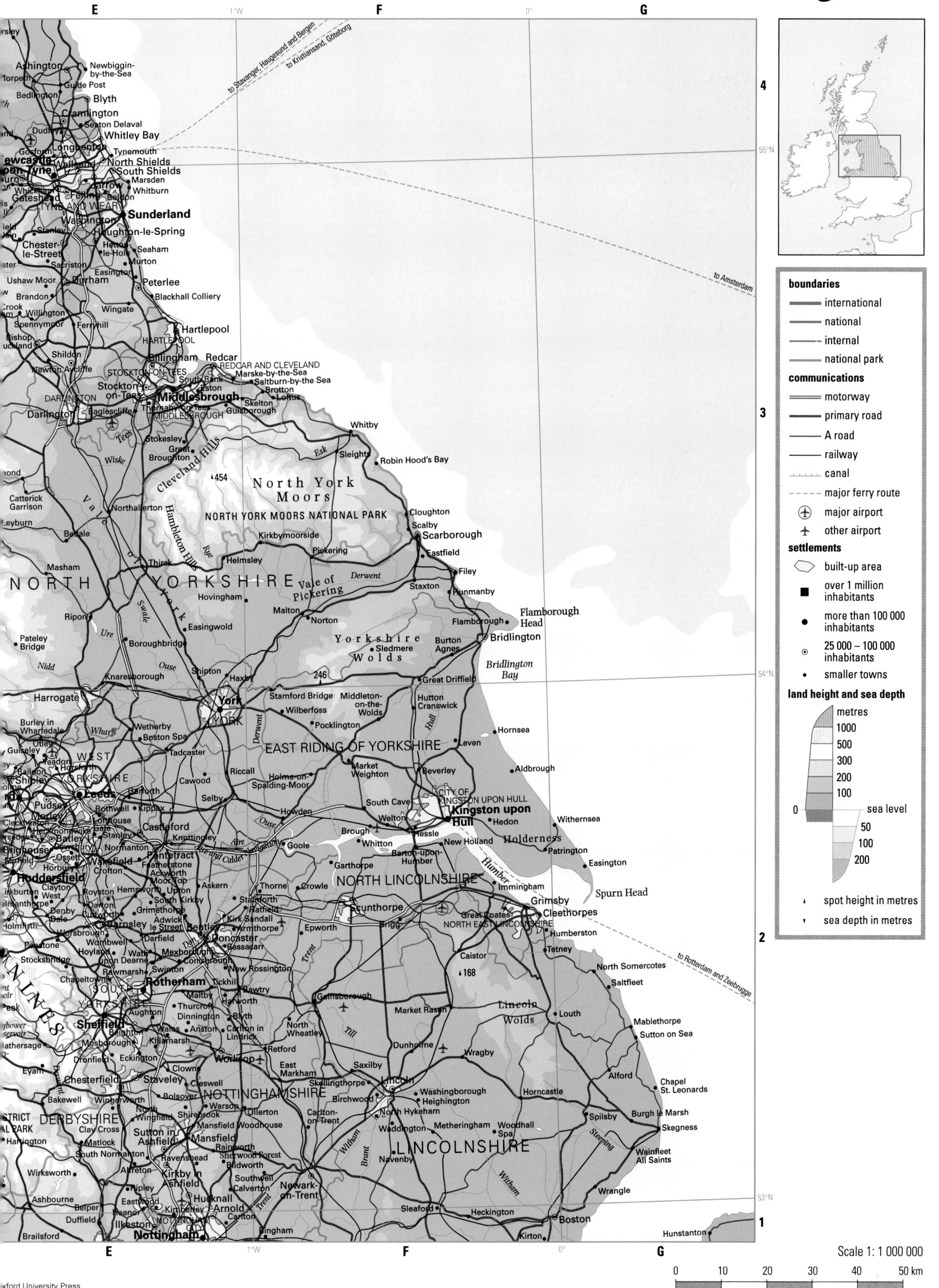

E
F
G
1°W
0°
4
3
2
1
55°N
54°N
53°N
to Stavanger, Haugesund and Bergen
to Kristiansand, Göteborg
to Amsterdam
to Rotterdam and Zeebrugge
boundaries
international
national
internal
national park
communications
motorway
primary road
A road
railway
canal
major ferry route
major airport
other airport
settlements
built-up area
over 1 million inhabitants
more than 100 000 inhabitants
25 000 – 100 000 inhabitants
smaller towns
land height and sea depth
metres
1000
500
300
200
100
0
sea level
50
100
200
spot height in metres
sea depth in metres
Scale 1: 1 000 000
0
10
20
30
40
50 km
Ashington
Newbiggin-by-the-Sea
Guide Post
Blyth
Cramlington
Seaton Delaval
Whitley Bay
Tynemouth
North Shields
South Shields
Marsden
Whitburn
Gateshead
TYNE AND WEAR
Sunderland
Washington
Houghton-le-Spring
Chester-le-Street
Seaham
Murton
Easington
Durham
Peterlee
Blackhall Colliery
Ushaw Moor
Brandon
Willington
Spennymoor
Ferryhill
Wingate
Hartlepool
HARTLEPOOL
Shildon
Newton Aycliffe
Billingham
Redcar
REDCAR AND CLEVELAND
Marske-by-the-Sea
Saltburn-by-the Sea
STOCKTON-ON-TEES
Stockton-on-Tees
Middlesbrough
MIDDLESBROUGH
Brotton
Loftus
Skelton
Guisborough
DARLINGTON
Darlington
Eaglescliffe
Thornaby-on-Tees
Whitby
Sleights
Robin Hood's Bay
Stokesley
Great Broughton
Cleveland Hills
454
North York Moors
NORTH YORK MOORS NATIONAL PARK
Catterick Garrison
Northallerton
Hambleton Hills
Cloughton
Scalby
Scarborough
Eastfield
Kirkbymoorside
Pickering
Helmsley
Bedale
Masham
Thirsk
NORTH YORKSHIRE
Vale of Pickering
Derwent
Filey
Staxton
Hunmanby
Hovingham
Malton
Norton
Flamborough
Flamborough Head
Bridlington
Ripon
Easingwold
Pateley Bridge
Boroughbridge
Yorkshire Wolds
Sledmere
Burton Agnes
Bridlington Bay
Knaresborough
Shipton
Haxby
246
Great Driffield
Harrogate
York
YORK
Stamford Bridge
Middleton-on-the-Wolds
Hutton Cranswick
Wilberfoss
Pocklington
Burley in Wharfedale
Wetherby
Boston Spa
Hornsea
Tadcaster
EAST RIDING OF YORKSHIRE
Leven
WEST YORKSHIRE
Market Weighton
Beverley
Aldbrough
Riccall
Cawood
Holme-on-Spalding-Moor
Leeds
Garforth
Selby
CITY OF KINGSTON UPON HULL
Kingston upon Hull
South Cave
Howden
Welton
Hedon
Withernsea
Castleford
Knottingley
Brough
Hessle
Goole
Whitton
New Holland
Holderness
Patrington
Pontefract
Wakefield
Barton-upon-Humber
Easington
Huddersfield
Garthorpe
NORTH LINCOLNSHIRE
Immingham
Askern
Thorne
Crowle
Spurn Head
Scunthorpe
Grimsby
Cleethorpes
Great Coates
NORTH EAST LINCOLNSHIRE
Brigg
Barnsley
Epworth
Humberston
Doncaster
Bessacarr
Tetney
Caistor
Mexborough
Conisbrough
New Rossington
168
North Somercotes
Saltfleet
Rotherham
SOUTH YORKSHIRE
Tickhill
Bawtry
Gainsborough
Harworth
Market Rasen
Lincoln Wolds
Louth
Sheffield
Dinnington
Blyth
North Wheatley
Mablethorpe
Sutton on Sea
Retford
Worksop
Dunholme
Wragby
Chesterfield
Staveley
Clowne
East Markham
Saxilby
Alford
Chapel St. Leonards
Bolsover
NOTTINGHAMSHIRE
Skellingthorpe
Lincoln
Washingborough
Heighington
Horncastle
Bakewell
North Hykeham
Birchwood
Burgh le Marsh
Skegness
DERBYSHIRE
Ollerton
Carlton-on-Trent
Spilsby
Mansfield
Mansfield Woodhouse
Metheringham
Woodhall Spa
Matlock
Sutton in Ashfield
Waddington
LINCOLNSHIRE
Wainfleet All Saints
Navenby
Rainworth
Sherwood Forest
Bildworth
Wirksworth
Kirkby in Ashfield
Southwell
Newark-on-Trent
Wrangle
Calverton
Ashbourne
Hucknall
Arnold
Sleaford
Heckington
Belper
Duffield
Ilkeston
Nottingham
Carlton
Bingham
Boston
Kirton
Hunstanton
Brailsford
Tees
Wiske
Swale
Ure
Nidd
Ouse
Wharfe
Rye
Esk
Hull
Aire
Trent
Till
Witham
Brant
Steeping
Humber

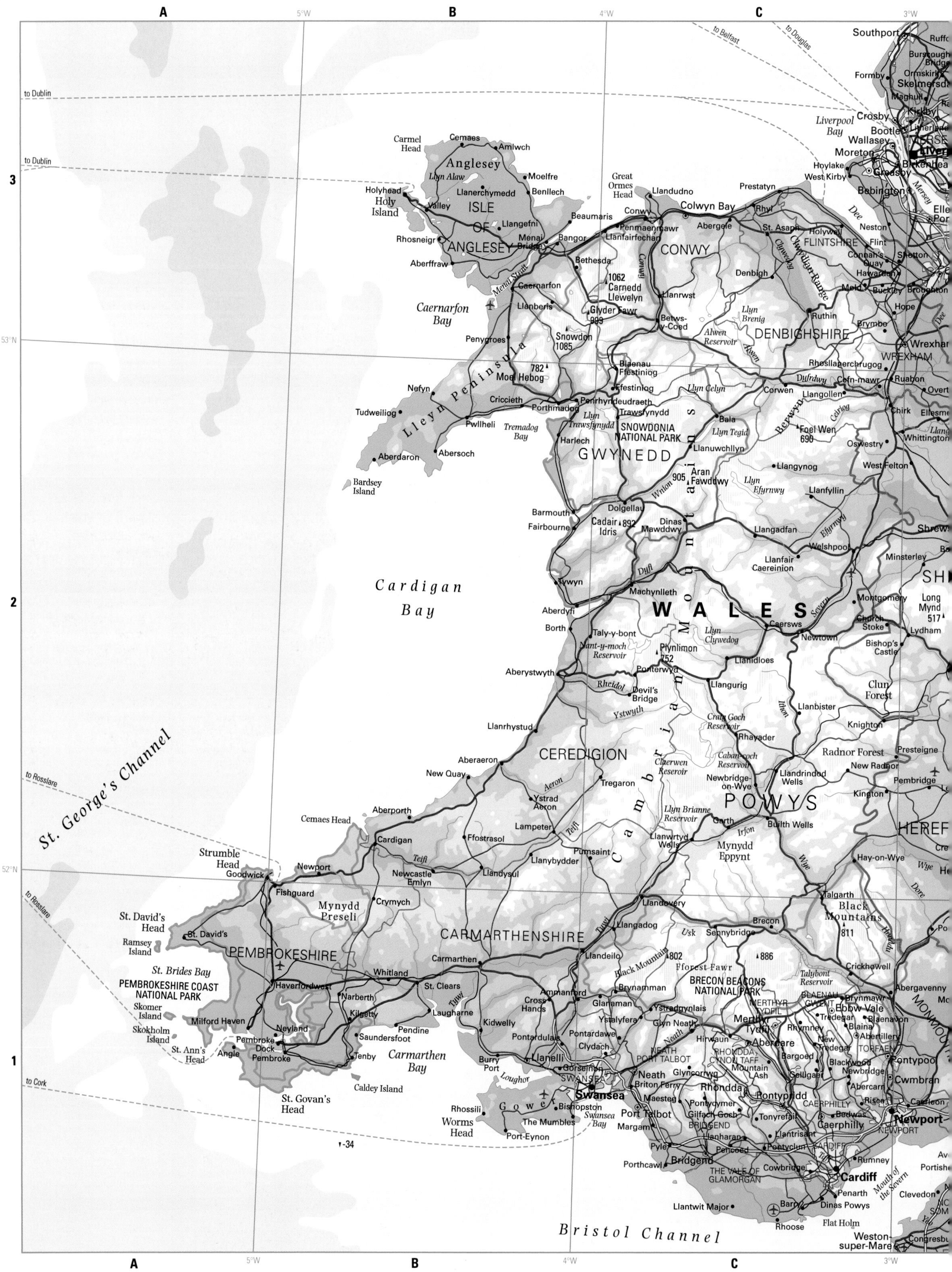

A          5°W          B          4°W          C          3°W

to Dublin

3

to Dublin

Carmel Head
Cemaes
Amlwch
Anglesey
Moelfre
Llyn Alaw
Benllech
Llanerchymedd
Great Ormes Head
Llandudno
Prestatyn
Southport
Ruffc
Burscough Bridge
Formby
Ormskirk
Skelmersd
Maghull
Kirkby
Crosby
Holyhead
Holy Island
Valley
Rhosneigr
ISLE OF ANGLESEY
Llangefni
Beaumaris
Bangor
Penmaenmawr
Llanfairfechan
Colwyn Bay
Rhyl
Abergele
St. Asaph
Holywell
CONWY
Liverpool Bay
Wallasey
Hoylake
West Kirby
Moreton
Bebington
Elle
Por
Liverpool
Birkenhead
Gladso
Neston
FLINTSHIRE
Flint
Connah's Quay
Shotton
Hawarden
Mold
Buckley
Broughton

Aberffraw
Menai Bridge
Caernarfon
Llanberis
1062 Carnedd Llewelyn
Glyder Fawr 999
Betws-y-Coed
Llanrwst
Denbigh
Ruthin
Brymbo
Hope
Wrexham
WREXHAM
Overt

Caernarfon Bay
Penygroes
Snowdon 1085
Llyn Brenig
Alwen Reservoir
DENBIGHSHIRE
Rhoslanerchrugog
Cefn-mawr
Ruabon
Chirk
Ellesme
Whittington

Nefyn
Moel Hebog 782
Blaenau Ffestiniog
Ffestiniog
Llyn Celyn
Corwen
Llangollen
Carrog
Foel Wen 690
Llan

Tudweiliog
Lleyn Peninsula
Criccieth
Porthmadog
Penrhyndeudraeth
Trawsfynydd
SNOWDONIA NATIONAL PARK
Bala
Llyn Tegid
Berwyn
West Felton
Oswestry

Pwllheli
Tremadog Bay
Llyn Trawsfynydd
Harlech
GWYNEDD
Llanuwchllyn
Aran Fawddwy 905
Llangynog
Llyn Efyrnwy
Llanfyllin

Aberdaron
Absroch
Barmouth
Fairbourne
Dolgellau
Cadair Idris 892
Dinas Mawddwy
Llangadfan
Welshpool
Minsterley
Shrew

Bardsey Island
Tywyn
Dyfi
Machynlleth
W A L E S
Severn
Montgomery
Church Stoke
Long Mynd 517
Lydham
SH

Cardigan Bay
Aberdyfi
Borth
Taly-y-bont
Nant-y-moch Reservoir
Plynlimon 752
Ponterwyd
Llyn Clywedog
Caersws
Newtown
Llanidloes
Bishop's Castle

2

Aberystwyth
Rheidol
Devil's Bridge
Ystwyth
Llangurig
Clun Forest

Llanrhystud
Craig Goch Reservoir
Rhayader
Llanbister
Knighton

Aberaeron
CEREDIGION
Cabancoch Reservoir
Llandrindod Wells
Radnor Forest
New Radnor
Presteigne
Pembridge
Le

New Quay
Aeron
Claerwen Reseroir
Newbridge-on-Wye
Kington
HEREF

Aberporth
Cemaes Head
Tregaron
Ystrad Aeron
Llyn Brianne Reservoir
Garth
POWYS
Builth Wells
Hay-on-Wye
Cre
He

St. David's Head
Ramsey Island
Strumble Head
Goodwick
Fishguard
Cardigan
Newcastle Emlyn
Ffostrasol
Teifi
Lampeter
Llanybydder
Pumsaint
Llanwrtyd Wells
Irfon
Mynydd Eppynt
Talgarth
Black Mountains 811
Crickhowell

52°N

to Rosslare
St. George's Channel
Newport
Crymych
Teifi
Llandysul
Brecon
Po

St. David's
Mynydd Preseli
Whitland
St. Clears
Carmarthen
Llandeilo
Usk
Llangadog
Llandovery
Sennybridge
Brecon
BRECON BEACONS NATIONAL PARK
886
Talybont Reservoir
Abergavenny
Mc

PEMBROKESHIRE
Haverfordwest
Narberth
Tywi
Laugharne
Black Mountain 1802
Fforest Fawr
MONMOU

St. Brides Bay
PEMBROKESHIRE COAST NATIONAL PARK
Kilgetty
Pendine
Kidwelly
Ammanford
Cross Hands
Brynamman
Glanaman
Ystradgynlais
Glyn Neath
Hirwaun
MERTHYR TYDFIL
Merthyr Tydfil
BLAENAU GWENT
Brynmawr
Ebbw Vale
Blaenavon
Blaina
Abertillery

Skomer Island
Milford Haven
Neyland
Saundersfoot
Ystalyfera
Pontardawe
NEATH PORT TALBOT
RHONDDA CYNON TAFF
New Tredegar
Blackwood
TORFAEN
Pontypool
Cwmbran

Skokholm Island
St. Ann's Head
Pembroke Dock
Pembroke
Angle
Tenby
Burry Port
Llanelli
Gorseinon
Pontardulais
Clydach
Glyncorrwg
Glyncorrwg Mountain
Aberdare
Rhymney
Bargoed
Tredegar
Abercarn
Risca
Pontypool

1

to Cork
Caldey Island
Carmarthen Bay
SWANSEA
Swansea
Neath
Briton Ferry
Port Talbot
Maesteg
Pontycymer
Rhondda
Pontypridd
CAERPHILLY
Caerphilly
Bedwas
Newport
NEWPORT

St. Govan's Head
Rhossili
Worms Head
Bishopston
The Mumbles
Swansea Bay
Margam
Gower
Port-Eynon
Gilfach Goch
Llanharan
Llantrisant
Pontyclun
Rumney
Risca
Cwmbran
Caefeon

-34
Porthcawl
Pyle
BRIDGEND
Bridgend
Pencoed
Cowbridge
THE VALE OF GLAMORGAN
CARDIFF
Cardiff
Penarth
Av
Portishe

Llantwit Major
Barry
Dinas Powys
MC
SOM

Bristol Channel
Rhoose
Flat Holm
Weston-super-Mare
Clevedon
Congresb

A          5°W          B          4°W          C          3°W

Transverse Mercator Projection          © Oxford University Pres

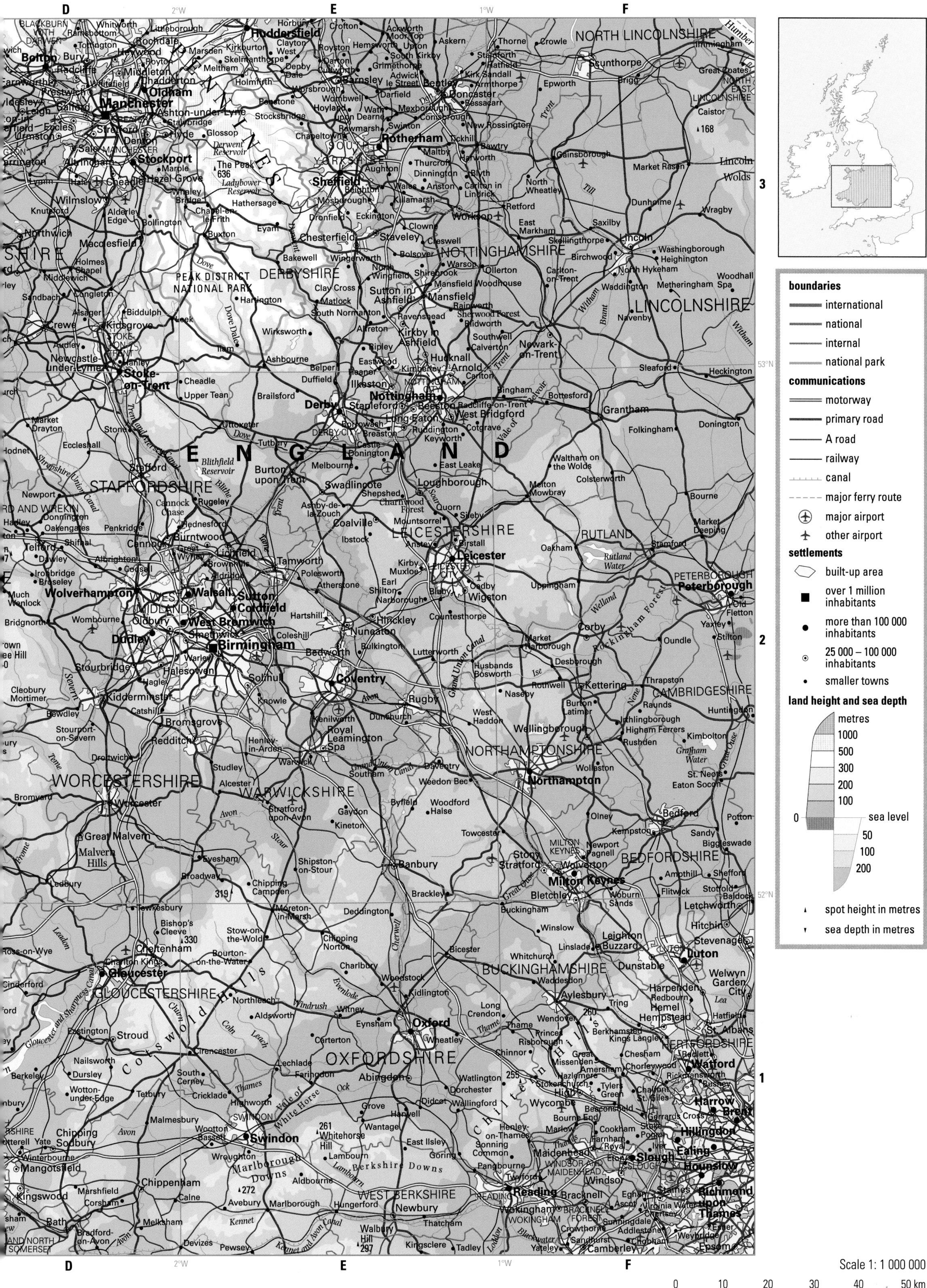

**boundaries**
international
national
internal
national park

**communications**
motorway
primary road
A road
railway
canal
major ferry route
✈ major airport
✈ other airport

**settlements**
⬡ built-up area
■ over 1 million inhabitants
● more than 100 000 inhabitants
◉ 25 000 – 100 000 inhabitants
• smaller towns

**land height and sea depth**
metres
1000
500
300
200
100
0 sea level
50
100
200
▲ spot height in metres
▼ sea depth in metres

Scale 1: 1 000 000

0   10   20   30   40   50 km

# South East England

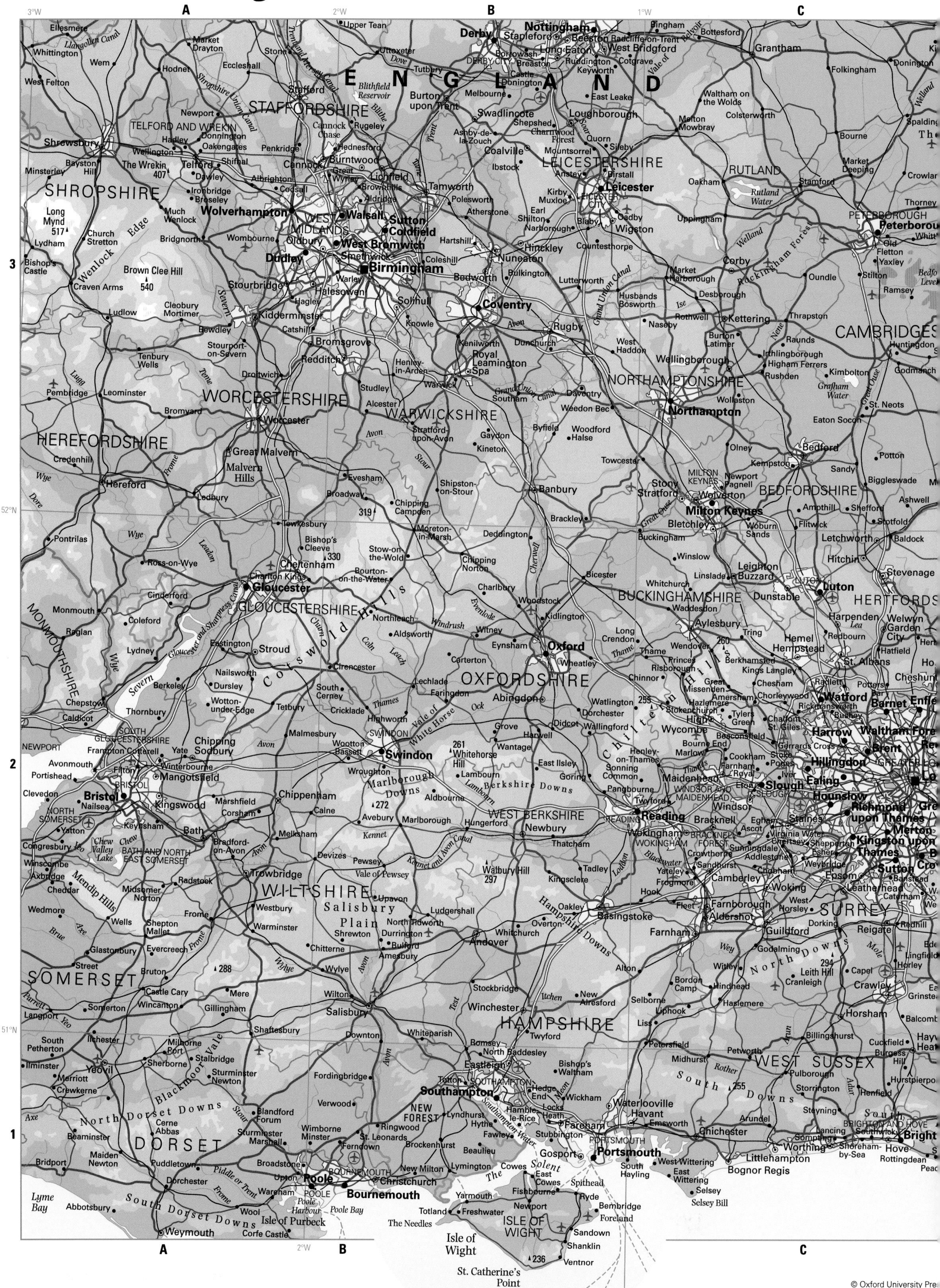

3°W
A
2°W
B
1°W
C
52°N
51°N
3
2
1
ENGLAND
SHROPSHIRE
TELFORD AND WREKIN
STAFFORDSHIRE
LEICESTERSHIRE
RUTLAND
WEST MIDLANDS
WORCESTERSHIRE
WARWICKSHIRE
HEREFORDSHIRE
NORTHAMPTONSHIRE
CAMBRIDGESHIRE
BEDFORDSHIRE
MILTON KEYNES
BUCKINGHAMSHIRE
HERTFORDSHIRE
GLOUCESTERSHIRE
OXFORDSHIRE
MONMOUTHSHIRE
SOUTH GLOUCESTERSHIRE
BRISTOL
NORTH SOMERSET
BATH AND NORTH EAST SOMERSET
SWINDON
WILTSHIRE
WEST BERKSHIRE
READING
WOKINGHAM
BRACKNELL FOREST
WINDSOR AND MAIDENHEAD
SLOUGH
GREATER LONDON
SURREY
HAMPSHIRE
SOUTHAMPTON
PORTSMOUTH
WEST SUSSEX
BRIGHTON AND HOVE
SOMERSET
DORSET
POOLE
BOURNEMOUTH
ISLE OF WIGHT
PETERBOROUGH
LUTON
DERBY CITY
LEICESTER CITY
Nottingham
Derby
Leicester
Birmingham
Wolverhampton
Walsall
West Bromwich
Dudley
Coventry
Northampton
Milton Keynes
Luton
Oxford
Gloucester
Cheltenham
Worcester
Hereford
Bristol
Bath
Swindon
Reading
Slough
Southampton
Portsmouth
Bournemouth
Poole
Peterborough
Brighton
Watford
Barnet
Harrow
Brent
Ealing
Hillingdon
Hounslow
Richmond upon Thames
Kingston upon Thames
Merton
Sutton
Waltham Forest
Shrewsbury
Telford
Stafford
Tamworth
Lichfield
Nuneaton
Rugby
Kettering
Corby
Bedford
Aylesbury
Banbury
Stratford-upon-Avon
Royal Leamington Spa
Warwick
Kidderminster
Redditch
Bromsgrove
Evesham
Tewkesbury
Stroud
Cirencester
Newbury
Basingstoke
Winchester
Salisbury
Andover
Guildford
Woking
Crawley
Horsham
Chichester
Worthing
Bognor Regis
Gosport
Fareham
Havant
Eastleigh
Yeovil
Dorchester
Weymouth
Isle of Purbeck
Isle of Wight
Newport
Ryde
Sandown
Shanklin
Ventnor
St. Catherine's Point
The Needles
Selsey Bill
Lyme Bay
Poole Bay
The Solent
Spithead
Southampton Water
Cotswold Hills
Chiltern Hills
North Downs
South Downs
Hampshire Downs
Berkshire Downs
Marlborough Downs
Salisbury Plain
Mendip Hills
Blackmoor Vale
North Dorset Downs
South Dorset Downs
Wenlock Edge
Long Mynd 517
Brown Clee Hill 540
Malvern Hills
Cannock Chase
Charnwood Forest
Rockingham Forest
New Forest
Rutland Water
Grafham Water
Blithfield Reservoir
Vale of White Horse
Vale of Pewsey
Vale of Belvoir
Severn
Avon
Thames
Trent
Wye
Kennet
Kennet and Avon Canal
Grand Union Canal
Gloucester and Sharpness Canal
Shropshire Union Canal
Trent and Mersey Canal
Llangollen Canal
Great Ouse
Nene
Welland
Cherwell
Evenlode
Windrush
Test
Itchen
Stour
Frome
Wey
Mole
Arun
Rother

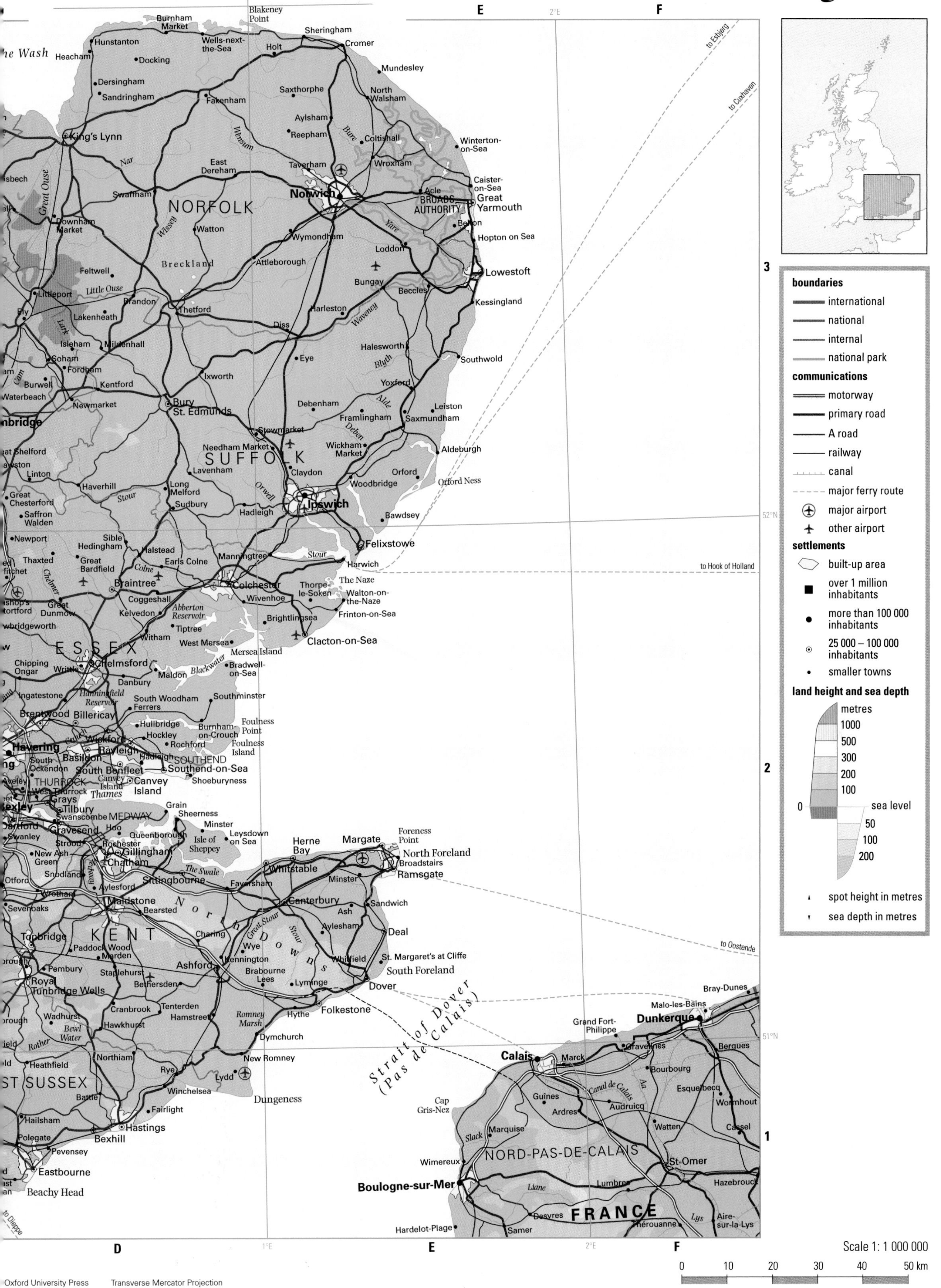
boundaries
international
national
internal
national park
communications
motorway
primary road
A road
railway
canal
major ferry route
major airport
other airport
settlements
built-up area
over 1 million inhabitants
more than 100 000 inhabitants
25 000 – 100 000 inhabitants
smaller towns
land height and sea depth
metres
1000
500
300
200
100
0
sea level
50
100
200
spot height in metres
sea depth in metres
Scale 1: 1 000 000
0
10
20
30
40
50 km
D
E
F
1
2
3
1°E
2°E
52°N
51°N
NORFOLK
SUFFOLK
ESSEX
KENT
NORD-PAS-DE-CALAIS
FRANCE
Strait of Dover (Pas de Calais)
The Wash
Hunstanton
Heacham
Burnham Market
Wells-next-the-Sea
Blakeney Point
Holt
Sheringham
Cromer
Mundesley
Docking
Dersingham
Sandringham
Fakenham
Saxthorpe
North Walsham
Aylsham
Reepham
Coltishall
Winterton-on-Sea
King's Lynn
Nar
East Dereham
Taverham
Wroxham
Caister-on-Sea
Great Yarmouth
Acle
BROADS AUTHORITY
Norwich
Swaffham
Downham Market
Watton
Wymondham
Belton
Hopton on Sea
Loddon
Breckland
Attleborough
Lowestoft
Feltwell
Bungay
Beccles
Kessingland
Littleport
Little Ouse
Brandon
Thetford
Harleston
Diss
Lakenheath
Ely
Isleham
Mildenhall
Halesworth
Soham
Fordham
Eye
Southwold
Burwell
Kentford
Ixworth
Yoxford
Newmarket
Bury St. Edmunds
Debenham
Leiston
Framlingham
Saxmundham
Stowmarket
Needham Market
Wickham Market
Aldeburgh
Shelford
Linton
Lavenham
Claydon
Orford
Orford Ness
Haverhill
Long Melford
Woodbridge
Great Chesterford
Saffron Walden
Sudbury
Hadleigh
Ipswich
Bawdsey
Newport
Sible Hedingham
Halstead
Felixstowe
Thaxted
Great Bardfield
Earls Colne
Manningtree
Harwich
to Hook of Holland
Braintree
Colchester
Thorpe-le-Soken
The Naze
Walton-on-the-Naze
Coggeshall
Wivenhoe
Great Dunmow
Kelvedon
Abberton Reservoir
Brightlingsea
Frinton-on-Sea
Tiptree
Witham
West Mersea
Clacton-on-Sea
Mersea Island
Chipping Ongar
Writtle
Chelmsford
Maldon
Blackwater
Bradwell-on-Sea
Danbury
Ingatestone
Hanningfield Reservoir
South Woodham Ferrers
Southminster
Brentwood
Billericay
Hullbridge
Hockley
Burnham-on-Crouch
Foulness Point
Foulness Island
Wickford
Rayleigh
Rochford
Havering
Basildon
Hadleigh
South Ockendon
South Benfleet
SOUTHEND
Southend-on-Sea
Shoeburyness
THURROCK
West Thurrock
Canvey Island
Grays
Thames
Tilbury
Grain
Swanscombe
MEDWAY
Sheerness
Minster
Gravesend
Hoo
Queenborough
Isle of Sheppey
Leysdown on Sea
Herne Bay
Margate
Foreness Point
North Foreland
Broadstairs
Ramsgate
Swanley
Strood
Rochester
Gillingham
Chatham
New Ash Green
Snodland
The Swale
Whitstable
Minster
Sittingbourne
Faversham
Aylesford
Canterbury
Sandwich
Maidstone
Bearsted
Ash
Sevenoaks
North Downs
Great Stour
Stour
Aylesham
Deal
Tonbridge
Charing
Paddock Wood
Marden
Wye
Whitfield
St. Margaret's at Cliffe
South Foreland
Pembury
Staplehurst
Ashford
Kennington
Brabourne Lees
Lyminge
Dover
Royal Tunbridge Wells
Bethersden
Cranbrook
Tenterden
Hythe
Folkestone
Wadhurst
Bewl Water
Hawkhurst
Hamstreet
Romney Marsh
Dymchurch
Rother
Heathfield
Northiam
New Romney
Rye
Lydd
Dungeness
Winchelsea
Battle
Fairlight
Hailsham
Hastings
Bexhill
Polegate
Pevensey
Eastbourne
Beachy Head
to Dieppe
to Esbjerg
to Cuxhaven
to Oostende
Bray-Dunes
Malo-les-Bains
Dunkerque
Grand Fort-Philippe
Gravelines
Bergues
Calais
Marck
Bourbourg
Aa
Canal de Calais
Guînes
Ardres
Audruicq
Esquelbecq
Wormhout
Cap Gris-Nez
Watten
Cassel
Marquise
Slack
Wimereux
St-Omer
Boulogne-sur-Mer
Liane
Lumbres
Hazebrouck
Desvres
Samer
Thérouanne
Lys
Aire-sur-la-Lys
Hardelot-Plage
Great Ouse
Wissey
Lark
Cam
Wensum
Bure
Yare
Waveney
Blyth
Alde
Deben
Orwell
Stour
Colne
Chelmer
Crouch
Medway
Oxford University Press
Transverse Mercator Projection

# South West England

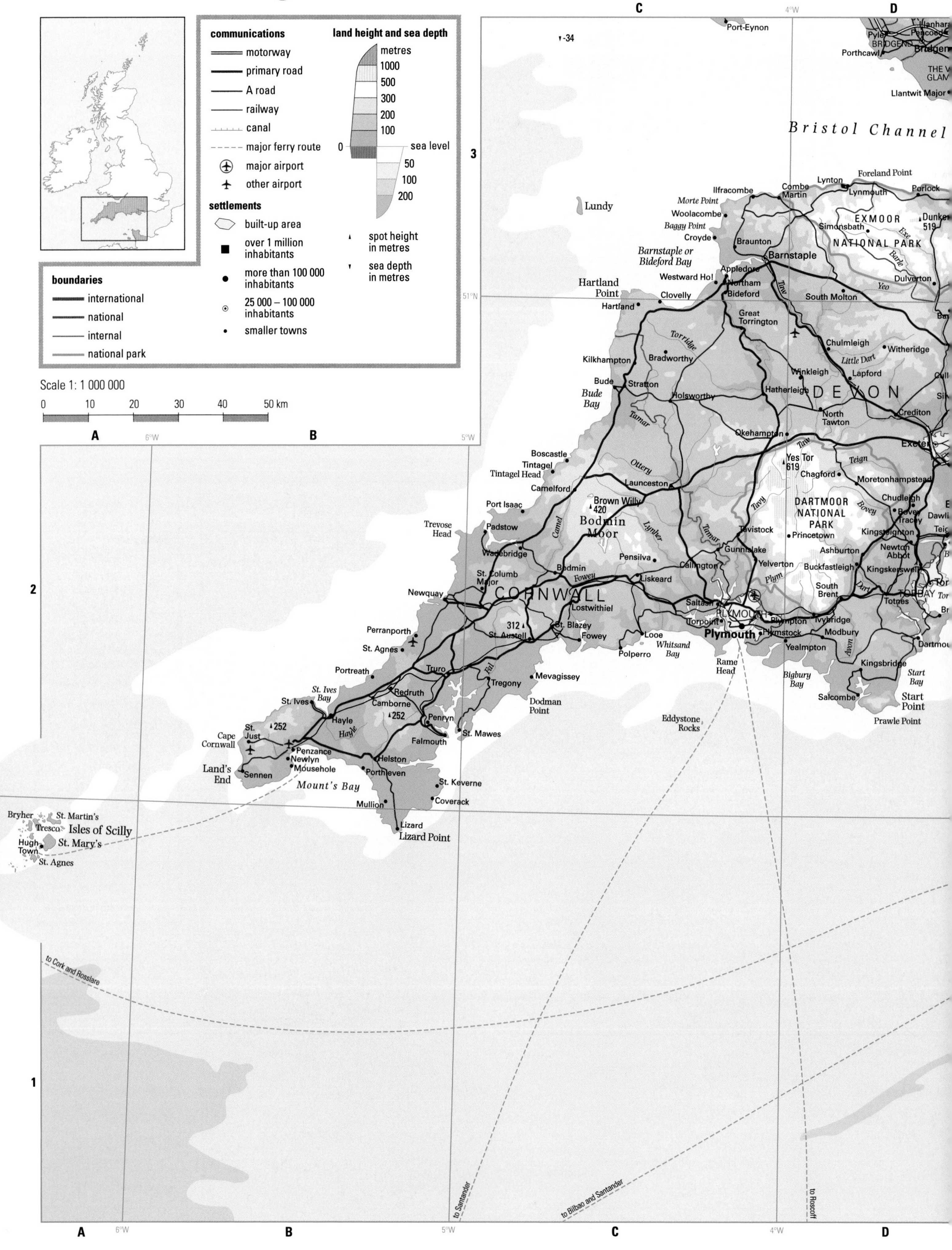
communications
motorway
primary road
A road
railway
canal
major ferry route
major airport
other airport
land height and sea depth
metres
1000
500
300
200
100
0
sea level
50
100
200
settlements
built-up area
over 1 million inhabitants
more than 100 000 inhabitants
25 000 – 100 000 inhabitants
smaller towns
spot height in metres
sea depth in metres
boundaries
international
national
internal
national park
Scale 1: 1 000 000
0 10 20 30 40 50 km
Bristol Channel
Lundy
EXMOOR NATIONAL PARK
DEVON
DARTMOOR NATIONAL PARK
CORNWALL
Bodmin Moor
Barnstaple
Plymouth
Exeter
Newquay
Truro
Penzance
Falmouth
St. Ives
St. Austell
Bideford
Ilfracombe
Okehampton
Launceston
Bude
Tavistock
Lands End
Lizard Point
Mount's Bay
Isles of Scilly
St. Mary's
Tresco
Bryher
St. Martin's
Hugh Town
St. Agnes
Yes Tor 619
Brown Willy 420
Barnstaple or Bideford Bay
Hartland Point
Start Point
Prawle Point
Eddystone Rocks
Whitsand Bay
to Cork and Rosslare
to Santander
to Bilbao and Santander
to Roscoff
51°N
6°W
5°W
4°W

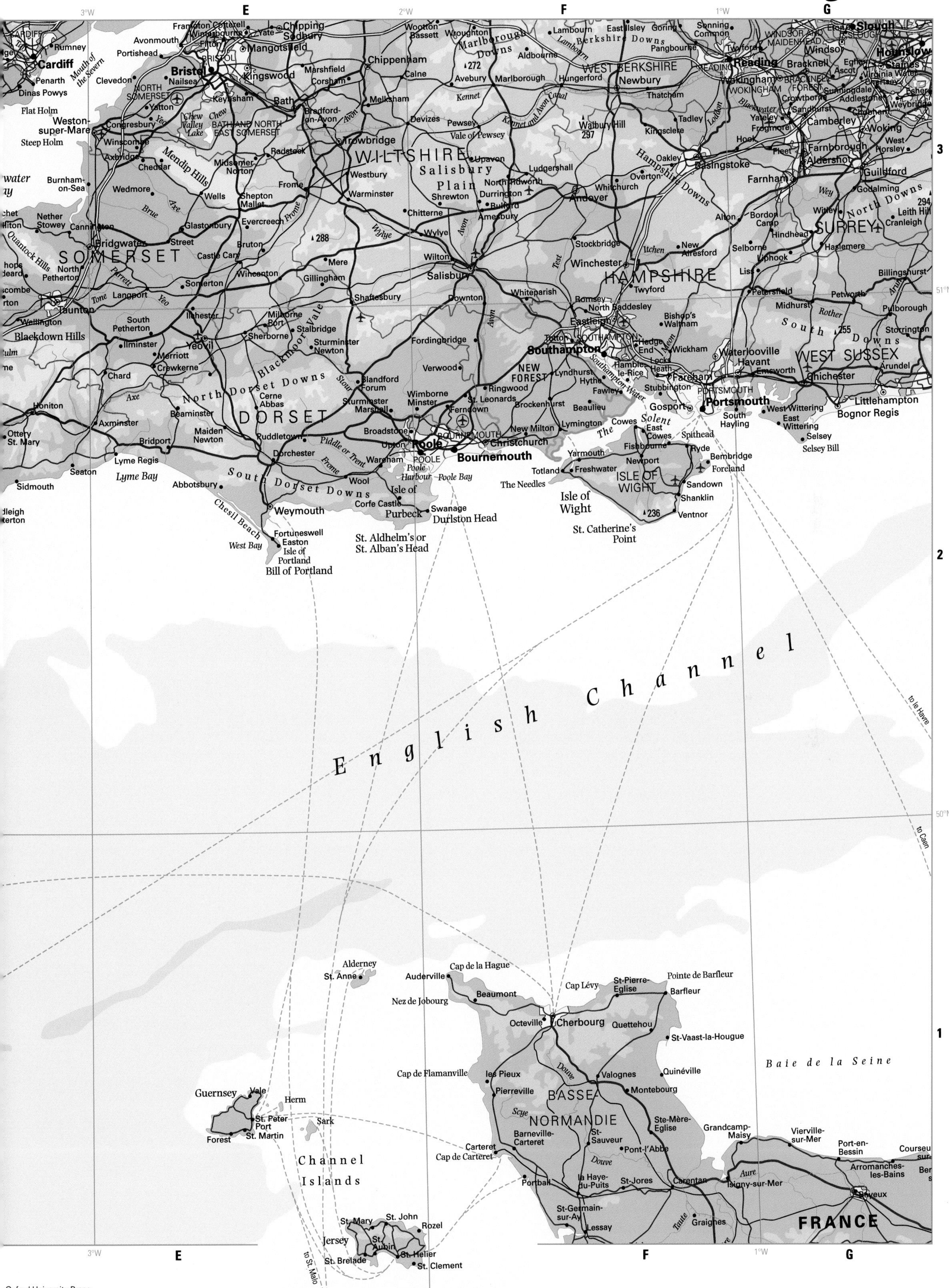
English Channel
SOMERSET
WILTSHIRE
DORSET
HAMPSHIRE
WEST BERKSHIRE
SURREY
WEST SUSSEX
ISLE OF WIGHT
NEW FOREST
BATH AND NORTH EAST SOMERSET
NORTH SOMERSET
BASSE-NORMANDIE
FRANCE
Channel Islands
Cardiff
Penarth
Dinas Powys
Rumney
Mouth of the Severn
Flat Holm
Steep Holm
Avonmouth
Portishead
Clevedon
Bristol
Nailsea
Kingswood
Mangotsfield
Keynsham
Bath
Yatton
Congresbury
Weston-super-Mare
Winscombe
Axbridge
Cheddar
Mendip Hills
Chew Valley Lake
Burnham-on-Sea
Wedmore
Wells
Midsomer Norton
Radstock
Frome
Shepton Mallet
Glastonbury
Street
Bridgwater
Nether Stowey
Cannington
Quantock Hills
North Petherton
Taunton
Wellington
Blackdown Hills
Langport
Somerton
Castle Cary
Bruton
Evercreech
Wincanton
Ilchester
South Petherton
Ilminster
Yeovil
Chard
Crewkerne
Merriott
Honiton
Axminster
Ottery St. Mary
Sidmouth
Seaton
Lyme Regis
Lyme Bay
Bridport
Beaminster
Maiden Newton
Cerne Abbas
Puddletown
Dorchester
Abbotsbury
Chesil Beach
West Bay
Weymouth
Fortuneswell
Easton
Isle of Portland
Bill of Portland
South Dorset Downs
North Dorset Downs
Blackmoor Vale
Sherborne
Milborne Port
Stalbridge
Sturminster Newton
Gillingham
Mere
Shaftesbury
Blandford Forum
Sturminster Marshall
Wimborne Minster
Ferndown
St. Leonards
Ringwood
Verwood
Fordingbridge
Broadstone
Upton
Poole
Poole Harbour
Poole Bay
Bournemouth
Christchurch
Wareham
Wool
Isle of Purbeck
Corfe Castle
Swanage
Durlston Head
St. Aldhelm's or St. Alban's Head
Chippenham
Calne
Corsham
Marshfield
Chipping Sodbury
Yate
Frampton Cotterell
Winterbourne
Filton
Melksham
Bradford-on-Avon
Trowbridge
Devizes
Westbury
Warminster
Chitterne
Wylye
Wilton
Salisbury
Salisbury Plain
Shrewton
Durrington
Bulford
Amesbury
Upavon
Pewsey
Vale of Pewsey
Marlborough
Marlborough Downs
Wootton Bassett
Wroughton
Aldbourne
Avebury
Ludgershall
North Tidworth
Andover
Whitchurch
Downton
Whiteparish
Romsey
North Baddesley
Eastleigh
Totton
Southampton
Southampton Water
Hedge End
Bishop's Waltham
Wickham
Fareham
Gosport
Portsmouth
South Hayling
Havant
Waterlooville
Emsworth
Chichester
West Wittering
East Wittering
Selsey
Selsey Bill
Bognor Regis
Littlehampton
Arundel
Storrington
Pulborough
Petworth
Midhurst
Petersfield
Billingshurst
Liss
Selborne
Liphook
Haslemere
Hindhead
Bordon Camp
Alton
Farnham
Aldershot
Farnborough
Fleet
Hook
Basingstoke
Oakley
Overton
Hampshire Downs
Winchester
Stockbridge
New Alresford
Twyford
Lyndhurst
Hythe
Fawley
Beaulieu
Brockenhurst
Lymington
New Milton
Hamble-le-Rice
Locks Heath
Stubbington
The Solent
Cowes
East Cowes
Spithead
Ryde
Fishbourne
Newport
Yarmouth
Freshwater
Totland
The Needles
Isle of Wight
Sandown
Shanklin
Ventnor
St. Catherine's Point
Bembridge
Foreland
Lambourn
East Ilsley
Goring
Pangbourne
Berkshire Downs
Sonning Common
Reading
Twyford
Wokingham
Bracknell
Crowthorne
Sandhurst
Yateley
Frogmore
Camberley
Woking
Guildford
Godalming
Witley
Cranleigh
Leith Hill
North Downs
South Downs
Newbury
Thatcham
Hungerford
Kingsclere
Tadley
Walbury Hill
Windsor
Slough
Ascot
Egham
Staines
Hounslow
Virginia Water
Sunningdale
Chertsey
Addlestone
Weybridge
Esher
West Horsley
Chobham
Alderney
St. Anne
Guernsey
Vale
Herm
Sark
St. Peter Port
St. Martin
Forest
Jersey
St. Mary
St. John
Rozel
St. Aubin
St. Helier
St. Brelade
St. Clement
to St. Malo
Cap de la Hague
Auderville
Nez de Jobourg
Beaumont
Octeville
Cherbourg
Cap Lévy
St-Pierre-Eglise
Pointe de Barfleur
Barfleur
Quettehou
St-Vaast-la-Hougue
Cap de Flamanville
les Pieux
Pierreville
Valognes
Quinéville
Montebourg
Ste-Mère-Eglise
St-Sauveur
Pont-l'Abbé
Barneville-Carteret
Carteret
Cap de Carteret
Portbail
la Haye-du-Puits
St-Jores
Carentan
Isigny-sur-Mer
Grandcamp-Maisy
Vierville-sur-Mer
Port-en-Bessin
Arromanches-les-Bains
Bayeux
St-Germain-sur-Ay
Lessay
Graignes
Baie de la Seine
to le Havre
to Caen
3°W
2°W
1°W
51°N
50°N

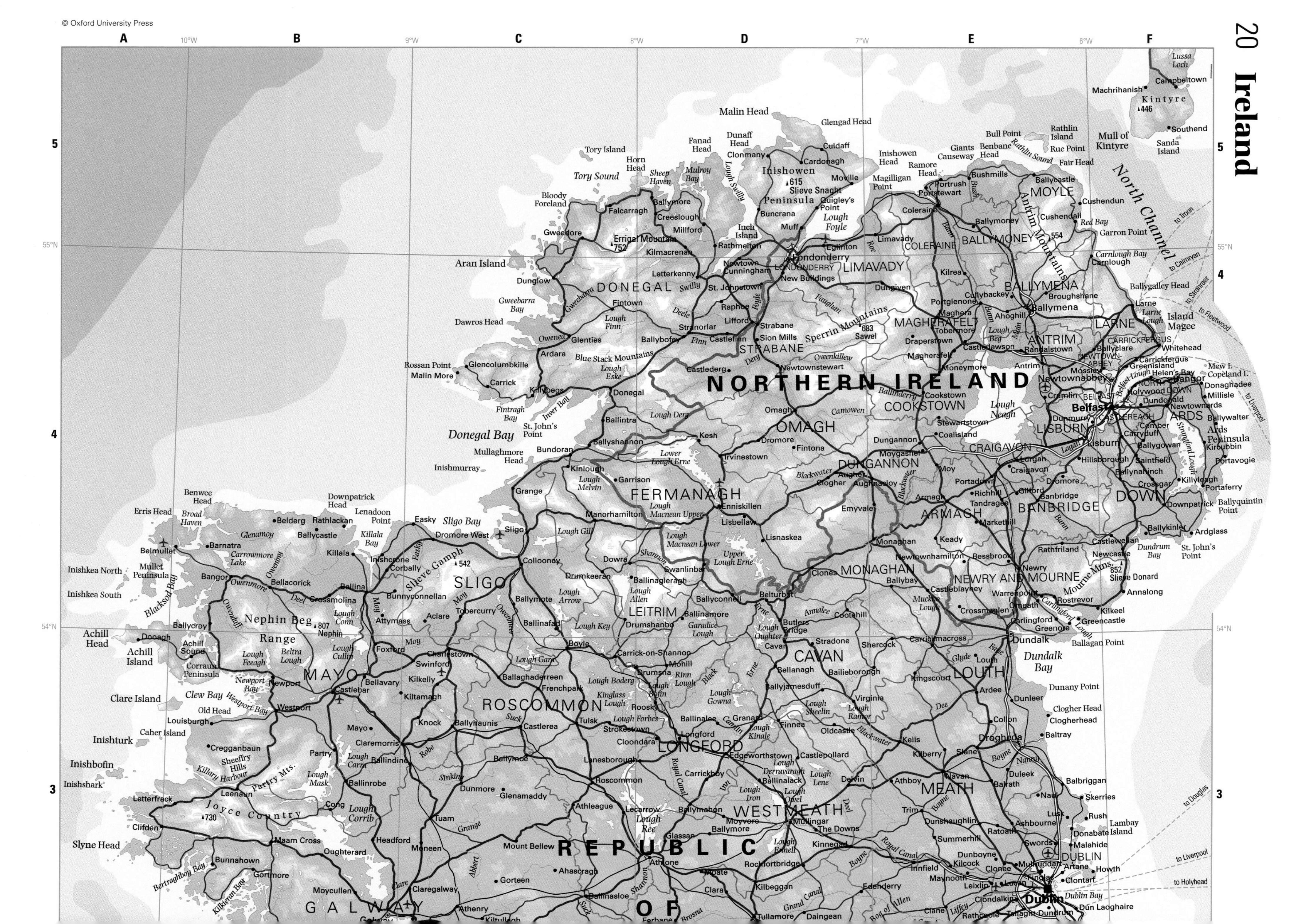
NORTHERN IRELAND
REPUBLIC OF
North Channel
Kintyre
Mull of Kintyre
Campbeltown
Machrihanish
Southend
Sanda Island
Rathlin Island
Fair Head
Antrim Mountains
Giants Causeway
Portrush
Coleraine
Ballymena
Larne
Belfast
Lisburn
Bangor
Newtownards
Strangford Lough
Lough Neagh
Dundalk
Dundalk Bay
Drogheda
DUBLIN
Dublin
Dublin Bay
Howth
Malahide
Lambay Island
Londonderry
Lough Foyle
Inishowen Peninsula
Malin Head
Lough Swilly
Letterkenny
DONEGAL
Donegal
Donegal Bay
Tory Island
Bloody Foreland
Aran Island
Sperrin Mountains
Omagh
Strabane
Enniskillen
Upper Lough Erne
Lower Lough Erne
FERMANAGH
ARMAGH
Armagh
DOWN
Mourne Mts.
Newry
MONAGHAN
Monaghan
CAVAN
Cavan
LOUTH
MEATH
WESTMEATH
Mullingar
LONGFORD
Longford
LEITRIM
SLIGO
Sligo
Sligo Bay
ROSCOMMON
Roscommon
Athlone
MAYO
Castlebar
Westport
Clew Bay
Achill Island
Achill Head
Nephin Beg Range
Ballina
Belmullet
Mullet Peninsula
Erris Head
Blacksod Bay
Inishkea North
Inishkea South
Clare Island
Inishturk
Inishbofin
Inishshark
Slyne Head
Clifden
Joyce Country
Lough Mask
Lough Corrib
GALWAY
Tuam
Ballinasloe
Shannon
Royal Canal
Grand Canal
Boyne
to Liverpool
to Holyhead
to Douglas
to Fleetwood
to Stranraer
to Cairnryan
to Troon

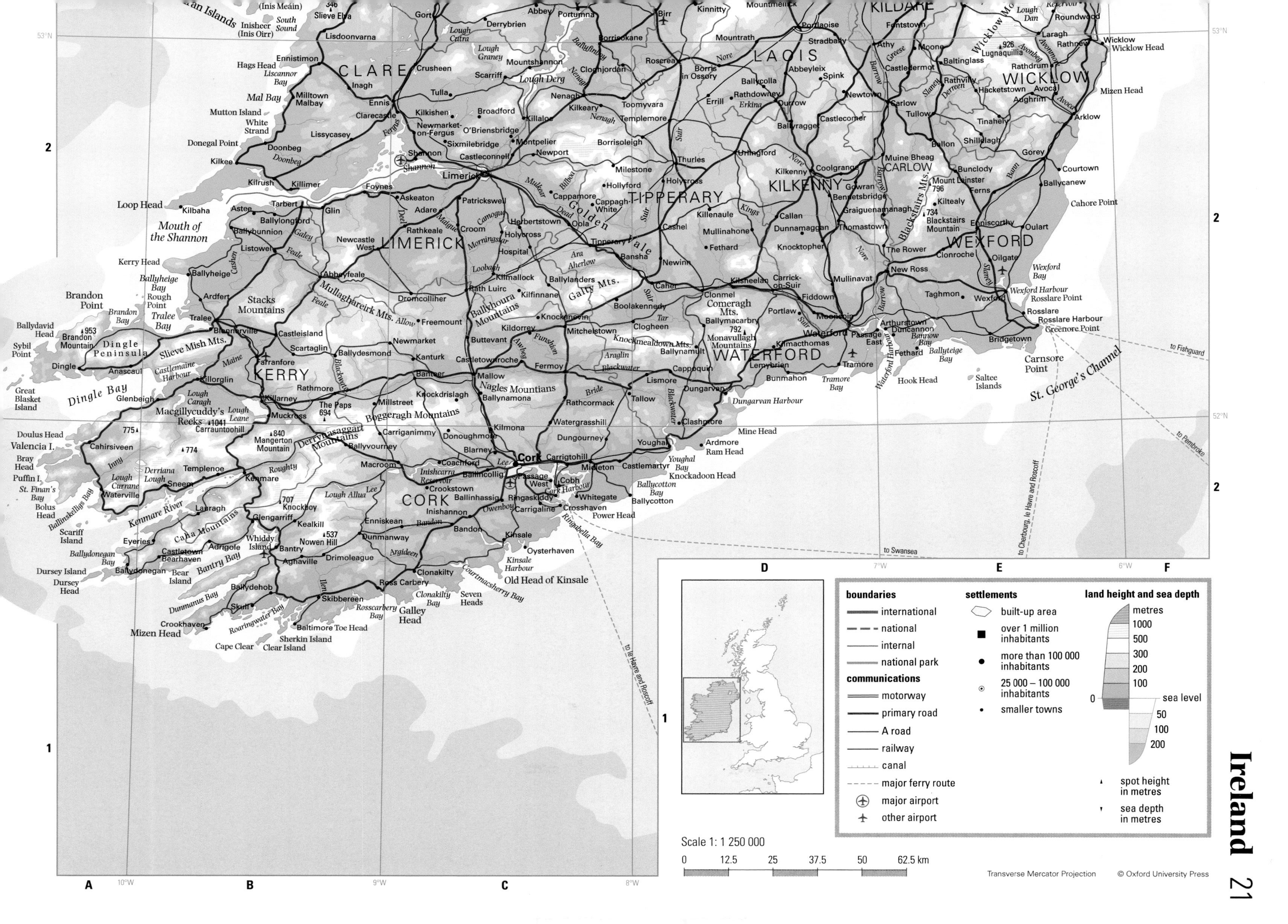
boundaries
international
national
internal
national park
communications
motorway
primary road
A road
railway
canal
major ferry route
major airport
other airport
settlements
built-up area
over 1 million inhabitants
more than 100 000 inhabitants
25 000 – 100 000 inhabitants
smaller towns
land height and sea depth
metres
1000
500
300
200
100
0
sea level
50
100
200
spot height in metres
sea depth in metres
Scale 1: 1 250 000
0
12.5
25
37.5
50
62.5 km
Transverse Mercator Projection
© Oxford University Press
CLARE
LIMERICK
KERRY
CORK
TIPPERARY
WATERFORD
KILKENNY
LAOIS
CARLOW
WEXFORD
WICKLOW
St. George's Channel
Dingle Bay
Mouth of the Shannon
Old Head of Kinsale
Cork
Limerick
Waterford
Wexford
Tralee
Killarney
Ennis
Clonmel
Kilkenny
Carlow
Wicklow
Dingle Peninsula
Slieve Mish Mts.
Macgillycuddy's Reeks
Galty Mts.
Comeragh Mts.
Blackstairs Mts.
Nagles Mountains
Boggeragh Mountains
Caha Mountains
Stacks Mountains
Loop Head
Brandon Point
Mizen Head
Carnsore Point
Wicklow Head
Valencia I.
Saltee Islands
to Fishguard
to Pembroke
to Swansea
to Cherbourg, le Havre and Roscoff
to le Havre and Roscoff
A
B
C
D
E
F
1
2

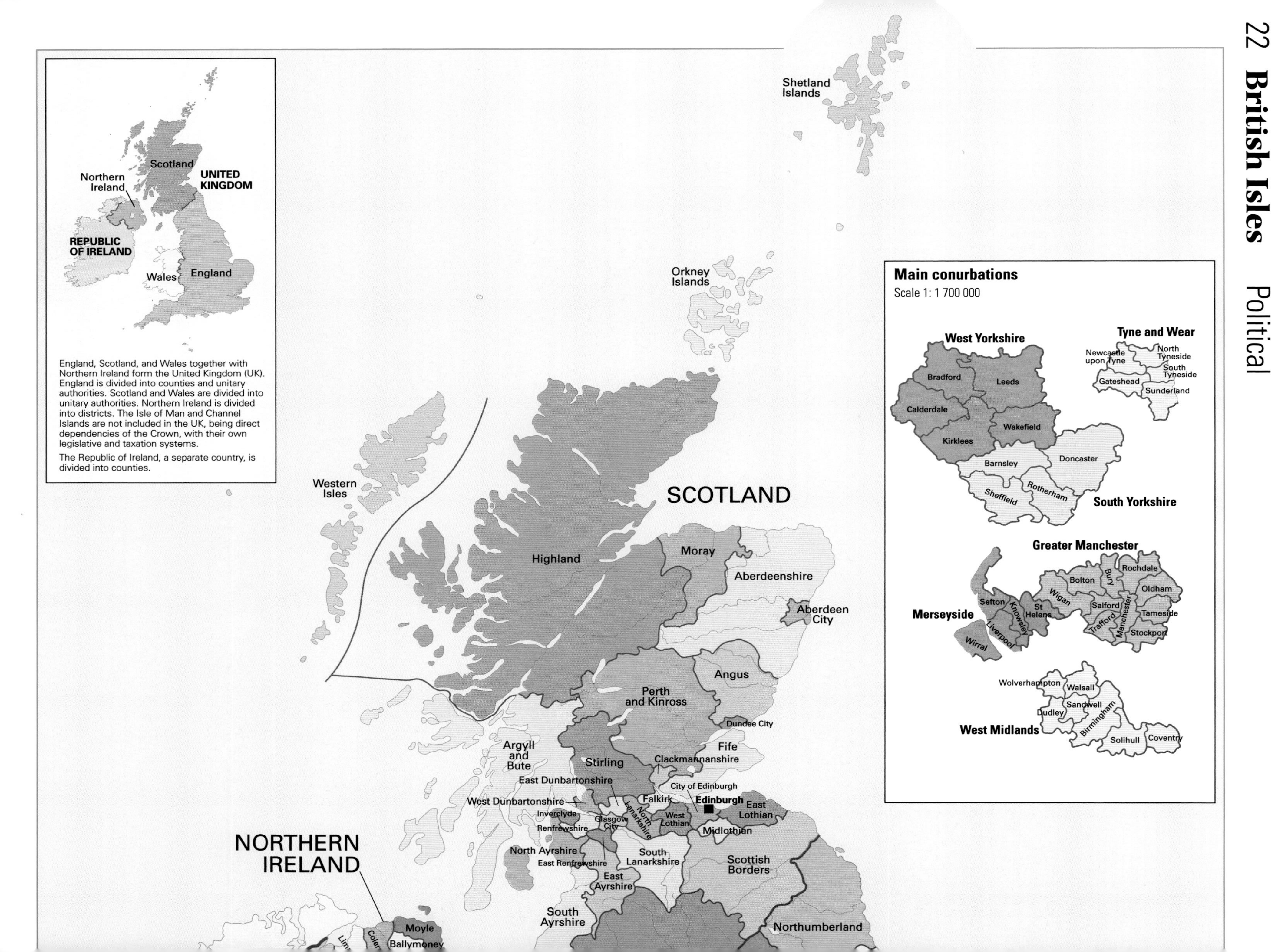

## Main conurbations
Scale 1: 1 700 000

**Tyne and Wear**
Newcastle upon Tyne
North Tyneside
South Tyneside
Gateshead
Sunderland

**South Yorkshire**
Doncaster
Barnsley
Rotherham
Sheffield

**West Yorkshire**
Leeds
Wakefield
Bradford
Calderdale
Kirklees

**Greater Manchester**
Rochdale
Oldham
Tameside
Stockport
Bury
Manchester
Salford
Trafford
Bolton
Wigan
St Helens
Knowsley

**Merseyside**
Sefton
Liverpool
Wirral

**West Midlands**
Wolverhampton
Walsall
Sandwell
Birmingham
Solihull
Coventry
Dudley

**SCOTLAND**

Shetland Islands
Orkney Islands
Western Isles
Highland
Moray
Aberdeenshire
Aberdeen City
Angus
Dundee City
Perth and Kinross
Fife
Clackmannanshire
City of Edinburgh
**Edinburgh**
East Lothian
Midlothian
Scottish Borders
Falkirk
West Lothian
North Lanarkshire
South Lanarkshire
Stirling
East Dunbartonshire
Glasgow City
Inverclyde
Renfrewshire
East Renfrewshire
East Ayrshire
North Ayrshire
South Ayrshire
Argyll and Bute
West Dunbartonshire
Northumberland

**NORTHERN IRELAND**
Moyle
Ballymoney
Coler
Lim

England, Scotland, and Wales together with Northern Ireland form the United Kingdom (UK). England is divided into counties and unitary authorities. Scotland and Wales are divided into unitary authorities. Northern Ireland is divided into districts. The Isle of Man and Channel Islands are not included in the UK, being direct dependencies of the Crown, with their own legislative and taxation systems.

The Republic of Ireland, a separate country, is divided into counties.

**UNITED KINGDOM**
Scotland
England
Wales
Northern Ireland
**REPUBLIC OF IRELAND**

Scale 1: 3 000 000 (main map)

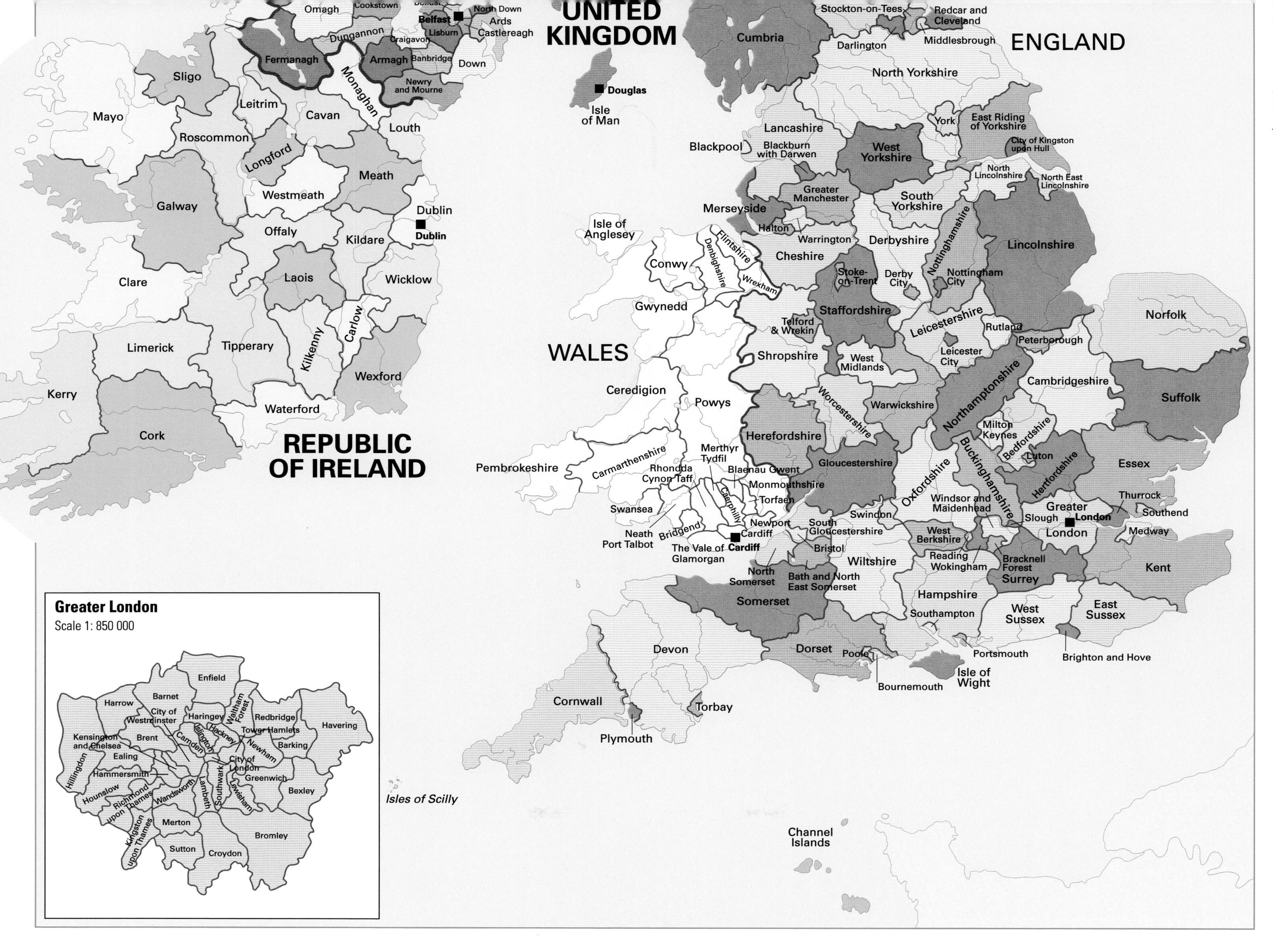

ENGLAND

UNITED KINGDOM

Redcar and Cleveland
Middlesbrough
Stockton-on-Tees
Darlington
North Yorkshire
East Riding of Yorkshire
City of Kingston upon Hull
North East Lincolnshire
North Lincolnshire
York
Lincolnshire
Norfolk
Suffolk
Essex
Thurrock
Southend
Medway
Kent
East Sussex
Brighton and Hove
Cumbria
Lancashire
West Yorkshire
South Yorkshire
Derbyshire
Nottinghamshire
Nottingham City
Leicestershire
Rutland
Leicester City
Peterborough
Cambridgeshire
Northamptonshire
Milton Keynes
Bedfordshire
Luton
Hertfordshire
Greater London
Slough
London
Bracknell Forest
Surrey
West Sussex
Portsmouth
Isle of Wight
Blackburn with Darwen
Blackpool
Greater Manchester
Warrington
Halton
Merseyside
Cheshire
Derby City
Stoke-on-Trent
Staffordshire
West Midlands
Warwickshire
Worcestershire
Telford & Wrekin
Shropshire
Herefordshire
Gloucestershire
Oxfordshire
Windsor and Maidenhead
West Berkshire
Reading
Wokingham
Hampshire
Southampton
Bournemouth
Swindon
South Gloucestershire
Bristol
Wiltshire
Bath and North East Somerset
North Somerset
Somerset
Dorset
Poole
Devon
Torbay
Torquay
Plymouth
Cornwall
Channel Islands

Douglas
Isle of Man

WALES

Flintshire
Wrexham
Denbighshire
Conwy
Gwynedd
Isle of Anglesey
Powys
Ceredigion
Pembrokeshire
Carmarthenshire
Swansea
Neath Port Talbot
Bridgend
The Vale of Glamorgan
Rhondda Cynon Taff
Merthyr Tydfil
Caerphilly
Blaenau Gwent
Torfaen
Monmouthshire
Newport
Cardiff

Isles of Scilly

North Down
Ards
Castlereagh
Down
Belfast
Lisburn
Banbridge
Newry and Mourne
Craigavon
Armagh
Cookstown
Dungannon
Omagh
Fermanagh
Monaghan

Dublin
Dublin

REPUBLIC OF IRELAND

Louth
Meath
Cavan
Leitrim
Longford
Westmeath
Offaly
Laois
Kildare
Wicklow
Wexford
Carlow
Kilkenny
Waterford
Tipperary
Sligo
Roscommon
Galway
Mayo
Clare
Limerick
Kerry
Cork

**Greater London**
Scale 1: 850 000

Havering
Redbridge
Barking
Newham
Bexley
Bromley
Greenwich
City of London
Lewisham
Waltham Forest
Hackney
Tower Hamlets
Islington
Southwark
Haringey
Camden
Lambeth
Wandsworth
Enfield
Barnet
Westminster
City of Westminster
Brent
Harrow
Ealing
Hammersmith
Hillingdon
Hounslow
Richmond upon Thames
Kingston upon Thames
Merton
Sutton
Croydon
Kensington and Chelsea

© Oxford University Press

# British Isles Physical

Scale 1: 4 500 000

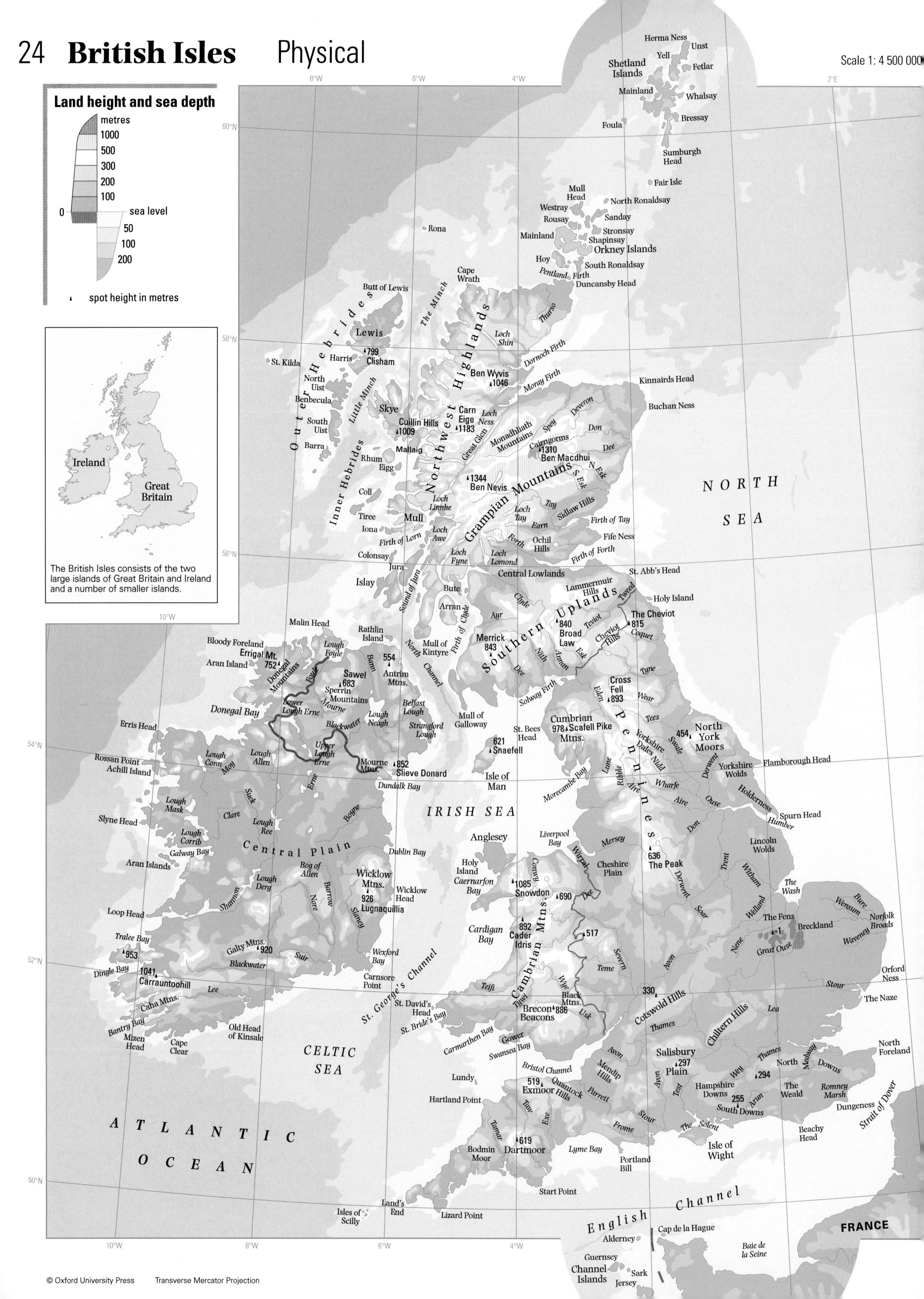

The British Isles consists of the two large islands of Great Britain and Ireland and a number of smaller islands.

 Transverse Mercator Projection

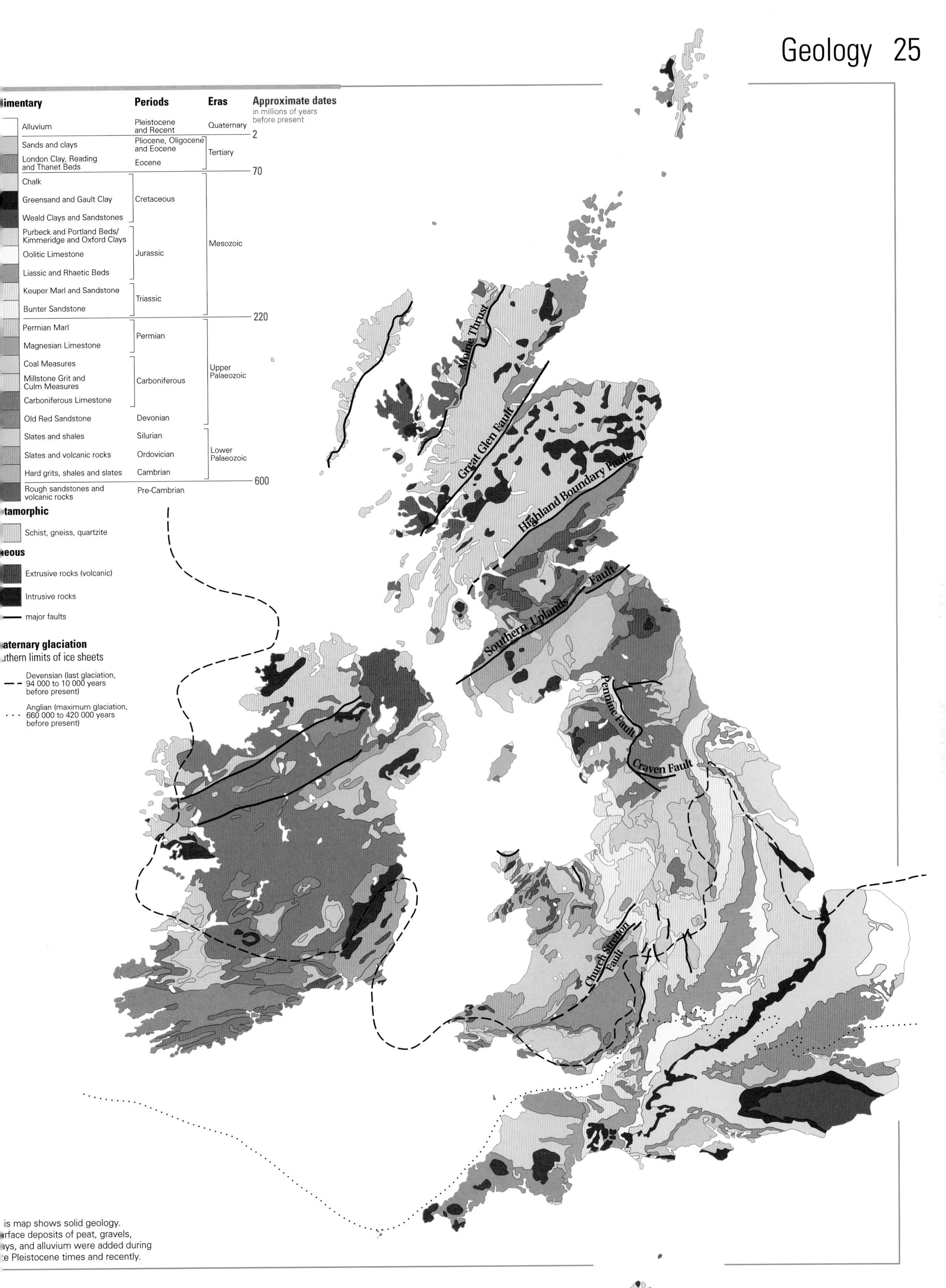

| imentary | Periods | Eras | Approximate dates in millions of years before present |
|---|---|---|---|
| Alluvium | Pleistocene and Recent | Quaternary | 2 |
| Sands and clays | Pliocene, Oligocene and Eocene | Tertiary | |
| London Clay, Reading and Thanet Beds | Eocene | | 70 |
| Chalk | Cretaceous | | |
| Greensand and Gault Clay | | | |
| Weald Clays and Sandstones | | | |
| Purbeck and Portland Beds/ Kimmeridge and Oxford Clays | Jurassic | Mesozoic | |
| Oolitic Limestone | | | |
| Liassic and Rhaetic Beds | | | |
| Keuper Marl and Sandstone | Triassic | | |
| Bunter Sandstone | | | 220 |
| Permian Marl | Permian | | |
| Magnesian Limestone | | | |
| Coal Measures | Carboniferous | Upper Palaeozoic | |
| Millstone Grit and Culm Measures | | | |
| Carboniferous Limestone | | | |
| Old Red Sandstone | Devonian | | |
| Slates and shales | Silurian | | |
| Slates and volcanic rocks | Ordovician | Lower Palaeozoic | |
| Hard grits, shales and slates | Cambrian | | 600 |
| Rough sandstones and volcanic rocks | Pre-Cambrian | | |

**tamorphic**

Schist, gneiss, quartzite

**neous**

Extrusive rocks (volcanic)

Intrusive rocks

— major faults

**aternary glaciation**
uthern limits of ice sheets

- - - Devensian (last glaciation, 94 000 to 10 000 years before present)

· · · Anglian (maximum glaciation, 660 000 to 420 000 years before present)

Moine Thrust

Great Glen Fault

Highland Boundary Fault

Southern Uplands Fault

Pennine Fault

Craven Fault

Church Stretton Fault

is map shows solid geology.
rface deposits of peat, gravels,
ys, and alluvium were added during
e Pleistocene times and recently.

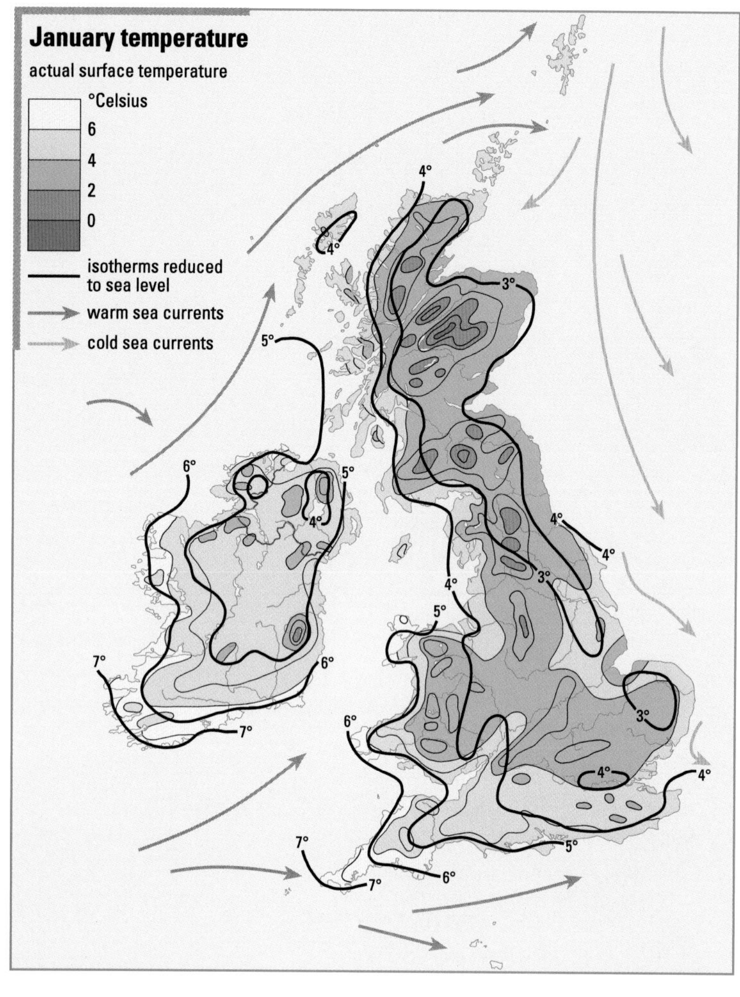

## January temperature

actual surface temperature

°Celsius
- 6
- 4
- 2
- 0

— isotherms reduced to sea level

→ warm sea currents

→ cold sea currents

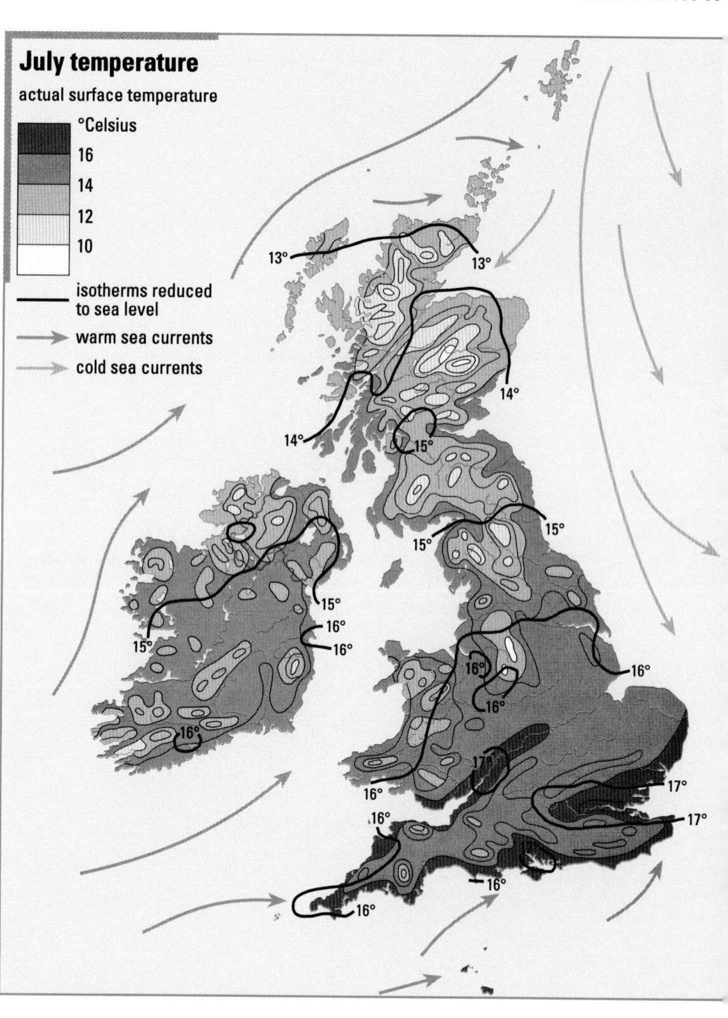

## July temperature

actual surface temperature

°Celsius
- 16
- 14
- 12
- 10

— isotherms reduced to sea level

→ warm sea currents

→ cold sea currents

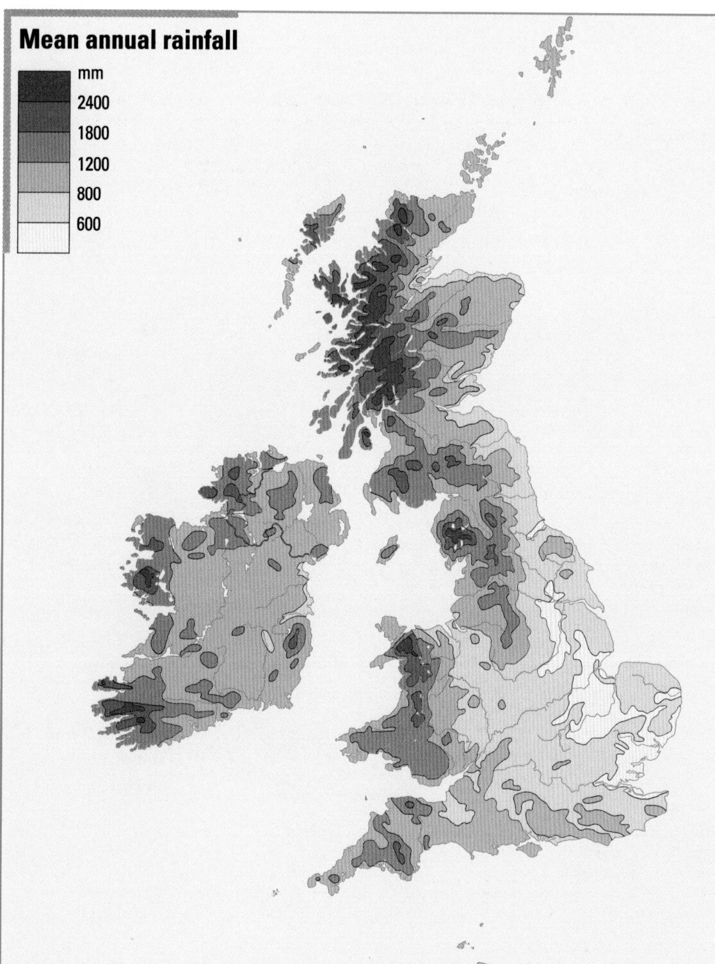

## Mean annual rainfall

mm
- 2400
- 1800
- 1200
- 800
- 600

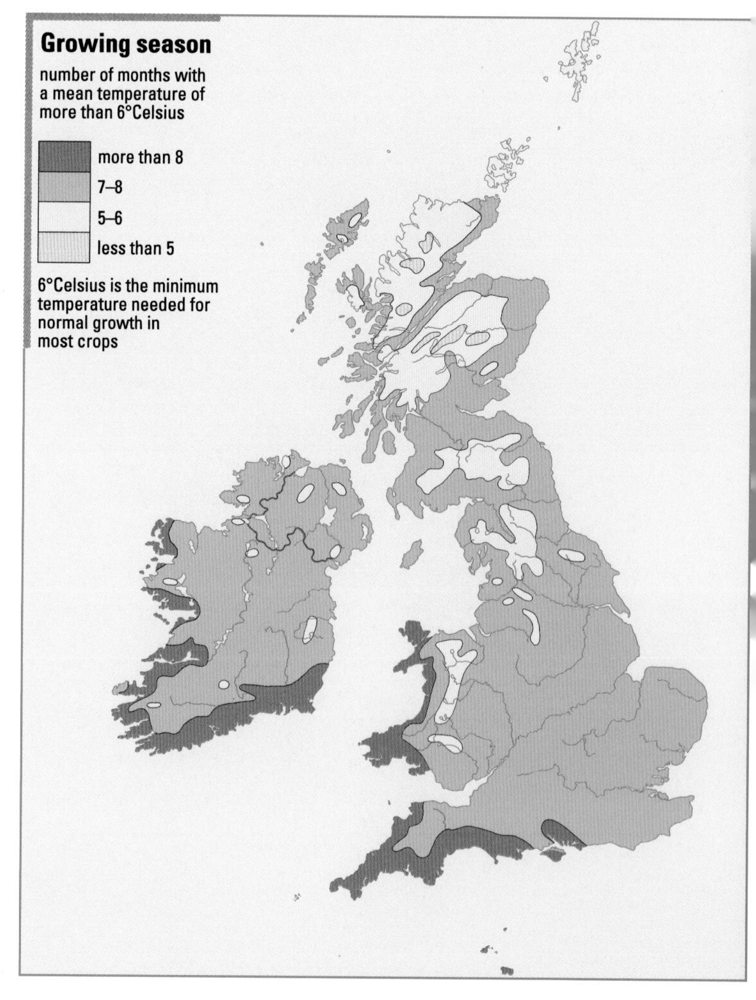

## Growing season

number of months with a mean temperature of more than 6°Celsius

- more than 8
- 7–8
- 5–6
- less than 5

6°Celsius is the minimum temperature needed for normal growth in most crops

Transverse Mercator Projection    © Oxford University Press

## Climate graphs for selected British stations

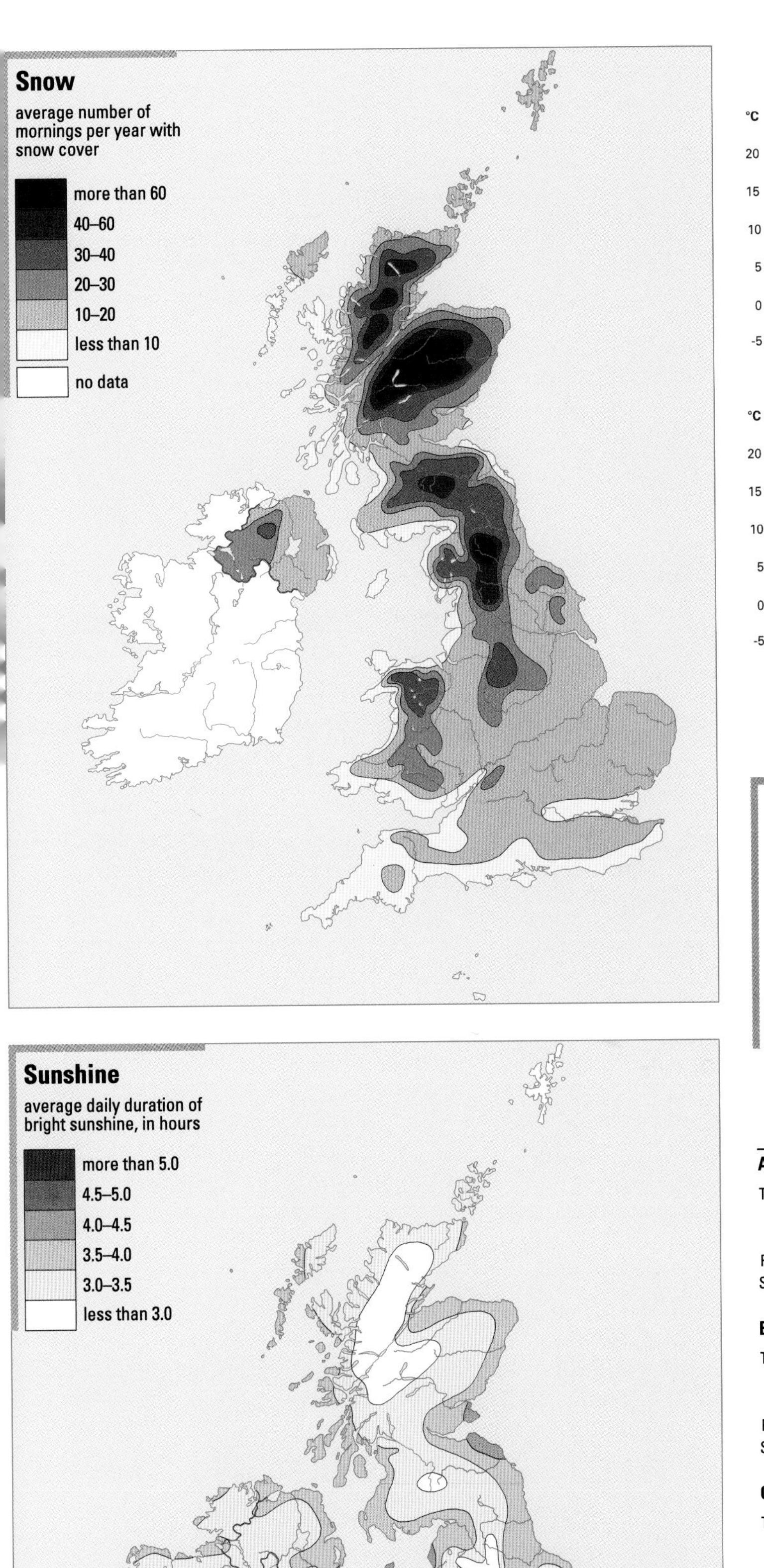

Belfast 68m · Edinburgh 26m · London 7m · St. Mawgan 103m · Stornoway 15m · Valencia 11m

## Climate stations

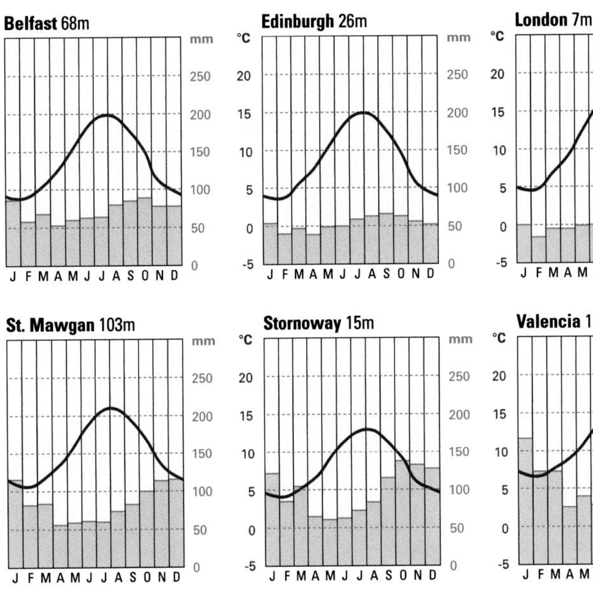

Stornoway · Braemar · Tiree · Edinburgh · Belfast · Anglesey · Valencia · Cambridge · St. Mawgan · London · Exeter

### Climate data
averages are for 1961–1990

**Anglesey (Valley)** 10m    climate station and its height above sea level

**Temperature (°C)**   high   average daily maximum temperature
      **mean**   average monthly temperature
      low   average daily minimum temperature

**Rainfall (mm)**    average monthly precipitation

**Sunshine (hours)**    average daily duration of bright sunshine

| | | Jan | Feb | Mar | Apr | May | Jun | Jul | Aug | Sep | Oct | Nov | Dec | YEAR |
|---|---|---|---|---|---|---|---|---|---|---|---|---|---|---|
| **Anglesey (Valley)** 10m | | | | | | | | | | | | | | |
| Temperature (°C) | high | 7.7 | 7.7 | 9.1 | 11.4 | 14.4 | 16.9 | 18.4 | 18.5 | 16.7 | 14.2 | 10.6 | 8.7 | 12.8 |
| | mean | 5.5 | 5.1 | 6.5 | 8.3 | 11.1 | 13.6 | 15.3 | 15.4 | 13.9 | 11.6 | 8.1 | 6.4 | 10.0 |
| | low | 3.2 | 2.5 | 3.8 | 5.1 | 7.7 | 10.3 | 12.2 | 12.3 | 11.0 | 8.9 | 5.6 | 4.1 | 7.2 |
| Rainfall (mm) | | 83 | 56 | 65 | 53 | 49 | 52 | 53 | 74 | 74 | 91 | 99 | 94 | 843 |
| Sunshine (hours) | | 1.8 | 3.0 | 4.0 | 5.9 | 7.2 | 7.0 | 6.4 | 6.0 | 4.7 | 3.3 | 2.2 | 1.6 | 4.4 |
| **Braemar** 339m | | | | | | | | | | | | | | |
| Temperature (°C) | high | 3.8 | 3.9 | 6.0 | 9.2 | 12.8 | 16.2 | 17.5 | 16.9 | 14.0 | 10.8 | 6.3 | 4.6 | 10.1 |
| | mean | 0.8 | 0.6 | 2.7 | 4.9 | 8.1 | 11.4 | 13.0 | 12.5 | 10.2 | 7.3 | 3.2 | 1.8 | 6.3 |
| | low | -2.2 | -2.7 | -0.7 | 0.6 | 3.4 | 6.5 | 8.4 | 8.0 | 6.3 | 3.8 | 0.1 | -1.0 | 2.5 |
| Rainfall (mm) | | 106 | 62 | 72 | 48 | 66 | 58 | 54 | 71 | 81 | 93 | 87 | 91 | 889 |
| Sunshine (hours) | | 0.8 | 2.0 | 3.1 | 4.6 | 5.2 | 5.6 | 5.1 | 4.8 | 3.5 | 2.2 | 1.2 | 0.6 | 3.2 |
| **Cambridge** 26m | | | | | | | | | | | | | | |
| Temperature (°C) | high | 6.4 | 6.8 | 9.7 | 12.5 | 16.4 | 19.6 | 21.5 | 21.5 | 18.8 | 14.9 | 9.7 | 7.3 | 13.7 |
| | mean | 3.7 | 3.9 | 6.0 | 8.2 | 11.6 | 14.6 | 16.6 | 16.5 | 14.3 | 11.0 | 6.6 | 4.6 | 9.8 |
| | low | 1.0 | 0.9 | 2.2 | 3.9 | 6.7 | 9.6 | 11.7 | 11.5 | 9.8 | 7.1 | 3.5 | 1.8 | 5.8 |
| Rainfall (mm) | | 43 | 32 | 42 | 43 | 49 | 50 | 44 | 53 | 46 | 49 | 51 | 49 | 551 |
| Sunshine (hours) | | 1.8 | 2.5 | 3.5 | 4.7 | 6.2 | 6.5 | 6.0 | 5.7 | 4.8 | 3.5 | 2.2 | 1.5 | 4.1 |
| **Exeter** 32m | | | | | | | | | | | | | | |
| Temperature (°C) | high | 8.0 | 8.0 | 10.2 | 12.7 | 15.8 | 19.1 | 21.0 | 20.8 | 18.4 | 15.0 | 11.0 | 9.0 | 14.0 |
| | mean | 5.0 | 5.0 | 6.6 | 8.6 | 11.5 | 14.6 | 16.5 | 16.3 | 14.2 | 11.4 | 7.8 | 6.0 | 10.2 |
| | low | 2.0 | 2.0 | 2.9 | 4.4 | 7.1 | 10.1 | 12.0 | 11.7 | 9.9 | 7.7 | 4.1 | 2.9 | 6.4 |
| Rainfall (mm) | | 93 | 71 | 61 | 50 | 54 | 47 | 45 | 54 | 57 | 73 | 72 | 87 | 764 |
| Sunshine (hours) | | 1.7 | 2.4 | 3.5 | 5.1 | 6.0 | 6.3 | 6.2 | 5.6 | 4.4 | 2.9 | 2.3 | 1.6 | 4.0 |
| **Tiree** 12m | | | | | | | | | | | | | | |
| Temperature (°C) | high | 7.3 | 7.1 | 8.3 | 10.1 | 12.6 | 14.7 | 15.8 | 16.0 | 14.5 | 12.5 | 9.5 | 8.1 | 11.3 |
| | mean | 5.1 | 4.9 | 5.8 | 7.3 | 9.7 | 11.9 | 13.3 | 13.5 | 12.2 | 10.3 | 7.2 | 5.6 | 8.9 |
| | low | 2.9 | 2.6 | 3.3 | 4.4 | 6.7 | 9.1 | 10.8 | 10.9 | 9.8 | 8.1 | 4.8 | 3.8 | 6.4 |
| Rainfall (mm) | | 127 | 79 | 96 | 59 | 59 | 61 | 78 | 95 | 129 | 140 | 122 | 120 | 1165 |
| Sunshine (hours) | | 1.3 | 2.4 | 3.5 | 5.7 | 6.9 | 6.5 | 5.1 | 5.0 | 3.7 | 2.5 | 1.6 | 1.0 | 3.8 |
| | | Jan | Feb | Mar | Apr | May | Jun | Jul | Aug | Sep | Oct | Nov | Dec | YEAR |

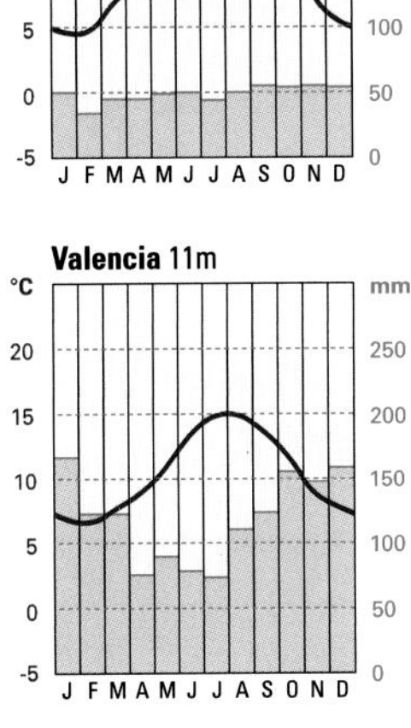

### Snow
average number of mornings per year with snow cover

- more than 60
- 40–60
- 30–40
- 20–30
- 10–20
- less than 10
- no data

### Sunshine
average daily duration of bright sunshine, in hours

- more than 5.0
- 4.5–5.0
- 4.0–4.5
- 3.5–4.0
- 3.0–3.5
- less than 3.0

Scale 1: 4 500 00

## Land use

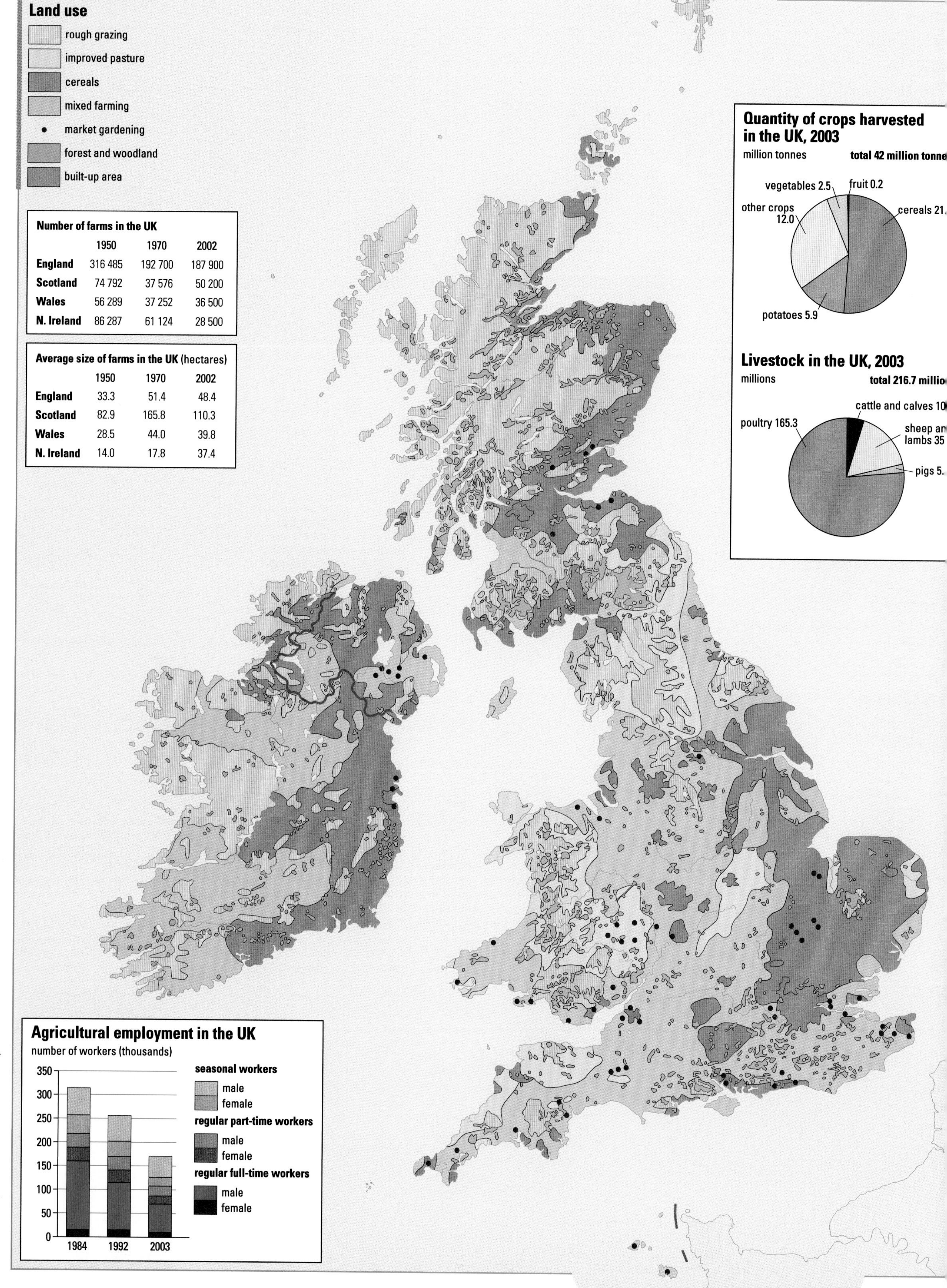

| | | |
|---|---|---|
| | | rough grazing |
| | | improved pasture |
| | | cereals |
| | | mixed farming |
| • | | market gardening |
| | | forest and woodland |
| | | built-up area |

### Number of farms in the UK

| | 1950 | 1970 | 2002 |
|---|---|---|---|
| **England** | 316 485 | 192 700 | 187 900 |
| **Scotland** | 74 792 | 37 576 | 50 200 |
| **Wales** | 56 289 | 37 252 | 36 500 |
| **N. Ireland** | 86 287 | 61 124 | 28 500 |

### Average size of farms in the UK (hectares)

| | 1950 | 1970 | 2002 |
|---|---|---|---|
| **England** | 33.3 | 51.4 | 48.4 |
| **Scotland** | 82.9 | 165.8 | 110.3 |
| **Wales** | 28.5 | 44.0 | 39.8 |
| **N. Ireland** | 14.0 | 17.8 | 37.4 |

### Quantity of crops harvested in the UK, 2003

million tonnes **total 42 million tonne**

vegetables 2.5  fruit 0.2
other crops 12.0
cereals 21.
potatoes 5.9

### Livestock in the UK, 2003

millions **total 216.7 millio**

poultry 165.3
cattle and calves 10
sheep an lambs 35
pigs 5.

### Agricultural employment in the UK

number of workers (thousands)

**seasonal workers**
male
female

**regular part-time workers**
male
female

**regular full-time workers**
male
female

| | 1984 | 1992 | 2003 |
|---|---|---|---|

(y-axis: 0, 50, 100, 150, 200, 250, 300, 350)

Transverse Mercator Projection

© Oxford University Press

Scale 1 : 12 500 000

### Wheat, 2003

percentage of farmland used for wheat

- over 40%
- 30–40%
- 20–30%
- 10–20%
- 0–10%

### Barley, 2003

percentage of farmland used for barley

- over 20%
- 15–20%
- 10–15%
- 5–10%
- 0–5%

### Potatoes, 2003

percentage of farmland used for potatoes

- over 3%
- 2–3%
- 1–2%
- 0.5–1%
- 0–0.5%

### Market gardening, 2003

percentage of farmland used for market gardening

- over 4%
- 3–4%
- 2–3%
- 1–2%
- under 1%

### Dairy and beef cattle, 2003

one dot represents 10 000 animals

- dairy cattle
- beef cattle

### Sheep and Pigs, 2003

one dot represents 100 000 animals

- sheep
- pigs

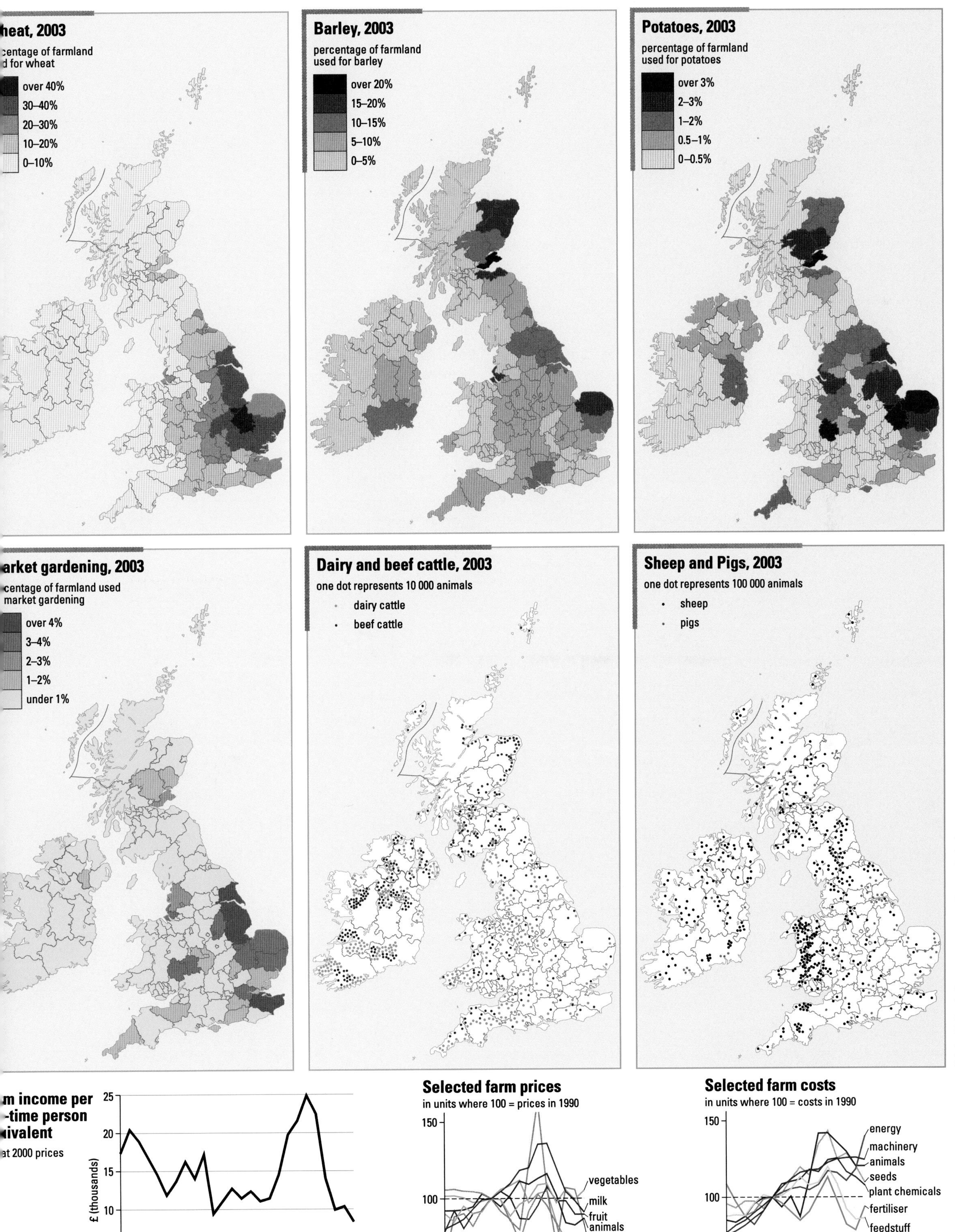

Transverse Mercator Projection

© Oxford University Press

### Farm income per full-time person equivalent

at 2000 prices

£ (thousands)

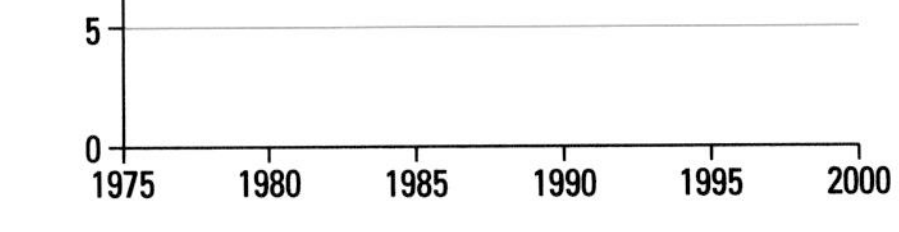

1975  1980  1985  1990  1995  2000

### Selected farm prices

in units where 100 = prices in 1990

- vegetables
- milk
- fruit
- animals
- root crops
- cereals
- wool

1985  1990  1995  2000

### Selected farm costs

in units where 100 = costs in 1990

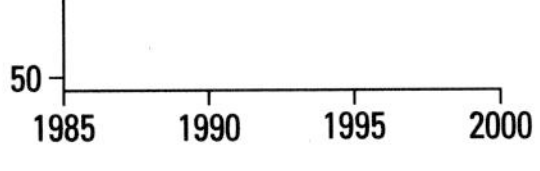

- energy
- machinery
- animals
- seeds
- plant chemicals
- fertiliser
- feedstuff

1985  1990  1995  2000

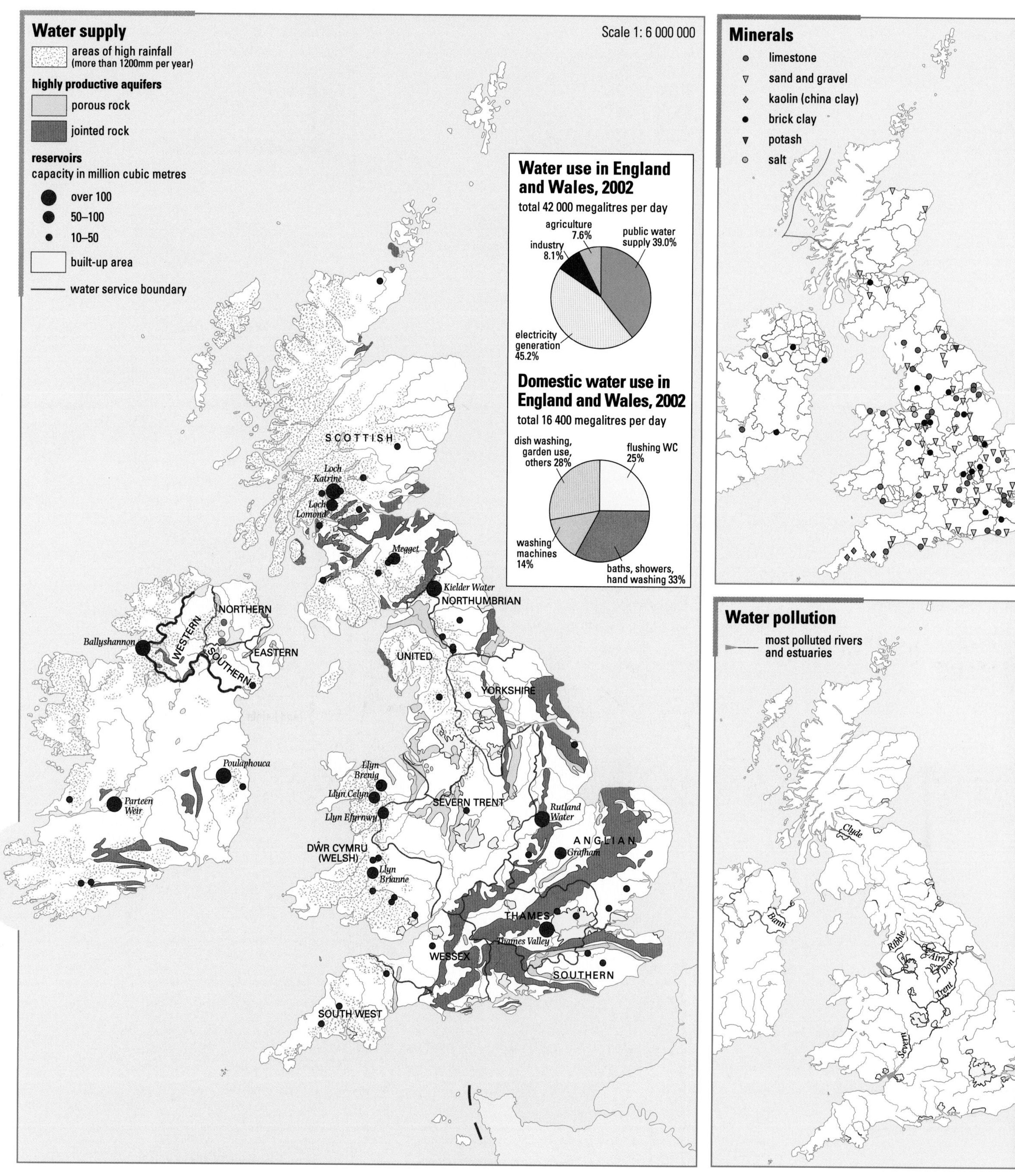

## Water supply

areas of high rainfall
(more than 1200mm per year)

**highly productive aquifers**

porous rock

jointed rock

**reservoirs**
capacity in million cubic metres

- over 100
- 50–100
- 10–50

built-up area

water service boundary

Scale 1: 6 000 000

## Minerals

- limestone
- ▽ sand and gravel
- ◆ kaolin (china clay)
- • brick clay
- ▼ potash
- ⊙ salt

### Water use in England and Wales, 2002
total 42 000 megalitres per day

- agriculture 7.6%
- public water supply 39.0%
- industry 8.1%
- electricity generation 45.2%

### Domestic water use in England and Wales, 2002
total 16 400 megalitres per day

- dish washing, garden use, others 28%
- flushing WC 25%
- washing machines 14%
- baths, showers, hand washing 33%

## Water pollution

most polluted rivers and estuaries

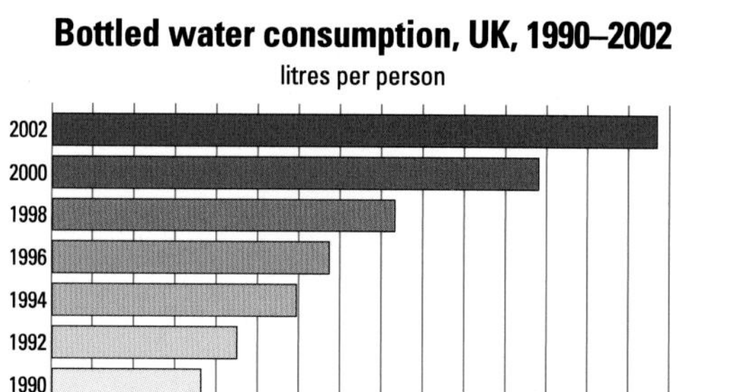

### Bottled water consumption, UK, 1990–2002
litres per person

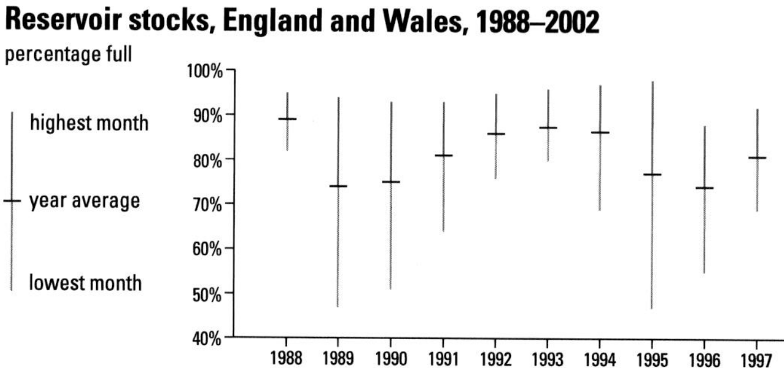

### Reservoir stocks, England and Wales, 1988–2002
percentage full

highest month

year average

lowest month

© Oxford University Press

cale 1: 6 000 000 (main map)

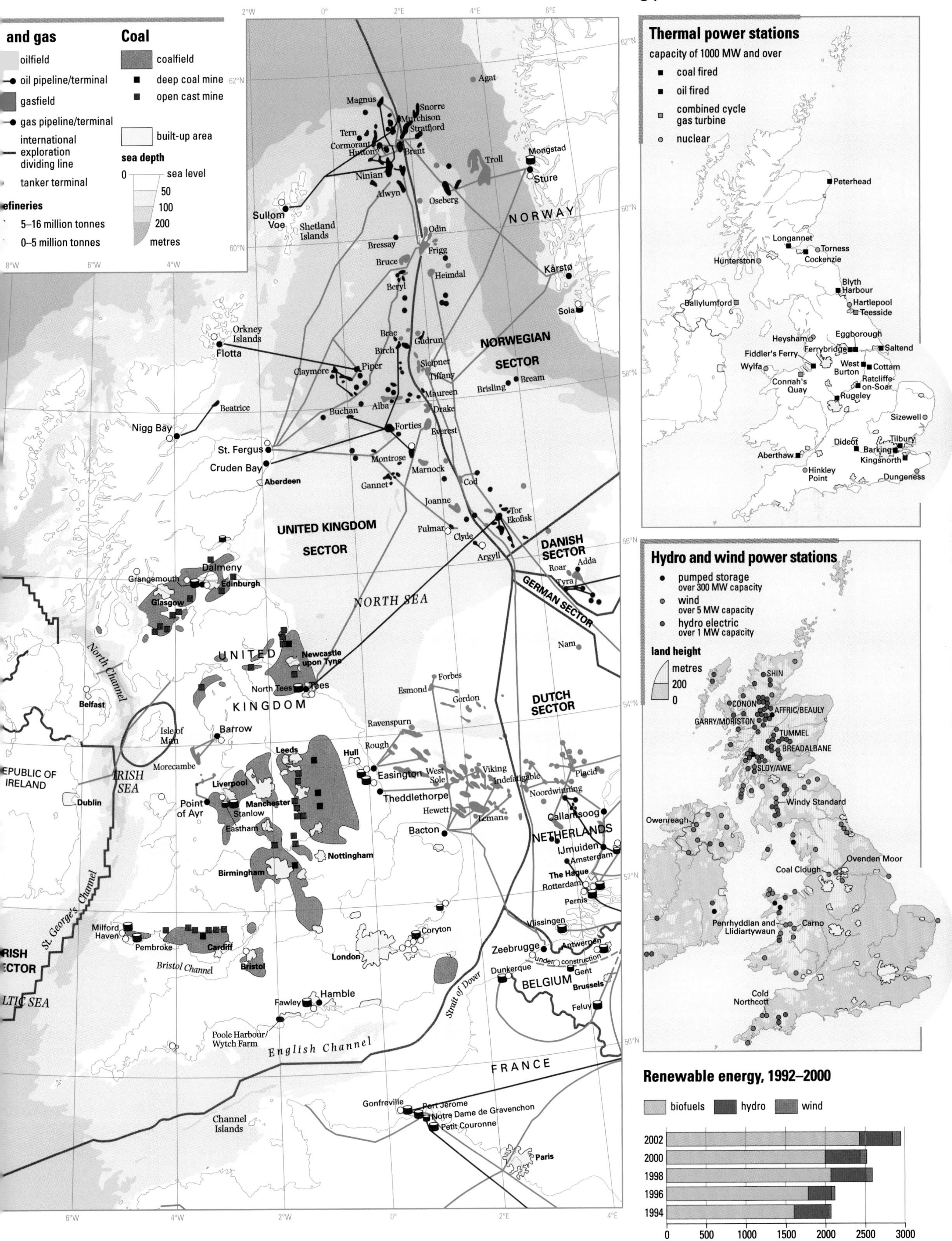

and gas
oilfield
oil pipeline/terminal
gasfield
gas pipeline/terminal
international exploration dividing line
tanker terminal
efineries
5–16 million tonnes
0–5 million tonnes
Coal
coalfield
deep coal mine
open cast mine
built-up area
sea depth
0 sea level
50
100
200
metres
NORTH SEA
UNITED KINGDOM SECTOR
NORWEGIAN SECTOR
DANISH SECTOR
GERMAN SECTOR
DUTCH SECTOR
NORWAY
UNITED KINGDOM
EPUBLIC OF IRELAND
NETHERLANDS
BELGIUM
FRANCE
IRISH SEA
English Channel
Bristol Channel
North Channel
St. George's Channel
Strait of Dover
Magnus
Snorre
Murchison
Statfjord
Tern
Cormorant
Hutton
Brent
Ninian
Alwyn
Troll
Oseberg
Mongstad
Sture
Agat
Sullom Voe
Shetland Islands
Odin
Bressay
Frigg
Bruce
Heimdal
Beryl
Kårstø
Sola
Orkney Islands
Flotta
Brae
Gudrun
Birch
Sleipner
Tiffany
Claymore
Piper
Maureen
Brisling
Bream
Alba
Buchan
Drake
Beatrice
Forties
Everest
Nigg Bay
St. Fergus
Cruden Bay
Aberdeen
Montrose
Marnock
Gannet
Cod
Joanne
Tor
Ekofisk
Fulmar
Clyde
Argyll
Roar
Adda
Tyra
Dalmeny
Grangemouth
Edinburgh
Glasgow
Nam
Newcastle upon Tyne
North Tees
Tees
Belfast
Forbes
Esmond
Gordon
Isle of Man
Barrow
Ravenspurn
Rough
Hull
Leeds
Morecambe
Easington
West Sole
Viking
Indefatigable
Placid
Noordwinning
Liverpool
Dublin
Point of Ayr
Manchester
Stanlow
Theddlethorpe
Hewett
Leman
Callantsoog
Eastham
Bacton
IJmuiden
Amsterdam
Nottingham
The Hague
Rotterdam
Birmingham
Pernis
Vlissingen
Milford Haven
Pembroke
Cardiff
Coryton
Bristol
London
Zeebrugge
Antwerpen
under construction
Gent
Dunkerque
Brussels
Feluy
Hamble
Fawley
Poole Harbour/ Wytch Farm
Channel Islands
Gonfreville
Port Jerome
Notre Dame de Gravenchon
Petit Couronne
Paris
RISH ECTOR
LTIC SEA
Thermal power stations
capacity of 1000 MW and over
coal fired
oil fired
combined cycle gas turbine
nuclear
Peterhead
Longannet
Torness
Cockenzie
Hunterston
Blyth Harbour
Hartlepool
Teesside
Ballylumford
Heysham
Eggborough
Ferrybridge
Saltend
Fiddler's Ferry
Wylfa
West Burton
Cottam
Connah's Quay
Ratcliffe-on-Soar
Rugeley
Sizewell
Didcot
Tilbury
Barking
Aberthaw
Kingsnorth
Hinkley Point
Dungeness
Hydro and wind power stations
pumped storage over 300 MW capacity
wind over 5 MW capacity
hydro electric over 1 MW capacity
land height
metres
200
0
SHIN
CONON
AFFRIC/BEAULY
GARRY/MORISTON
TUMMEL
BREADALBANE
SLOY/AWE
Windy Standard
Owenreagh
Ovenden Moor
Coal Clough
Penrhyddlan and Llidiartywaun
Carno
Cold Northcott
Renewable energy, 1992–2000
biofuels
hydro
wind
2002
2000
1998
1996
1994
0
500
1000
1500
2000
2500
3000
thousand tonnes of oil equivalent

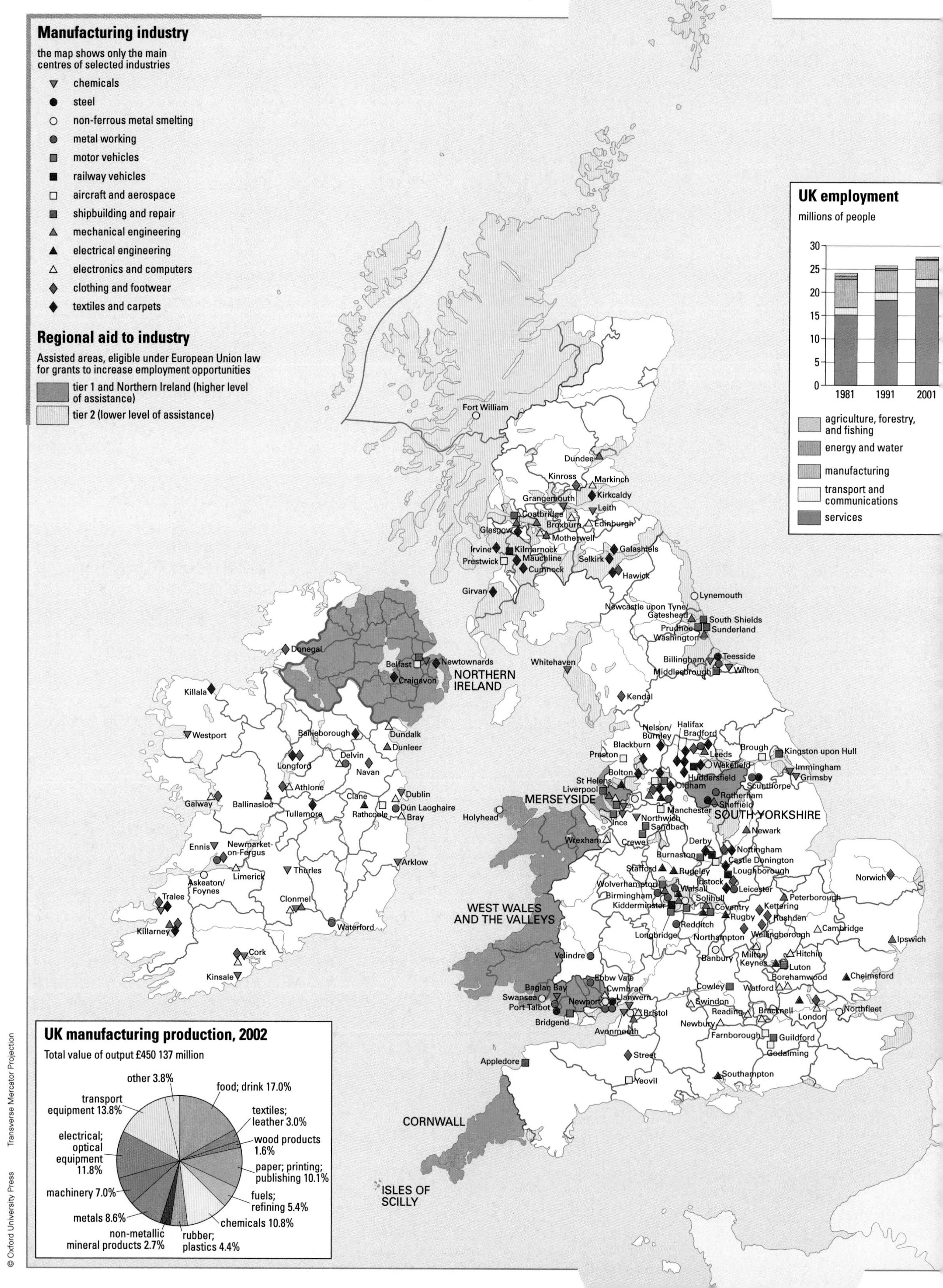

## Manufacturing industry

the map shows only the main centres of selected industries

- ▽ chemicals
- ● steel
- ○ non-ferrous metal smelting
- ◉ metal working
- ▣ motor vehicles
- ▪ railway vehicles
- ☐ aircraft and aerospace
- ▨ shipbuilding and repair
- △ mechanical engineering
- ▲ electrical engineering
- △ electronics and computers
- ◆ clothing and footwear
- ◆ textiles and carpets

## Regional aid to industry

Assisted areas, eligible under European Union law for grants to increase employment opportunities

- tier 1 and Northern Ireland (higher level of assistance)
- tier 2 (lower level of assistance)

### UK employment
millions of people

agriculture, forestry, and fishing
energy and water
manufacturing
transport and communications
services

### UK manufacturing production, 2002

Total value of output £450 137 million

- other 3.8%
- transport equipment 13.8%
- food; drink 17.0%
- textiles; leather 3.0%
- wood products 1.6%
- electrical; optical equipment 11.8%
- paper; printing; publishing 10.1%
- machinery 7.0%
- fuels; refining 5.4%
- metals 8.6%
- chemicals 10.8%
- non-metallic mineral products 2.7%
- rubber; plastics 4.4%

Transverse Mercator Projection

© Oxford University Press

Scale 1: 12 500 000

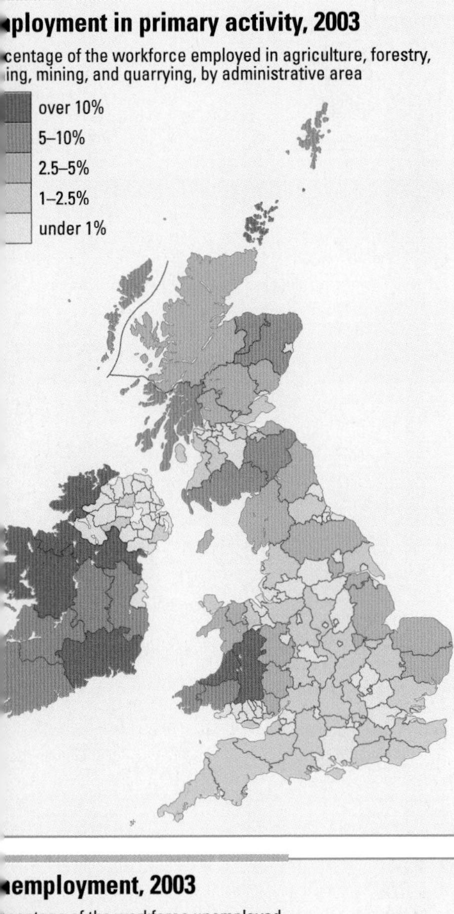

### ...ployment in primary activity, 2003

...centage of the workforce employed in agriculture, forestry,
...ing, mining, and quarrying, by administrative area

- over 10%
- 5–10%
- 2.5–5%
- 1–2.5%
- under 1%

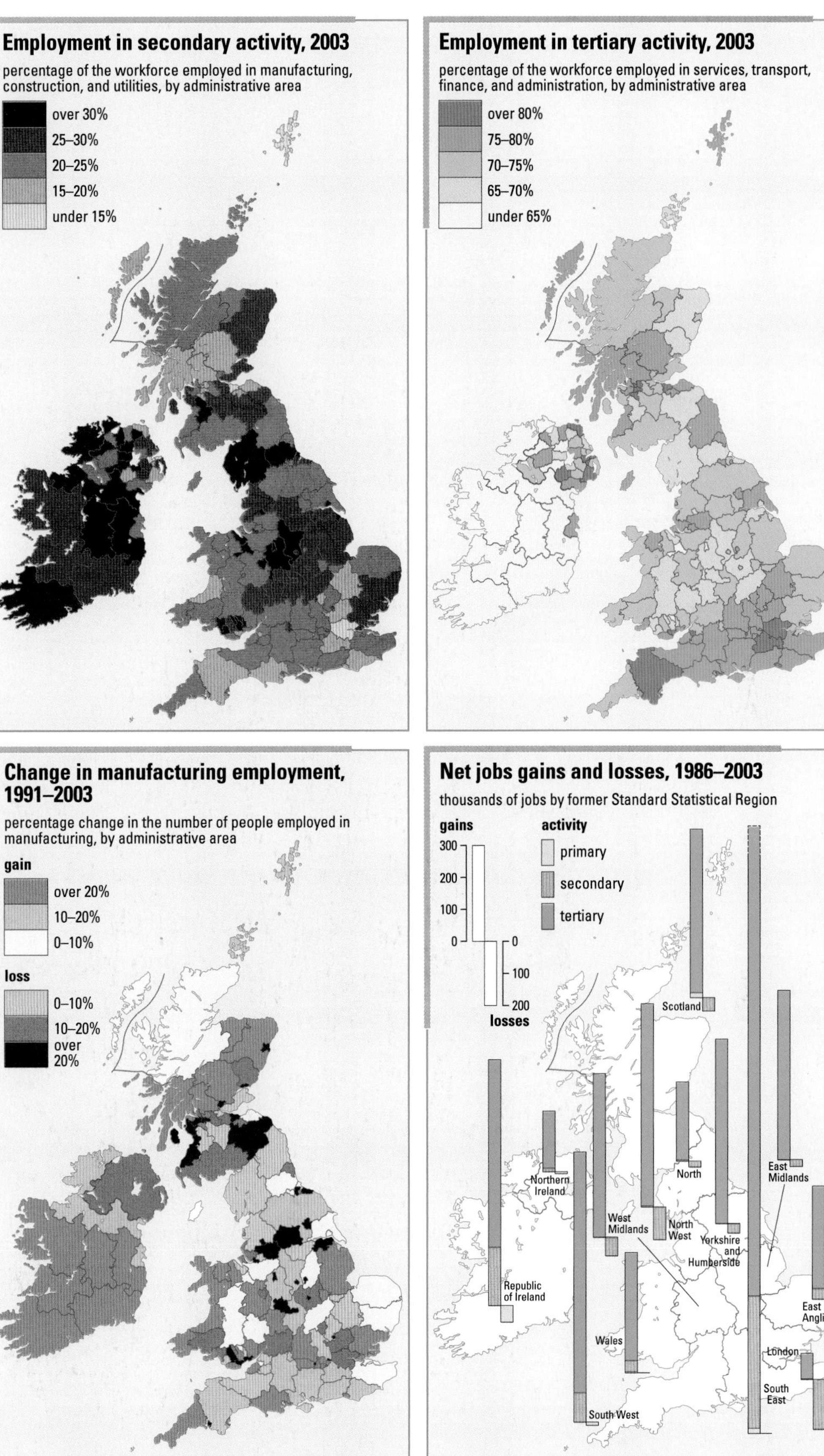

### Employment in secondary activity, 2003

percentage of the workforce employed in manufacturing,
construction, and utilities, by administrative area

- over 30%
- 25–30%
- 20–25%
- 15–20%
- under 15%

### Employment in tertiary activity, 2003

percentage of the workforce employed in services, transport,
finance, and administration, by administrative area

- over 80%
- 75–80%
- 70–75%
- 65–70%
- under 65%

### ...employment, 2003

...centage of the workforce unemployed,
...administrative area

- over 7%
- 6–7%
- 5–6%
- 4–5%
- 3–4%
- under 3%

### Change in manufacturing employment, 1991–2003

percentage change in the number of people employed in
manufacturing, by administrative area

**gain**
- over 20%
- 10–20%
- 0–10%

**loss**
- 0–10%
- 10–20%
- over 20%

### Net jobs gains and losses, 1986–2003

thousands of jobs by former Standard Statistical Region

**gains**
300
200
100
0
–100
–200
**losses**

**activity**
- primary
- secondary
- tertiary

Scotland
Northern Ireland
North
East Midlands
West Midlands
North West
Yorkshire and Humberside
Republic of Ireland
East Anglia
Wales
London
South East
South West

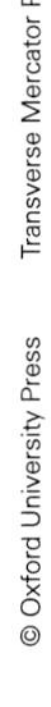

Transverse Mercator Projection

© Oxford University Press

### UK workforce structure, 2003

Total workforce 29 455 000

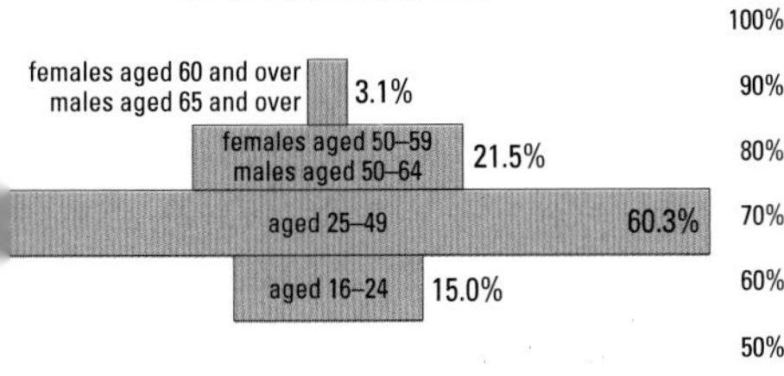

- females aged 60 and over, males aged 65 and over — 3.1%
- females aged 50–59, males aged 50–64 — 21.5%
- aged 25–49 — 60.3%
- aged 16–24 — 15.0%

### UK employment rates, 1960–2002

percentage of people of working age

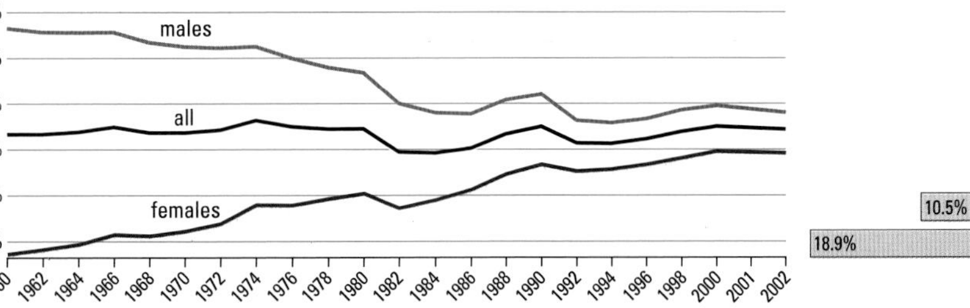

- males
- all
- females

### UK unemployment structure, 2003

percentage of all economically active people

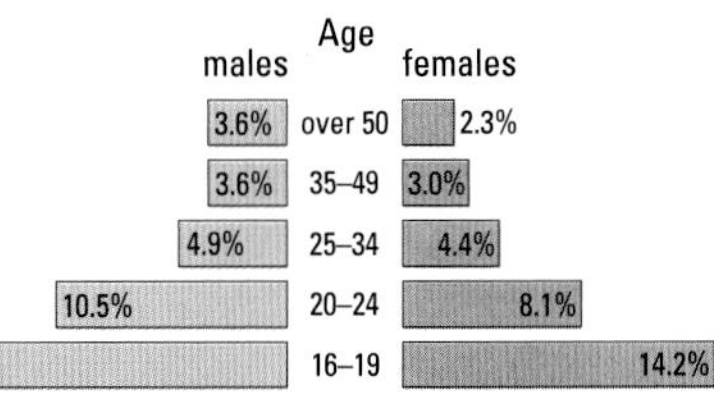

| males | Age | females |
|---|---|---|
| 3.6% | over 50 | 2.3% |
| 3.6% | 35–49 | 3.0% |
| 4.9% | 25–34 | 4.4% |
| 10.5% | 20–24 | 8.1% |
| 18.9% | 16–19 | 14.2% |

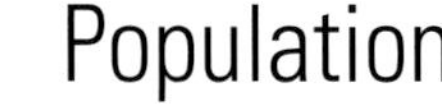

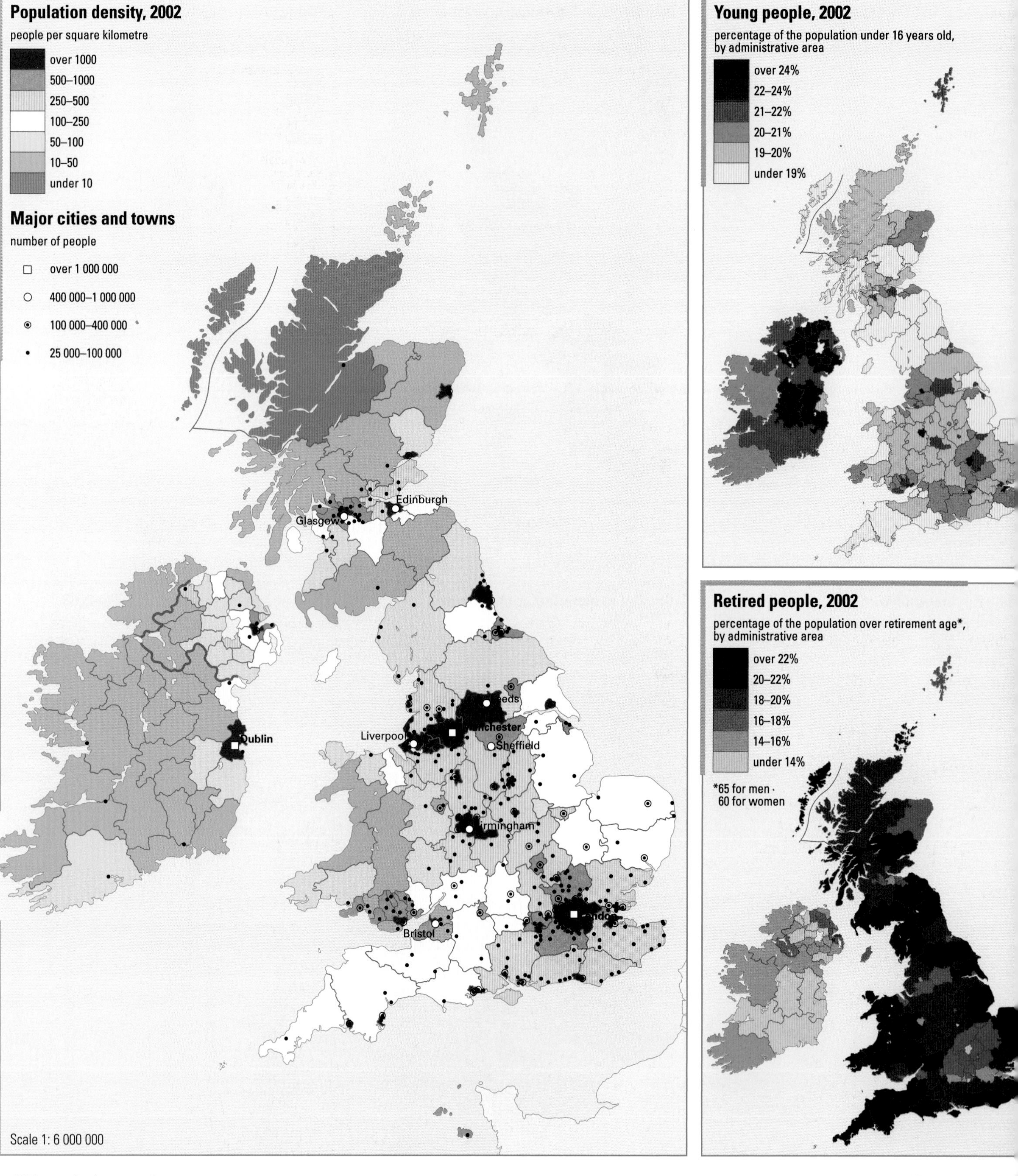

## Population density, 2002

people per square kilometre

- over 1000
- 500–1000
- 250–500
- 100–250
- 50–100
- 10–50
- under 10

## Major cities and towns

number of people

- □ over 1 000 000
- ○ 400 000–1 000 000
- ◉ 100 000–400 000
- • 25 000–100 000

## Young people, 2002

percentage of the population under 16 years old, by administrative area

- over 24%
- 22–24%
- 21–22%
- 20–21%
- 19–20%
- under 19%

## Retired people, 2002

percentage of the population over retirement age*, by administrative area

- over 22%
- 20–22%
- 18–20%
- 16–18%
- 14–16%
- under 14%

*65 for men
60 for women

Scale 1: 6 000 000

© Oxford University Press

| UK population trends | 1901 | 1911 | 1921 | 1931 | 1941 | 1951 | 1961 | 1971 | 1981 | 1991 | 2001 | 2011 | 2021 |
|---|---|---|---|---|---|---|---|---|---|---|---|---|---|
| Total population (millions) | 38.24 | 42.08 | 44.03 | 46.04 | 48.22 | 50.23 | 52.81 | 55.93 | 56.35 | 57.65 | 59.62 | 60.93 | 63.64 |
| Infant mortality (deaths per 1000 live births) | 138.0 | 110.0 | 76.0 | 62.0 | 50.0 | 27.0 | 21.0 | 17.9 | 11.0 | 7.4 | 5.6 | 5.5 | 5.5 |
| Birth rate (births per 1000 people) | 28.6 | 24.5 | 22.8 | 16.3 | 14.4 | 15.9 | 17.9 | 16.1 | 13.0 | 13.8 | 12.0 | 11.5 | 11.5 |
| Death rate (deaths per 1000 people) | 16.5 | 14.3 | 11.9 | 12.5 | 13.0 | 12.6 | 12.0 | 11.5 | 11.6 | 11.3 | 10.5 | 10.0 | 10.3 |
| Life expectancy (years) | 47.0 | 52.2 | 57.3 | 60.0 | 61.0 | 68.5 | 70.9 | 71.9 | 73.8 | 76.0 | 77.5 | 79.5 | 80.5 |

*projected*

Scale 1: 12 500 000 (smallest maps)

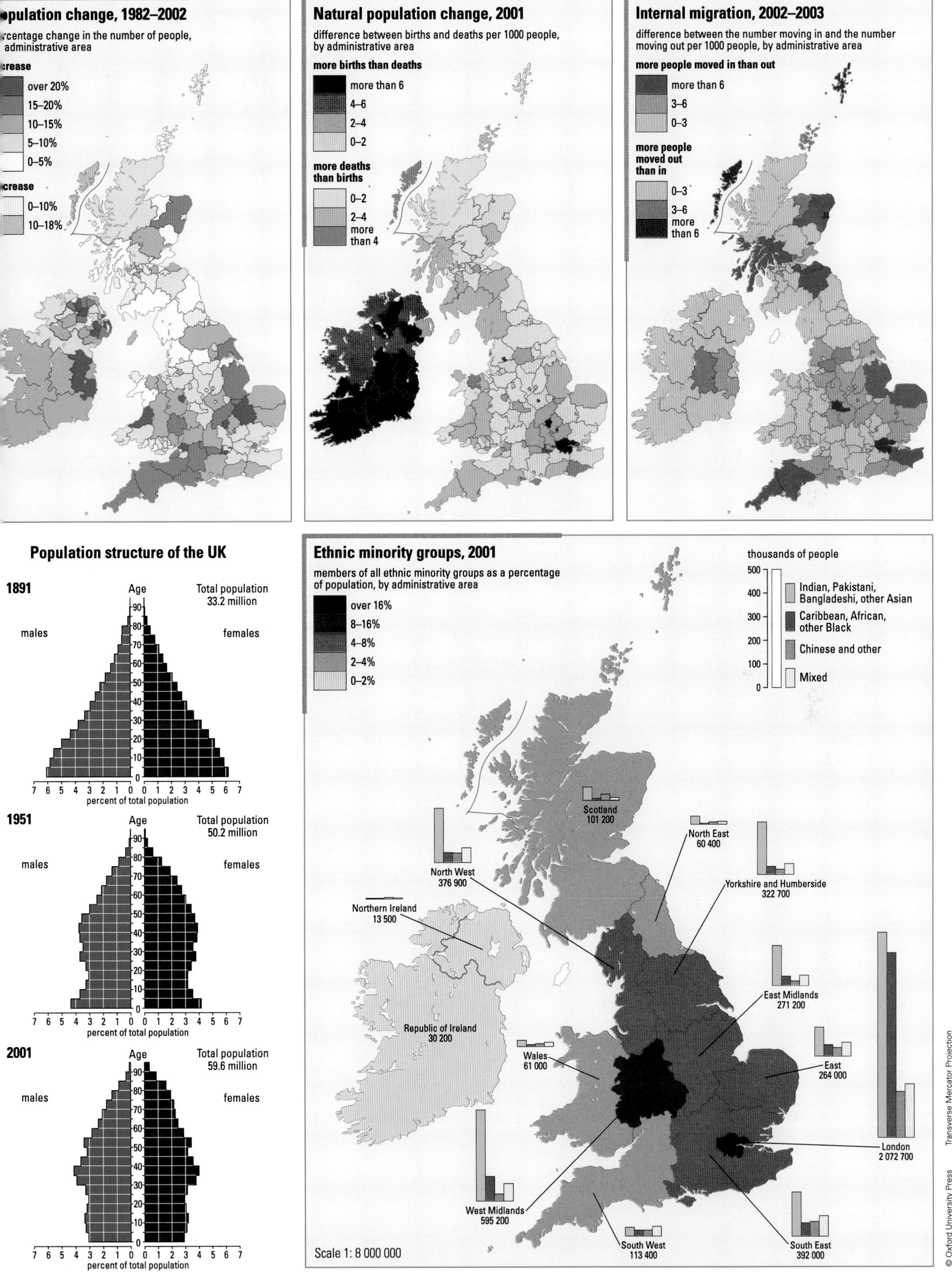

### pulation change, 1982–2002

rcentage change in the number of people, administrative area

**crease**
- over 20%
- 15–20%
- 10–15%
- 5–10%
- 0–5%

**crease**
- 0–10%
- 10–18%

### Natural population change, 2001

difference between births and deaths per 1000 people, by administrative area

**more births than deaths**
- more than 6
- 4–6
- 2–4
- 0–2

**more deaths than births**
- 0–2
- 2–4
- more than 4

### Internal migration, 2002–2003

difference between the number moving in and the number moving out per 1000 people, by administrative area

**more people moved in than out**
- more than 6
- 3–6
- 0–3

**more people moved out than in**
- 0–3
- 3–6
- more than 6

### Population structure of the UK

**1891**  Age  Total population 33.2 million
males  females
7 6 5 4 3 2 1 0 0 1 2 3 4 5 6 7
percent of total population

**1951**  Age  Total population 50.2 million
males  females
7 6 5 4 3 2 1 0 0 1 2 3 4 5 6 7
percent of total population

**2001**  Age  Total population 59.6 million
males  females
7 6 5 4 3 2 1 0 0 1 2 3 4 5 6 7
percent of total population

### Ethnic minority groups, 2001

members of all ethnic minority groups as a percentage of population, by administrative area
- over 16%
- 8–16%
- 4–8%
- 2–4%
- 0–2%

thousands of people
500
400
300
200
100
0

- Indian, Pakistani, Bangladeshi, other Asian
- Caribbean, African, other Black
- Chinese and other
- Mixed

Scotland
101 200

North East
60 400

Yorkshire and Humberside
322 700

North West
376 900

Northern Ireland
13 500

East Midlands
271 200

Republic of Ireland
30 200

Wales
61 000

East
264 000

London
2 072 700

West Midlands
595 200

South West
113 400

South East
392 000

Scale 1: 8 000 000

Transverse Mercator Projection

© Oxford University Press

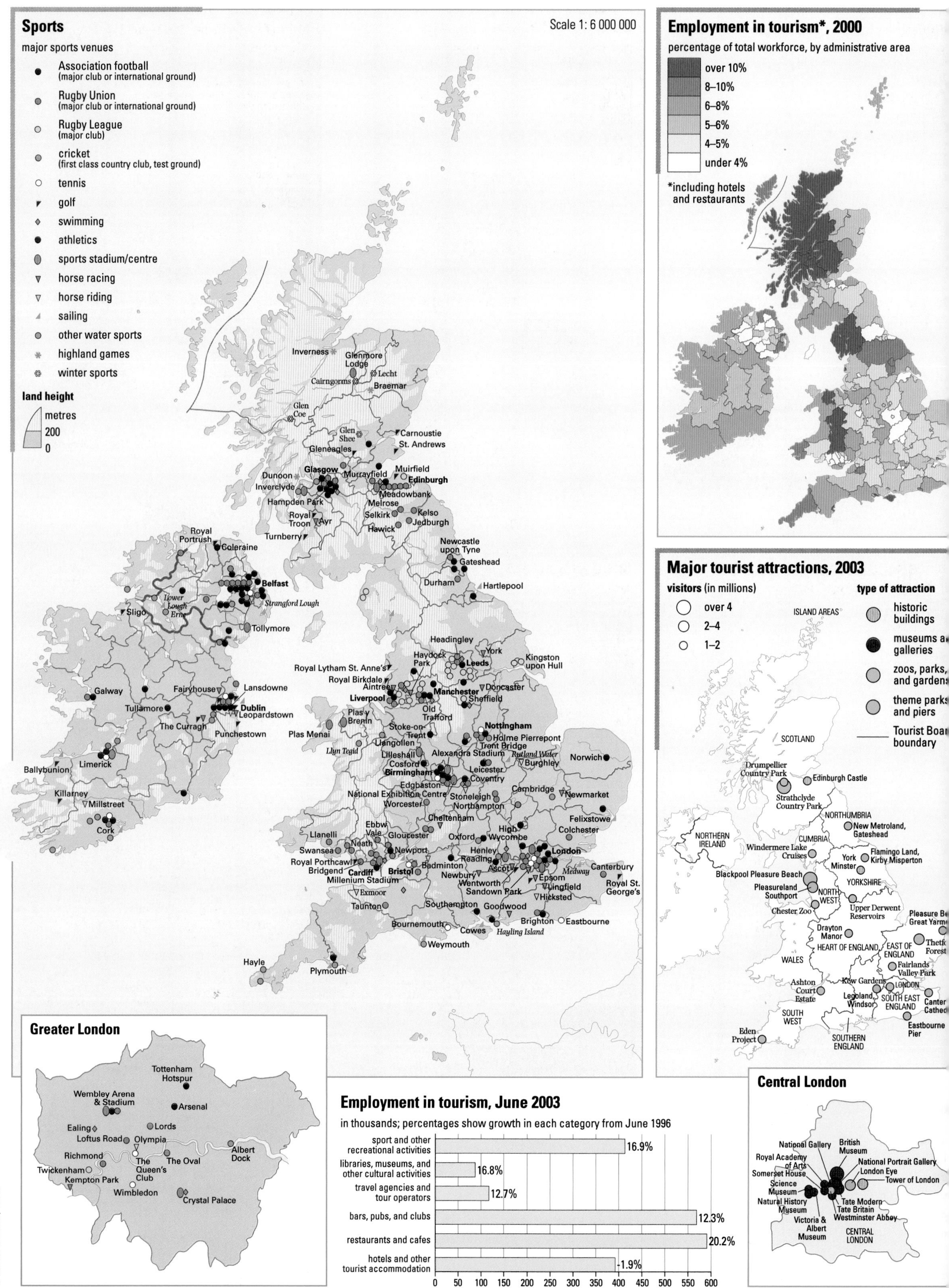

## Sports

major sports venues

- Association football (major club or international ground)
- Rugby Union (major club or international ground)
- Rugby League (major club)
- cricket (first class country club, test ground)
- tennis
- golf
- swimming
- athletics
- sports stadium/centre
- horse racing
- horse riding
- sailing
- other water sports
- highland games
- winter sports

**land height**

metres
200
0

Scale 1: 6 000 000

## Employment in tourism*, 2000

percentage of total workforce, by administrative area

- over 10%
- 8–10%
- 6–8%
- 5–6%
- 4–5%
- under 4%

*including hotels and restaurants

## Major tourist attractions, 2003

visitors (in millions)
- over 4
- 2–4
- 1–2

type of attraction
- historic buildings
- museums and galleries
- zoos, parks and gardens
- theme parks and piers
- Tourist Board boundary

ISLAND AREAS

SCOTLAND

Drumpellier Country Park
Edinburgh Castle
Strathclyde Country Park
NORTHUMBRIA
New Metroland, Gateshead
NORTHERN IRELAND
CUMBRIA
Windermere Lake Cruises
Flamingo Land, Kirby Misperton
York Minster
YORKSHIRE
Blackpool Pleasure Beach
Pleasureland Southport
NORTH WEST
Pleasure Beach Great Yarmouth
Chester Zoo
Upper Derwent Reservoirs
Drayton Manor
EAST OF ENGLAND
Thetford Forest
HEART OF ENGLAND
WALES
Fairlands Valley Park
Ashton Court Estate
Kew Gardens
LONDON
Legoland, Windsor
SOUTH EAST ENGLAND
Canterbury Cathedral
SOUTH WEST
Eden Project
SOUTHERN ENGLAND
Eastbourne Pier

### Greater London

Tottenham Hotspur
Wembley Arena & Stadium
Arsenal
Ealing
Lords
Loftus Road
Olympia
Richmond
Albert Dock
The Oval
Twickenham
The Queen's Club
Kempton Park
Wimbledon
Crystal Palace

### Employment in tourism, June 2003

in thousands; percentages show growth in each category from June 1996

| | |
|---|---|
| sport and other recreational activities | 16.9% |
| libraries, museums, and other cultural activities | 16.8% |
| travel agencies and tour operators | 12.7% |
| bars, pubs, and clubs | 12.3% |
| restaurants and cafes | 20.2% |
| hotels and other tourist accommodation | -1.9% |

0  50  100  150  200  250  300  350  400  450  500  550  600

### Central London

National Gallery
British Museum
Royal Academy of Arts
Somerset House
National Portrait Gallery
London Eye
Science Museum
Tower of London
Natural History Museum
Tate Modern
Tate Britain
Westminster Abbey
Victoria & Albert Museum
CENTRAL LONDON

© Oxford University Press

Scale 1: 12 500 000

## come, 2003

rage gross weekly earnings of workers in
-time employment, by administrative area

- over £475
- £425–£475
- £400–£425
- £375–£400
- £350–£375
- under £350
- no data

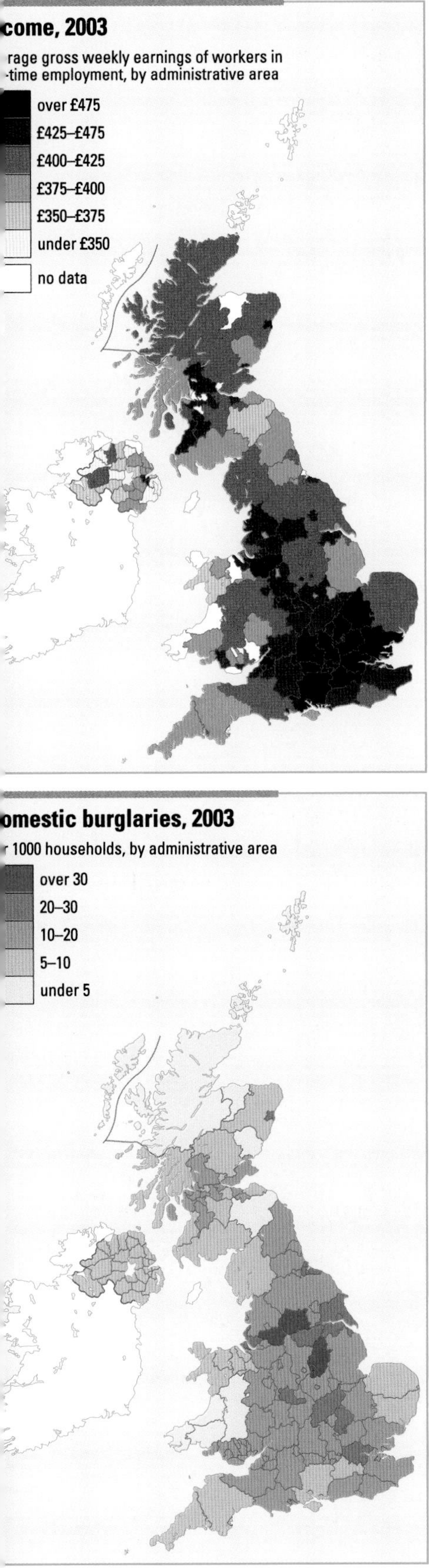

## Education, 2002

percentage of 16 year olds entering further
or higher education, by administrative area

- over 90%
- 85–90%
- 80–85%
- 75–80%
- 70–75%
- under 70%

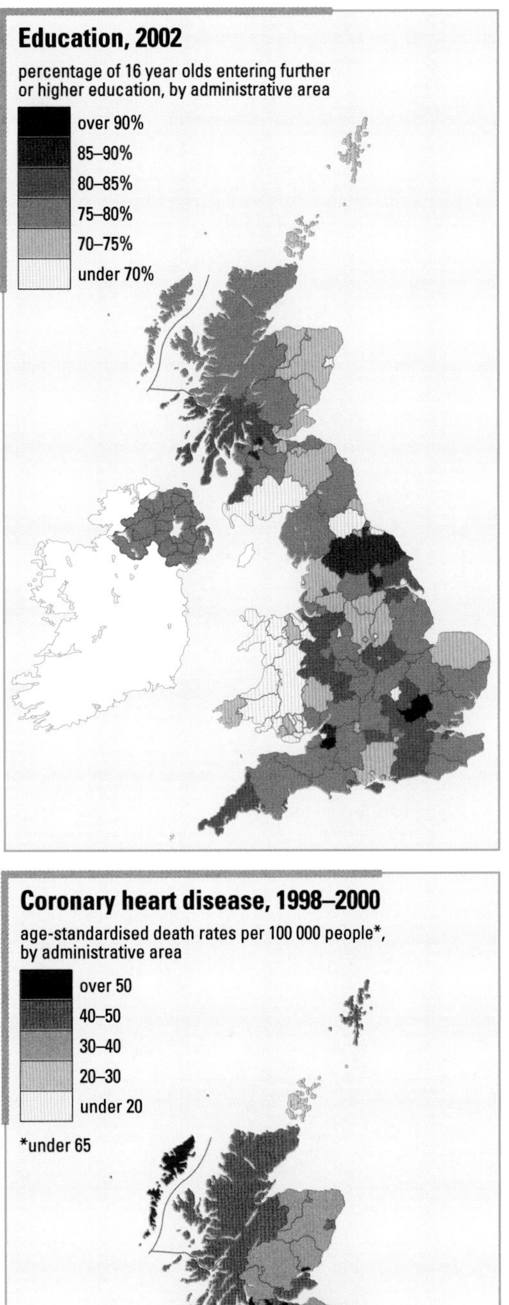

## Index of Multiple Deprivation (IMD), 2000

IMD is calculated from a number of indicators including
low income, unemployment, poor health, disability,
lack of education, unsatisfactory housing,
and poor access to services. The map
shows the 10% most deprived areas
within each part of the UK.

- England
- Wales
- Scotland
- Northern Ireland

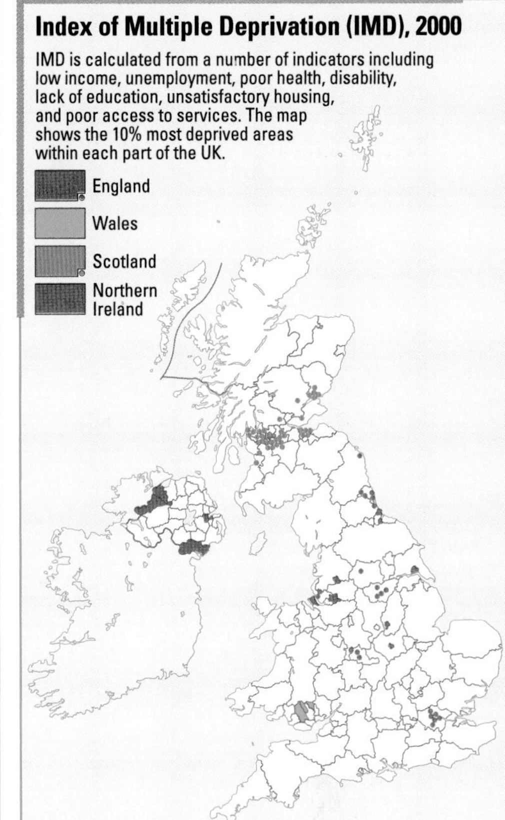

## omestic burglaries, 2003

1000 households, by administrative area

- over 30
- 20–30
- 10–20
- 5–10
- under 5

## Coronary heart disease, 1998–2000

age-standardised death rates per 100 000 people*,
by administrative area

- over 50
- 40–50
- 30–40
- 20–30
- under 20

*under 65

## House prices, 2004

comparative prices for similar size and style of house
in similar neighbourhoods

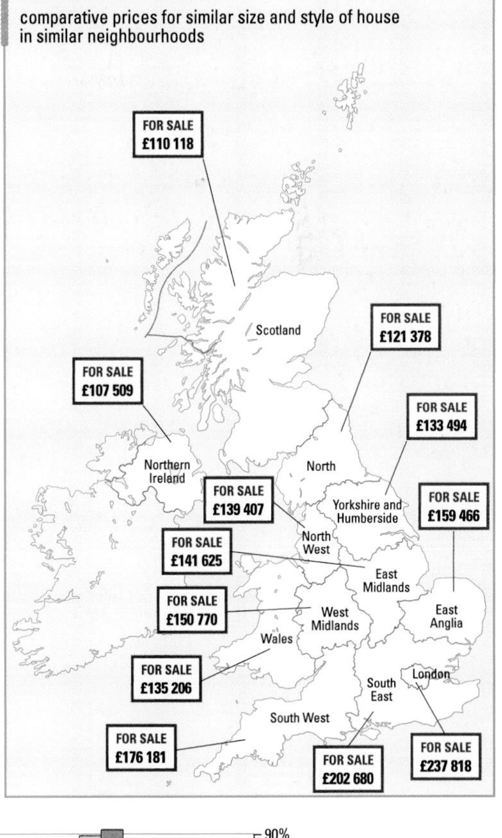

FOR SALE £110 118

FOR SALE £121 378

Scotland

FOR SALE £107 509

FOR SALE £133 494

Northern Ireland

North

FOR SALE £139 407

Yorkshire and Humberside

FOR SALE £159 466

North West

FOR SALE £141 625

East Midlands

FOR SALE £150 770

West Midlands

East Anglia

Wales

FOR SALE £135 206

South East

London

South West

FOR SALE £176 181

FOR SALE £202 680

FOR SALE £237 818

Transverse Mercator Projection

© Oxford University Press

## nsumer goods, 1970–2002

ent of UK households having use of
product

- car
- central heating
- washing machine
- dishwasher
- microwave oven
- video
- PC
- CD player
- ooo no data

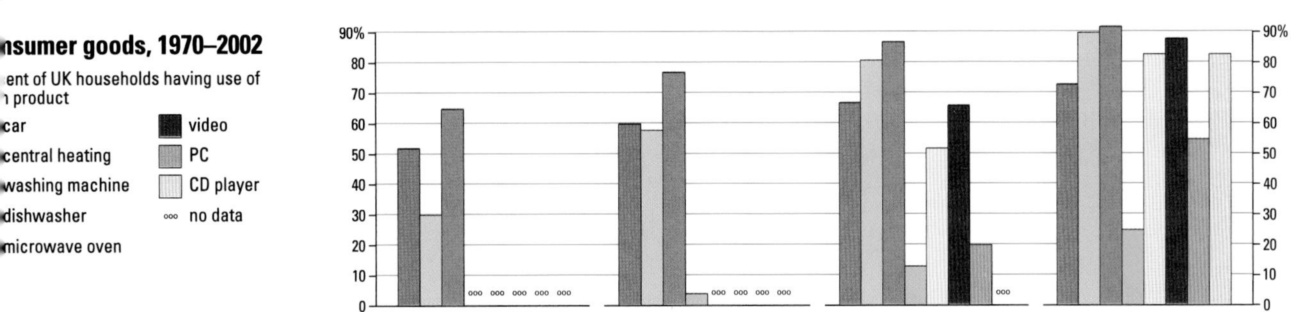

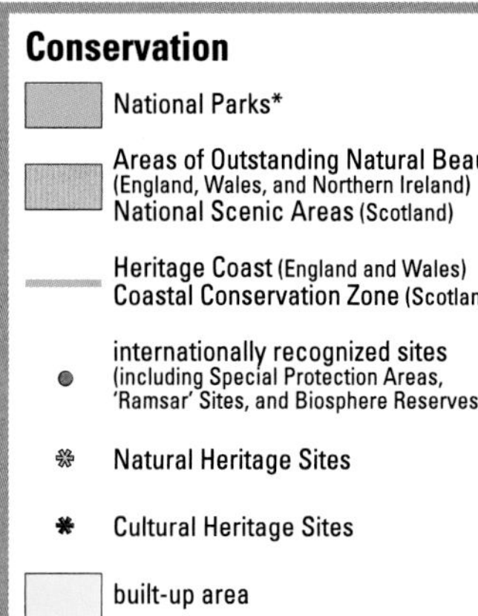

## Conservation

- National Parks*
- Areas of Outstanding Natural Beauty (England, Wales, and Northern Ireland) National Scenic Areas (Scotland)
- Heritage Coast (England and Wales) Coastal Conservation Zone (Scotland)
- internationally recognized sites (including Special Protection Areas, 'Ramsar' Sites, and Biosphere Reserves)
- ✿ Natural Heritage Sites
- ✻ Cultural Heritage Sites
- built-up area

*National parks are designated to conserve the natural beauty and cultural heritage of areas of outstanding landscape value. There are 10 national parks in England and Wales, which were designated in the 1950's following the National Parks and Access to the Countryside Act, 1949. The Broads is not officially a national park but is considered as such by the government and has had its own authority since 1989. Proposals to establish the New Forest and South Downs as national parks are still taking place. The National Parks (Scotland) Act was passed in July 2000, and the Cairngorms, and Loch Lomond and The Trossachs became Scotland's first national parks in 2002/3.

### Great Britain Countryside

percentage of broad habitats, Countryside Survey 200

one small square represents 1%

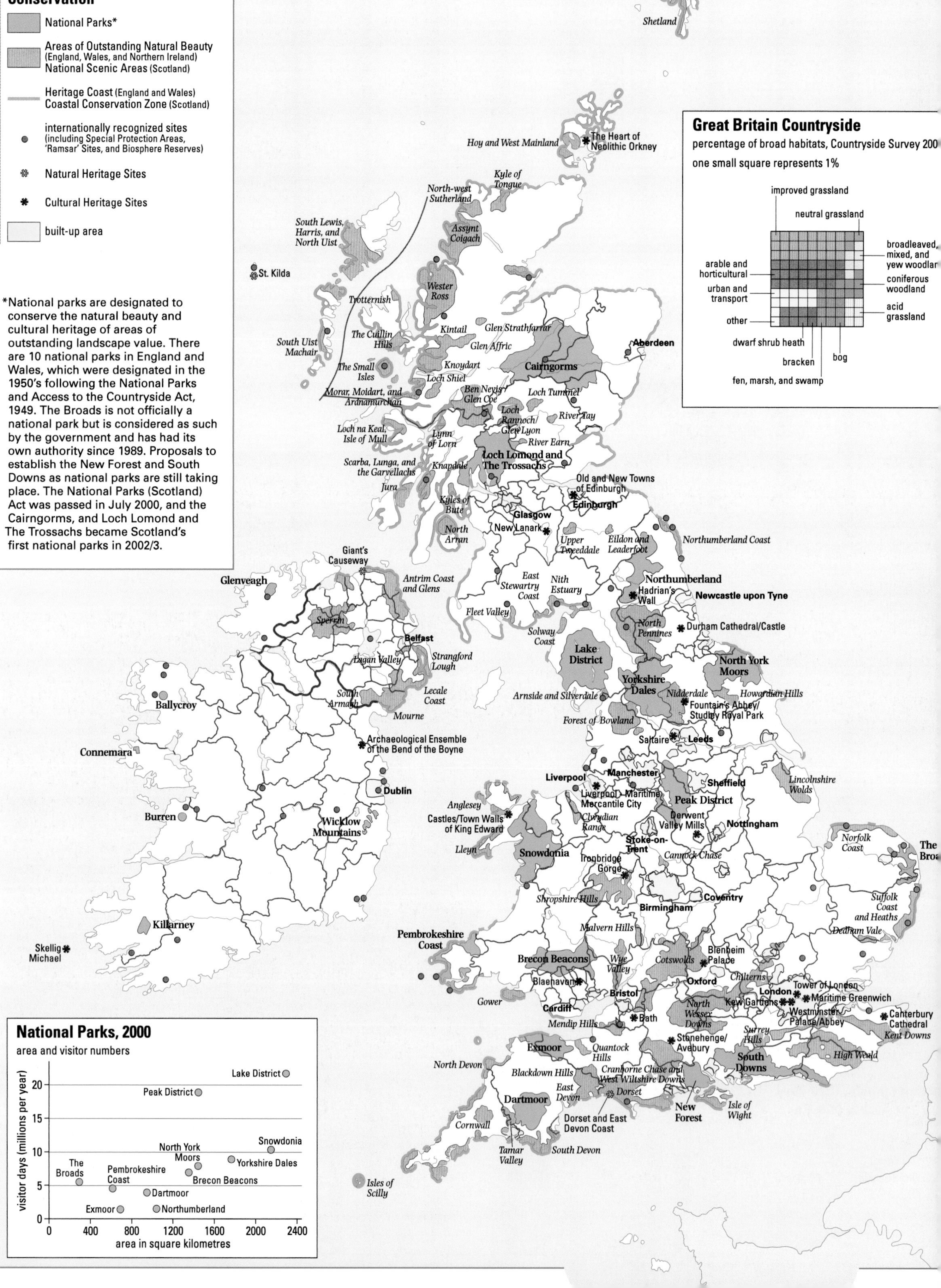

improved grassland
neutral grassland
broadleaved, mixed, and yew woodlan
arable and horticultural
coniferous woodland
urban and transport
acid grassland
other
dwarf shrub heath
bog
bracken
fen, marsh, and swamp

### National Parks, 2000

area and visitor numbers

Lake District
Peak District
Snowdonia
North York Moors
Yorkshire Dales
The Broads
Pembrokeshire Coast
Brecon Beacons
Dartmoor
Exmoor
Northumberland

visitor days (millions per year) — 0, 5, 10, 15, 20
area in square kilometres — 0, 400, 800, 1200, 1600, 2000, 2400

## Acid rain

Environmental damage is more likely where acid
deposition is high and soils (particularly those
that are already acid) are more sensitive.

areas where potential damage to vegetation
from nitrogen in acid rain is

- very high
- high
- moderate
- low

## Ozone

Number of days when ozone concentration exceeded
50 parts per billion, used to assess the potential for
effects on human health.

days per year

- over 45
- 35–45
- 30–35
- 25–30
- under 25

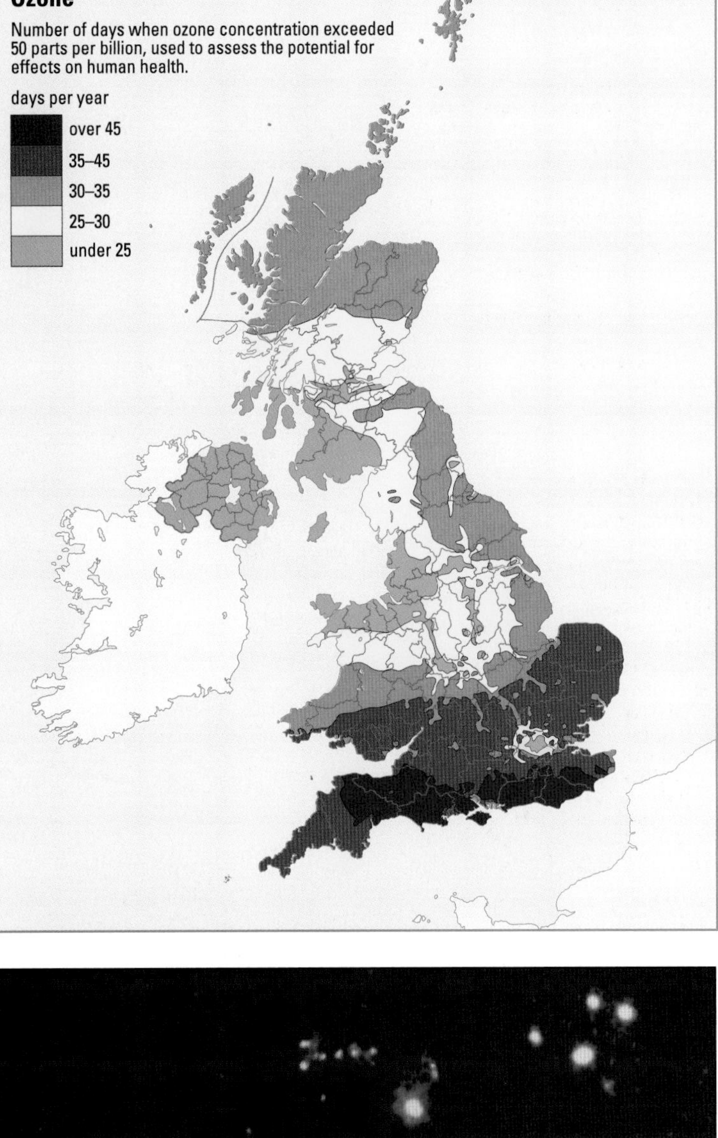

## Coastal and offshore pollution

bathing beaches heavily
polluted by sewage

**oil spills within UK waters**
tonnes

- over 5000
- 50–5000
- 0–50

**Braer 86 248 tonnes**
5 January 1993

*ATLANTIC OCEAN*

*NORTH SEA*

**Sea Empress 72 000 tonnes**
15 February 1996

*English Channel*

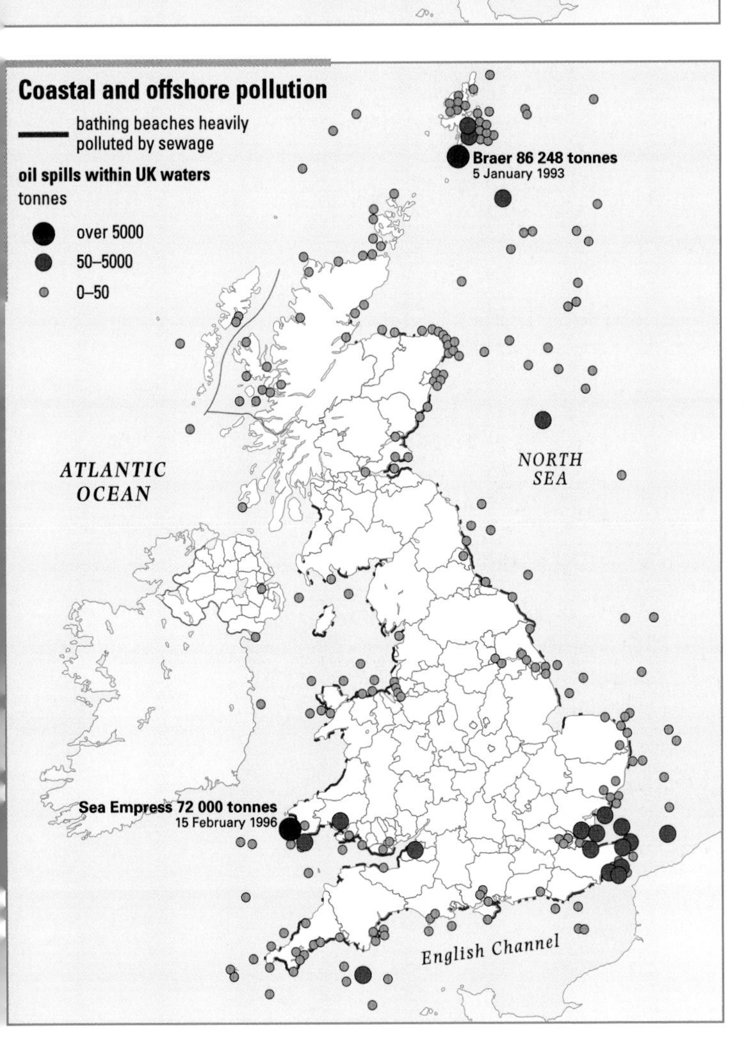

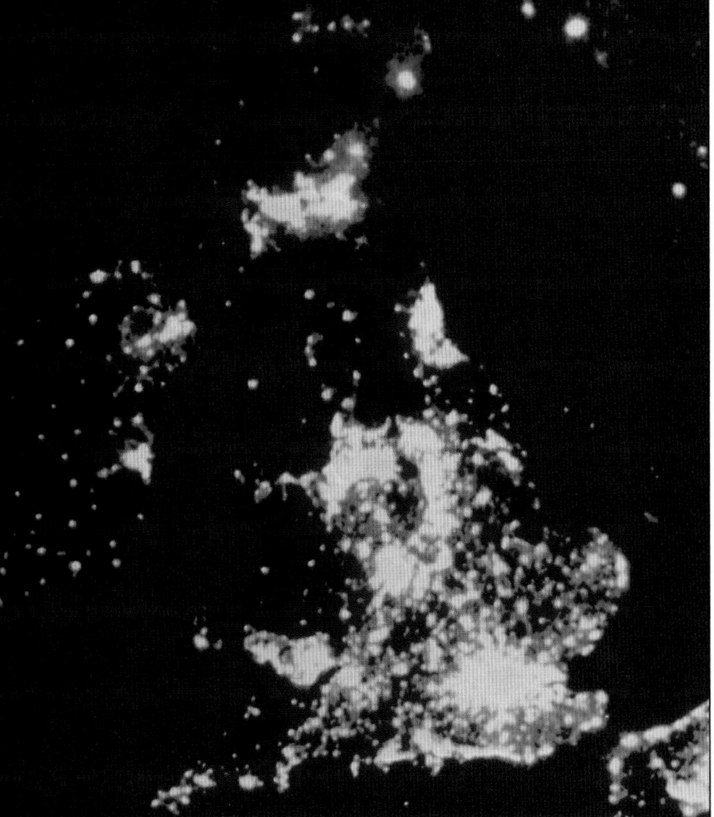

## Light pollution

Image of the British Isles at night showing city lights. The patches of light in the
North Sea are flares from oil rigs.

## Roads, airports, ferries

═══ motorway
─── major road
─── major ferry route

**airports, 2003**
passengers

✈ over 10 million
✈ 1–10 million
• 100–1 million

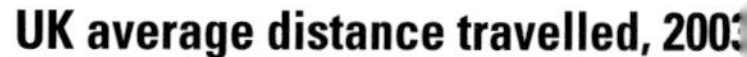

### UK average distance travelled, 2003

| | miles per person per year |
|---|---|
| **walking** | 192 |
| **bicycle** | 34 |
| **car** | 5 252 |
| **motorcycle** | 36 |
| **local bus** | 213 |
| **rail** | 347 |
| **taxi** | 49 |
| **air and ferry** | 96 |

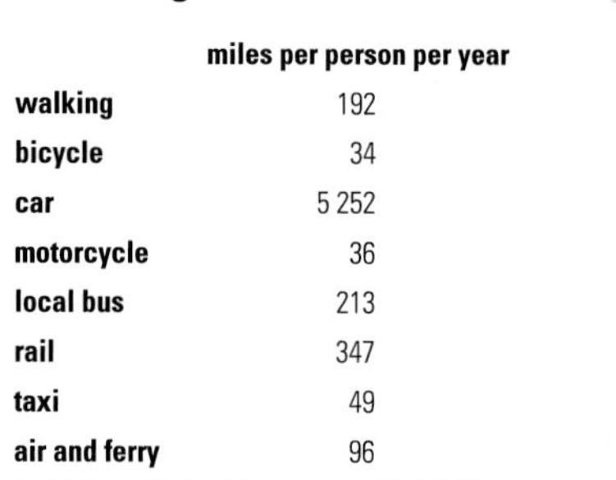

### Average distance travelled, 1993–200...

percentage change per person per year, UK

air an
ferry 13...

taxi 29%

rail 16%

motorcycle
12%

car 6%

local bus
−2%

walking
−4%

bicycle
−10%

### Rail network, por...

─── principal railway

• terminal or
major junction

▢ built-up area

**ports, 2003**
cargo handled, tonnes

⬤ over 40 million
● 10–40 million
• 50 000–10 million

**land height**

metres
200
0

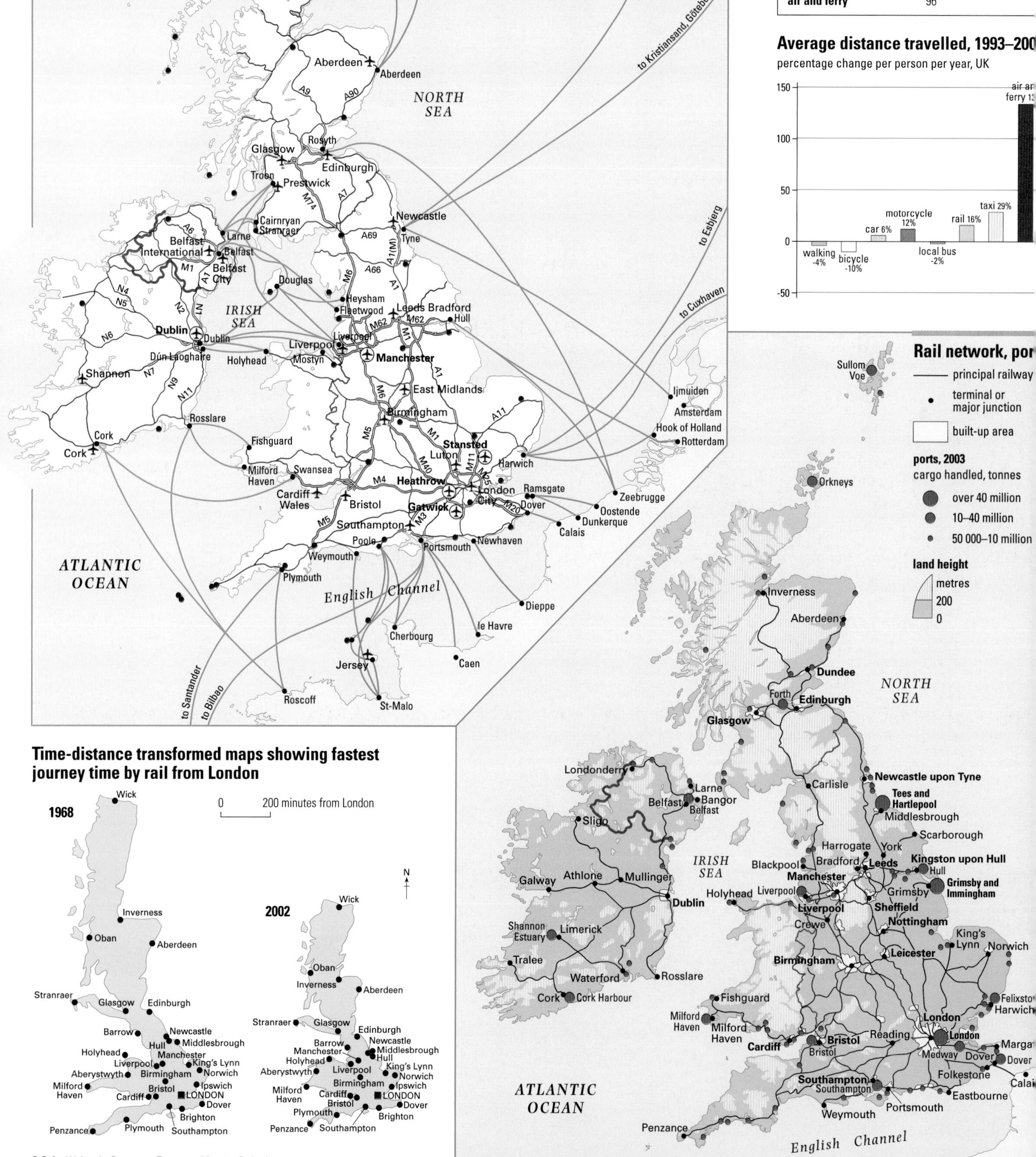

## Time-distance transformed maps showing fastest journey time by rail from London

0 ——— 200 minutes from London

**1968**

**2002**

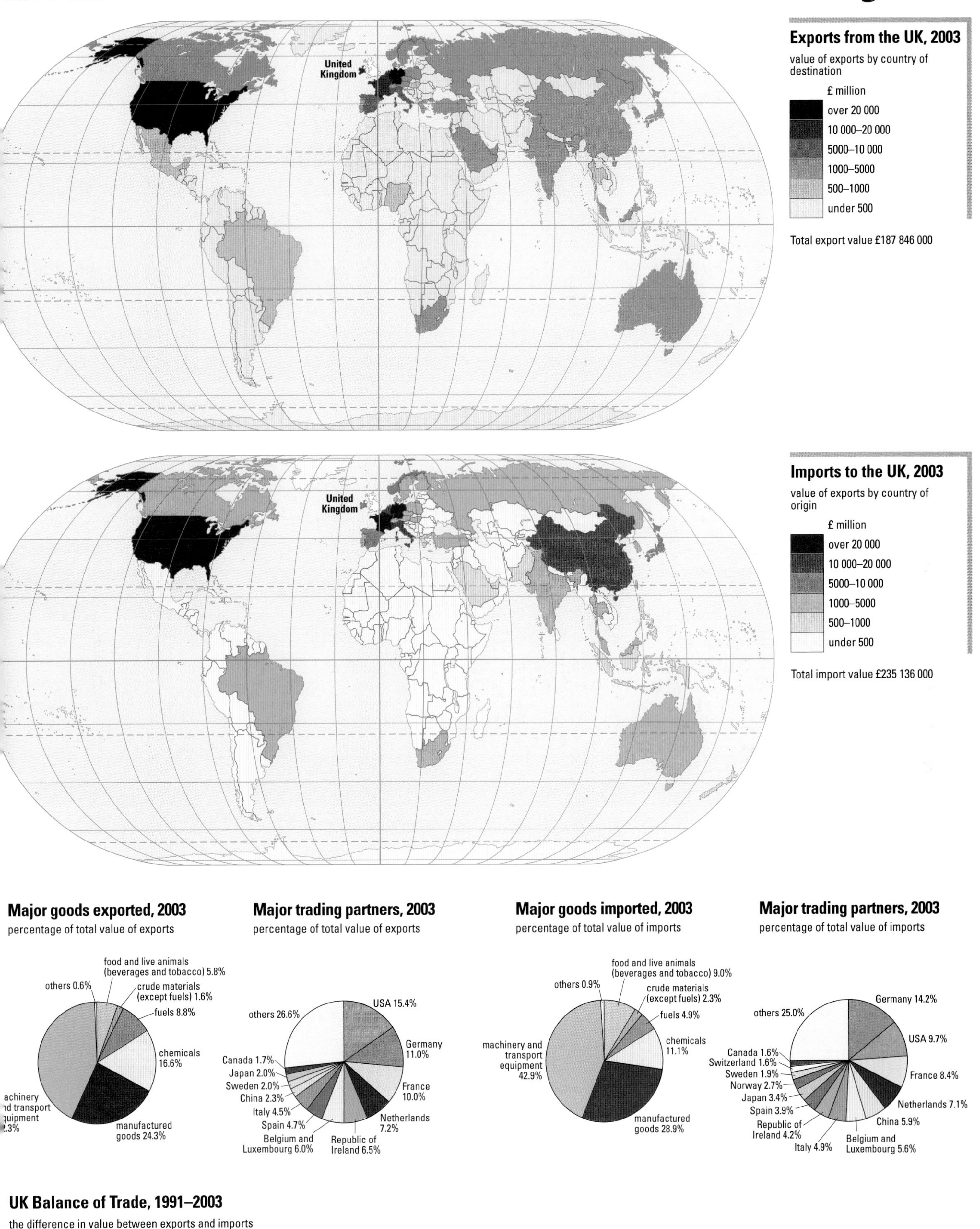

**Exports from the UK, 2003**

value of exports by country of destination

£ million

- over 20 000
- 10 000–20 000
- 5000–10 000
- 1000–5000
- 500–1000
- under 500

Total export value £187 846 000

**Imports to the UK, 2003**

value of exports by country of origin

£ million

- over 20 000
- 10 000–20 000
- 5000–10 000
- 1000–5000
- 500–1000
- under 500

Total import value £235 136 000

## Major goods exported, 2003

percentage of total value of exports

- others 0.6%
- food and live animals (beverages and tobacco) 5.8%
- crude materials (except fuels) 1.6%
- fuels 8.8%
- chemicals 16.6%
- machinery and transport equipment 42.3%
- manufactured goods 24.3%

## Major trading partners, 2003

percentage of total value of exports

- others 26.6%
- USA 15.4%
- Germany 11.0%
- France 10.0%
- Netherlands 7.2%
- Republic of Ireland 6.5%
- Belgium and Luxembourg 6.0%
- Spain 4.7%
- Italy 4.5%
- China 2.3%
- Sweden 2.0%
- Japan 2.0%
- Canada 1.7%

## Major goods imported, 2003

percentage of total value of imports

- others 0.9%
- food and live animals (beverages and tobacco) 9.0%
- crude materials (except fuels) 2.3%
- fuels 4.9%
- chemicals 11.1%
- machinery and transport equipment 42.9%
- manufactured goods 28.9%

## Major trading partners, 2003

percentage of total value of imports

- others 25.0%
- Germany 14.2%
- USA 9.7%
- France 8.4%
- Netherlands 7.1%
- China 5.9%
- Belgium and Luxembourg 5.6%
- Italy 4.9%
- Republic of Ireland 4.2%
- Spain 3.9%
- Japan 3.4%
- Norway 2.7%
- Sweden 1.9%
- Switzerland 1.6%
- Canada 1.6%

## UK Balance of Trade, 1991–2003

the difference in value between exports and imports

| | 1991 | 1992 | 1993 | 1994 | 1995 | 1996 | 1997 | 1998 | 1999 | 2000 | 2001 | 2002 | 2003 |
|---|---|---|---|---|---|---|---|---|---|---|---|---|---|
| Value of exports (£ million) | 103 939 | 107 863 | 122 039 | 135 260 | 153 577 | 167 196 | 171 923 | 164 056 | 166 198 | 188 085 | 190 055 | 186 517 | 187 846 |
| Value of imports (£ million) | 114 162 | 120 913 | 135 358 | 146 351 | 165 600 | 180 918 | 184 265 | 185 869 | 193 722 | 218 108 | 230 703 | 233 192 | 235 136 |

Eckert IV Projection     © Oxford University Press

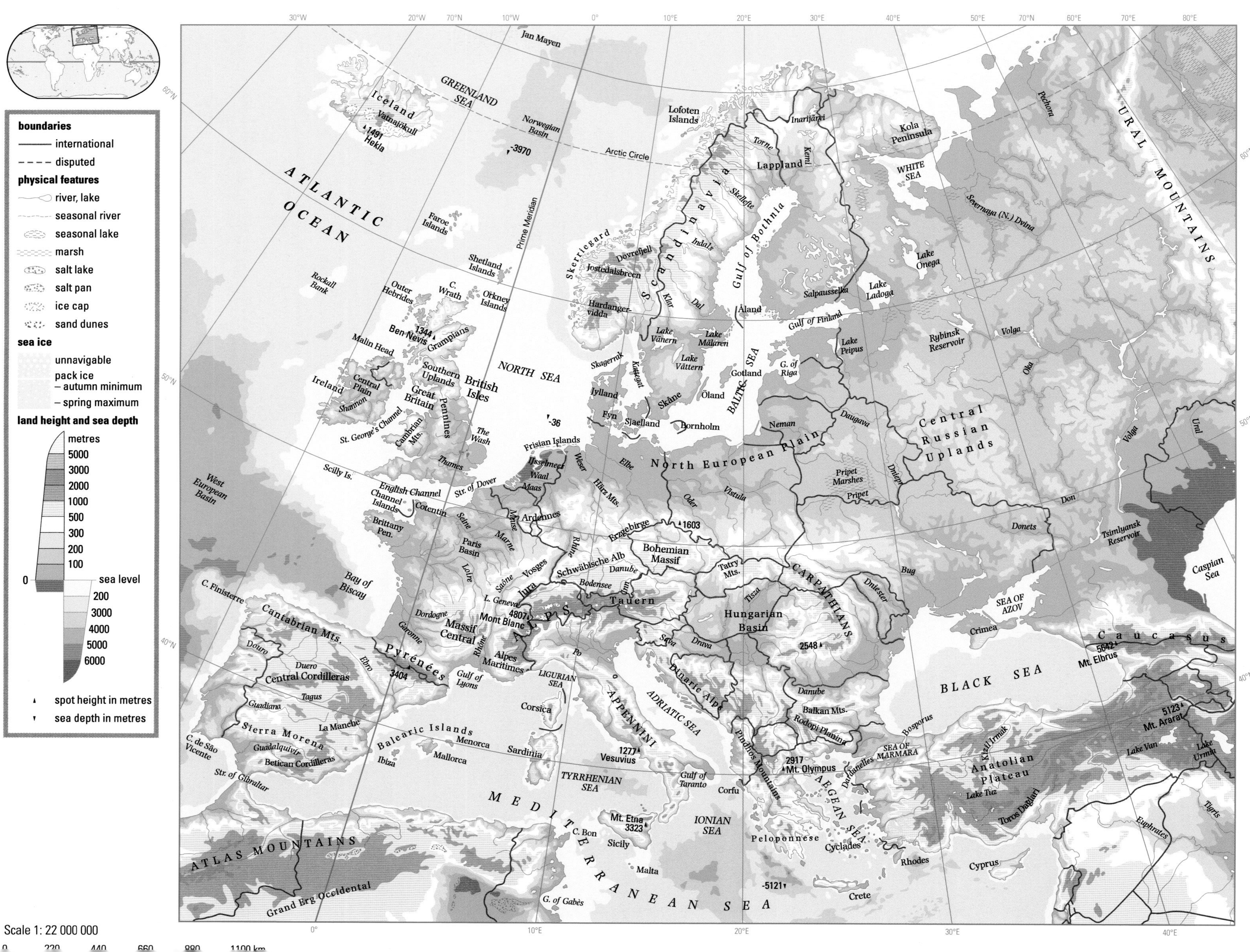
Jan Mayen
GREENLAND SEA
Iceland
Vatnajökull
1491
Hekla
Norwegian Basin
-3970
Arctic Circle
ATLANTIC OCEAN
Faroe Islands
Prime Meridian
Rockall Bank
Shetland Islands
Outer Hebrides
C. Wrath
Orkney Islands
1344
Ben Nevis
Grampians
Malin Head
Ireland
Central Plain
Shannon
Southern Uplands
British Isles
Great Britain
Pennines
St. George's Channel
Cambrian Mts.
The Wash
Thames
NORTH SEA
-36
West European Basin
Scilly Is.
English Channel
Channel Islands
Cotentin
Str. of Dover
Brittany Pen.
Lofoten Islands
Skerriegard
Dovrefjell
Jostedalsbreen
Hardanger-vidda
Scandinavia
Klar
Dal
Indals
Skellefte
Torne
Lappland
Kemi
Inarijärvi
Gulf of Bothnia
Åland
Lake Vänern
Lake Mälaren
Lake Vättern
Skagerrak
Kattegat
Jylland
Fyn
Sjaelland
Skåne
Öland
Gotland
BALTIC SEA
Bornholm
G. of Riga
Gulf of Finland
Salpausselkä
Kola Peninsula
WHITE SEA
Lake Onega
Lake Ladoga
Lake Peipus
Severnaya (N.) Dvina
Pechora
URAL MOUNTAINS
Rybinsk Reservoir
Volga
Oka
Ural
Central Russian Uplands
Neman
Daugava
North European Plain
Frisian Islands
Ijsselmeer
Waal
Maas
Weser
Elbe
Härz Mts.
Oder
Vistula
Pripet Marshes
Pripet
Dnieper
Don
Donets
Tsimlyansk Reservoir
Caspian Sea
Ardennes
Meuse
Seine
Marne
Paris Basin
Loire
Rhine
Erzgebirge
1603
Bohemian Massif
Schwäbische Alb
Danube
Bodensee
Inn
Tauern
Vosges
Saône
Jura
L. Geneva
4807
Mont Blanc
ALPS
Tatry Mts.
CARPATHIANS
Bug
Dniester
SEA OF AZOV
Crimea
Caucasus
5642
Mt. Elbrus
Bay of Biscay
C. Finisterre
Cantabrian Mts.
Douro
Duero
Ebro
Central Cordilleras
Tagus
Guadiana
La Mancha
Sierra Morena
Guadalquivir
Betican Cordilleras
C. de São Vicente
Str. of Gibraltar
Dordogne
Garonne
Massif Central
Rhône
Pyrénées
3404
Alpes Maritimes
Gulf of Lyons
LIGURIAN SEA
Corsica
Sardinia
Balearic Islands
Menorca
Mallorca
Ibiza
Po
APPENNINI
ADRIATIC SEA
Sava
Drava
Tisza
Hungarian Basin
2548
Dinaric Alps
Danube
Balkan Mts.
Rodopi Planina
BLACK SEA
Bosporus
SEA OF MARMARA
Dardanelles
Kizil Irmak
Anatolian Plateau
Lake Tuz
Toros Dağlari
5123
Mt. Ararat
Lake Van
Lake Urmia
Tigris
Euphrates
1277
Vesuvius
TYRRHENIAN SEA
Gulf of Taranto
Corfu
Pindhos Mountains
2917
Mt. Olympus
AEGEAN SEA
Mt. Etna
3323
C. Bon
Sicily
IONIAN SEA
Peloponnese
Cyclades
Rhodes
Cyprus
Malta
-5121
Crete
MEDITERRANEAN SEA
G. of Gabès
ATLAS MOUNTAINS
Grand Erg Occidental
boundaries
international
disputed
physical features
river, lake
seasonal river
seasonal lake
marsh
salt lake
salt pan
ice cap
sand dunes
sea ice
unnavigable
pack ice
– autumn minimum
– spring maximum
land height and sea depth
metres
5000
3000
2000
1000
500
300
200
100
0
sea level
200
3000
4000
5000
6000
spot height in metres
sea depth in metres
Scale 1: 22 000 000

Scale 1: 50 000 000

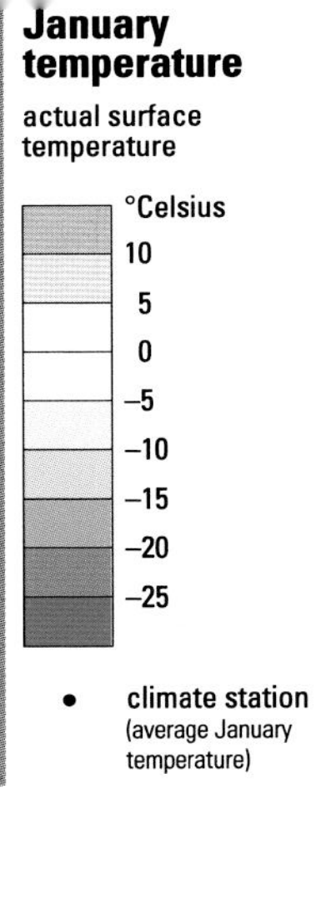

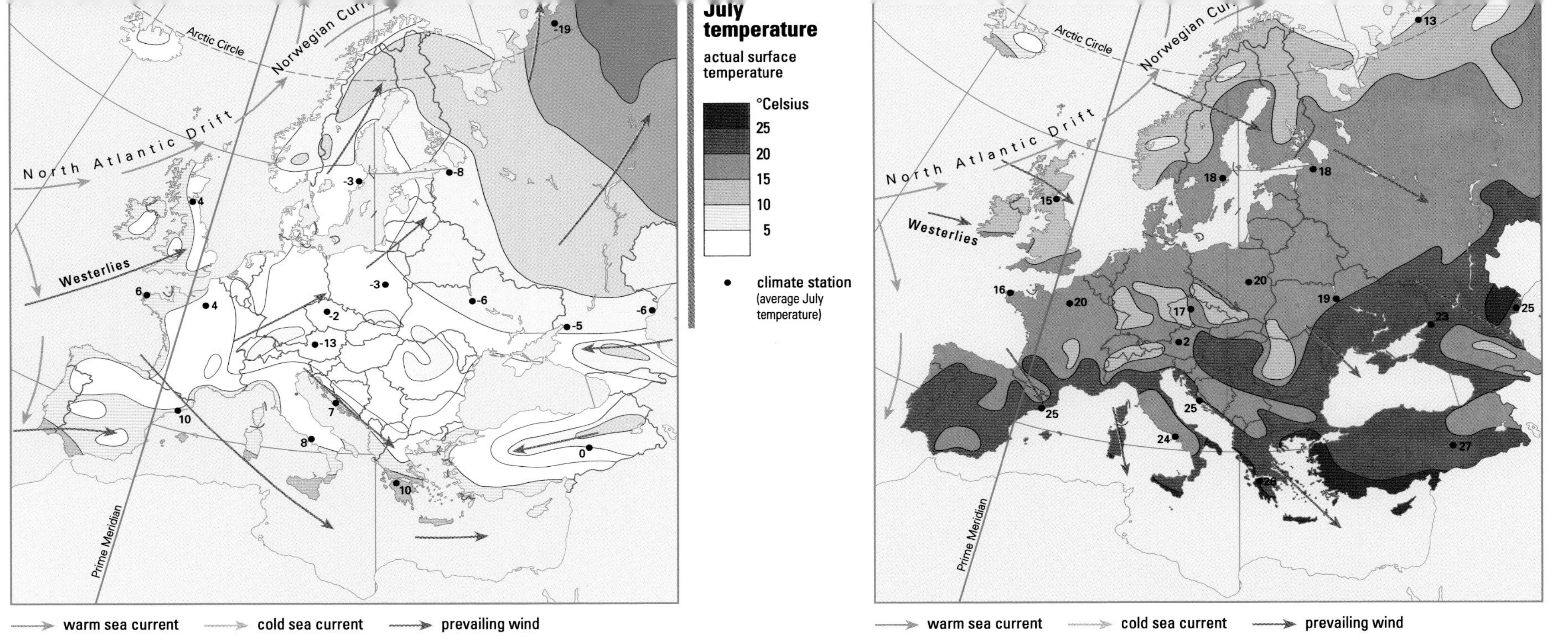

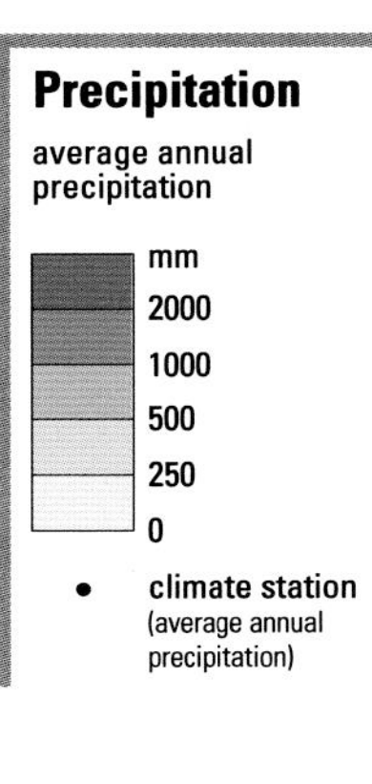

**July temperature**

actual surface temperature

°Celsius
25
20
15
10
5

• climate station (average July temperature)

**Ecosystems**

- coniferous forest
- mixed forest
- evergreens and shrubs
- temperate grasslands
- semi-desert
- tundra
- ice
- mountains

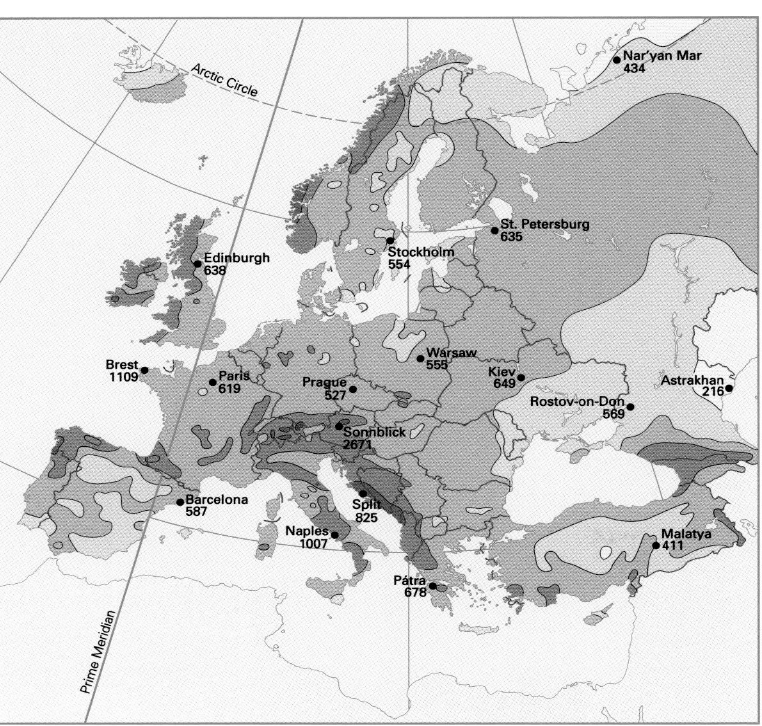

→ prevailing wind
→ cold sea current
↑ warm sea current

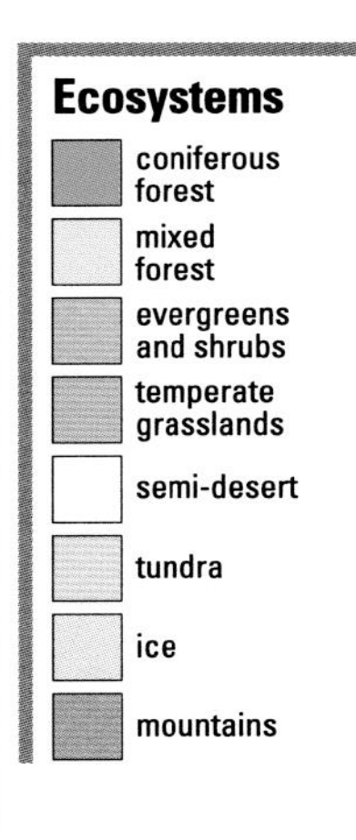

**January temperature**

actual surface temperature

°Celsius
10
5
0
−5
−10
−15
−20
−25

• climate station (average January temperature)

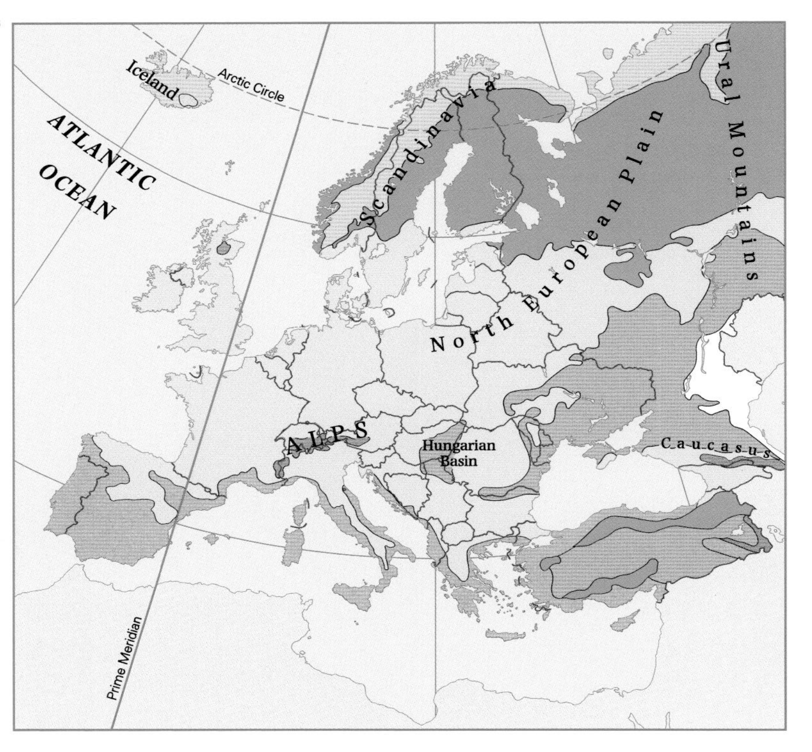

**Precipitation**

average annual precipitation

mm
2000
1000
500
250
0

• climate station (average annual precipitation)

© Oxford University Press

Conical Orthomorphic Projection

## Political

**boundaries**

— international

--- disputed

**settlements**

■ capital city

● other important city

### The European Union

**Brussels:** Headquarters

**Strasbourg:** European Parliament

**Luxembourg:** European Court of Justice

### Headquarters of other European and World Organisations

**The Hague:** International Court of Justice

**Geneva:** World Health Organisation (WHO)

**Paris:** United National Education, Scientific and Cultural Organisation (UNESCO)

Organisation for Economic Cooperation and Development (OECD)

**Rome:** Food and Agricultural Organisation of the United Nations (FAO)

Scale 1: 22 000 000 (main map)

Conical Orthomorphic Projection

### The European Union

date of joining

- 1957
- 1973
- 1981
- 1986
- 1990
- 1995
- 2004
- negotiating membership
- ★ headquarters

**Population Growth**
millions of people

2004
1995
1990
1986
1981
1973
1957

550 500 450 400 350 300 250 200 150 100 50 0

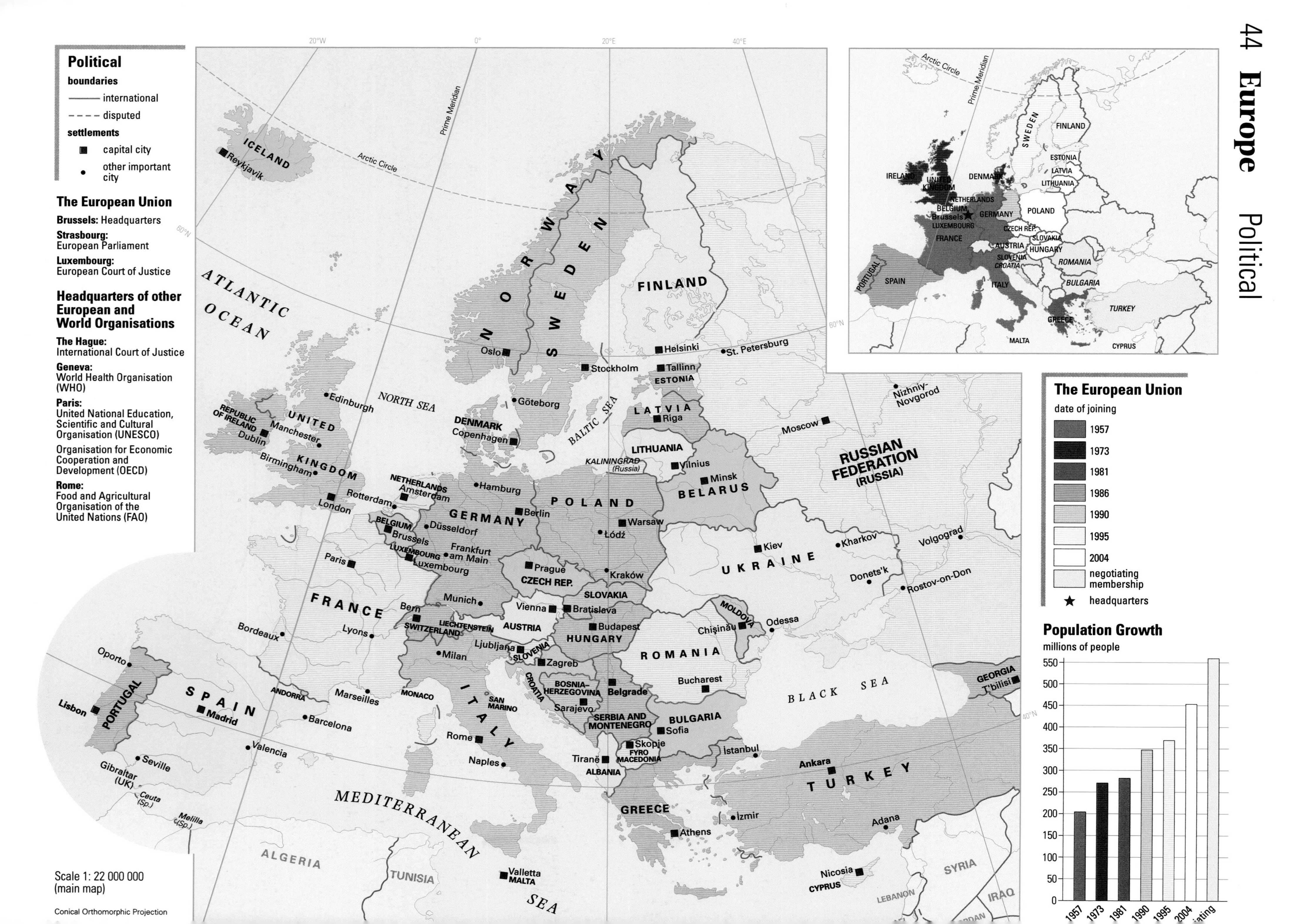

Scale 1: 50 000 000

**Population
density**

people per square
kilometre

- over 200
- 100–200
- 10–100
- 1–10
- under 1

**Major cities**

population in millions

- ■ over 3
- □ 1–3
- ● 0.5–1
- · 0.1–0.5

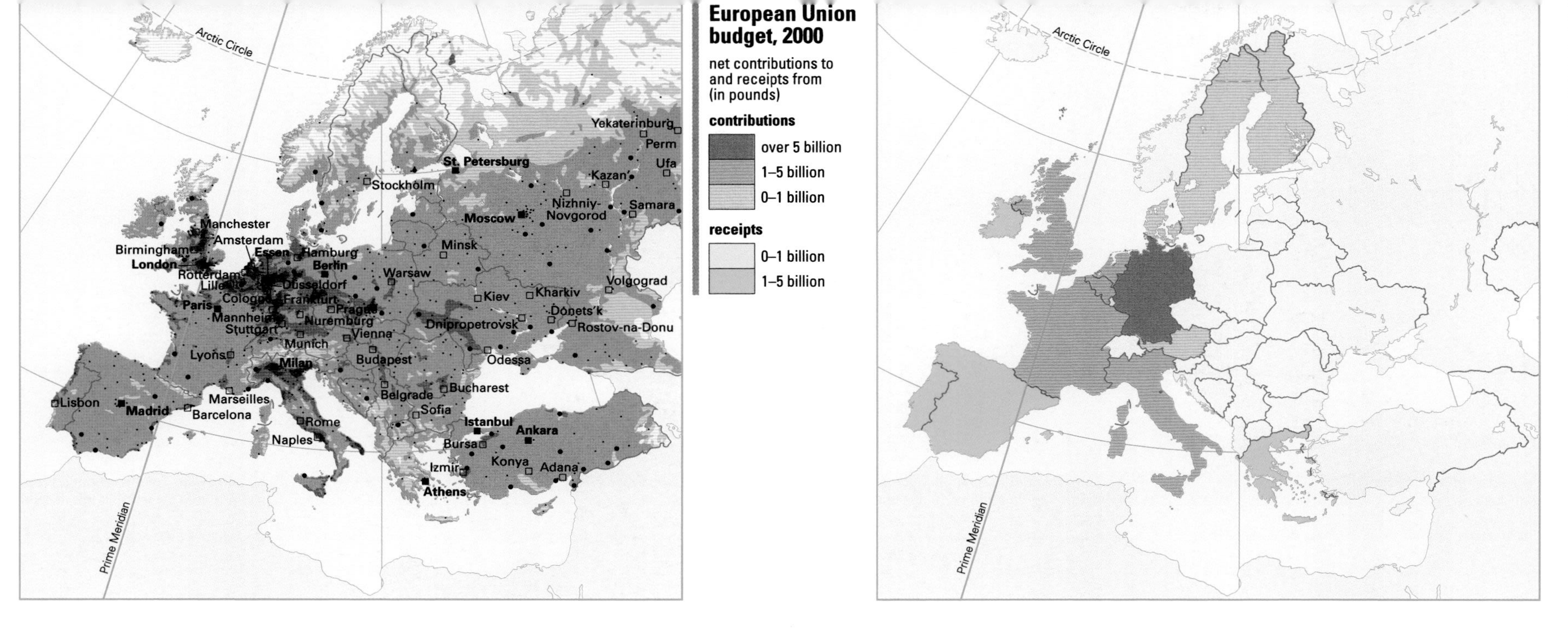

**European Union
budget, 2000**

net contributions to
and receipts from
(in pounds)

**contributions**

- over 5 billion
- 1–5 billion
- 0–1 billion

**receipts**

- 0–1 billion
- 1–5 billion

**Population
change,
1999–2004**

percentage change in
the number of people

**increase**

- over 4%
- 2–4%
- 1–2%
- 0–1%

**decrease**

- 0–1%
- 1–2%
- 2–4%
- over 4%

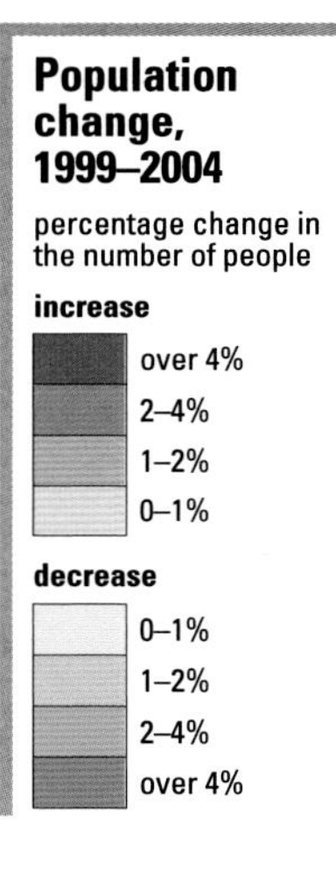

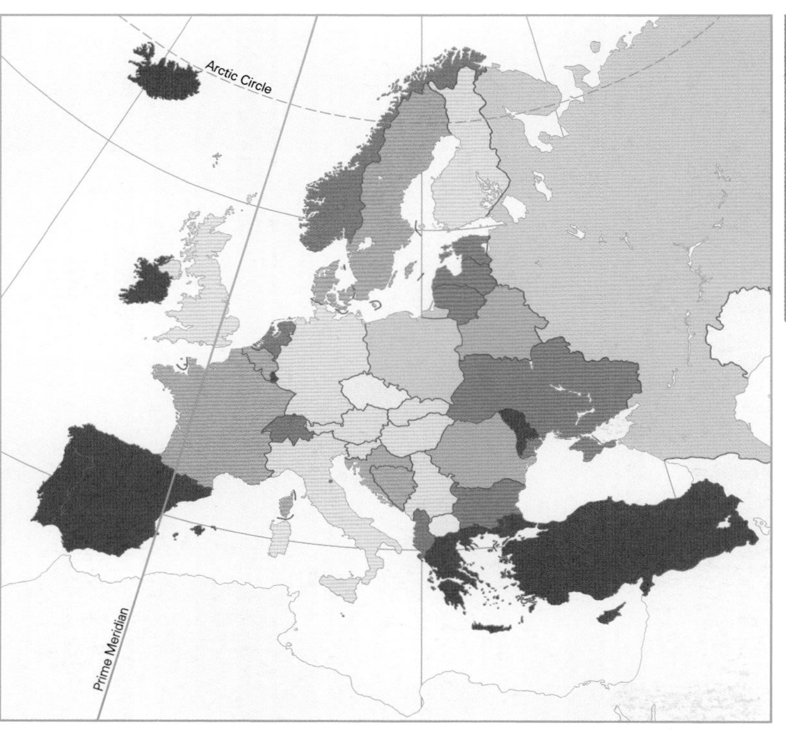

**Unemployment,
2003**

percentage of the work
force out of work

- over 20%
- 15–20%
- 10–15%
- 5–10%
- under 5%
- no data

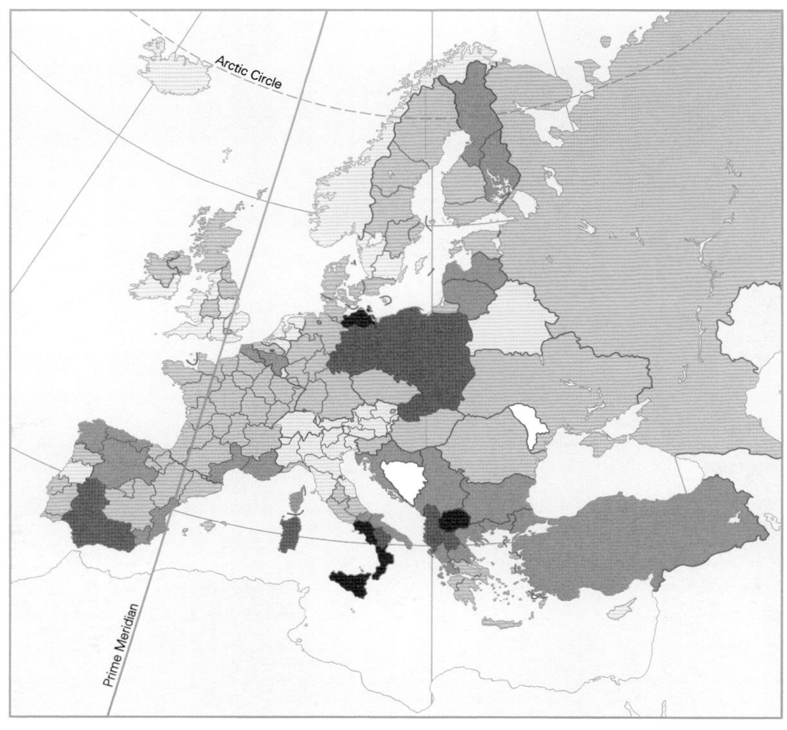

Conical Orthomorphic Projection

Scale 1: 22 000 000

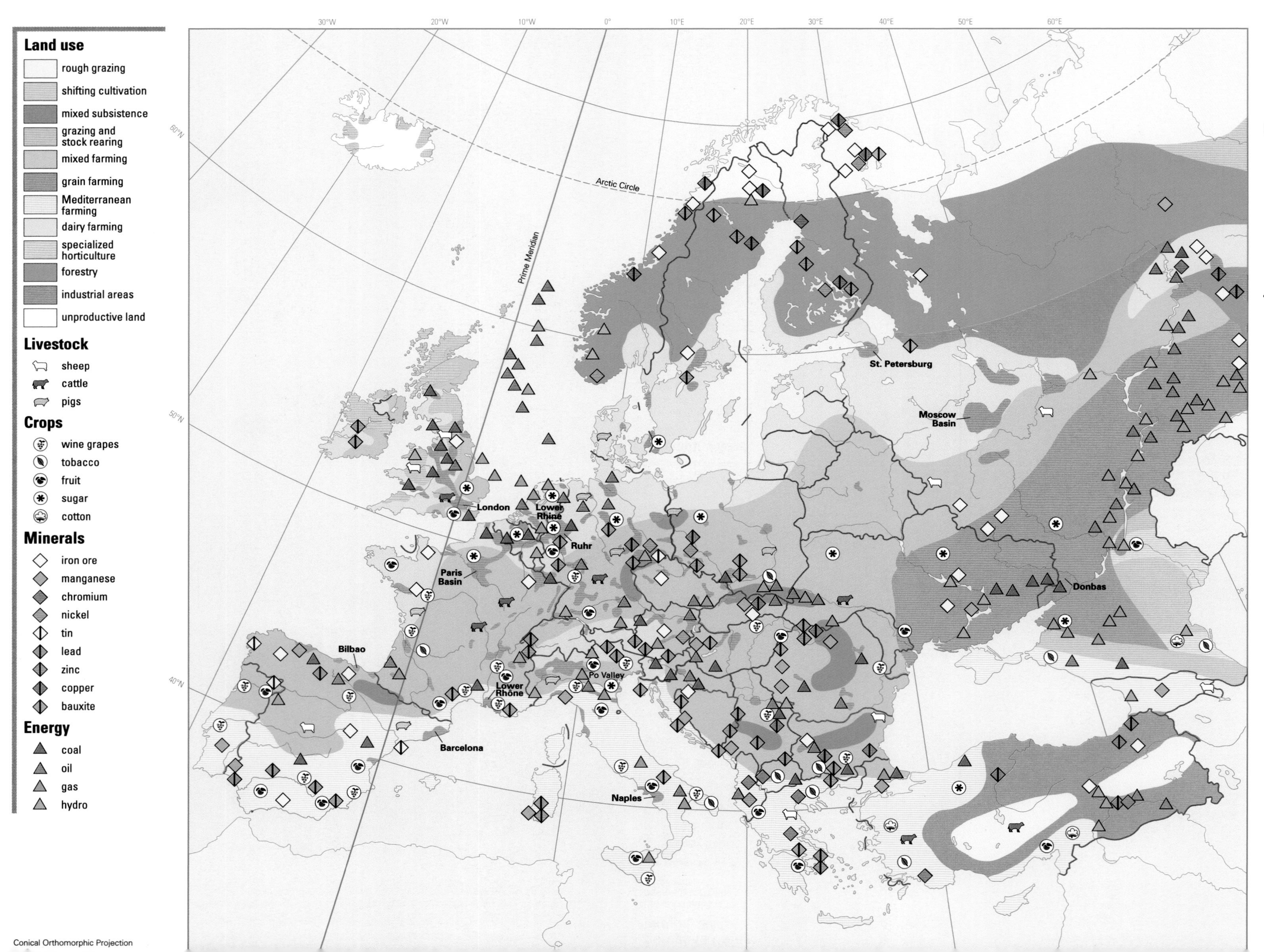

St. Petersburg

Moscow Basin

Donbas

London

Paris Basin

Lower Rhine

Ruhr

Po Valley

Lower Rhône

Barcelona

Naples

Bilbao

Arctic Circle

Prime Meridian

**Land use**
rough grazing
shifting cultivation
mixed subsistence
grazing and stock rearing
mixed farming
grain farming
Mediterranean farming
dairy farming
specialized horticulture
forestry
industrial areas
unproductive land

**Livestock**
sheep
cattle
pigs

**Crops**
wine grapes
tobacco
fruit
sugar
cotton

**Minerals**
iron ore
manganese
chromium
nickel
tin
lead
zinc
copper
bauxite

**Energy**
coal
oil
gas
hydro

Conical Orthomorphic Projection

Scale 1: 17 000 000

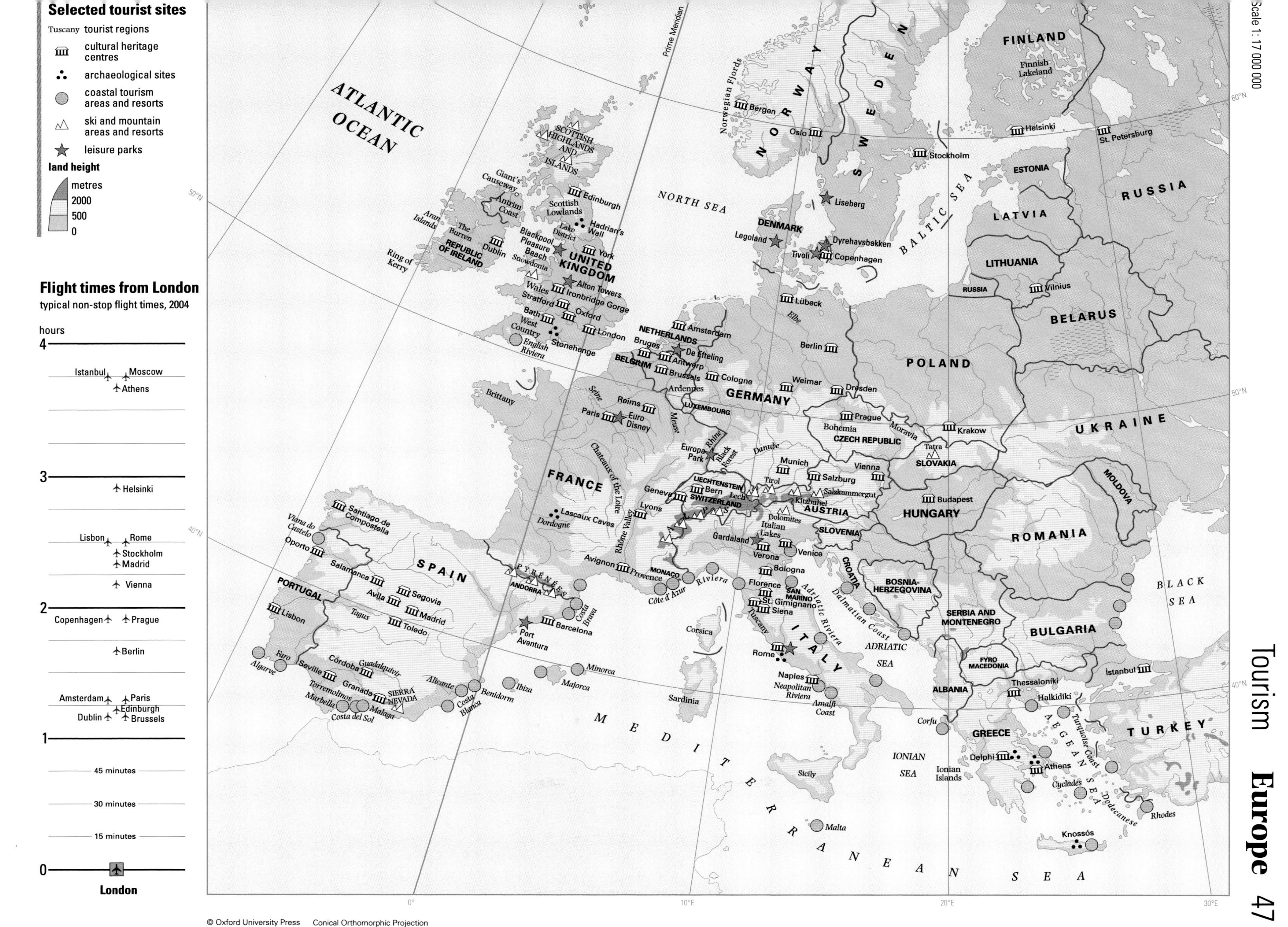

**Selected tourist sites**

Tuscany tourist regions

- 𝍌 cultural heritage centres
- •:• archaeological sites
- ◯ coastal tourism areas and resorts
- △ ski and mountain areas and resorts
- ★ leisure parks

**land height**

metres
2000
500
0

**Flight times from London**
typical non-stop flight times, 2004

hours

4 ── Istanbul ✈ Moscow
   ── ✈ Athens

3 ── ✈ Helsinki

   ── Lisbon ✈ Rome
   ── ✈ Stockholm
   ── ✈ Madrid
2 ── ✈ Vienna
   ── Copenhagen ✈ ✈ Prague
   ── ✈ Berlin

   ── Amsterdam ✈ ✈ Paris
1 ── Dublin ✈ ✈ Edinburgh
   ── ✈ Brussels

   ── 45 minutes

   ── 30 minutes

   ── 15 minutes

0 ── ✈ London

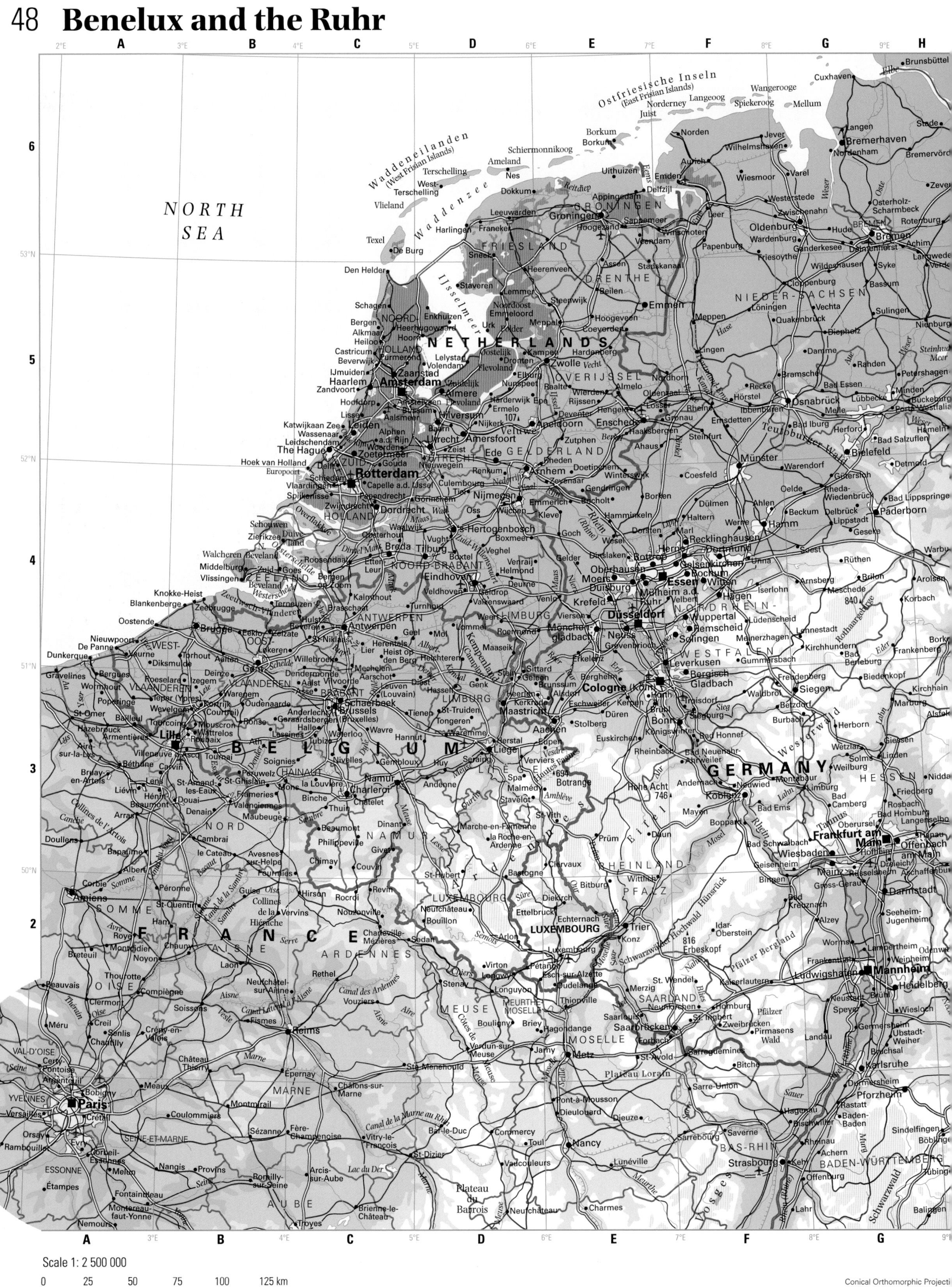

NORTH SEA

*Waddeneilanden (West Frisian Islands)*

NETHERLANDS

BELGIUM

GERMANY

FRANCE

LUXEMBOURG

Paris

Amsterdam

Rotterdam

Brussels

Cologne

Essen

Düsseldorf

Frankfurt am Main

Luxembourg

Scale 1: 2 500 000

0    25    50    75    100    125 km

Conical Orthomorphic Projection

© Oxford University Press

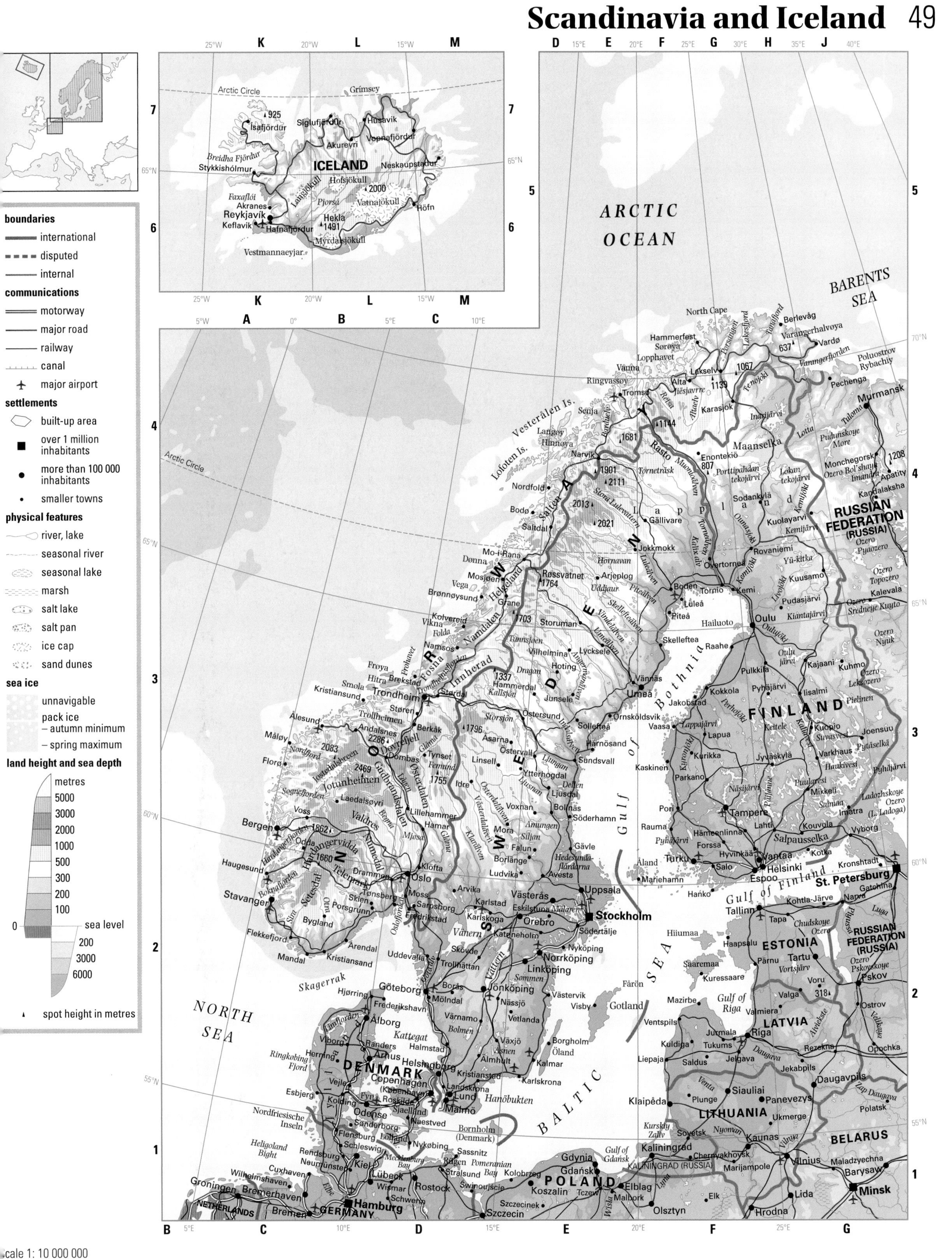

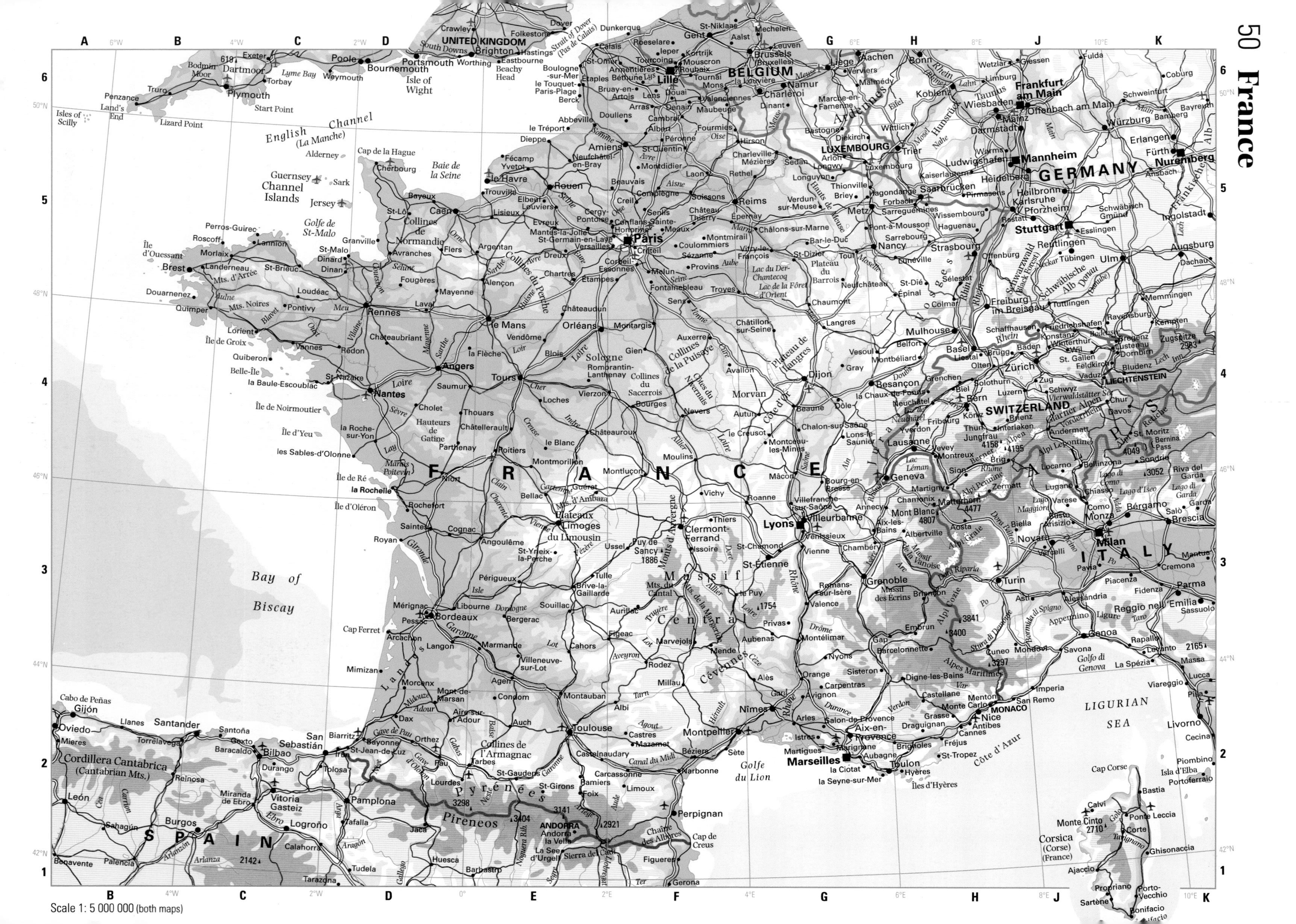
FRANCE
UNITED KINGDOM
BELGIUM
LUXEMBOURG
GERMANY
SWITZERLAND
LIECHTENSTEIN
ITALY
MONACO
ANDORRA
SPAIN
English Channel (La Manche)
Channel Islands
Bay of Biscay
Golfe du Lion
LIGURIAN SEA
Corsica (Corse) (France)
Massif Central
Pyrénées
Pireneos
Cordillera Cantabrica (Cantabrian Mts.)
Paris
Lyons
Marseilles
Bordeaux
Toulouse
Nantes
Lille
Strasbourg
Scale 1: 5 000 000 (both maps)

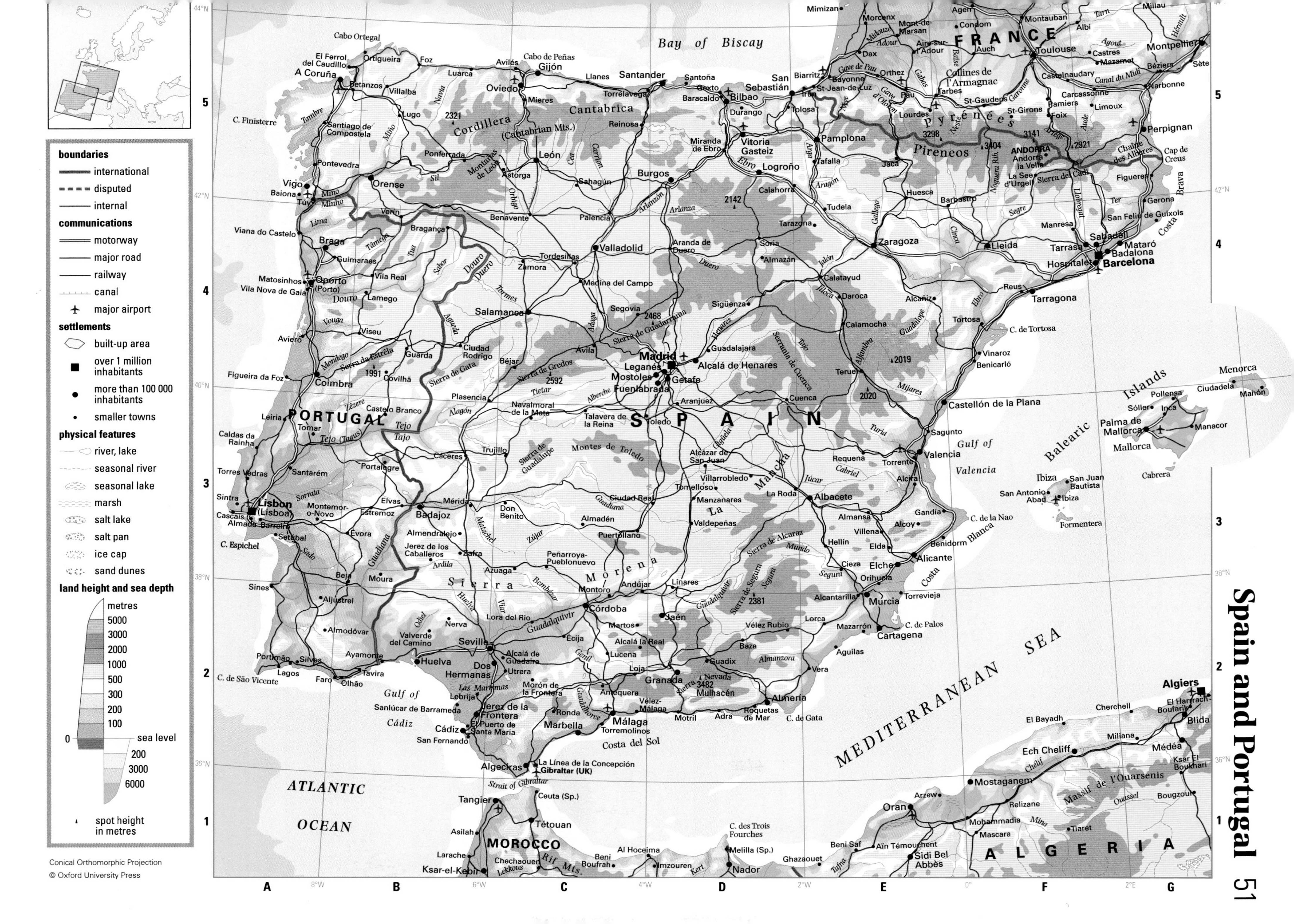
boundaries
international
disputed
internal
communications
motorway
major road
railway
canal
major airport
settlements
built-up area
over 1 million inhabitants
more than 100 000 inhabitants
smaller towns
physical features
river, lake
seasonal river
seasonal lake
marsh
salt lake
salt pan
ice cap
sand dunes
land height and sea depth
metres
5000
3000
2000
1000
500
300
200
100
0
sea level
200
3000
6000
spot height in metres
Conical Orthomorphic Projection
© Oxford University Press
Bay of Biscay
MEDITERRANEAN SEA
ATLANTIC OCEAN
SPAIN
PORTUGAL
FRANCE
ALGERIA
MOROCCO
Balearic Islands
Cordillera Cantabrica (Cantabrian Mts.)
Pyrénées
Pireneos
Sierra Morena
La Mancha
Madrid
Lisbon (Lisboa)
Barcelona
Valencia
Seville
Zaragoza
Málaga
Oporto (Porto)
Bilbao
Gulf of Valencia
Gulf of Cádiz
Strait of Gibraltar
Gibraltar (UK)
Ceuta (Sp.)
Melilla (Sp.)
Algiers
Oran
Menorca
Mallorca
Ibiza
Formentera

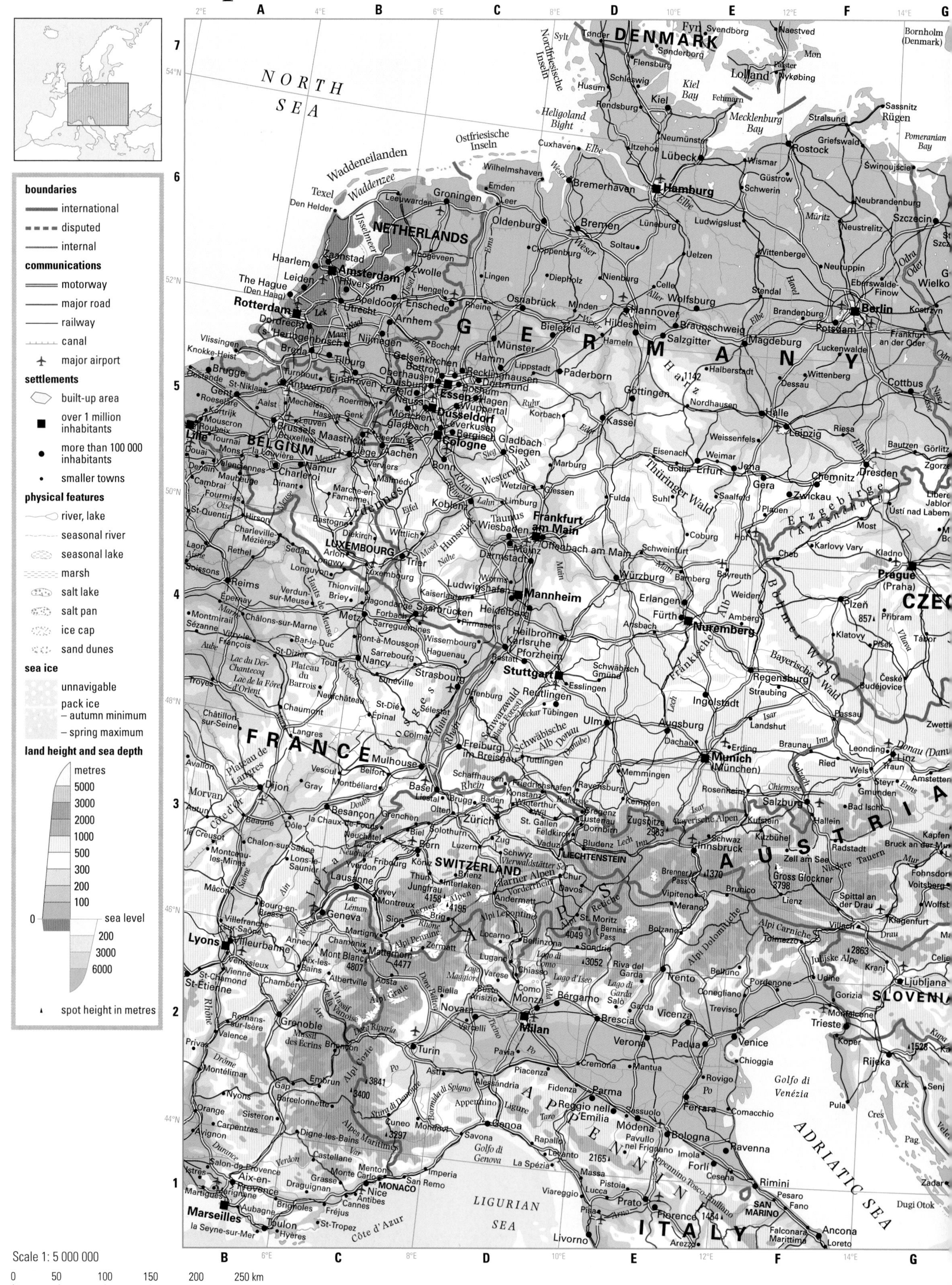

Conical Orthomorphic Projection    © Oxford University Press

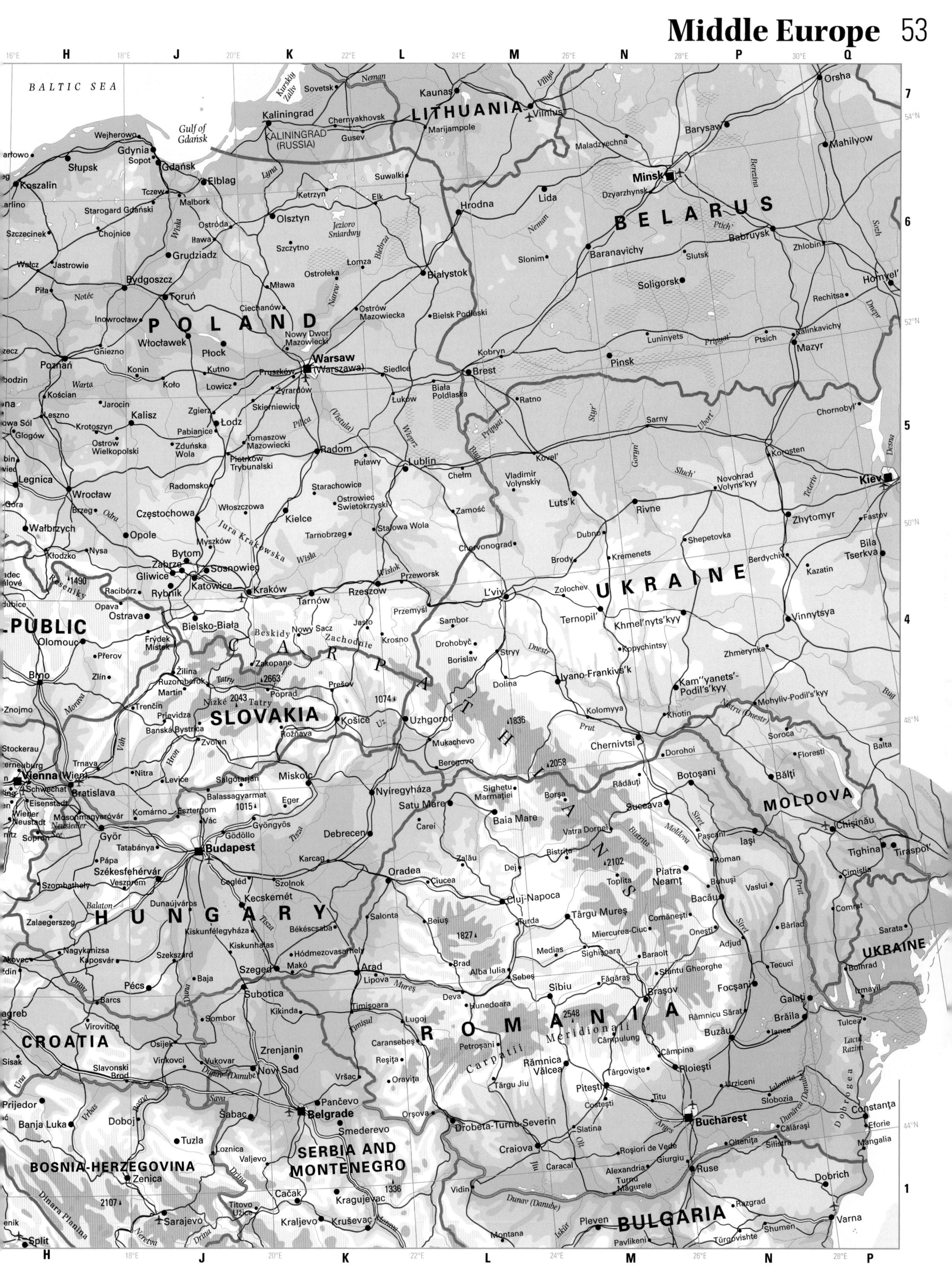
BALTIC SEA
POLAND
LITHUANIA
BELARUS
UKRAINE
SLOVAKIA
HUNGARY
ROMANIA
MOLDOVA
CROATIA
BOSNIA-HERZEGOVINA
SERBIA AND MONTENEGRO
BULGARIA
Warsaw (Warszawa)
Minsk
Kiev
Budapest
Bucharest
Belgrade
Vilnius
Chişinău
Vienna (Wien)
Bratislava
Sarajevo

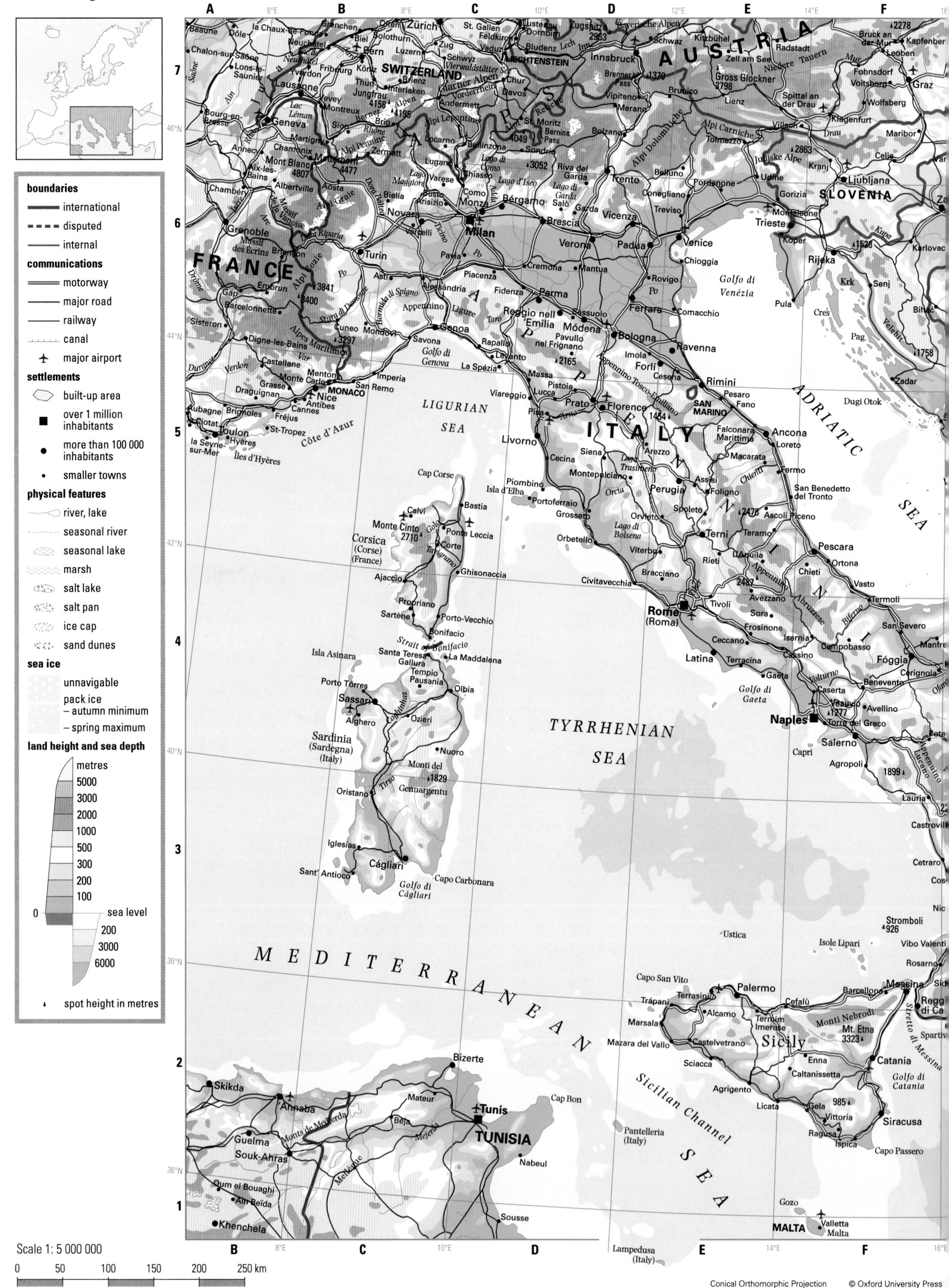

boundaries
international
disputed
internal
communications
motorway
major road
railway
canal
major airport
settlements
built-up area
over 1 million inhabitants
more than 100 000 inhabitants
smaller towns
physical features
river, lake
seasonal river
seasonal lake
marsh
salt lake
salt pan
ice cap
sand dunes
sea ice
unnavigable
pack ice
– autumn minimum
– spring maximum
land height and sea depth
metres
5000
3000
2000
1000
500
300
200
100
0
sea level
200
3000
6000
spot height in metres
Scale 1: 5 000 000
0
50
100
150
200
250 km
A
B
C
D
E
F
1
2
3
4
5
6
7
6°E
8°E
10°E
12°E
14°E
16°E
46°N
44°N
42°N
40°N
38°N
36°N
AUSTRIA
SWITZERLAND
LIECHTENSTEIN
FRANCE
ITALY
SLOVENIA
MONACO
SAN MARINO
TUNISIA
MALTA
ALPS
APENNINES
LIGURIAN SEA
TYRRHENIAN SEA
MEDITERRANEAN
ADRIATIC SEA
Sicilian Channel
SEA
Golfo di Venézia
Golfo di Genova
Golfo di Gaeta
Golfo di Cágliari
Golfo di Catania
Strait of Bonifacio
Stretto di Messina
Côte d'Azur
Îles d'Hyères
Corsica (Corse) (France)
Sardinia (Sardegna) (Italy)
Sicily
Capri
Ustica
Isole Lipari
Stromboli 926
Pantelleria (Italy)
Lampedusa (Italy)
Isla d'Elba
Isla Asinara
Cap Corse
Cap Bon
Capo Carbonara
Capo San Vito
Capo Passero
Monte Cinto 2710
Monti del Gennargentu 1829
Mt. Etna 3323
Monti Nebrodi
Mont Blanc 4807
Matterhorn 4477
Jungfrau 4158
Gross Glockner 3798
Zugspitze 2963
Bernina Pass 4049
Brenner Pass 1370
Vesuvio 1277
Milan
Rome (Roma)
Naples
Tunis
Turin
Genoa
Venice
Florence
Bologna
Verona
Palermo
Catania
Cágliari
Trieste
Ljubljana
Zürich
Bern
Geneva
Lausanne
Innsbruck
Graz
Klagenfurt
Nice
Toulon
Grenoble
Ajaccio
Bastia
Sassari
Bizerte
Annaba
Skikda
Sousse
Valletta
Ancona
Pescara
Bari
Livorno
Pisa
Parma
Padua
Trento
Bolzano
Brescia
Bergamo
Messina
Siracusa
Salerno
Fóggia
Perugia
Rimini
Ravenna
Rijeka
Conical Orthomorphic Projection
© Oxford University Press

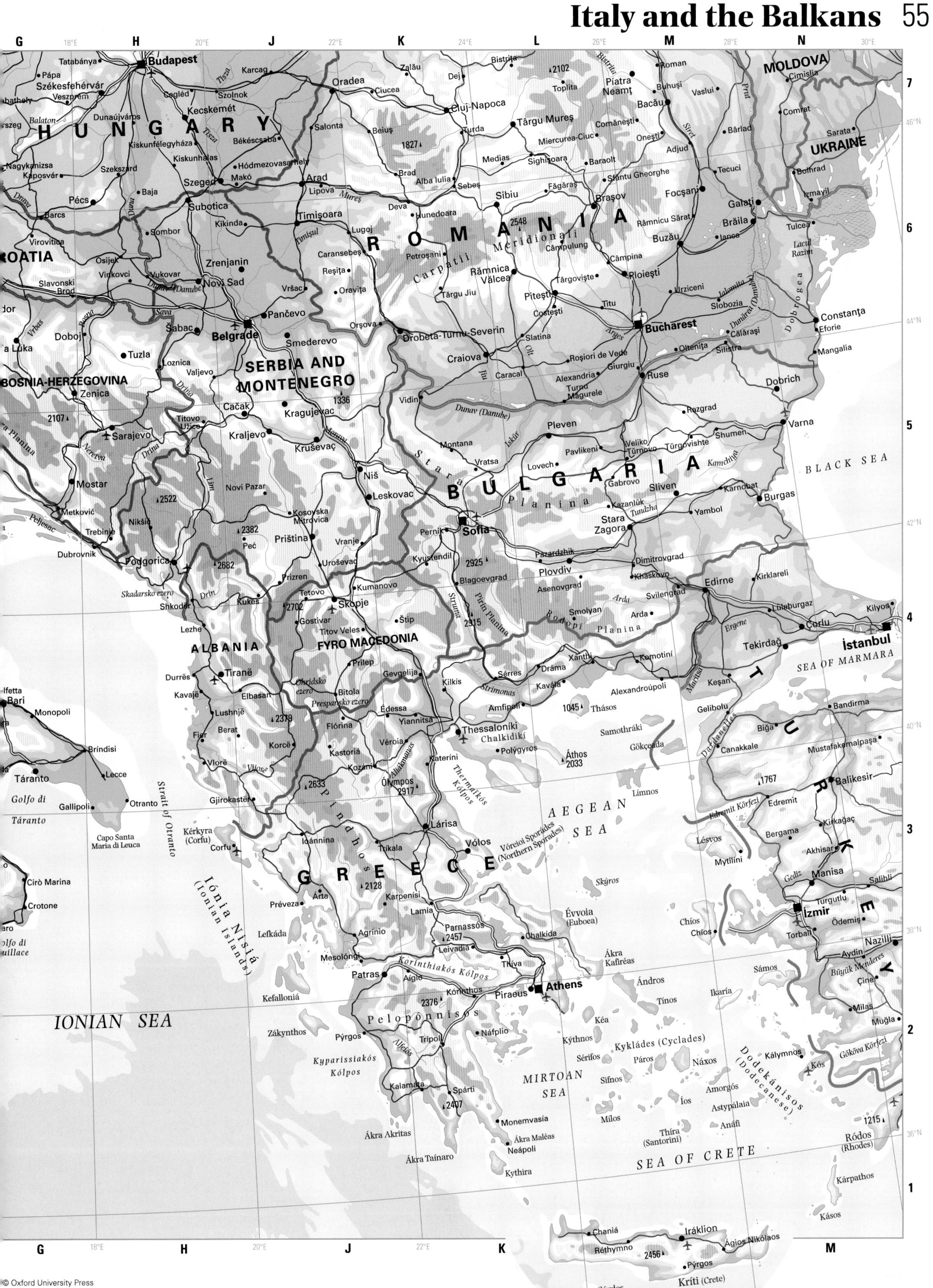

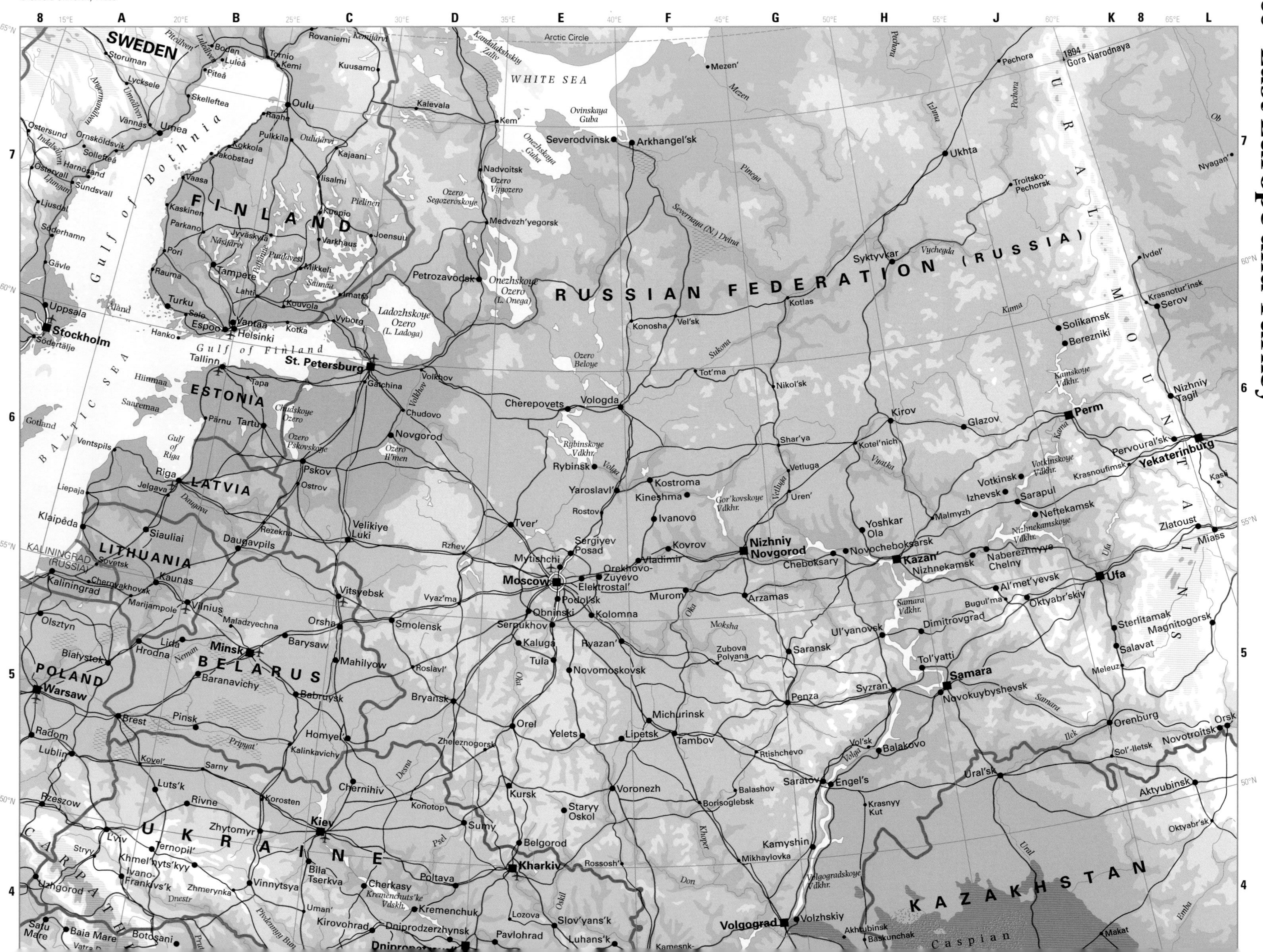
RUSSIAN FEDERATION (RUSSIA)
URAL MOUNTAINS
KAZAKHSTAN
UKRAINE
BELARUS
LITHUANIA
LATVIA
ESTONIA
FINLAND
SWEDEN
POLAND
KALININGRAD (RUSSIA)
WHITE SEA
BALTIC SEA
Gulf of Bothnia
Gulf of Finland
Gulf of Riga
Arctic Circle
Caspian
CARPATHIANS
Moscow
St. Petersburg
Nizhniy Novgorod
Kazan'
Samara
Ufa
Perm
Yekaterinburg
Volgograd
Kharkiv
Kiev
Minsk
Warsaw
Stockholm
Helsinki
Tallinn
Riga
Vilnius
Arkhangel'sk
Severodvinsk
Petrozavodsk
Vologda
Yaroslavl'
Tver'
Smolensk
Bryansk
Kursk
Voronezh
Saratov
Penza
Orenburg
Izhevsk
Kirov
Syktyvkar
Ukhta
Pechora
Kotlas
Novgorod
Pskov
Kaunas
Klaipėda
Daugavpils
Homyel
Mahilyow
Vitsyebsk
Hrodna
Brest
Pinsk
L'viv
Rivne
Zhytomyr
Vinnytsya
Poltava
Chernihiv
Tula
Ryazan'
Lipetsk
Tambov
Ul'yanovsk
Tol'yatti
Engel's
Aktyubinsk
Ural'sk
Orsk
Zlatoust
Miass
Ob
Vychegda
Mezen'
Pinega
Sukhona
Volga
Oka
Kama
Don
Khoper
Desna
Pripyat'
Daugava
Neman
Ladozhskoye Ozero (L. Ladoga)
Onezhskoye Ozero (L. Onega)
Rybinskoye Vdkhr.
Gor'kovskoye Vdkhr.
Volgogradskoye Vdkhr.
1894 Gora Narodnaya
7
6
5
4
8
A
B
C
D
E
F
G
H
J
K
L
65°N
60°N
55°N
50°N
15°E
20°E
25°E
30°E
35°E
40°E
45°E
50°E
55°E
60°E
65°E

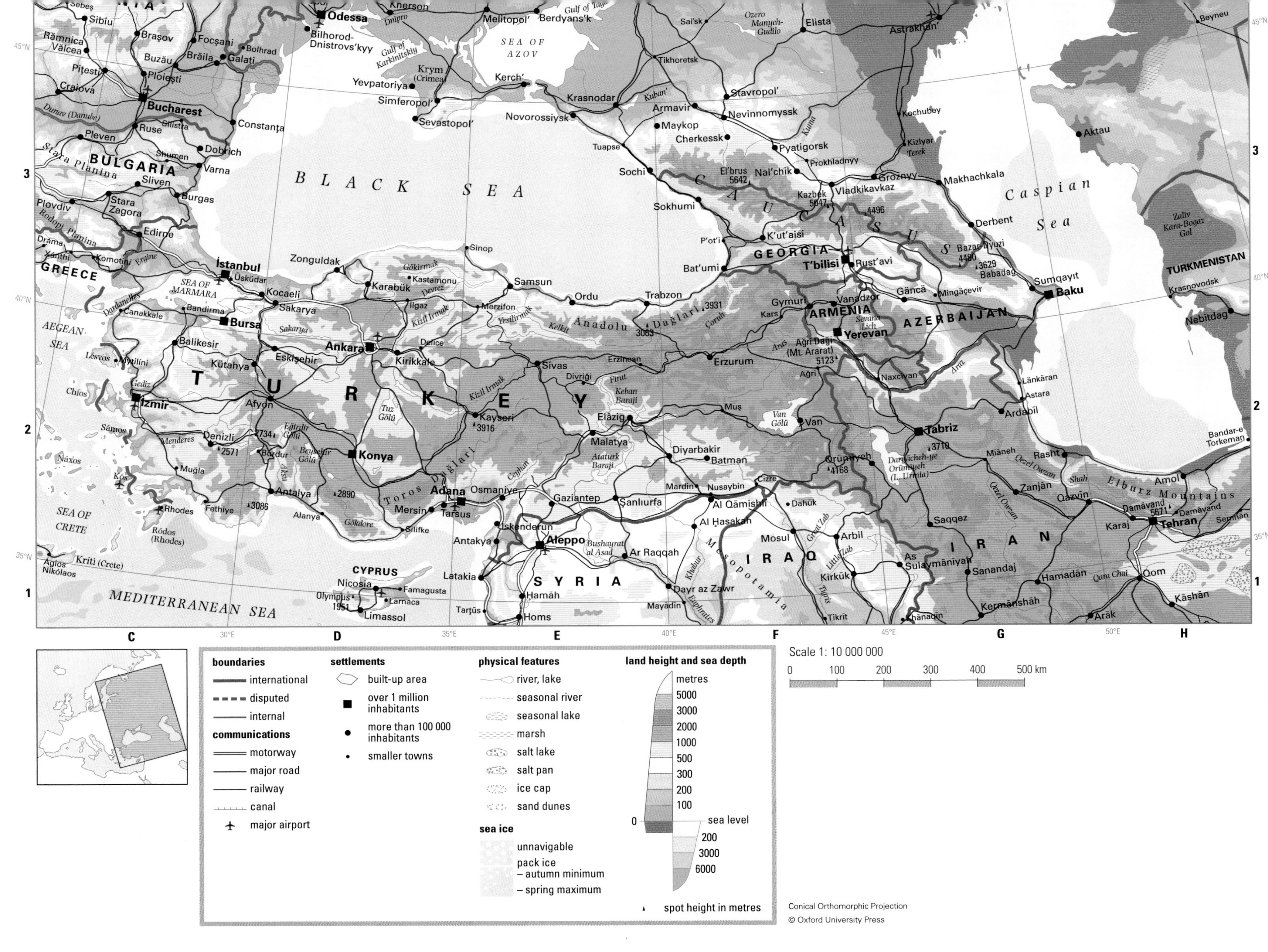

Scale 1 : 10 000 000

0    100    200    300    400    500 km

Conical Orthomorphic Projection
© Oxford University Press

**boundaries**
international
disputed
internal

**communications**
motorway
major road
railway
canal
✈ major airport

**settlements**
built-up area
over 1 million inhabitants
more than 100 000 inhabitants
smaller towns

**physical features**
river, lake
seasonal river
seasonal lake
marsh
salt lake
salt pan
ice cap
sand dunes

**sea ice**
unnavigable
pack ice
– autumn minimum
– spring maximum

**land height and sea depth**
metres
5000
3000
2000
1000
500
300
200
100
sea level
200
3000
6000
▲ spot height in metres

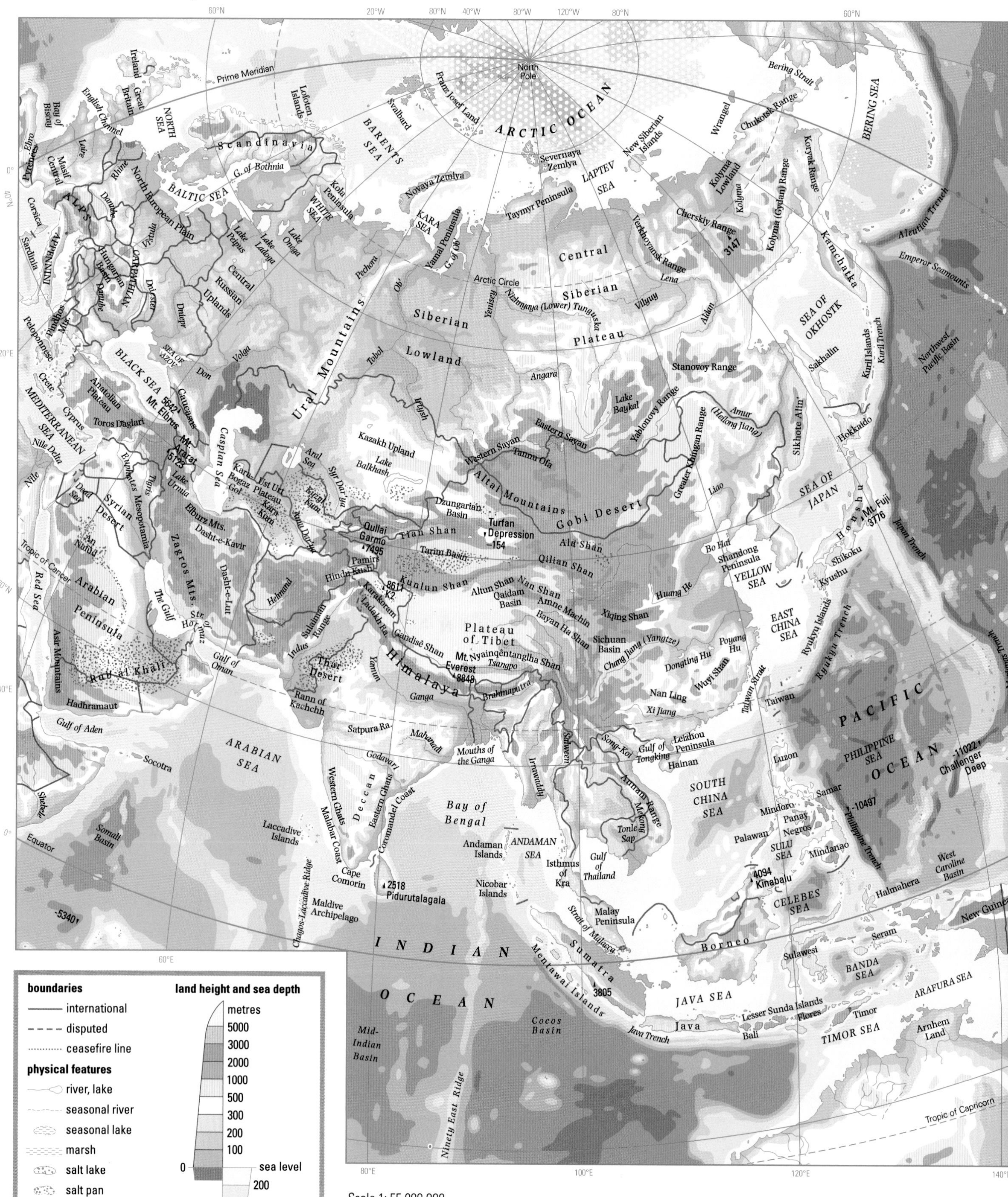

Zenithal equal Area Projection

**boundaries**
| — | international |
| -- -- | disputed |
| ..... | ceasefire line |

**physical features**
river, lake
seasonal river
seasonal lake
marsh
salt lake
salt pan
ice cap
sand dunes

**sea ice**
unnavigable
pack ice
– autumn minimum
– spring maximum

**land height and sea depth**
metres
5000
3000
2000
1000
500
300
200
100
0     sea level
200
3000
4000
5000
6000

▲ spot height in metres
▼ sea depth in metres

Scale 1: 55 000 000

| 0 | 550 | 1100 | 1650 | 2200 | 2750 km |

le 1: 55 000 000

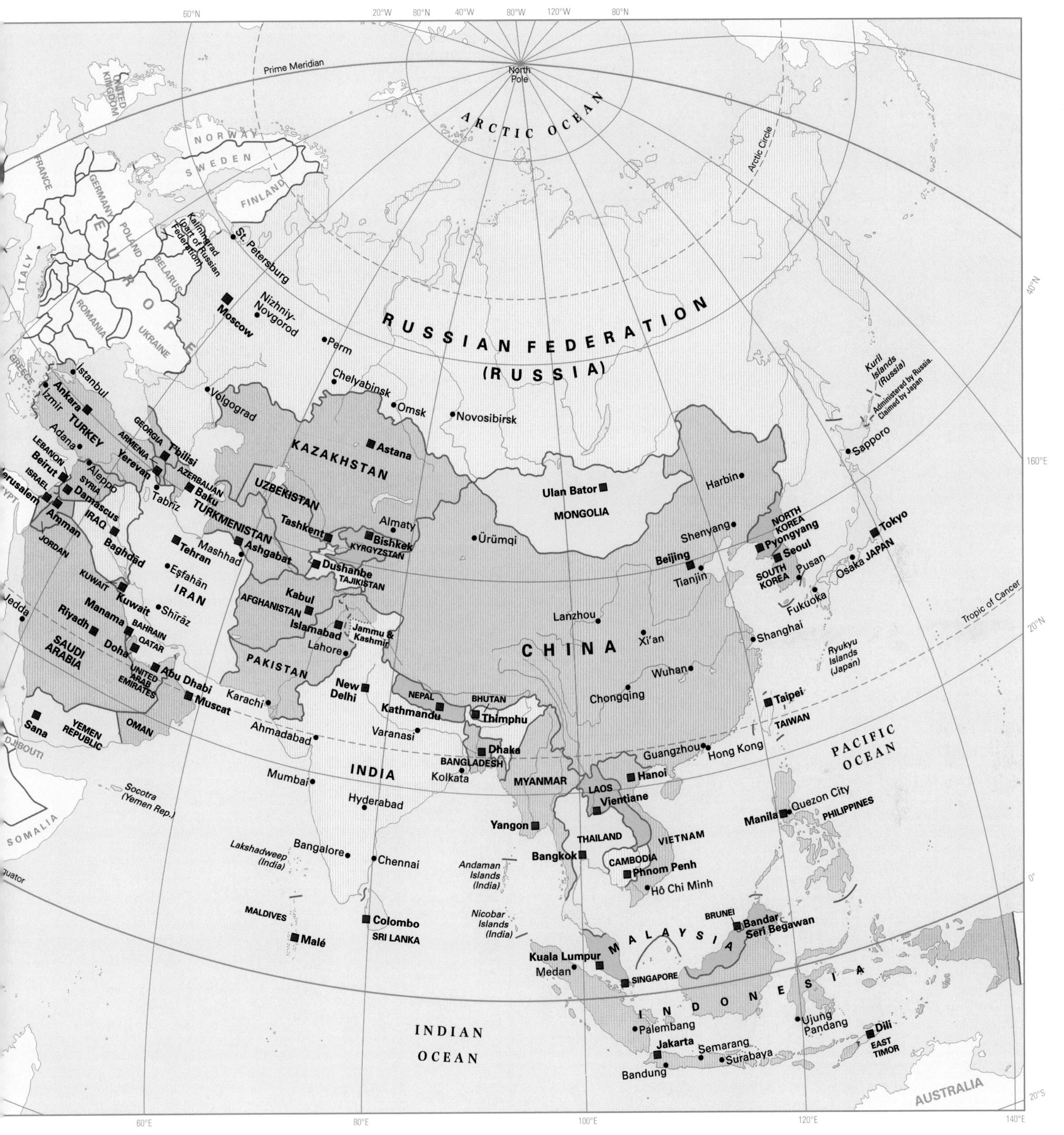

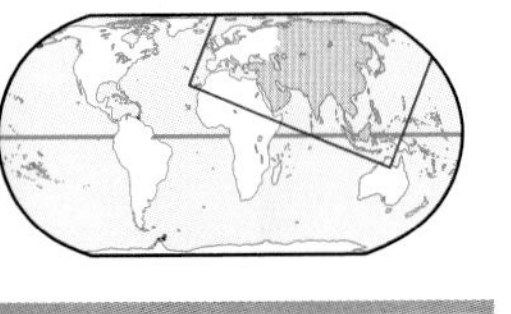

**Asian urban and rural population, 2003**

percentage of
total population

rural    urban

500 000–1 million

1–5 million

5–10 million

percentage of urban
population by city size

less than
500 000

over
10 million

international boundary

disputed boundary

ceasefire line

■ capital city

• other important city

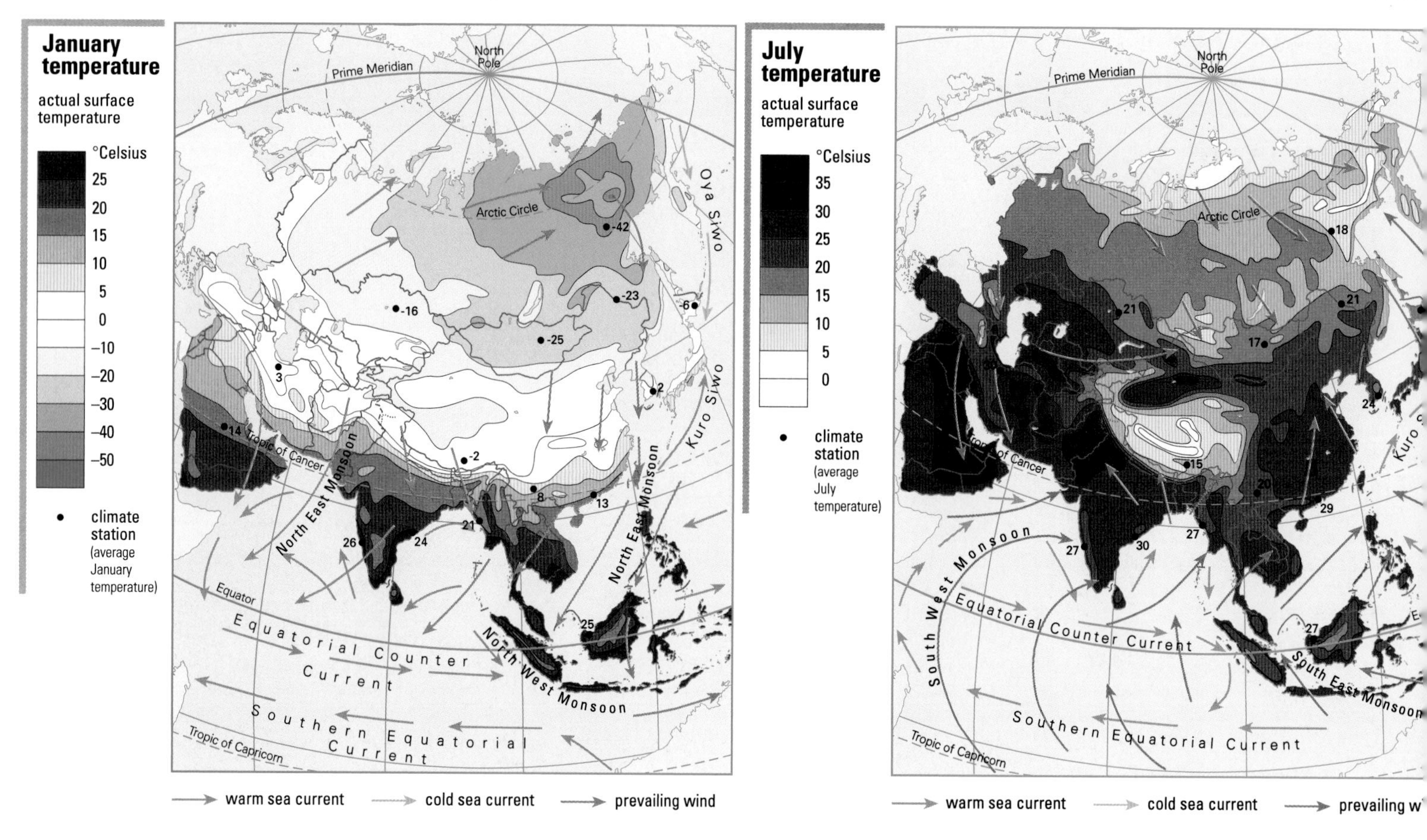

## January temperature

actual surface temperature

°Celsius

25
20
15
10
5
0
−10
−20
−30
−40
−50

● climate station (average January temperature)

## July temperature

actual surface temperature

°Celsius

35
30
25
20
15
10
5
0

● climate station (average July temperature)

⟶ warm sea current ⟶ cold sea current ⟶ prevailing wind

⟶ warm sea current ⟶ cold sea current ⟶ prevailing w

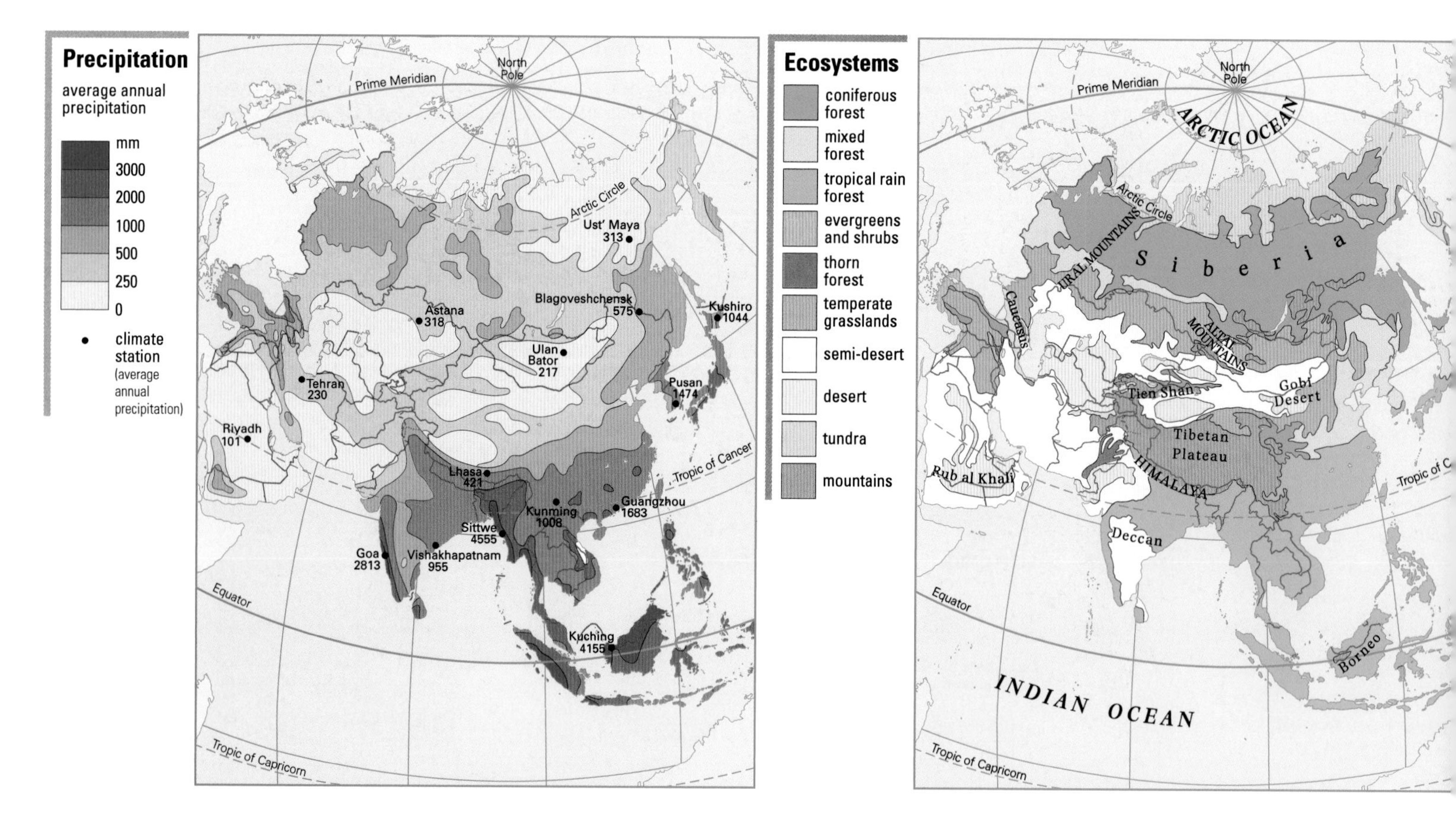

## Precipitation

average annual precipitation

mm

3000
2000
1000
500
250
0

● climate station (average annual precipitation)

Ust' Maya 313

Astana 318
Blagoveshchensk 575
Kushiro 1044
Tehran 230
Ulan Bator 217
Pusan 1474
Riyadh 101
Lhasa 421
Kunming 1008
Guangzhou 1683
Sittwe 4555
Goa 2813
Vishakhapatnam 955
Kuching 4155

## Ecosystems

- coniferous forest
- mixed forest
- tropical rain forest
- evergreens and shrubs
- thorn forest
- temperate grasslands
- semi-desert
- desert
- tundra
- mountains

ARCTIC OCEAN

URAL MOUNTAINS
Caucasus
Siberia
ALTAI MOUNTAINS
Tien Shan
Gobi Desert
Tibetan Plateau
Rub al Khali
HIMALAYA
Deccan
Borneo

Equator

INDIAN OCEAN

Tropic of Capricorn

cale 1: 75 000 000

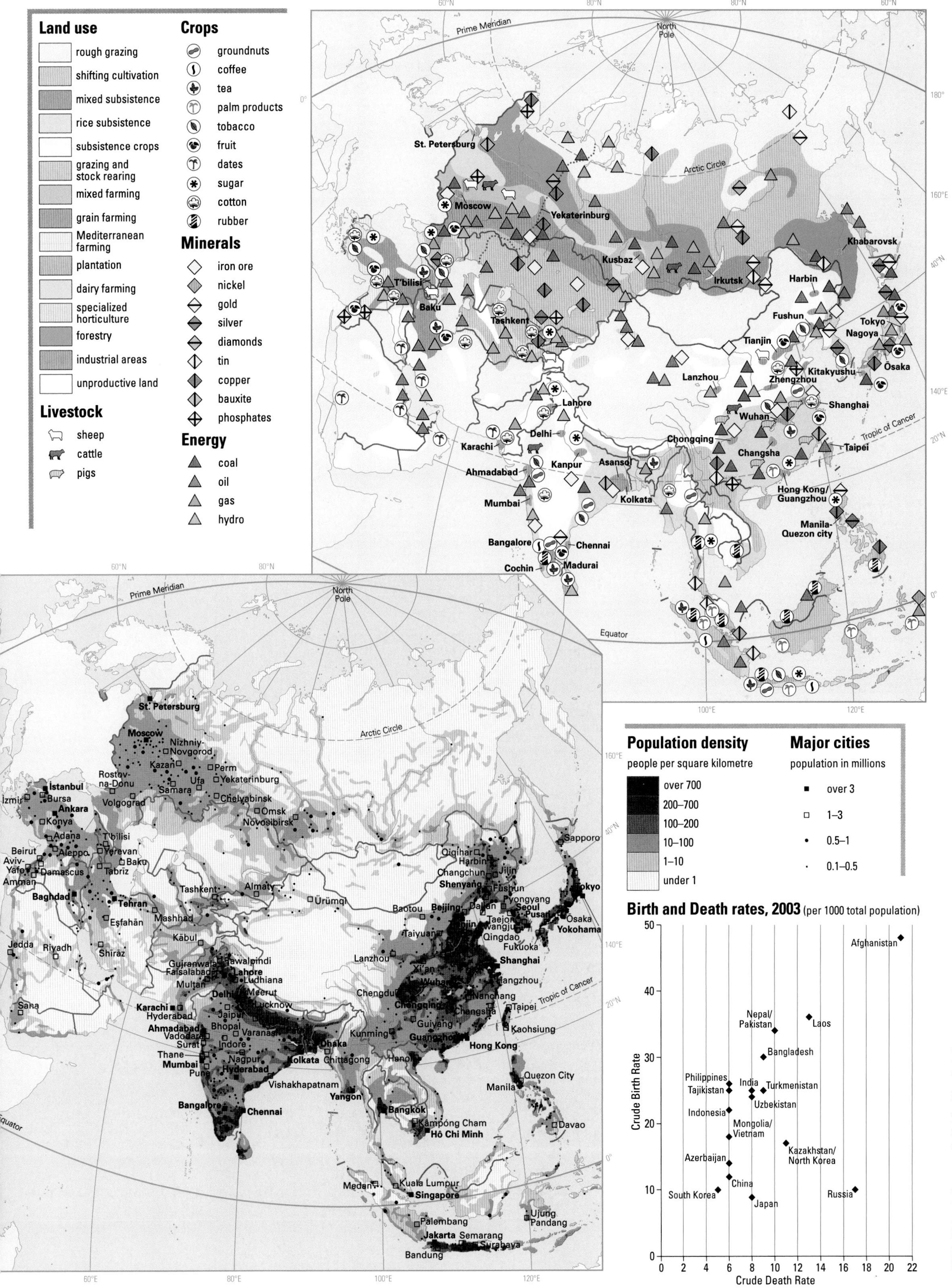

## Land use

| | |
|---|---|
| | rough grazing |
| | shifting cultivation |
| | mixed subsistence |
| | rice subsistence |
| | subsistence crops |
| | grazing and stock rearing |
| | mixed farming |
| | grain farming |
| | Mediterranean farming |
| | plantation |
| | dairy farming |
| | specialized horticulture |
| | forestry |
| | industrial areas |
| | unproductive land |

## Livestock

- sheep
- cattle
- pigs

## Crops

- groundnuts
- coffee
- tea
- palm products
- tobacco
- fruit
- dates
- sugar
- cotton
- rubber

## Minerals

- iron ore
- nickel
- gold
- silver
- diamonds
- tin
- copper
- bauxite
- phosphates

## Energy

- coal
- oil
- gas
- hydro

## Population density

people per square kilometre

| | |
|---|---|
| | over 700 |
| | 200–700 |
| | 100–200 |
| | 10–100 |
| | 1–10 |
| | under 1 |

## Major cities

population in millions

- ■ over 3
- □ 1–3
- • 0.5–1
- · 0.1–0.5

## Birth and Death rates, 2003 (per 1000 total population)

Afghanistan

Nepal/Pakistan    Laos

Bangladesh

Philippines    India    Turkmenistan
Tajikistan    Uzbekistan
Indonesia    Mongolia/Vietnam
Azerbaijan    Kazakhstan/North Korea
South Korea    China    Russia
Japan

Crude Birth Rate
Crude Death Rate

© Oxford University Press

Zenithal Equal Area Projection

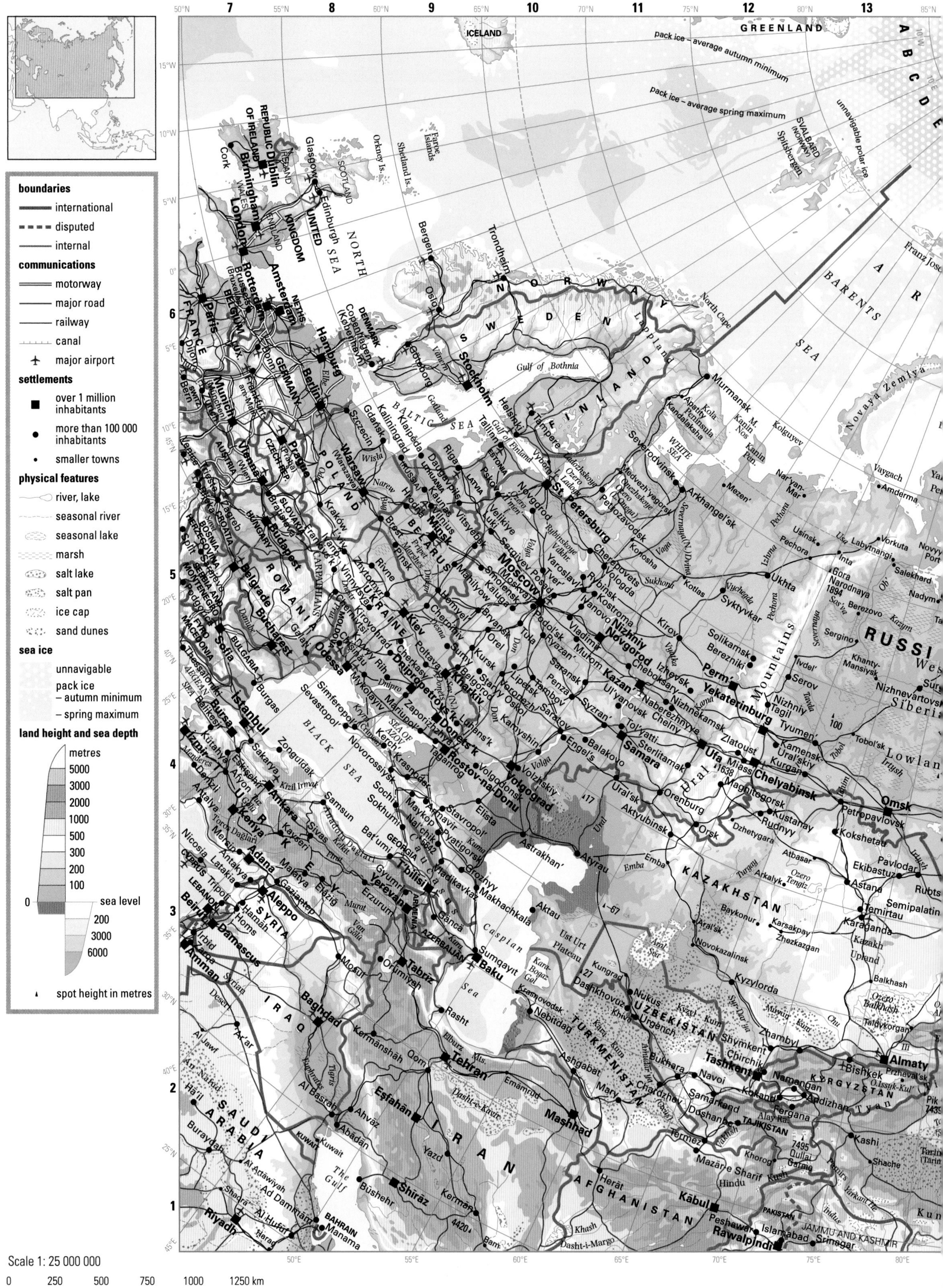

Conical Orthomorphic Projection

Scale 1: 25 000 000

0    250    500    750    1000    1250 km

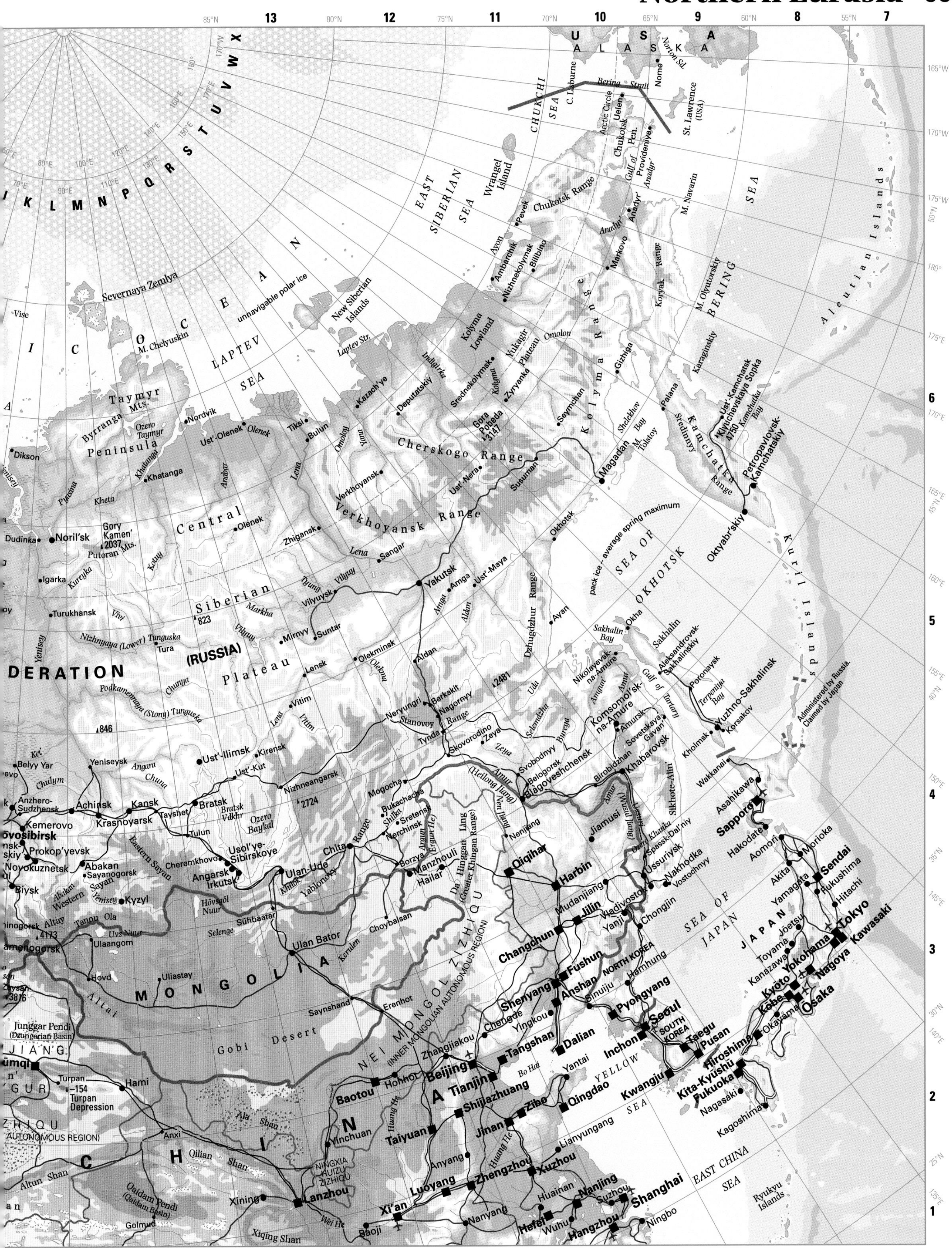

85°N
13
80°N
12
75°N
11
70°N
10
65°N
9
60°N
8
55°N
7
U S A
A L A S K A
Norton Sd.
Nome
Bering Strait
St. Lawrence (USA)
CHUKCHI SEA
C. Lisburne
Arctic Circle
Uelen
Chukotsk Pen.
Provideniya
Gulf of Anadyr'
Anadyr'
Wrangel Island
EAST SIBERIAN SEA
Pevek
Chukotsk Range
M. Navarin
BERING SEA
Aleutian Islands
Ayon
Ambarchik
Nizhnekolymsk
Bilibino
Markovo
Koryak Range
M. Olyutorskiy
Karaginskiy
Kolyma Lowland
Kolyma Range
Yukagir Plateau
Omolon
Gizhiga
Palana
Ust'-Kamchatsk
Klyuchevskaya Sopka 4750
Kamchatka Bay
Petropavlovsk-Kamchatskiy
Kamchatka Range
Sredinnyy
Shelekhov Bay
M. Tolstoy
New Siberian Islands
Laptev Str.
Indigirka
Srednekolymsk
Zyryanka
Seymchan
Gora Pobeda 3147
Severnaya Zemlya
unnavigable polar ice
Vise
M. Chelyuskin
LAPTEV SEA
Kazach'ye
Deputatskiy
Taymyr Mts.
Ozero Taymyr
Byrranga Mts.
Taymyr Peninsula
Nordvik
Ust'-Olenek
Olenek
Tiksi
Bulun
Omoloy
Yana
Cherskogo Range
Dikson
Khatanga
Anabar
Lena
Verkhoyansk
Ust'-Nera
Susuman
Magadan
Verkhoyansk Range
Pyasina
Kheta
Central Siberian Plateau
Olenek
Zhigansk
Okhotsk
pack ice – average spring maximum
SEA OF OKHOTSK
Oktyabr'skiy
Kuril Islands
Dudinka
Noril'sk
Gory Kamen' 2037
Putoran Mts.
Kotuy
Sangar
Igarka
Kureyka
Yakutsk
Amga
Ust'-Maya
Vilyuysk
Vilyuy
Tyung
Turukhansk
Markha
Aldan
Dzhugdzhur Range
Ayan
823
Mirnyy
Suntar
Okha
Sakhalin Bay
Sakhalin
Nizhnyaya (Lower) Tunguska
Tura
(RUSSIA)
Olekminsk
Aldan
Yenisey
DERATION
Lensk
Plateau
Olekma
2481
Nikolayevsk-na-Amure
Aleksandrovsk-Sakhalinskiy
Gulf of Tartary
Poronaysk
Terpeniya Bay
Yuzhno-Sakhalinsk
Administered by Russia. Claimed by Japan
Podkamennaya (Stony) Tunguska
Chunya
Vitim
Neryungri
Berkakit
Nagornyy
Stanovoy Range
Uda
Komsomol'sk-na-Amure
Amursk
Sovetskaya Gavan'
Korsakov
Kholmsk
846
Tynda
Skovorodino
Zeya
Selemdzha
Bureya
Ket'
Belyy Yar
Chulym
Yeniseysk
Angara
Ust'-Ilimsk
Kirensk
Ust'-Kut
Chuna
Svobodnyy
Belogorsk
Blagoveshchensk
Birobidzhan
Khabarovsk
Wakkanai
Asahikawa
Anzhero-Sudzhensk
Achinsk
Kansk
Tayshet
Bratsk
Bratsk Vdkhr
Nizhneangarsk
2724
Mogocha
Bukachacha
Shilka
Sretensk
Nerchinsk
Amur (Heilong Jiang)
Nen Jiang
Ussuri (Wusuli Jiang)
Sikhote-Alin'
Sapporo
Kemerovo
Krasnoyarsk
Ozero Baykal
Tulun
Da Hinggan Ling (Greater Khingan Range)
Argun (Ergun He)
Nenjiang
Jiamusi
Ozero Khanka
Spassk-Dal'niy
Hakodate
Aomori
Morioka
Novosibirsk
Prokop'yevsk
Eastern Sayan
Cheremkhovo
Usol'ye-Sibirskoye
Chita
Borzya
Manzhouli
Hailar
Qiqihar
Ussuriysk
Nakhodka
Vostochnyy
Akita
Sendai
Fukushima
Novokuznetsk
Abakan
Sayanogorsk
Angarsk
Irkutsk
Ulan-Ude
Khilok
Yablonovy
Harbin
Mudanjiang
Vladivostok
Yamagata
Hitachi
Biysk
Western Sayan
Yenisey
Kyzyl
Hövsgöl Nuur
Sühbaatar
Selenge
Choybalsan
Kerulen
Jilin
Yanji
Chongjin
SEA OF JAPAN
Joetsu
Tokyo
Altay
Tannu Ola
Uvs Nuur
Ulaangom
4173
Ulan Bator
Changchun
Toyama
Yokohama
Kawasaki
3
Fushun
NORTH KOREA
Hamhung
Kanazawa
Nagoya
Hovd
Uliastay
MONGOLIA
Shenyang
Anshan
Sinuiju
Pyongyang
Kyoto
Kobe
Osaka
Junggar Pendi (Dzungarian Basin)
Altai
Saynshand
Erenhot
NEI MONGOL ZIZHIQU (INNER MONGOLIAN AUTONOMOUS REGION)
Chengde
Yingkou
Seoul
SOUTH KOREA
Taegu
Pusan
Hiroshima
Okayama
Gobi Desert
Zhangjiakou
Tangshan
Dalian
Inchon
Beijing
Bo Hai
Yantai
YELLOW SEA
Kwangju
Kita-Kyushu
Fukuoka
Turpan
-154
Turpan Depression
Hami
Baotou
Hohhot
Tianjin
Shijiazhuang
Zibo
Qingdao
Nagasaki
Kagoshima
AUTONOMOUS REGION
Anxi
Ala Shan
Yinchuan
Huang He
Taiyuan
Jinan
Lianyungang
CHINA
Qilian Shan
NINGXIA HUIZU ZIZHIQU
Anyang
Xuzhou
EAST CHINA SEA
Altun Shan
Qaidam Pendi (Qaidam Basin)
Xining
Lanzhou
Luoyang
Zhengzhou
Huainan
Nanjing
Suzhou
Shanghai
Ryukyu Islands
Golmud
Xi'an
Nanyang
Hefei
Wuhu
Hangzhou
Ningbo
Baoji
Wei He
Xiqing Shan
90°E
95°E
100°E
105°E
110°E
115°E
120°E
125°E
130°E
135°E
140°E
145°E
150°E
155°E
160°E
165°E
170°E
175°E
180°
175°W
170°W
165°W
1
2
3
4
5
6

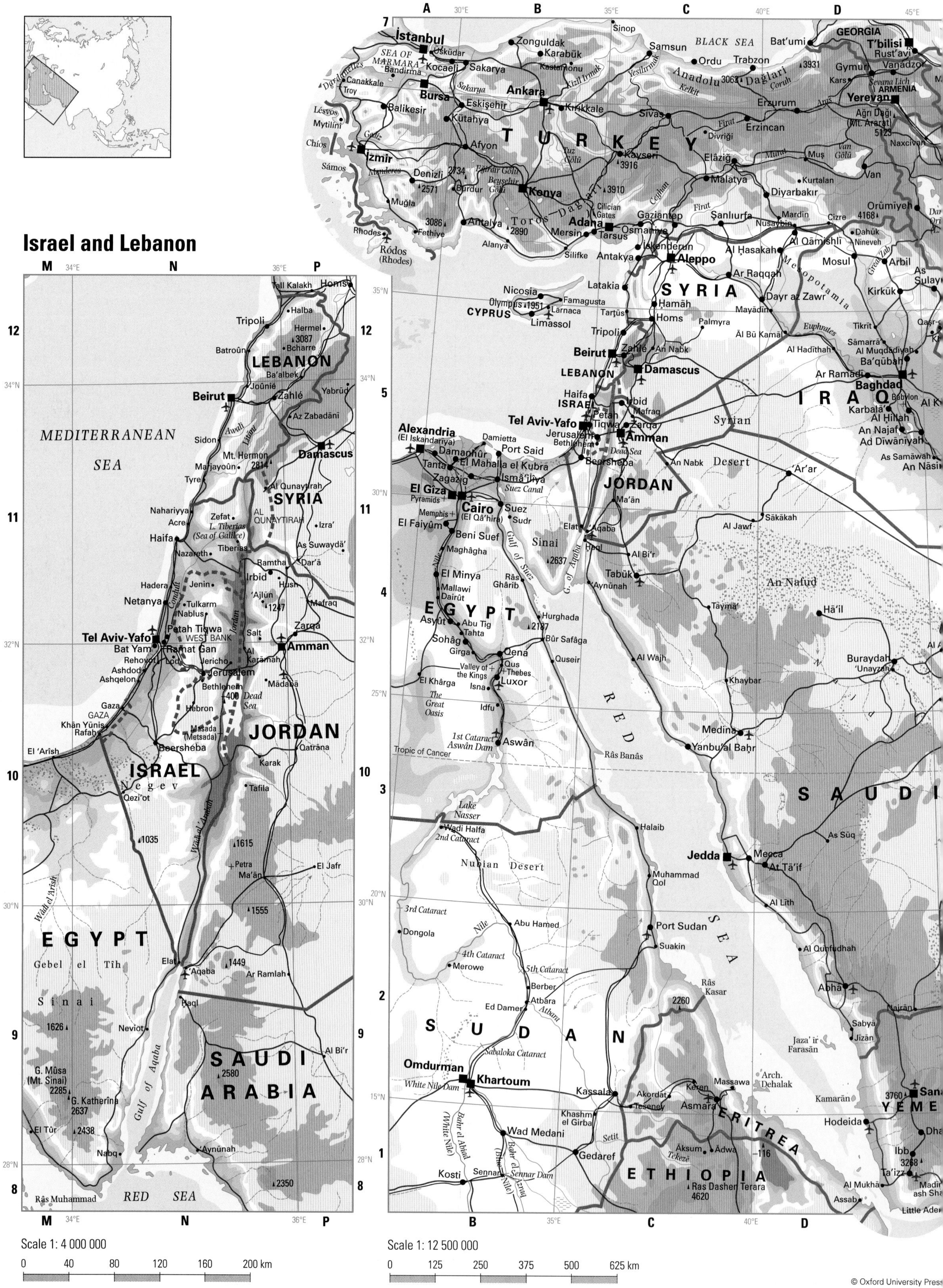

## Israel and Lebanon

Scale 1: 4 000 000

| 0 | 40 | 80 | 120 | 160 | 200 km |

Scale 1: 12 500 000

| 0 | 125 | 250 | 375 | 500 | 625 km |

© Oxford University Press

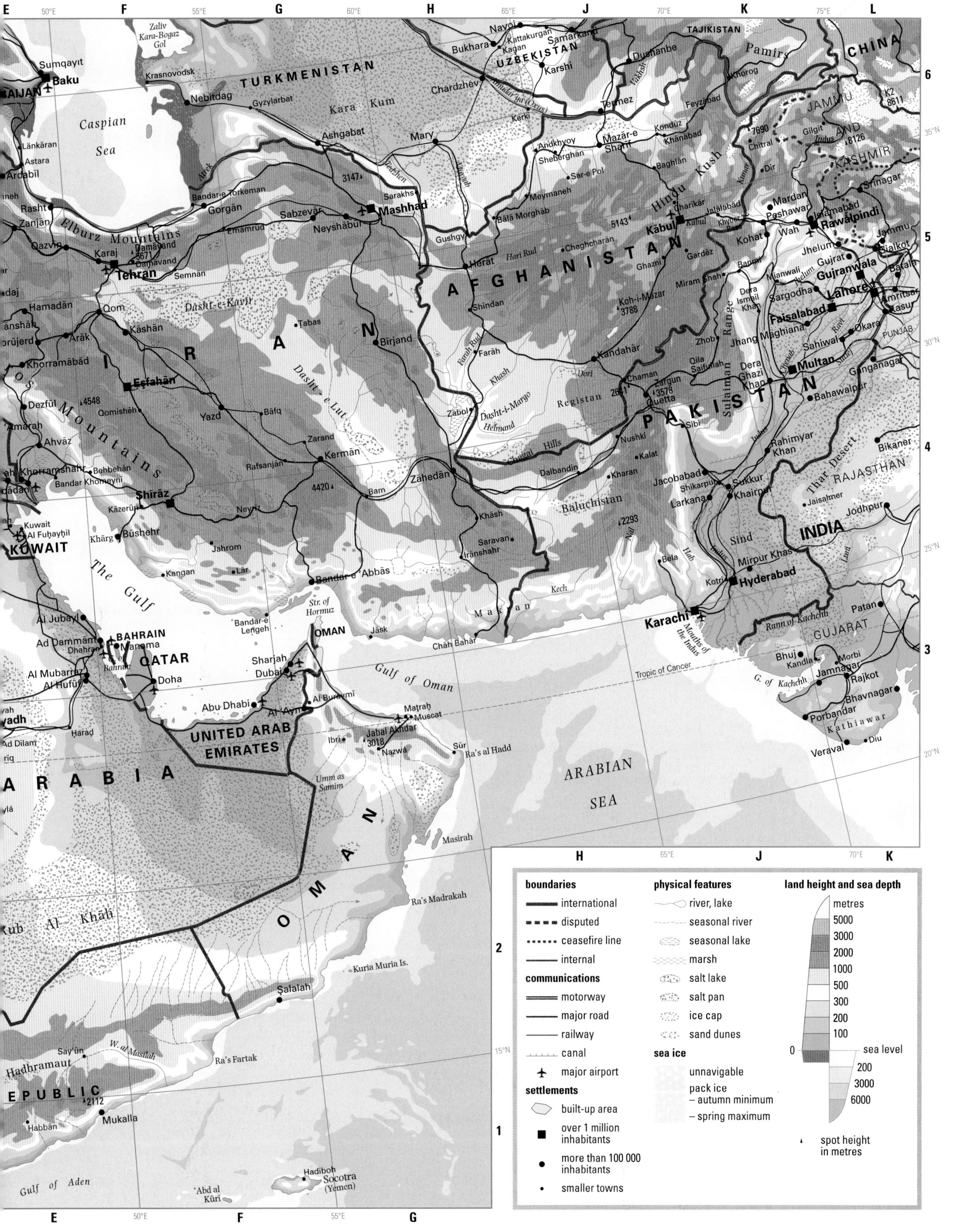

E  50°E  F  55°E  G  60°E  H  65°E  J  70°E  K  75°E  L

**6**

Sumqayıt
Baku
AIJAN
Länkäran
Astara
Ardabīl
neh
Rasht
Zanjān
Qazvīn
Karaj
Tehran
Hamadān
nshah
rüjerd
Khorramābād
Arāk
Qom
Kāshān
Dezfūl
Esfahān
Qomisheh
Ahvāz
Behbehān
Khorramshahr
Bandar Khomeynī
Shīrāz
Kāzerūn
Kuwait
Al Fuḥayḥil
Khārg
Būshehr
KUWAIT
Kangan
Jahrom
Lār
Neyrīz
Al Jubayl
The Gulf
Bandar-e Lengeh
Ad Dammām
BAHRAIN
Dhahran
Manama
Al Mubarraz
Bahrain
QATAR
Al Hufūf
Doha
Abu Dhabi
Al Aymī
Sharjah
Dubai
OMAN
Al Buraymī
 yadh
Ad Dilam
Harad
UNITED ARAB
EMIRATES
Matraḥ
Muscat
riq
Ibrī
Jabal Akhdar
3018
Nazwa
Ra's al Hadd
A R A B I A
Umm as Samīm
Sūr
Masīrah

Caspian Sea

Zaliv Kara-Bogaz Gol
Krasnovodsk
Nebitdag
Gyzylarbat
Kara Kum
TURKMENISTAN
Ashgabat
Atrek
Gorgān
Bandar-e Torkeman
Bandar-e Torkeman
Damāvand
5671
Semnān
Dasht-e Kavīr
Tabas
4548
Yazd
Bāfq
Zarand
Rafsanjān
4420
Bam
Kermān
Khāsh

Mary
Tejen
3147
Sarakhs
Sabzevār
Neyshābūr
Mashhad
Herat
Hari Rud
Bīrjand
Dasht-e Lut
Farah Rud
Khash
Farāh
Zābol
Dasht-i-Margo
Helmand
Zāhedān
Saravan
Īrānshahr
M a k r a n
Jāsk
Chāh Bahār

Nebitdag

Chardzhev
Kerki
Gushgy
Bālā Morghāb
Chaghcharān
Shindand
A  F  G  H  A  N  I  S  T  A  N
Koh-i-Mazar
3788
Kandahār
Dori
Chaman
Registan
Chagai Hills
Dalbandin
Baluchistan
2293
Kech
Bela
Hab

Bukhara
Kagan
Chardzhev
UZBEKISTAN
Karshi
Termez
Andkhvoy
Sheberghān
Sar-e Pol
Meymaneh
5143
Gardēz
Ghaznī
Miram Shah
Qila Saifullah
Zargun
3578
2611
Quetta
Sibi
Nushki
Kalat
Kharan
Khuzdar
Jacobabad
Shikarpur
Larkana
Khairpur
Sukkur
Sind
Mirpur Khas
Mirpur Khas
Kotri
Hyderabad
Karachi
Months of the Indus

Navoi
Kattakurgan
Samarkand
Kagan
Amudar'ya (Oxus)
TAJIKISTAN
Dushanbe
Khorog
Termez
Konduz
Khānabad
Mazar-e Sharif
Baghlān
Charikār
Jalālābād
Kabul
Kabul
Khyber
Kohat
Bannu
Dera Ismail Khan
Zhob
Dera Ghazi Khan
Jhang Maghiana
Multan
Bahawalpur
Rahimyar Khan
P A K I S T A N
Sibi
Thar Desert
Jaisalmer
Bikaner
Jodhpur
RAJASTHAN
INDIA

Pamirs
CHINA
JAMMU
7690
Feyzabad
Gilgit
Chitral
Dir
Hindu Kush
Mardan
Peshawar
Wah
Islamabad
Rawalpindi
Jhelum
Gujrat
Gujranwala
Lahore
Faisalabad
Sahiwal
Okara
Bahawalnagar
PUNJAB

K2
8611
JAMMU
AND
KASHMIR
8126
Srīnagar
Kohat
Kotli
Mianwali
Sargodha
Wah
Sialkot
Batala
Amritsar
Kasu
Ganganagar

**5**

**4**

**3**

**2**

**1**

Str. of Hormuz
Bandar-e Abbas
Gulf of Oman
ARABIAN
SEA

O M A N
Ra's Madrakah
Kub Al Khali
yla
Ra's Fartak
Salālah
W. al Maṣīlah
Say'ūn
Hadhramaut
2112
Ḥabbān
Mukalla
EPUBLIC
Gulf of Aden

Tropic of Cancer

Kuria Muria Is.
Hadīboh
Socotra (Yemen)
'Abd al Kūrī
Ra's Fartak

H  65°E  J  70°E  K

© Oxford University Press    Conical Orthomorphic projection

**boundaries**
— international
--- disputed
···· ceasefire line
— internal

**communications**
═ motorway
— major road
— railway
⊥⊥⊥ canal
✈ major airport

**settlements**
⬡ built-up area
■ over 1 million inhabitants
● more than 100 000 inhabitants
• smaller towns

**physical features**
〜 river, lake
----- seasonal river
〰 seasonal lake
≈ marsh
⬭ salt lake
∴ salt pan
⬭ ice cap
⬭ sand dunes

**sea ice**
unnavigable
pack ice
– autumn minimum
– spring maximum

**land height and sea depth**

metres
5000
3000
2000
1000
500
300
200
100
0  sea level
200
3000
6000

▲ spot height in metres

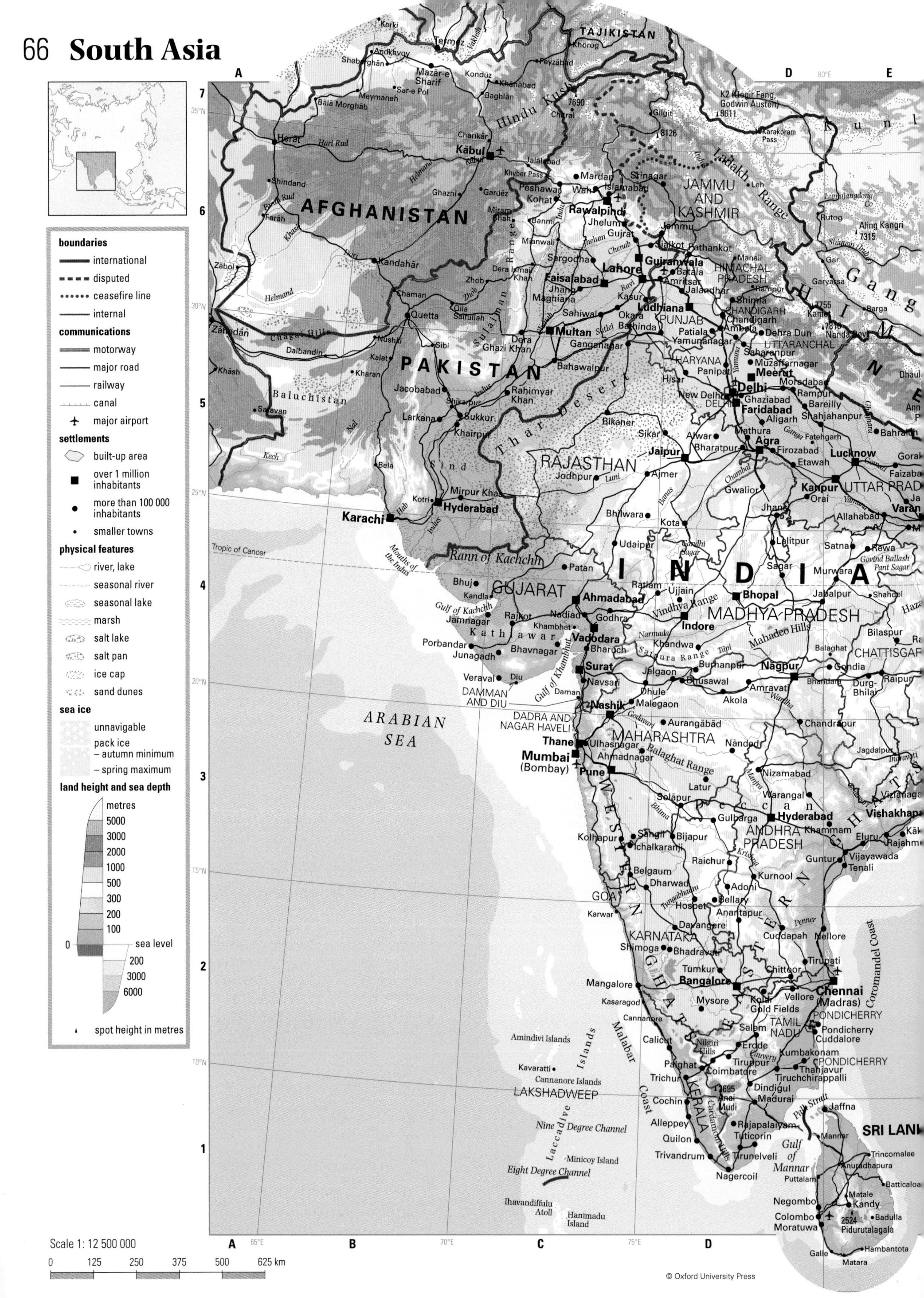
boundaries
international
disputed
ceasefire line
internal
communications
motorway
major road
railway
canal
major airport
settlements
built-up area
over 1 million inhabitants
more than 100 000 inhabitants
smaller towns
physical features
river, lake
seasonal river
seasonal lake
marsh
salt lake
salt pan
ice cap
sand dunes
sea ice
unnavigable
pack ice
– autumn minimum
– spring maximum
land height and sea depth
metres
5000
3000
2000
1000
500
300
200
100
0
sea level
200
3000
6000
spot height in metres
Scale 1: 12 500 000
0
125
250
375
500
625 km
AFGHANISTAN
PAKISTAN
INDIA
TAJIKISTAN
SRI LANKA
JAMMU AND KASHMIR
HIMACHAL PRADESH
PUNJAB
HARYANA
UTTARANCHAL
DELHI
RAJASTHAN
UTTAR PRADESH
GUJARAT
MADHYA PRADESH
CHATTISGARH
DAMMAN AND DIU
DADRA AND NAGAR HAVELI
MAHARASHTRA
ANDHRA PRADESH
GOA
KARNATAKA
KERALA
TAMIL NADU
PONDICHERRY
LAKSHADWEEP
ARABIAN SEA
Hindu Kush
Karakoram Pass
Ladakh Range
K2 (Qogir Feng, Godwin Austen) 8611
Thar Desert
Baluchistan
Sind
Rann of Kachchh
Gulf of Kachchh
Gulf of Khambhat
Mouths of the Indus
Tropic of Cancer
Western Ghats
Eastern Ghats
Deccan
Malabar Coast
Coromandel Coast
Palk Strait
Gulf of Mannar
Laccadive Islands
Amindivi Islands
Cannanore Islands
Nine Degree Channel
Eight Degree Channel
Minicoy Island
Ihavandiffulu Atoll
Hanimadu Island
Kabul
Herat
Kandahar
Quetta
Mazar-e Sharif
Islamabad
Rawalpindi
Peshawar
Lahore
Faisalabad
Multan
Gujranwala
Karachi
Hyderabad
Srinagar
Amritsar
Ludhiana
Chandigarh
Delhi
New Delhi
Faridabad
Meerut
Agra
Jaipur
Jodhpur
Lucknow
Kanpur
Ahmadabad
Vadodara
Surat
Rajkot
Indore
Bhopal
Nagpur
Nashik
Thane
Mumbai (Bombay)
Pune
Hyderabad
Bangalore
Chennai (Madras)
Madurai
Coimbatore
Cochin
Trivandrum
Colombo
Kandy
Jaffna
2524 Pidurutalagala
© Oxford University Press

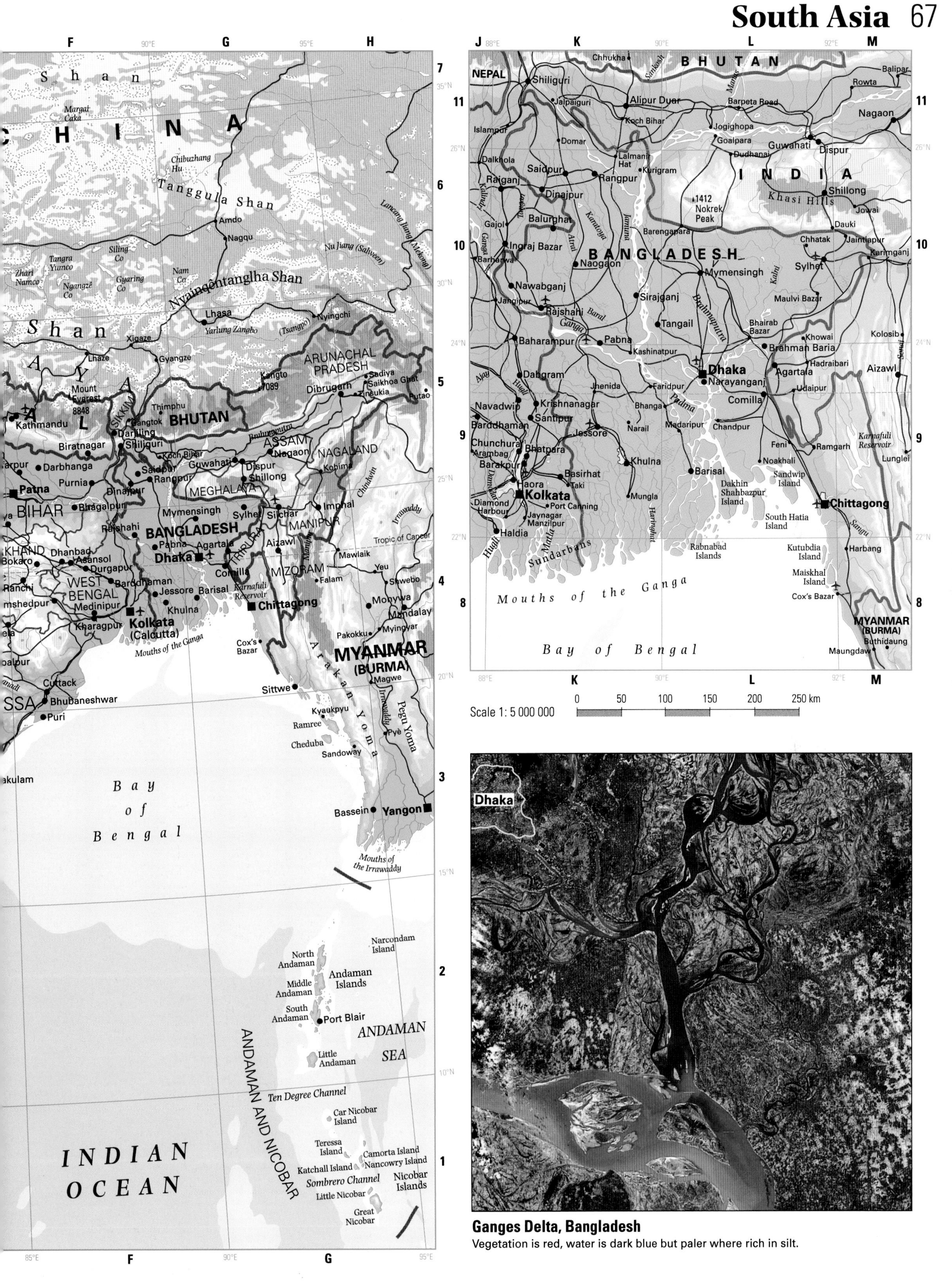

**Ganges Delta, Bangladesh**
Vegetation is red, water is dark blue but paler where rich in silt.

Dhaka

Scale 1: 5 000 000

0   50   100   150   200   250 km

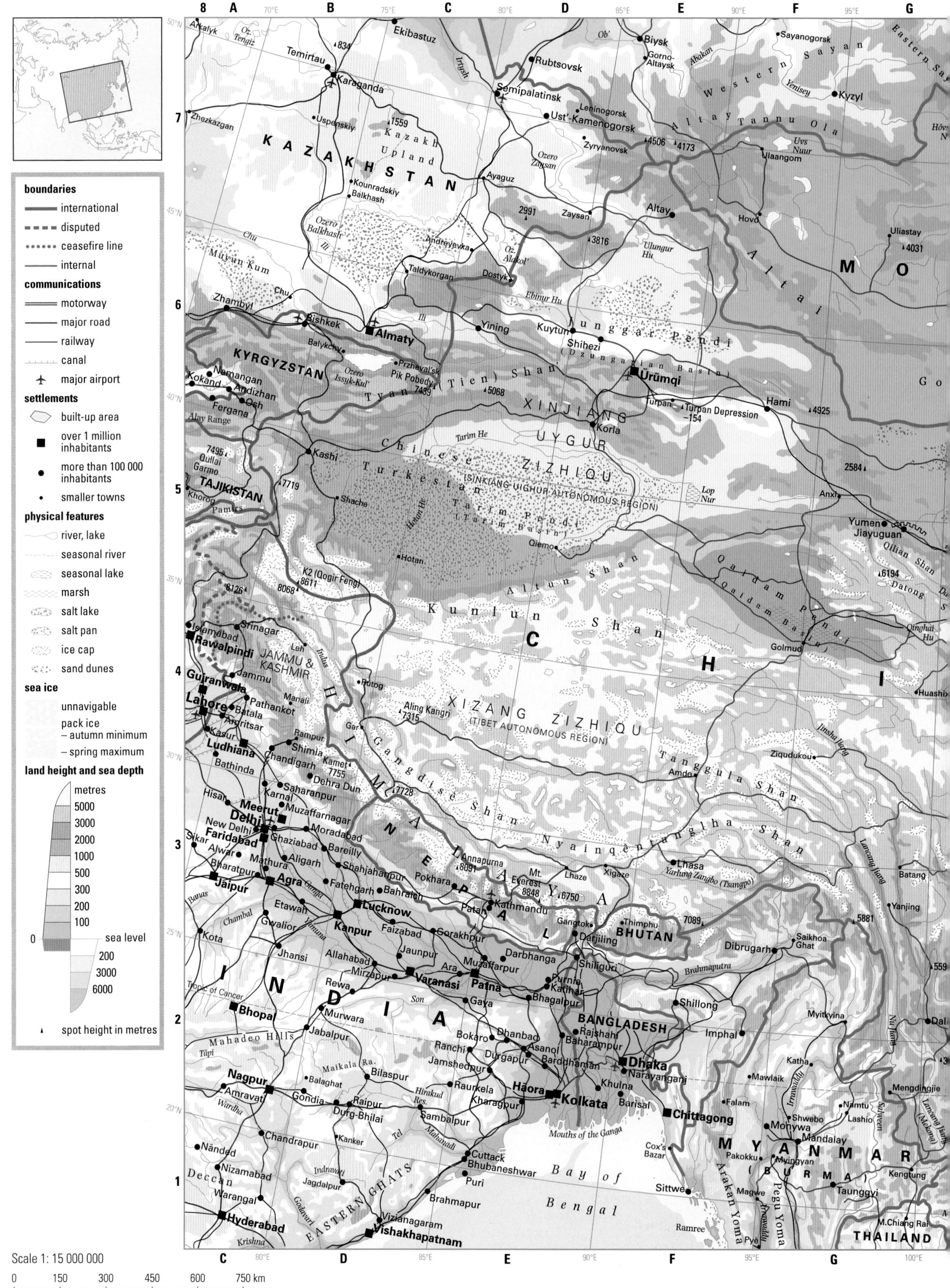
boundaries
international
disputed
ceasefire line
internal
communications
motorway
major road
railway
canal
major airport
settlements
built-up area
over 1 million inhabitants
more than 100 000 inhabitants
smaller towns
physical features
river, lake
seasonal river
seasonal lake
marsh
salt lake
salt pan
ice cap
sand dunes
sea ice
unnavigable
pack ice – autumn minimum
– spring maximum
land height and sea depth
metres
5000
3000
2000
1000
500
300
200
100
0
sea level
200
3000
6000
spot height in metres
KAZAKHSTAN
Kazakh Upland
KYRGYZSTAN
TAJIKISTAN
XINJIANG UYGUR ZIZHIQU
(SINKIANG UIGHUR AUTONOMOUS REGION)
Tarim Pendi (Tarim Basin)
Junggar Pendi (Dzungarian Basin)
Tyan Shan (Tien Shan)
Kunlun Shan
Altun Shan
Qaidam Pendi
XIZANG ZIZHIQU
(TIBET AUTONOMOUS REGION)
Tanggula Shan
Gangdisê Shan
Nyainqêntanglha Shan
JAMMU & KASHMIR
HIMALAYA
NEPAL
BHUTAN
BANGLADESH
INDIA
MYANMAR (BURMA)
THAILAND
Bay of Bengal
CHINA
Western Sayan
Tannu Ola
Altai
Ürümqi
Almaty
Bishkek
Lhasa
Kathmandu
Delhi
Dhaka
Kolkata
Mt. Everest 8848
K2 (Qogir Feng) 8611
Turpan Depression –154
Scale 1: 15 000 000
0 150 300 450 600 750 km

Conical Orthomorphic Projection

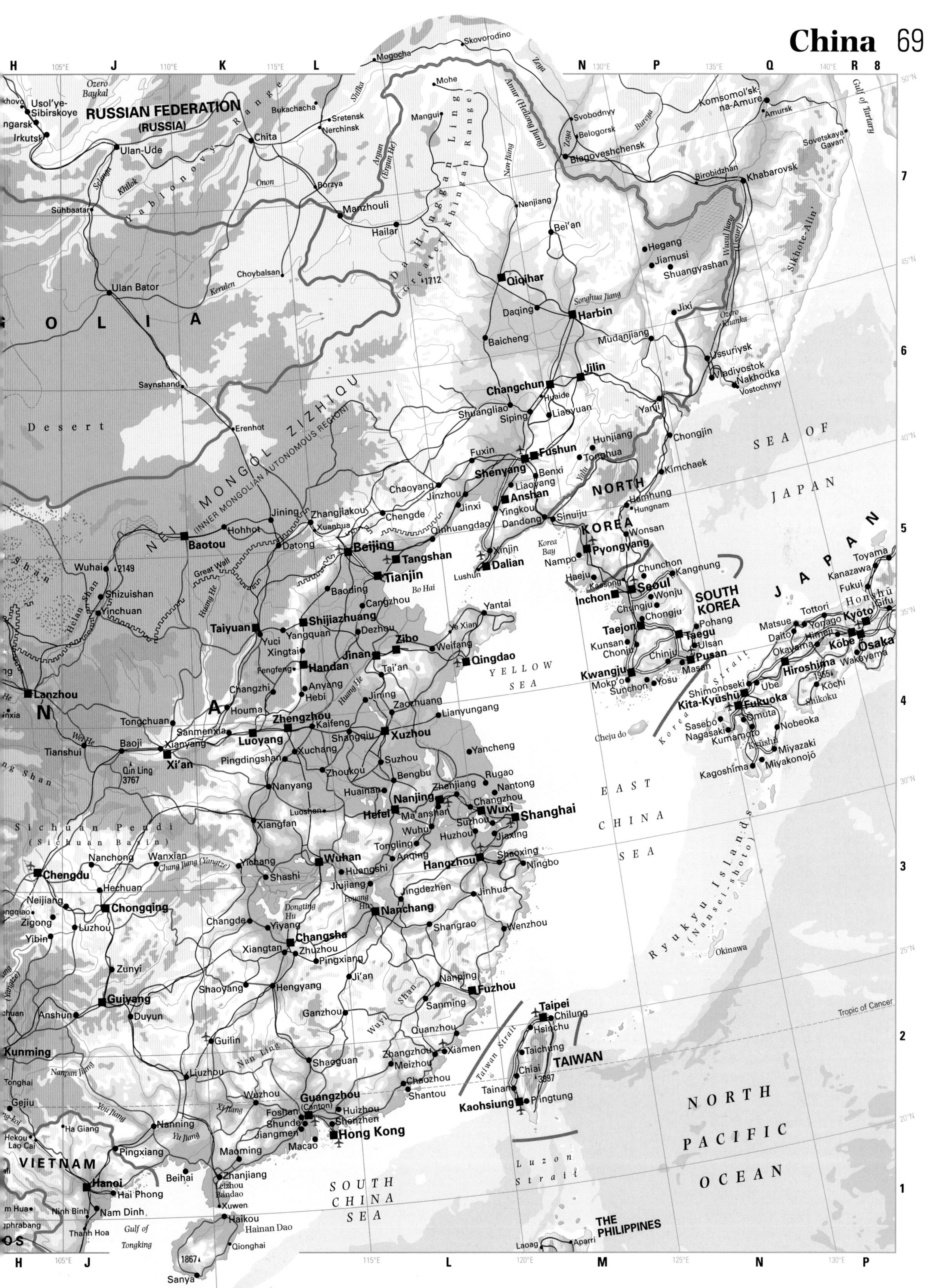
RUSSIAN FEDERATION
(RUSSIA)
Ozero Baykal
Usol'ye-Sibirskoye
Irkutsk
Ulan-Ude
Chita
Yablonovy Range
Bukachacha
Sretensk
Nerchinsk
Shilka
Mogocha
Skovorodino
Mohe
Mangui
Argun (Ergun He)
Borzya
Manzhouli
Hailar
Onon
Sühbaatar
Ulan Bator
Choybalsan
Kerulen
Da Hinggan Ling
Greater Khingan Range
Zeya
Amur (Heilong Jiang)
Svobodnyy
Belogorsk
Blagoveshchensk
Bureya
Komsomol'sk-na-Amure
Amursk
Sovetskaya Gavan'
Gulf of Tartary
Birobidzhan
Khabarovsk
Sikhote-Alin'
Nenjiang
Nen Jiang
Bei'an
Hegang
Jiamusi
Shuangyashan
Qiqihar
Songhua Jiang
Harbin
Daqing
Jixi
Ozero Khanka
Baicheng
Mudanjiang
Ussuriysk
Vladivostok
Nakhodka
Vostochnyy
Jilin
Changchun
Huaide
Liaoyuan
Siping
Shuangliao
Yanji
Chongjin
Hunjiang
Tonghua
Fuxin
Fushun
Shenyang
Benxi
Liaoyang
Anshan
Yingkou
Dandong
Sinuiju
Kimchaek
Hamhung
Hungnam
Wonsan
Chaoyang
Jinzhou
Jinxi
Chengde
Qinhuangdao
Zhangjiakou
Xuanhua
Jining
Hohhot
Baotou
Datong
Erenhot
Saynshand
Desert
O L I A
NEI MONGOL ZIZHIQU
(INNER MONGOLIAN AUTONOMOUS REGION)
Great Wall
Huang He
Wuhai
2149
Shizuishan
Yinchuan
Helan Shan
Beijing
Tangshan
Tianjin
Bo Hai
Baoding
Cangzhou
Shijiazhuang
Taiyuan
Yangquan
Dezhou
Yuci
Xingtai
Zibo
Jinan
Weifang
Ye Xian
Yantai
Qingdao
Xinjin
Dalian
Lushun
Korea Bay
Nampo
Pyongyang
NORTH KOREA
Yalu
Haeju
Kaesong
Seoul
Inchon
Chunchon
Kangnung
Wonju
Chungju
Chongju
SOUTH KOREA
Taejon
Pohang
Taegu
Ulsan
Kunsan
Chonju
Chinju
Pusan
Masan
Kwangju
Mokp'o
Sunchon
Yosu
Cheju do
Korea Strait
SEA OF JAPAN
JAPAN
Toyama
Kanazawa
Fukui
Honshu
Gifu
Kyoto
Osaka
Kobe
Himeji
Okayama
Wakayama
Tottori
Yonago
Matsue
Daito
Hiroshima
Shikoku
Kochi
1955
Shimonoseki
Ube
Kita-Kyushu
Fukuoka
Sasebo
Nagasaki
Omuta
Kumamoto
Kyushu
Nobeoka
Miyazaki
Miyakonojo
Kagoshima
Fengfeng
Handan
Anyang
Hebi
Changzhi
Houma
Tai'an
Jining
Zaozhuang
Lianyungang
YELLOW SEA
Lanzhou
Tongchuan
Sanmenxia
Zhengzhou
Kaifeng
Luoyang
Xuzhou
Wei He
Baoji
Xianyang
Xi'an
Tianshui
Qin Ling
3767
Pingdingshan
Xuchang
Shangqiu
Yancheng
Zhoukou
Suzhou
Bengbu
Rugao
Nantong
Nanyang
Zhenjiang
Huainan
Nanjing
Changzhou
Luoshan
Hefei
Ma'anshan
Wuxi
Shanghai
Suzhou
Jiaxing
Wuhu
Huzhou
Xiangfan
EAST CHINA SEA
Sichuan Pendi
(Sichuan Basin)
Nanchong
Wanxian
Chang Jiang (Yangtze)
Yichang
Wuhan
Huangshi
Tongling
Anqing
Hangzhou
Shaoxing
Ningbo
Chengdu
Shashi
Hechuan
Jiujiang
Poyang Hu
Jingdezhen
Jinhua
Neijiang
Chongqing
Dongting Hu
Nanchang
Zigong
Luzhou
Yibin
Changde
Yiyang
Changsha
Shangrao
Wenzhou
Xiangtan
Zhuzhou
Pingxiang
Ryukyu Islands
(Nansei-shoto)
Okinawa
Zunyi
Ji'an
Shaoyang
Hengyang
Nanping
Fuzhou
Guiyang
Wuyi Shan
Sanming
Anshun
Duyun
Ganzhou
Taipei
Chilung
Hsinchu
Taichung
TAIWAN
Chiai
3997
Tainan
Pingtung
Kaohsiung
Taiwan Strait
Quanzhou
Xiamen
Zhangzhou
Meizhou
Chaozhou
Shantou
Guilin
Nan Ling
Shaoguan
Kunming
Nanpan Jiang
Liuzhou
Tonghai
Gejiu
Xi Jiang
Wuzhou
Guangzhou
(Canton)
Huizhou
Foshan
Shunde
Shenzhen
Jiangmen
Hong Kong
Macao
You Jiang
Nanning
Yu Jiang
Ha Giang
Hekou
Lao Cai
Pingxiang
VIETNAM
Hanoi
Hai Phong
Nam Dinh
Ninh Binh
Thanh Hoa
Maoming
Beihai
Zhanjiang
Leizhou Bandao
Xuwen
Haikou
Hainan Dao
Qionghai
1867
Sanya
Gulf of Tongking
SOUTH CHINA SEA
Luzon Strait
THE PHILIPPINES
Laoag
Aparri
NORTH PACIFIC OCEAN
Tropic of Cancer
105°E
110°E
115°E
120°E
125°E
130°E
135°E
140°E
50°N
45°N
40°N
35°N
30°N
25°N
20°N
H J K L M N P Q R
1 2 3 4 5 6 7 8

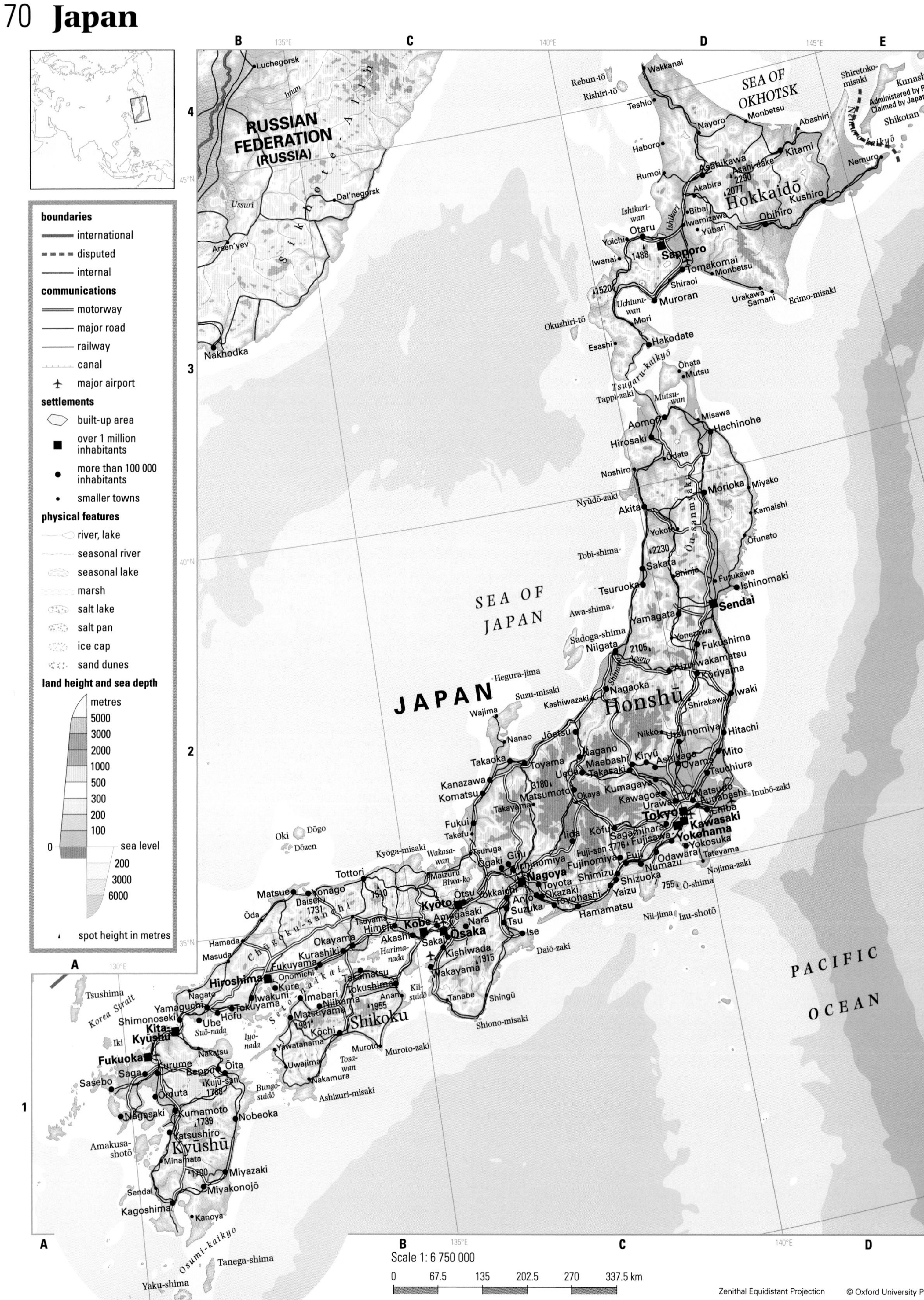

**boundaries**
- —— international
- ---- disputed
- —— internal

**communications**
- ═══ motorway
- —— major road
- —— railway
- ┼┼┼ canal
- ✈ major airport

**settlements**
- ⬡ built-up area
- ■ over 1 million inhabitants
- ● more than 100 000 inhabitants
- • smaller towns

**physical features**
- river, lake
- seasonal river
- seasonal lake
- marsh
- salt lake
- salt pan
- ice cap
- sand dunes

**land height and sea depth**

metres
5000
3000
2000
1000
500
300
200
100
0   sea level
200
3000
6000

▲ spot height in metres

Scale 1: 6 750 000

0    67.5    135    202.5    270    337.5 km

Zenithal Equidistant Projection

RUSSIAN FEDERATION (RUSSIA)

SEA OF OKHOTSK

SEA OF JAPAN

JAPAN

PACIFIC OCEAN

Honshū

Hokkaidō

Shikoku

Kyūshū

Korea Strait

Administered by Russia. Claimed by Japan.

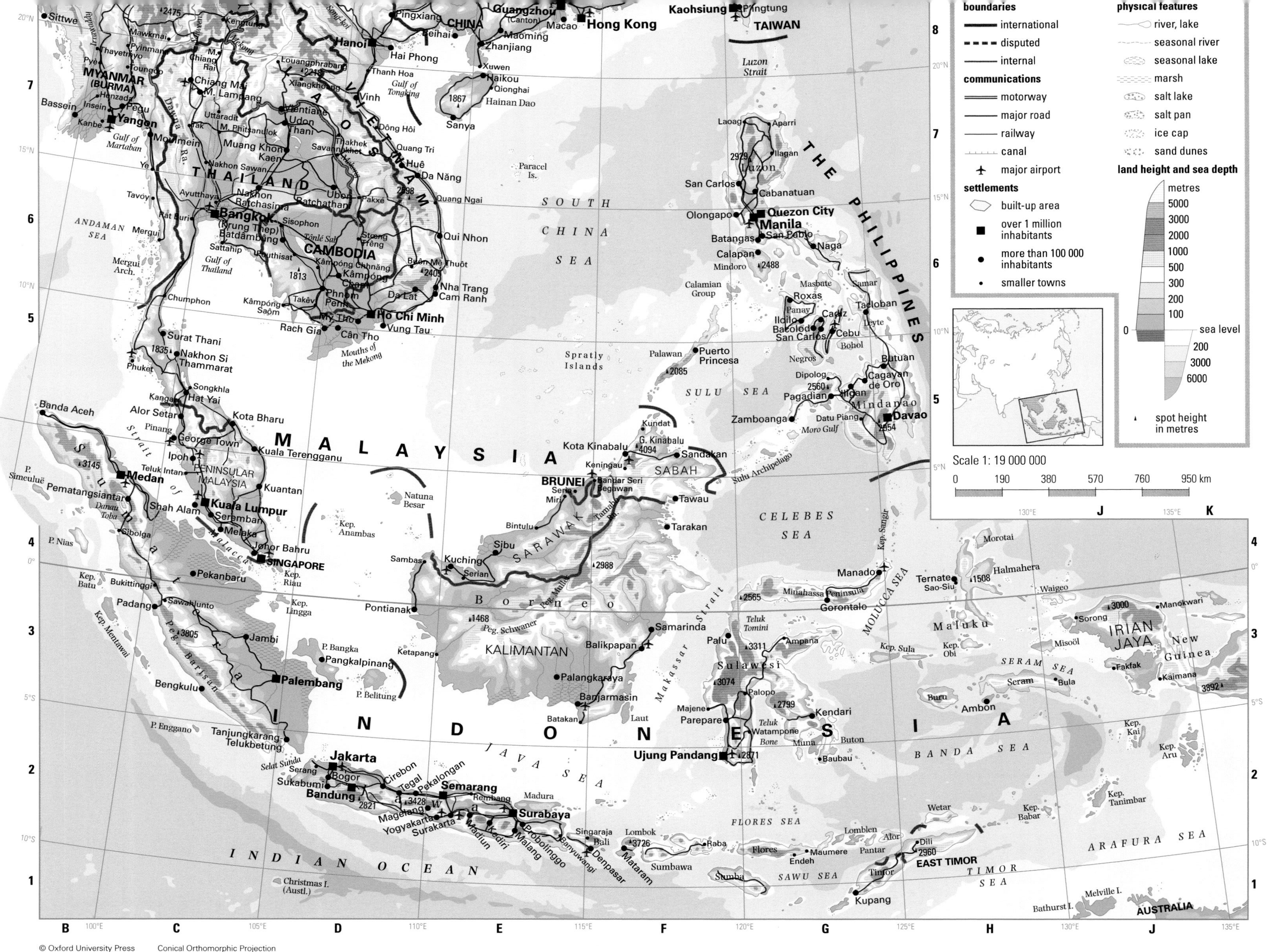
boundaries
international
disputed
internal
communications
motorway
major road
railway
canal
major airport
settlements
built-up area
over 1 million inhabitants
more than 100 000 inhabitants
smaller towns
physical features
river, lake
seasonal river
seasonal lake
marsh
salt lake
salt pan
ice cap
sand dunes
land height and sea depth
metres
5000
3000
2000
1000
500
300
200
100
0
sea level
200
3000
6000
spot height in metres
Scale 1: 19 000 000
0
190
380
570
760
950 km
TAIWAN
THE PHILIPPINES
CHINA
VIETNAM
LAOS
THAILAND
CAMBODIA
MYANMAR (BURMA)
MALAYSIA
SINGAPORE
BRUNEI
INDONESIA
EAST TIMOR
AUSTRALIA
SOUTH CHINA SEA
SULU SEA
CELEBES SEA
JAVA SEA
BANDA SEA
FLORES SEA
SAWU SEA
TIMOR SEA
ARAFURA SEA
SERAM SEA
MOLUCCA SEA
ANDAMAN SEA
INDIAN OCEAN
Gulf of Thailand
Gulf of Tongking
Gulf of Martaban
Luzon Strait
Makassar Strait
Strait of Malacca
Borneo
KALIMANTAN
SARAWAK
SABAH
PENINSULAR MALAYSIA
Sumatra
Jawa
Sulawesi
Maluku
IRIAN JAYA
New Guinea
Hong Kong
Macao
Guangzhou (Canton)
Kaohsiung
Hanoi
Hai Phong
Da Nang
Ho Chi Minh
Yangon
Bangkok (Krung Thep)
Vientiane
Phnom Penh
Kuala Lumpur
Manila
Quezon City
Davao
Cebu
Jakarta
Bandung
Semarang
Surabaya
Medan
Palembang
Ujung Pandang
Kupang
Dili
Denpasar
Mataram
Manado
Ambon
Kota Kinabalu
Bandar Seri Begawan
Kuching
Pontianak
Balikpapan
Samarinda
© Oxford University Press
Conical Orthomorphic Projection

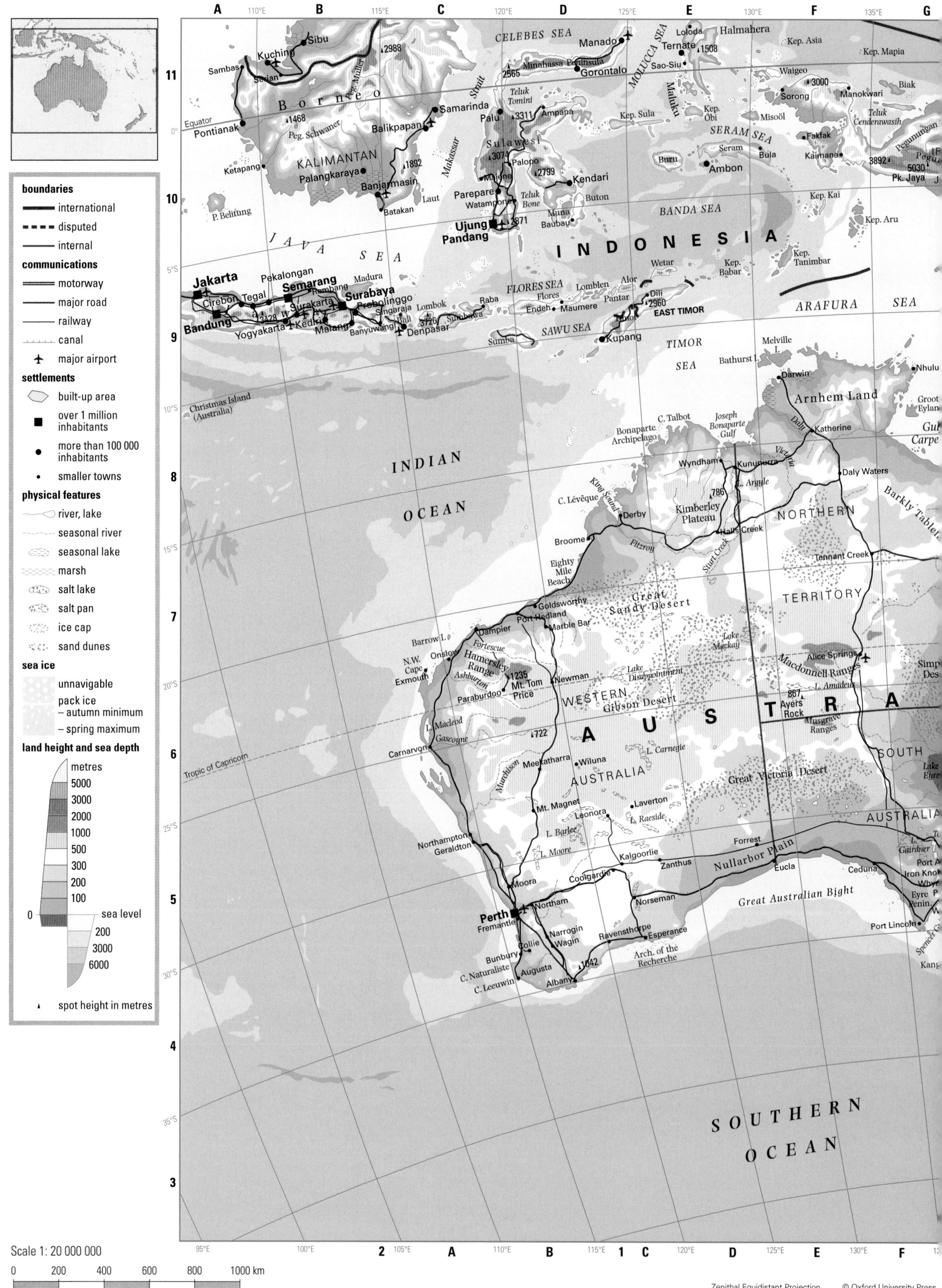

**boundaries**
international
disputed
internal

**communications**
motorway
major road
railway
canal
major airport

**settlements**
built-up area
over 1 million inhabitants
more than 100 000 inhabitants
smaller towns

**physical features**
river, lake
seasonal river
seasonal lake
marsh
salt lake
salt pan
ice cap
sand dunes

**sea ice**
unnavigable
pack ice
– autumn minimum
– spring maximum

**land height and sea depth**
metres
5000
3000
2000
1000
500
300
200
100
0
sea level
200
3000
6000

spot height in metres

Scale 1: 20 000 000

0   200   400   600   800   1000 km

Zenithal Equidistant Projection

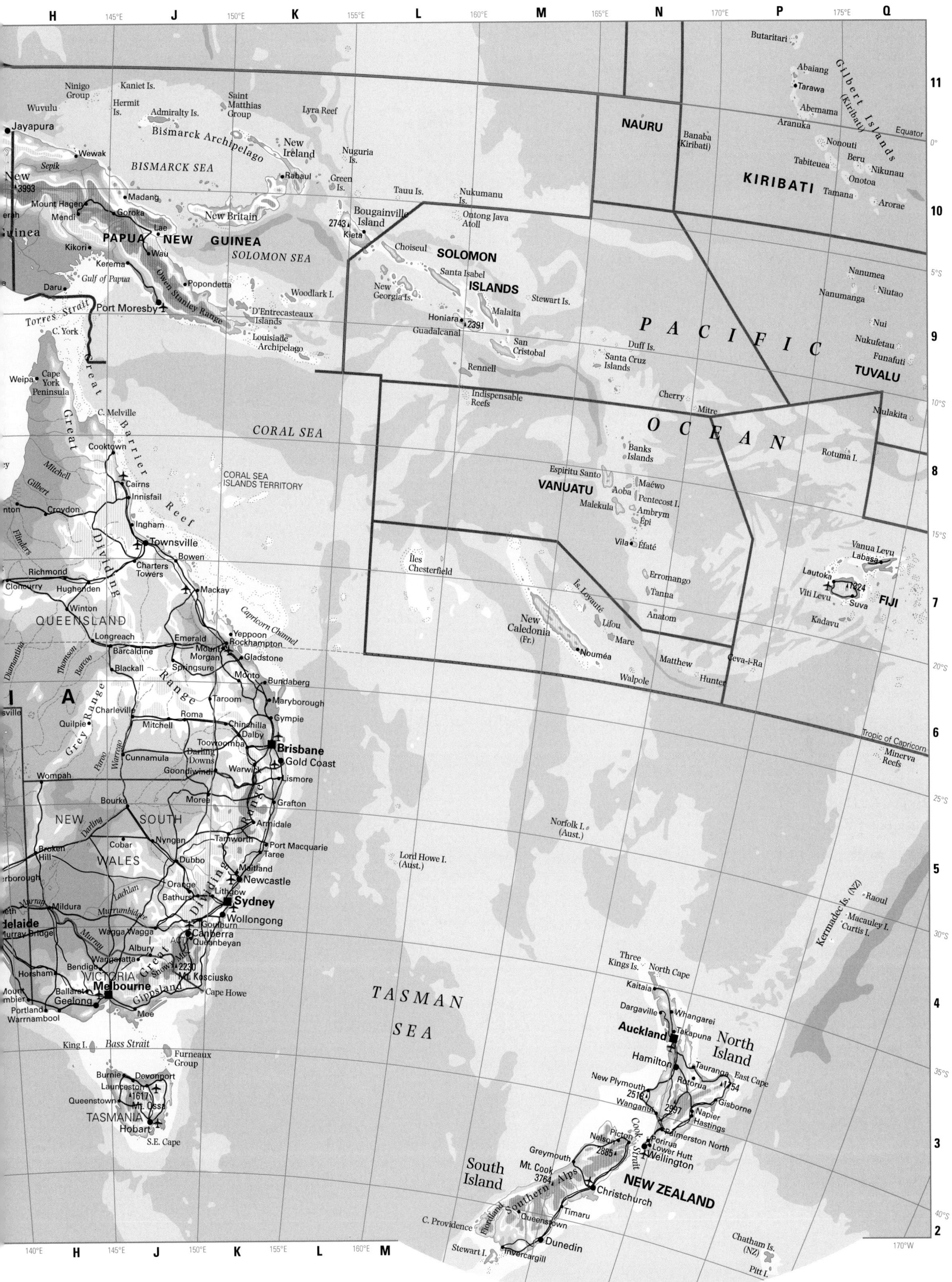

Butaritari
Abaiang
Tarawa
Abemama
Aranuka
Nonouti
Beru
Nikunau
Onotoa
Arorae
Tamana

NAURU
Banaba
(Kiribati)

KIRIBATI

Gilbert Islands (Kiribati)

Equator

Ninigo Group
Kaniet Is.
Hermit Is.
Admiralty Is.
Saint Matthias Group
Lyra Reef

New Ireland

Nuguria Is.

Wuvulu

Jayapura

Wewak

BISMARCK SEA

Bismarck Archipelago

New Britain

Rabaul

Green Is.

Tauu Is.

Nukumanu Is.

Ontong Java Atoll

New 3993

Mount Hagen
Madang
Goroka

Mendi

PAPUA NEW GUINEA

Lae

Bougainville Island

2743

Kieta

Choiseul

Santa Isabel

SOLOMON

Stewart Is.

Nanumea
Niutao

Nanumanga

Guinea

Kikori

Wau

SOLOMON SEA

Kerema

New Georgia Is.

ISLANDS

Malaita

Nui

Daru

Gulf of Papua

Popondetta

Woodlark I.

Honiara

2391

Nukufetau

Daru

Owen Stanley Range

Port Moresby

D'Entrecasteaux Islands

Guadalcanal

San Cristobal

Funafuti

TUVALU

Torres Strait

C. York

Louisiade Archipelago

Rennell

Duff Is.

PACIFIC

Weipa

Cape York Peninsula

Santa Cruz Islands

Cherry

Mitre

Rotuma I.

Mulakita

C. Melville

Indispensable Reefs

OCEAN

5°S

Cooktown

CORAL SEA

Banks Islands

0°

Cairns
Innisfail

CORAL SEA ISLANDS TERRITORY

Espiritu Santo

Maéwo

Mitchell

Ingham

Aoba

Pentecost I.

VANUATU

Malekula

Ambrym

Epi

10°S

Vanua Levu
Labasa

Gilbert

Croydon

Townsville

Bowen

Íles Chesterfield

Vila

Éfaté

Lautoka
1324

Richmond

Charters Towers

Viti Levu

Suva

FIJI

Clonourry

Hughenden

Mackay

Erromango

Kadavu

15°S

QUEENSLAND

Winton

Longreach

Emerald

Yeppoon
Rockhampton

Capricorn Channel

New Caledonia (Fr.)

Tanna

Lifou

Anatom

Barcaldine

Mount Morgan

Gladstone

Nouméa

Mare

Ceva-i-Ra

Blackall

Springsure

Monto

Matthew

Bundaberg

Walpole

Hunter

20°S

Charleville

Roma

Taroom

Maryborough

Gympie

Quilpie

Mitchell

Chinchilla
Dalby

Tropic of Capricorn

Minerva Reefs

Cunnamulla

Toowoomba

Brisbane

25°S

Goondiwindi

Warwick

Gold Coast

Norfolk I. (Aust.)

Bourke

Moree

Lismore

NEW SOUTH

Grafton

Cobar

Armidale

Lord Howe I. (Aust.)

WALES

Nyngan

Tamworth

Port Macquarie

Broken Hill

Dubbo

Taree

Kermadec Is. (NZ)

Raoul

30°S

Orange

Maitland

Bathurst

Newcastle

Macauley I.
Curtis I.

Lithgow

Sydney

Goulburn

Wollongong

Three Kings Is.

North Cape

VICTORIA

Canberra
Queanbeyan

TASMAN

Kaitaia

Mt. Kosciusko

2230

Dargaville

Whangarei

Melbourne

Cape Howe

SEA

Auckland

Takapuna

North Island

King I.

Bass Strait

Furneaux Group

Hamilton

Tauranga
East Cape

New Plymouth

Rotorua

1754

Burnie
Devonport

2518

Gisborne

Launceston

1617

Wanganui

2957

Napier

Hastings

Queenstown

Mt. Ossa

TASMANIA

Palmerston North

35°S

Hobart

Picton

Cook Strait

Porirua
Lower Hutt

Wellington

S.E. Cape

Nelson

2885

Greymouth

Mt. Cook

NEW ZEALAND

South Island

3764

Southern Alps

Christchurch

C. Providence

Fiordland

Queenstown

Timaru

40°S

Stewart I.

Invercargill

Dunedin

Chatham Is. (NZ)

Pitt I.

170°W

Scale 1: 75 000 000

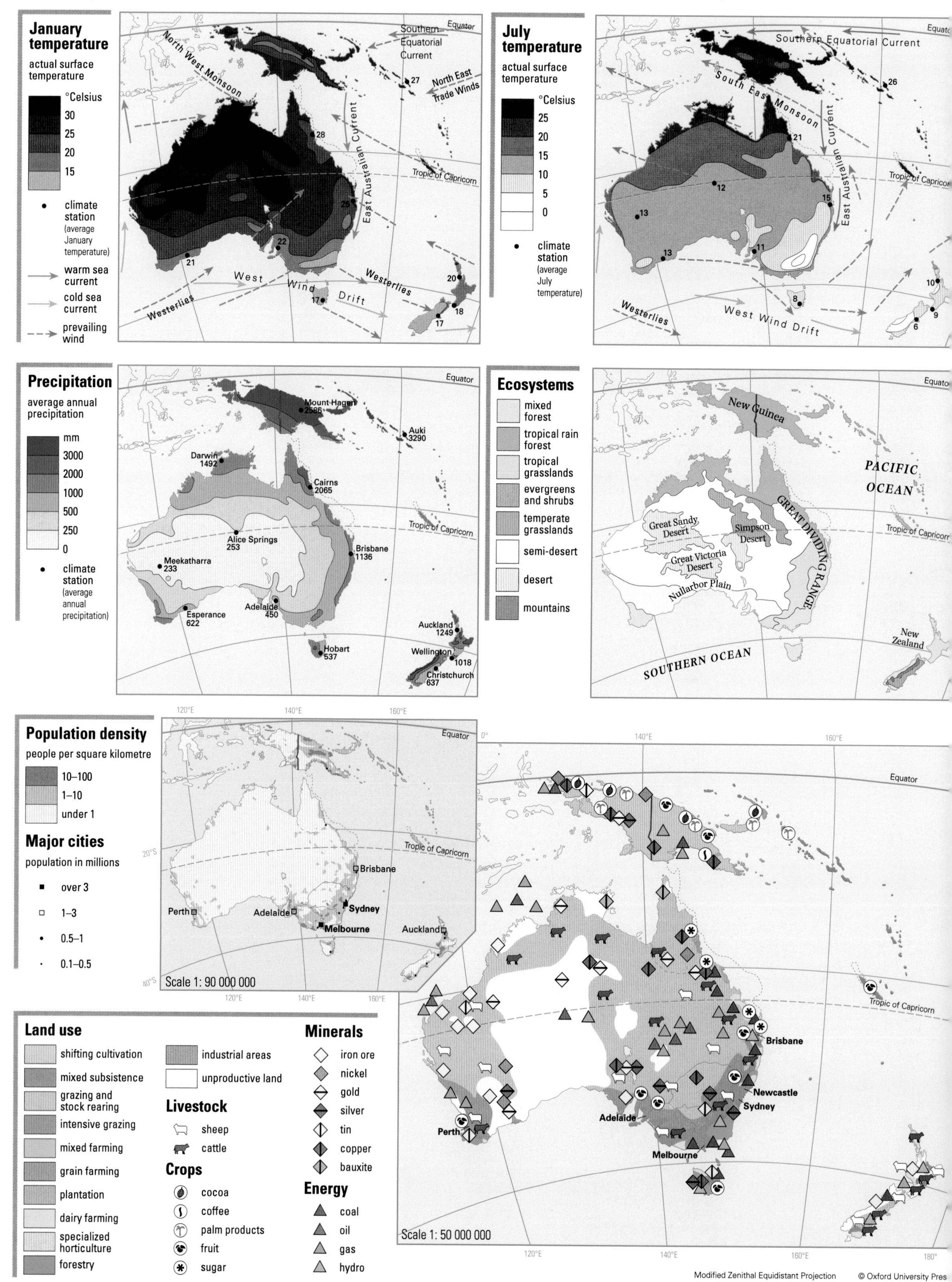

## January temperature

actual surface temperature

°Celsius
30
25
20
15

• climate station (average January temperature)

→ warm sea current

→ cold sea current

--→ prevailing wind

North West Monsoon

Southern Equatorial Current

Equator

North East Trade Winds

East Australian Current

Tropic of Capricorn

West Wind Drift

Westerlies

Westerlies

27
28
26
22
21
20
17
18

## July temperature

actual surface temperature

°Celsius
25
20
15
10
5
0

• climate station (average July temperature)

Southern Equatorial Current

Equator

South East Monsoon

East Australian Current

Tropic of Capricorn

Westerlies

West Wind Drift

26
21
12
13
13
15
11
8
10
9
6

## Precipitation

average annual precipitation

mm
3000
2000
1000
500
250
0

• climate station (average annual precipitation)

Mount Hagen 2586
Auki 3290
Darwin 1492
Cairns 2065
Alice Springs 253
Brisbane 1136
Meekatharra 233
Esperance 622
Adelaide 450
Auckland 1249
Hobart 537
Wellington 1018
Christchurch 637

Equator
Tropic of Capricorn

## Ecosystems

mixed forest
tropical rain forest
tropical grasslands
evergreens and shrubs
temperate grasslands
semi-desert
desert
mountains

New Guinea
Equator
PACIFIC OCEAN
Great Sandy Desert
Simpson Desert
GREAT DIVIDING RANGE
Tropic of Capricorn
Great Victoria Desert
Nullarbor Plain
New Zealand
SOUTHERN OCEAN

## Population density

people per square kilometre

10–100
1–10
under 1

## Major cities

population in millions

■ over 3
□ 1–3
• 0.5–1
· 0.1–0.5

120°E  140°E  160°E
Equator
20°S
Tropic of Capricorn
Brisbane
Perth  Adelaide  Sydney
Melbourne
Auckland
40°S
Scale 1: 90 000 000
120°E  140°E  160°E

## Land use

shifting cultivation
mixed subsistence
grazing and stock rearing
intensive grazing
mixed farming
grain farming
plantation
dairy farming
specialized horticulture
forestry

industrial areas
unproductive land

### Livestock

🐑 sheep
🐂 cattle

### Crops

cocoa
coffee
palm products
fruit
sugar

## Minerals

◇ iron ore
◆ nickel
◆ gold
◆ silver
◊ tin
◊ copper
◊ bauxite

## Energy

▲ coal
▲ oil
▲ gas
▲ hydro

0°  140°E  160°E
Equator
Tropic of Capricorn
Brisbane
Newcastle
Sydney
Adelaide
Perth
Melbourne
Scale 1: 50 000 000
120°E  140°E  160°E  180°

Modified Zenithal Equidistant Projection

© Oxford University Press

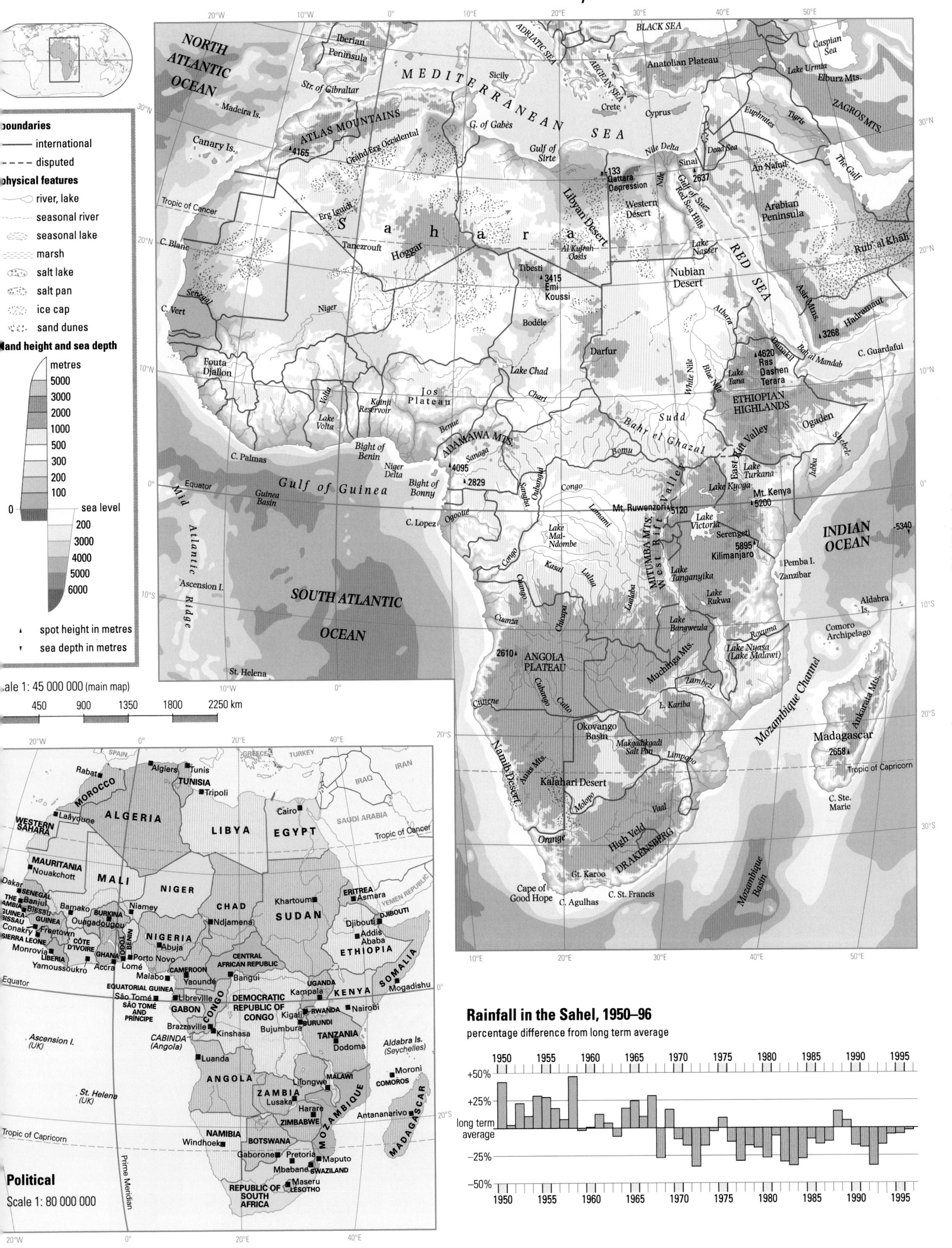

### Rainfall in the Sahel, 1950–96
percentage difference from long term average

© Oxford University Press Zenithal Equal Area Projection

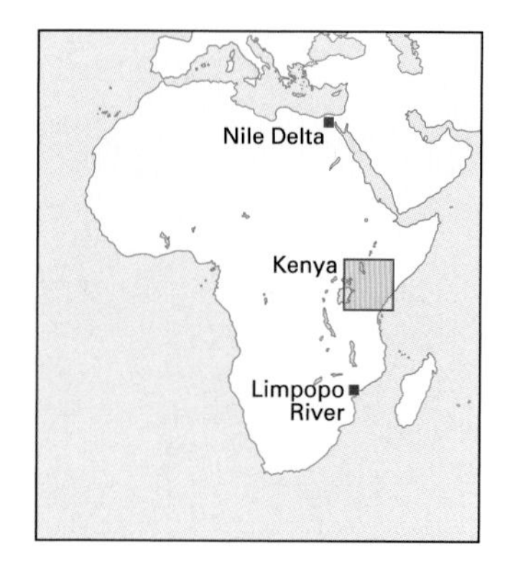

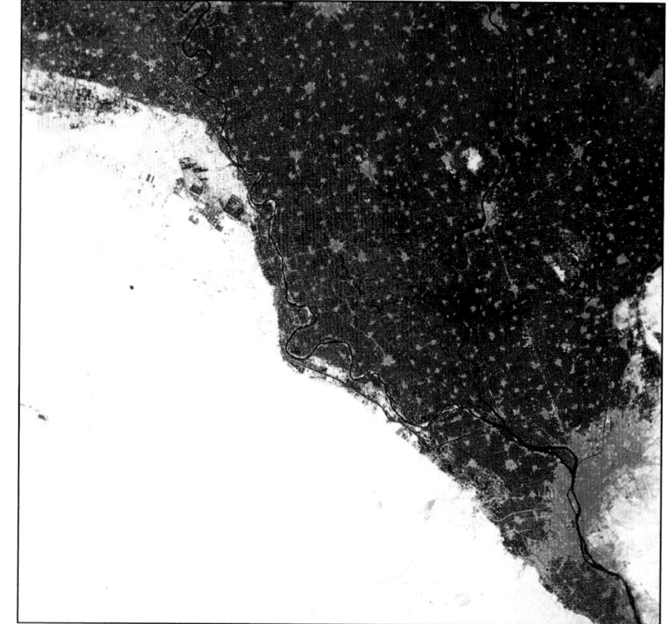

### Nile River Delta

Cairo is shown by the blue/grey area to lower right of the image. The city has grown in size from 1.5 million people in 1947 to more than 6 million in 1991. Other blue areas show rapid urban development in the delta. Yellow areas at top left of the image show the spread of agriculture in the desert, assisted by centre pivot irrigation.

### Kenya crop cover

Remote sensing can be used to predict food shortages. Dark green areas on the satellite image of Kenya for April, 2000 show the newly sown 'long rains' cereal crop. However, gaps in the dark green pattern indicate a poor harvest and in June, $88 million of international food aid was agreed. By August, low rainfall had led to widespread crop failure in the south, shown as light green, and spread of bare soil in the north, shown as orange and yellow.

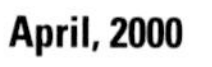

**April, 2000**

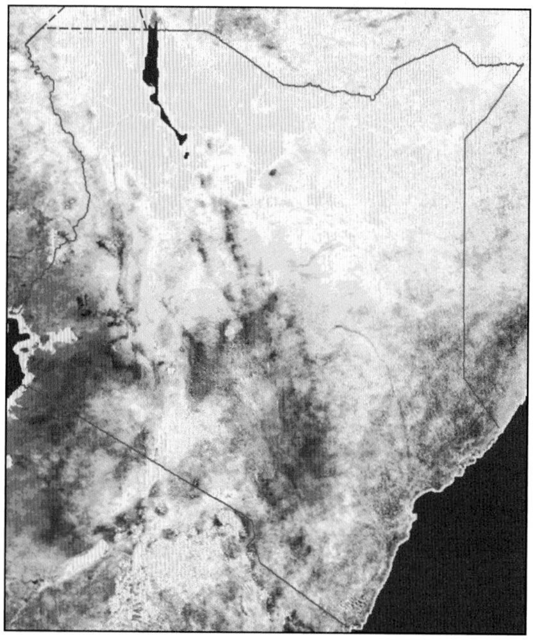

**August, 2000**

### Mozambique floods

**Before flooding, August, 1999.**

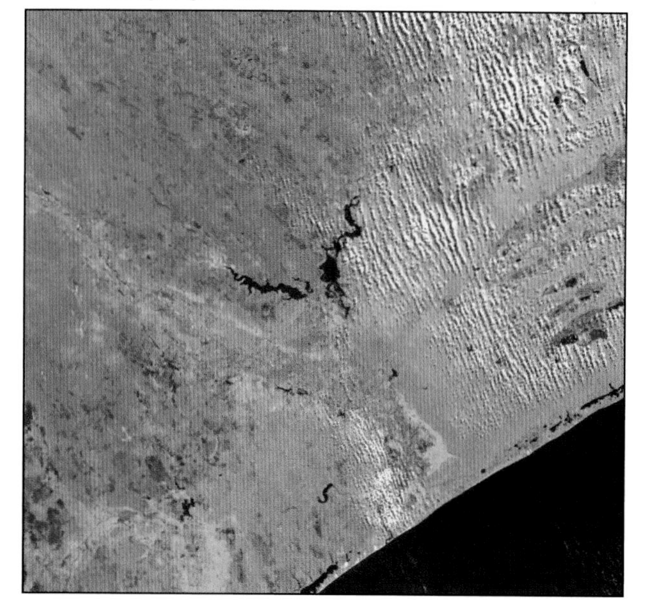

**After flooding, March 2000.**

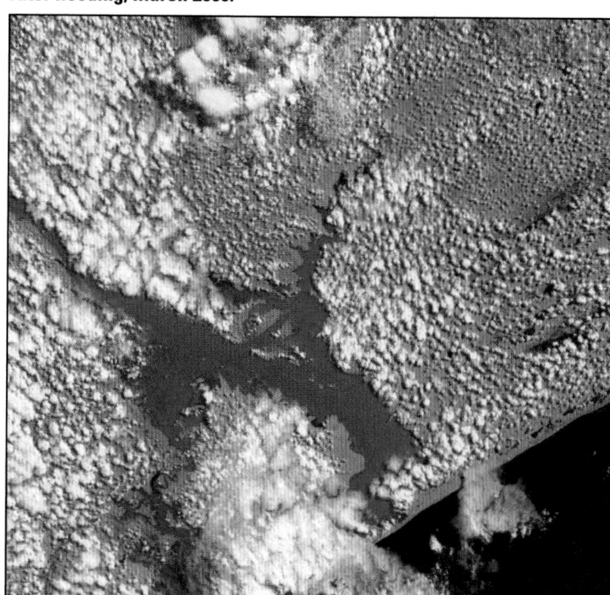

These images from the Landsat 7 satellite show the Limpopo river before and after flooding. Torrential rain between 4 and 7 February, 2000 added to already high levels of seasonal rainfall. Tropical cyclone Eline hit the southern coast of Mozambique on 21 February bringing even more rain. Over a million people were made homeless and 100 000 hectares of agricultural land flooded. 620 miles of roads were swept away.

Scale 1: 90 000 000

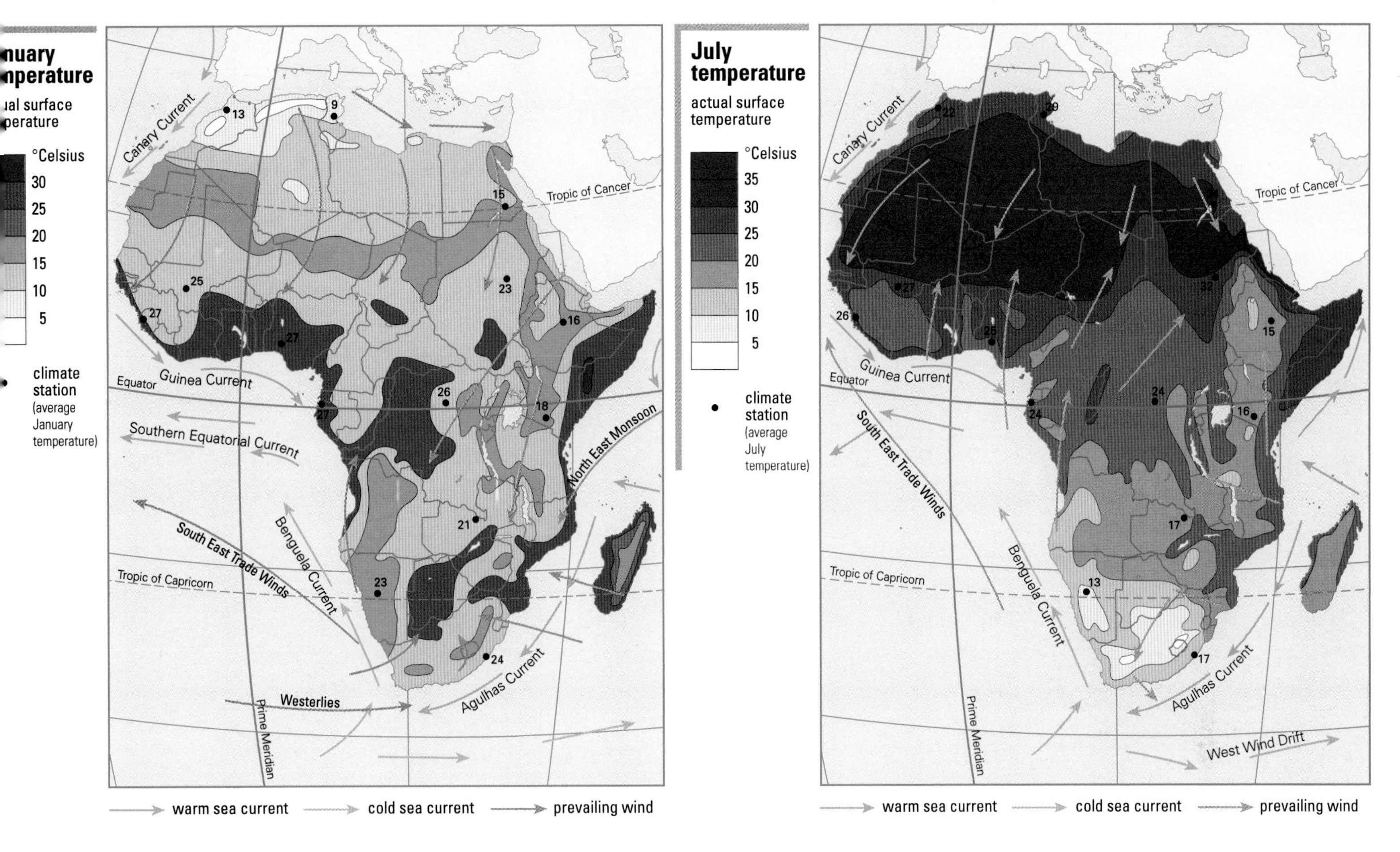

## January temperature

actual surface temperature

°Celsius
- 30
- 25
- 20
- 15
- 10
- 5

● climate station (average January temperature)

Canary Current
● 13
● 9
● 15
Tropic of Cancer
● 25
● 27
● 27
● 23
● 16
Guinea Current
Equator
● 27
● 26
● 18
Southern Equatorial Current
North East Monsoon
Benguela Current
● 21
South East Trade Winds
Tropic of Capricorn
● 23
Agulhas Current
● 24
Westerlies
Prime Meridian

→ warm sea current   → cold sea current   → prevailing wind

## July temperature

actual surface temperature

°Celsius
- 35
- 30
- 25
- 20
- 15
- 10
- 5

● climate station (average July temperature)

Canary Current
● 22
● 29
Tropic of Cancer
● 27
● 32
● 26
● 15
Guinea Current
Equator
● 28
● 24
● 16
● 24
South East Trade Winds
● 17
Benguela Current
Tropic of Capricorn
● 13
Agulhas Current
● 17
Prime Meridian
West Wind Drift

→ warm sea current   → cold sea current   → prevailing wind

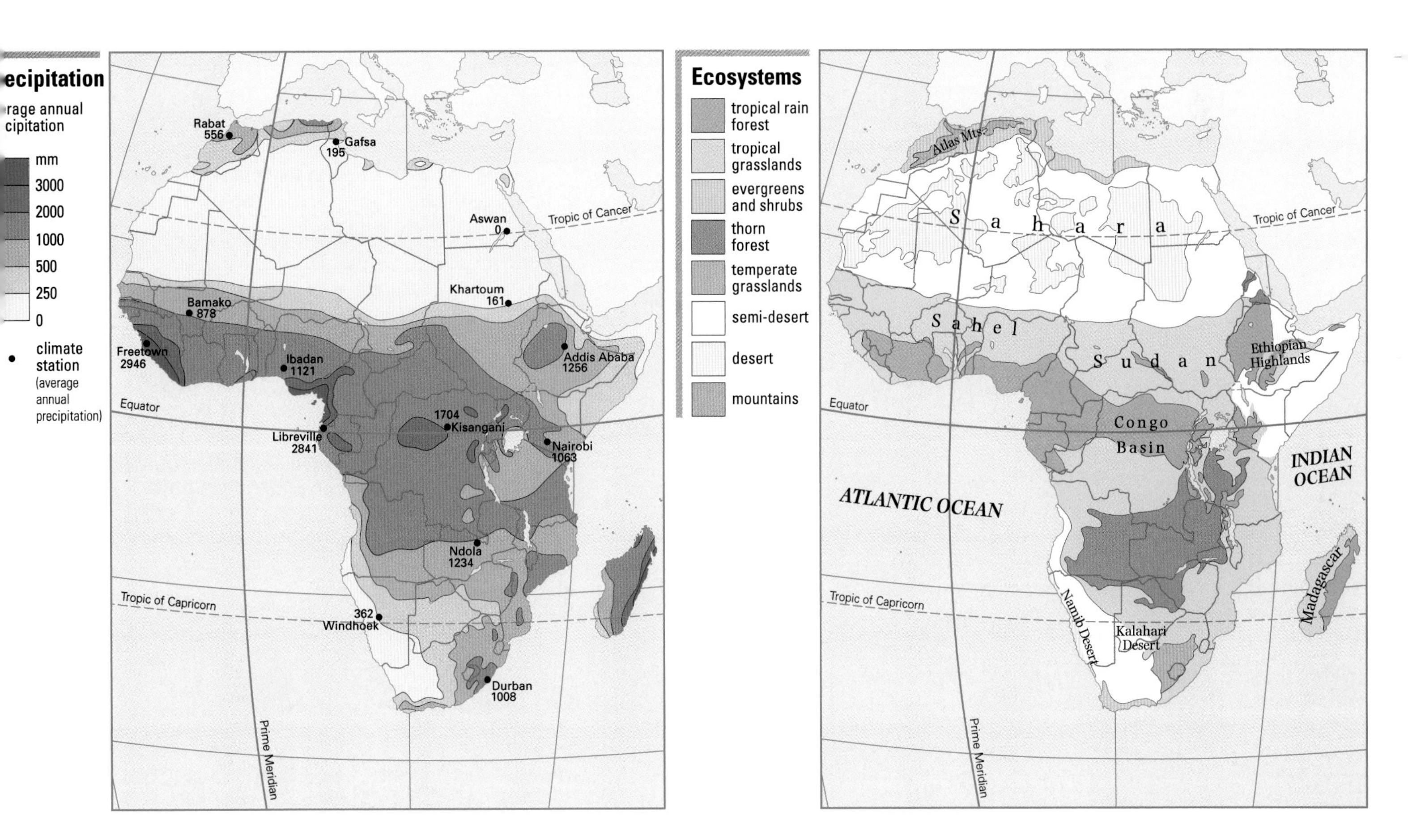

## Precipitation

average annual precipitation

mm
- 3000
- 2000
- 1000
- 500
- 250
- 0

● climate station (average annual precipitation)

Rabat 556
Gafsa 195
Aswan 0
Tropic of Cancer
Bamako 878
Khartoum 161
Freetown 2946
Ibadan 1121
Addis Ababa 1256
Equator
Libreville 2841
Kisangani 1704
Nairobi 1063
Ndola 1234
Tropic of Capricorn
Windhoek 362
Durban 1008
Prime Meridian

## Ecosystems

- tropical rain forest
- tropical grasslands
- evergreens and shrubs
- thorn forest
- temperate grasslands
- semi-desert
- desert
- mountains

Atlas Mts.
S  a  h  a  r  a
Tropic of Cancer
S a h e l
S u d a n
Ethiopian Highlands
Equator
Congo Basin
ATLANTIC OCEAN
INDIAN OCEAN
Tropic of Capricorn
Namib Desert
Kalahari Desert
Madagascar
Prime Meridian

Scale 1: 55 000 00

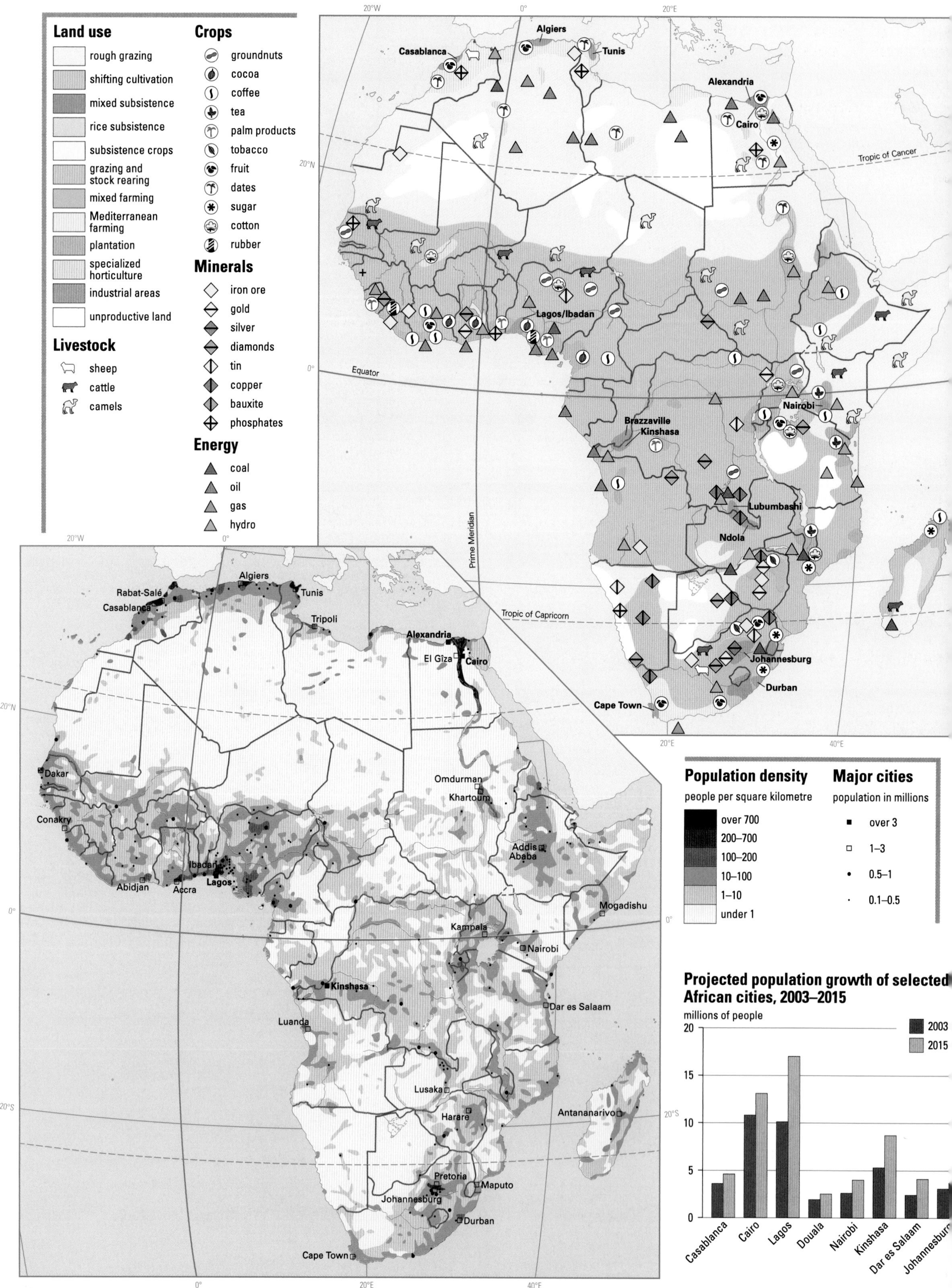

## Land use

- rough grazing
- shifting cultivation
- mixed subsistence
- rice subsistence
- subsistence crops
- grazing and stock rearing
- mixed farming
- Mediterranean farming
- plantation
- specialized horticulture
- industrial areas
- unproductive land

## Livestock

- sheep
- cattle
- camels

## Crops

- groundnuts
- cocoa
- coffee
- tea
- palm products
- tobacco
- fruit
- dates
- sugar
- cotton
- rubber

## Minerals

- iron ore
- gold
- silver
- diamonds
- tin
- copper
- bauxite
- phosphates

## Energy

- coal
- oil
- gas
- hydro

## Population density

people per square kilometre

- over 700
- 200–700
- 100–200
- 10–100
- 1–10
- under 1

## Major cities

population in millions

- over 3
- 1–3
- 0.5–1
- 0.1–0.5

## Projected population growth of selected African cities, 2003–2015

millions of people

2003
2015

Casablanca, Cairo, Lagos, Douala, Nairobi, Kinshasa, Dar es Salaam, Johannesburg

Zenithal Equal Area Projection

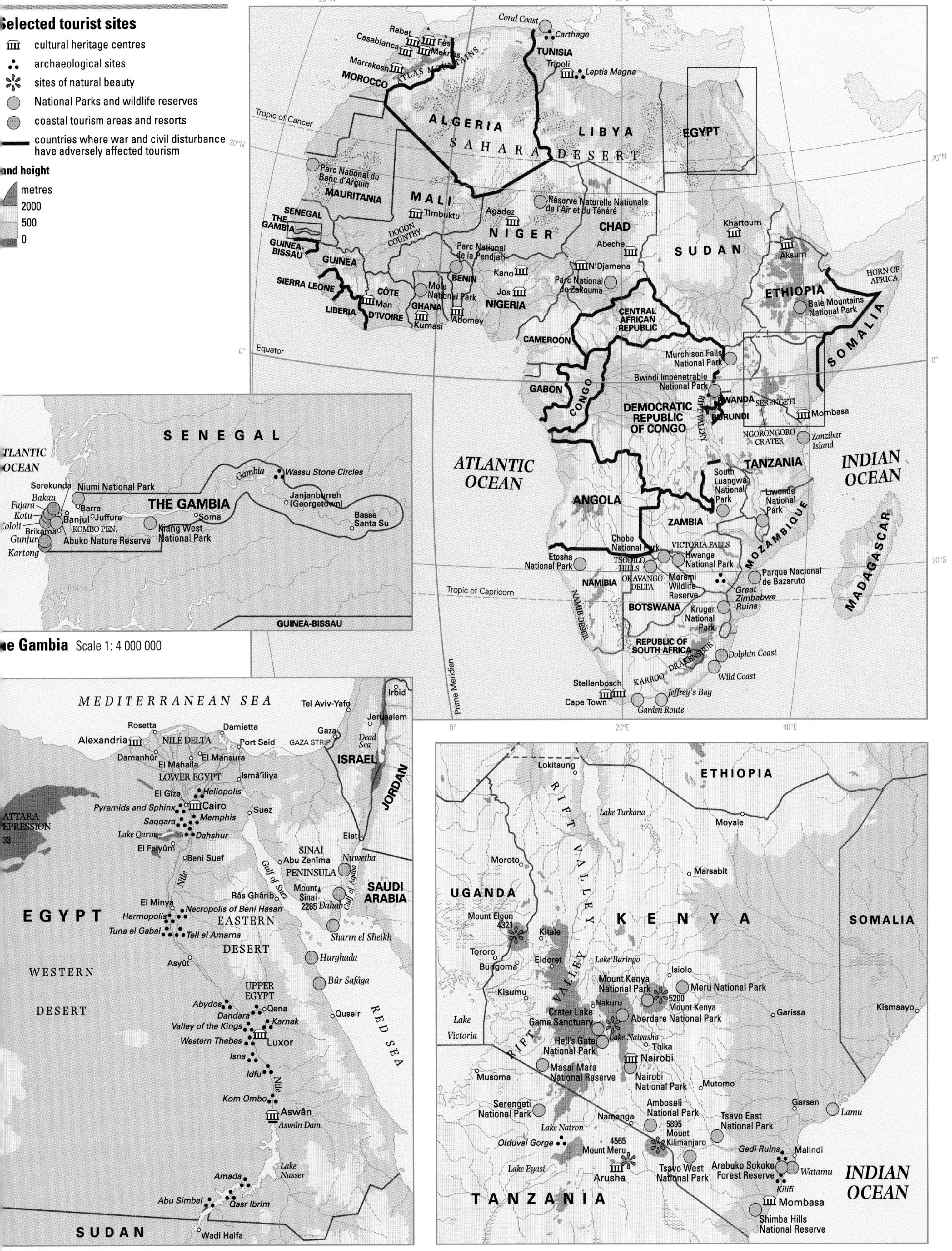

cale 1: 55 000 000 (main map)

**Selected tourist sites**

- ⛩ cultural heritage centres
- ⸪ archaeological sites
- ✳ sites of natural beauty
- ⬤ National Parks and wildlife reserves
- ⬤ coastal tourism areas and resorts
- ▬ countries where war and civil disturbance have adversely affected tourism

**and height**

metres

2000
500
0



Coral Coast
Carthage
Rabat
Casablanca · Fès
Meknes
TUNISIA
Tripoli
Marrakesh
· Leptis Magna
MOROCCO
ATLAS MOUNTAINS
LIBYA
EGYPT
Tropic of Cancer
ALGERIA
SAHARA DESERT
20°N
Parc National du Banc d'Arguin
MAURITANIA
MALI
NIGER
CHAD
SUDAN
Khartoum
SENEGAL
Timbuktu
Agadez
Réserve Naturelle Nationale de l'Aïr et du Ténéré
THE GAMBIA
GUINEA-BISSAU
DOGON COUNTRY
Abeche
Aksum
HORN OF AFRICA
GUINEA
Parc National de la Pendjari
Kano
N'Djamena
SIERRA LEONE
CÔTE D'IVOIRE
Mole National Park
BENIN
Jos
Parc National de Zakouma
ETHIOPIA
Bale Mountains National Park
LIBERIA
Man
GHANA
Abomey
NIGERIA
CENTRAL AFRICAN REPUBLIC
SOMALIA
Equator
Kumasi
CAMEROON
Murchison Falls National Park
Bwindi Impenetrable National Park
RWANDA
SERENGETI
Mombasa
GABON
CONGO
DEMOCRATIC REPUBLIC OF CONGO
BURUNDI
RIFT VALLEY
NGORONGORO CRATER
Zanzibar Island
ATLANTIC OCEAN
TANZANIA
INDIAN OCEAN
South Luangwa National Park
Liwonde National Park
ANGOLA
ZAMBIA
MOZAMBIQUE
MADAGASCAR
Chobe National Park
VICTORIA FALLS
Etosha National Park
TSODILO HILLS
Hwange National Park
Parque Nacional de Bazaruto
NAMIBIA
OKAVANGO DELTA
Moremi Wildlife Reserve
Great Zimbabwe Ruins
NAMIB DESERT
BOTSWANA
Kruger National Park
REPUBLIC OF SOUTH AFRICA
DRAKENSBERG
Dolphin Coast
Tropic of Capricorn
Stellenbosch
KARROO
Wild Coast
Cape Town
Jeffrey's Bay
Garden Route

**The Gambia inset** (Scale 1: 4 000 000)

SENEGAL
ATLANTIC OCEAN
Gambia
Wassu Stone Circles
Serekunda
Niumi National Park
Bakau
Barra
Juffure
Janjanbureh (Georgetown)
Fajara
Kotu
Banjul
Soma
Kololi
Brikama
KOMBO PEN.
Basse Santa Su
Gunjur
Abuko Nature Reserve
THE GAMBIA
Kiang West National Park
Kartong
GUINEA-BISSAU

**The Gambia** Scale 1: 4 000 000

**Nile Valley and Eastern Egypt inset** (Scale 1: 10 000 000)

MEDITERRANEAN SEA
Tel Aviv-Yafo
Irbid
Rosetta
Damietta
Gaza
Jerusalem
Alexandria
NILE DELTA
Port Said
GAZA STRIP
Dead Sea
Damanhûr
El Mahalla
ISRAEL
El Mansura
Ismâ'iliya
JORDAN
LOWER EGYPT
El Giza
Heliopolis
Pyramids and Sphinx
Cairo
Suez
Saqqara
Memphis
ATTARA DEPRESSION
Dahshur
Lake Qarun
SINAI
33
El Fayûm
Abu Zenîma
Nuweiba
EGYPT
Beni Suef
PENINSULA
Elat
SAUDI ARABIA
El Minya
Râs Ghârib
Mount Sinai 2285
Dahab
Hermopolis
EASTERN
Tuna el Gabal
Tell el Amarna
Sharm el Sheikh
WESTERN
Asyût
DESERT
Hurghada
DESERT
Bûr Safâga
UPPER EGYPT
Abydos
Qena
Dandara
Karnak
Quseir
RED SEA
Valley of the Kings
Western Thebes
Luxor
Isna
Idfu
Nile
Kom Ombo
Aswân
Aswân Dam
Lake Nasser
Amada
Abu Simbel
Qasr Ibrim
SUDAN
Wadi Halfa

**Nile Valley and Eastern Egypt** Scale 1: 10 000 000

**Kenya inset** (Scale 1: 10 000 000)

Lokitaung
ETHIOPIA
RIFT VALLEY
Lake Turkana
Moyale
Moroto
Marsabit
UGANDA
KENYA
SOMALIA
Mount Elgon 4321
Kitale
Tororo
Isiolo
Meru National Park
Bungoma
Eldoret
Lake Baringo
Garissa
Kisumu
Mount Kenya National Park
Nakuru
Mount Kenya 5200
Kismaayo
Crater Lake Game Sanctuary
Aberdare National Park
Lake Victoria
Hell's Gate National Park
Lake Naivasha
Thika
Nairobi
Musoma
Masai Mara National Reserve
Nairobi National Park
Mutomo
Lamu
Serengeti National Park
Amboseli National Park
Tsavo East National Park
Garsen
Lake Natron
Namanga
5895 Mount Kilimanjaro
Gedi Ruins
Malindi
Olduvai Gorge
4565 Mount Meru
Tsavo West National Park
Arabuko Sokoke Forest Reserve
Watamu
Lake Eyasi
Arusha
Kilifi
INDIAN OCEAN
TANZANIA
Mombasa
Shimba Hills National Reserve

**Kenya** Scale 1: 10 000 000

Zenithal Equal Area Projection     © Oxford University Press

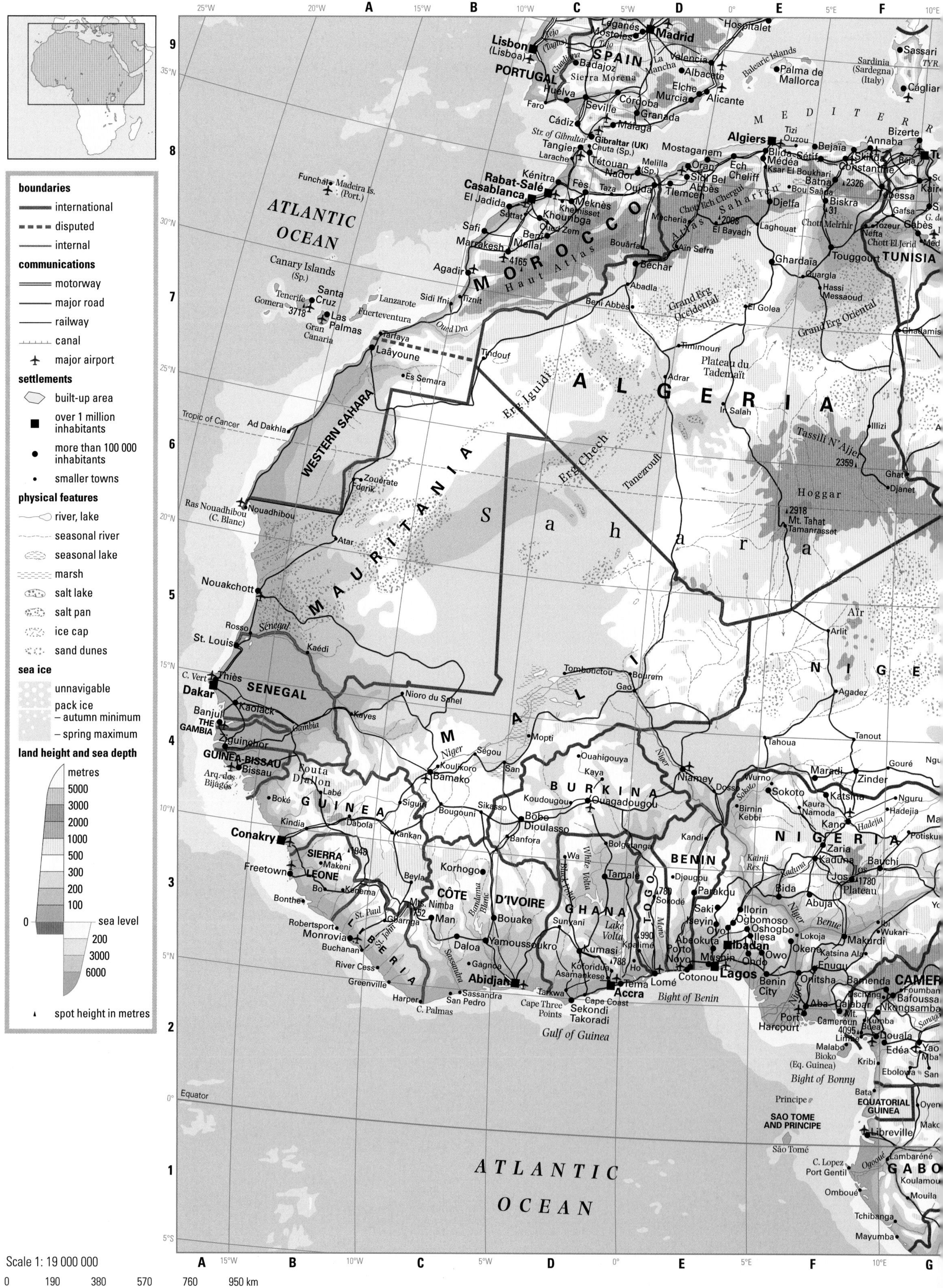

Scale 1: 19 000 000

Zenithal Equal Area Projection

## boundaries
—— international
- - - disputed
—— internal

## communications
===== motorway
—— major road
—— railway
+++++ canal
✈ major airport

## settlements
⬡ built-up area
■ over 1 million inhabitants
● more than 100 000 inhabitants
• smaller towns

## physical features
river, lake
seasonal river
seasonal lake
marsh
salt lake
salt pan
ice cap
sand dunes

## sea ice
unnavigable
pack ice
– autumn minimum
– spring maximum

## land height and sea depth
metres
5000
3000
2000
1000
500
300
200
100
sea level
200
3000
6000

▲ spot height in metres

0  190  380  570  760  950 km

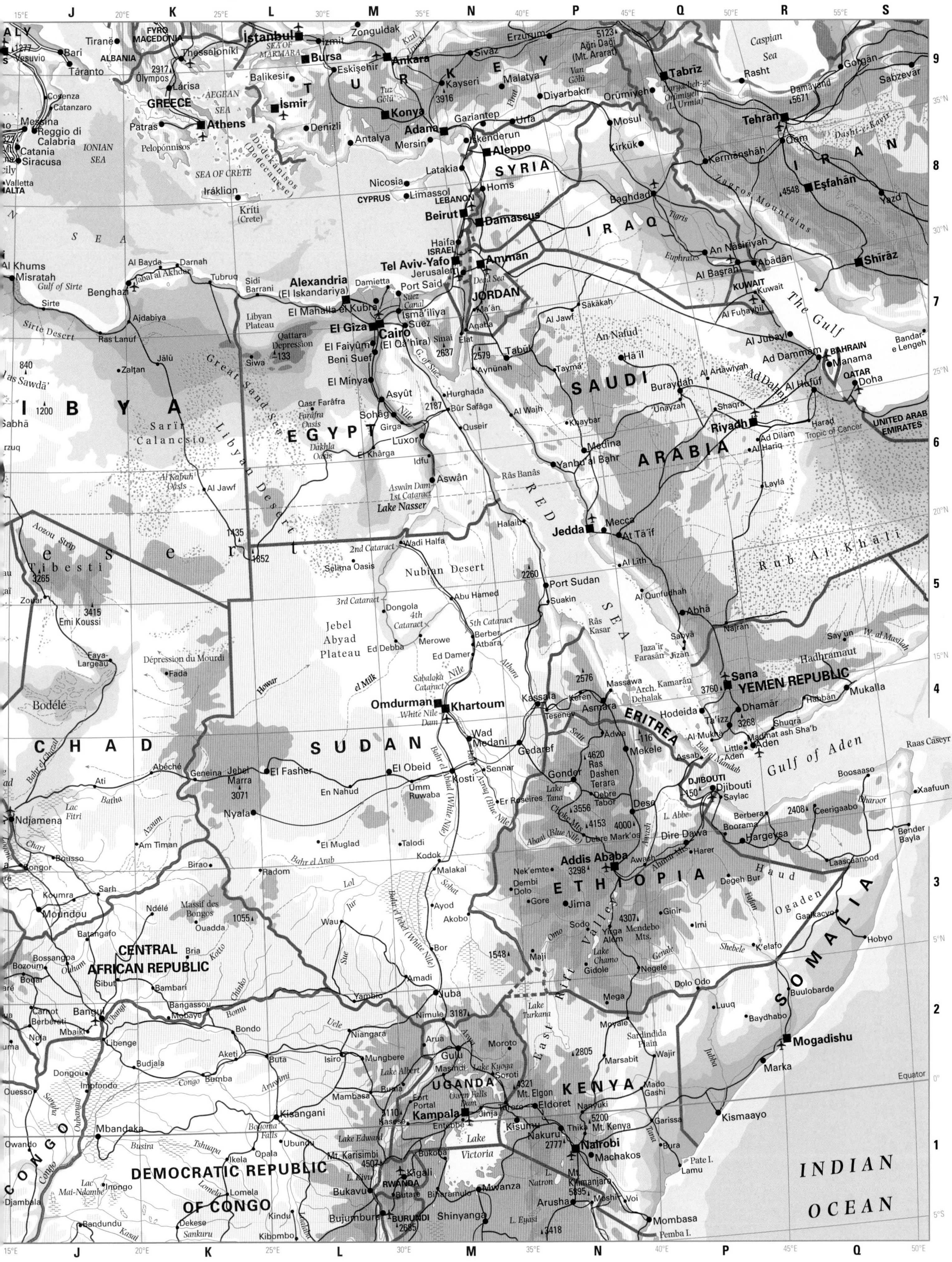
TURKEY
SYRIA
IRAQ
IRAN
JORDAN
ISRAEL
LEBANON
CYPRUS
KUWAIT
BAHRAIN
QATAR
UNITED ARAB EMIRATES
SAUDI ARABIA
YEMEN REPUBLIC
EGYPT
LIBYA
CHAD
SUDAN
ERITREA
DJIBOUTI
ETHIOPIA
SOMALIA
KENYA
UGANDA
RWANDA
BURUNDI
CENTRAL AFRICAN REPUBLIC
DEMOCRATIC REPUBLIC OF CONGO
CONGO
GREECE
ALBANIA
FYRO MACEDONIA
İstanbul
Ankara
İzmir
Konya
Bursa
Adana
Athens
Thessaloniki
Aleppo
Damascus
Beirut
Amman
Tel Aviv-Yafo
Jerusalem
Baghdad
Tehran
Eşfahān
Shīrāz
Tabrīz
Riyadh
Jedda
Mecca
Medina
Sana
Aden
Alexandria (El Iskandariya)
Cairo (El Qâ'hira)
El Giza
Port Said
Suez
Aswân
Luxor
Khartoum
Omdurman
Port Sudan
El Obeid
Addis Ababa
Djibouti
Asmara
Mogadishu
Nairobi
Kampala
Kigali
Mombasa
Kisangani
Bangui
Ndjamena
Benghazi
Misratah
Mediterranean Sea
Caspian Sea
RED SEA
Gulf of Aden
The Gulf
INDIAN OCEAN
Lake Victoria
Lake Nasser
Lake Turkana
Lake Tana
Nile
Great Sand Sea
Libyan Desert
Nubian Desert
Rub Al Khali
An Nafud
Tibesti
East Rift Valley
Tropic of Cancer
Equator

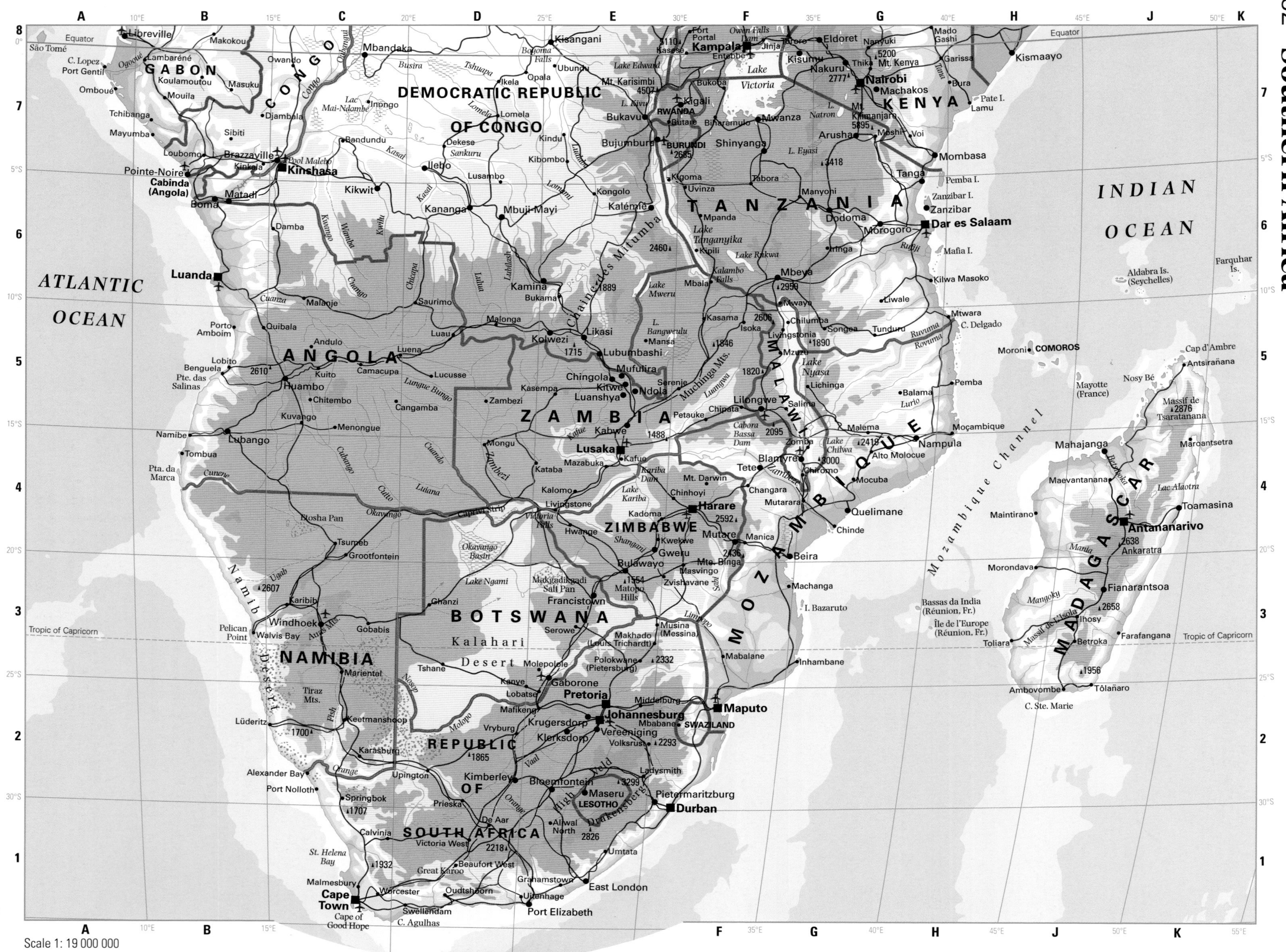
INDIAN OCEAN
ATLANTIC OCEAN
Mozambique Channel
DEMOCRATIC REPUBLIC OF CONGO
CONGO
GABON
ANGOLA
ZAMBIA
TANZANIA
KENYA
MALAWI
MOZAMBIQUE
MADAGASCAR
ZIMBABWE
BOTSWANA
NAMIBIA
REPUBLIC OF SOUTH AFRICA
LESOTHO
SWAZILAND
RWANDA
BURUNDI
COMOROS
Kalahari Desert
Namib Desert
Tropic of Capricorn
Equator
Scale 1: 19 000 000

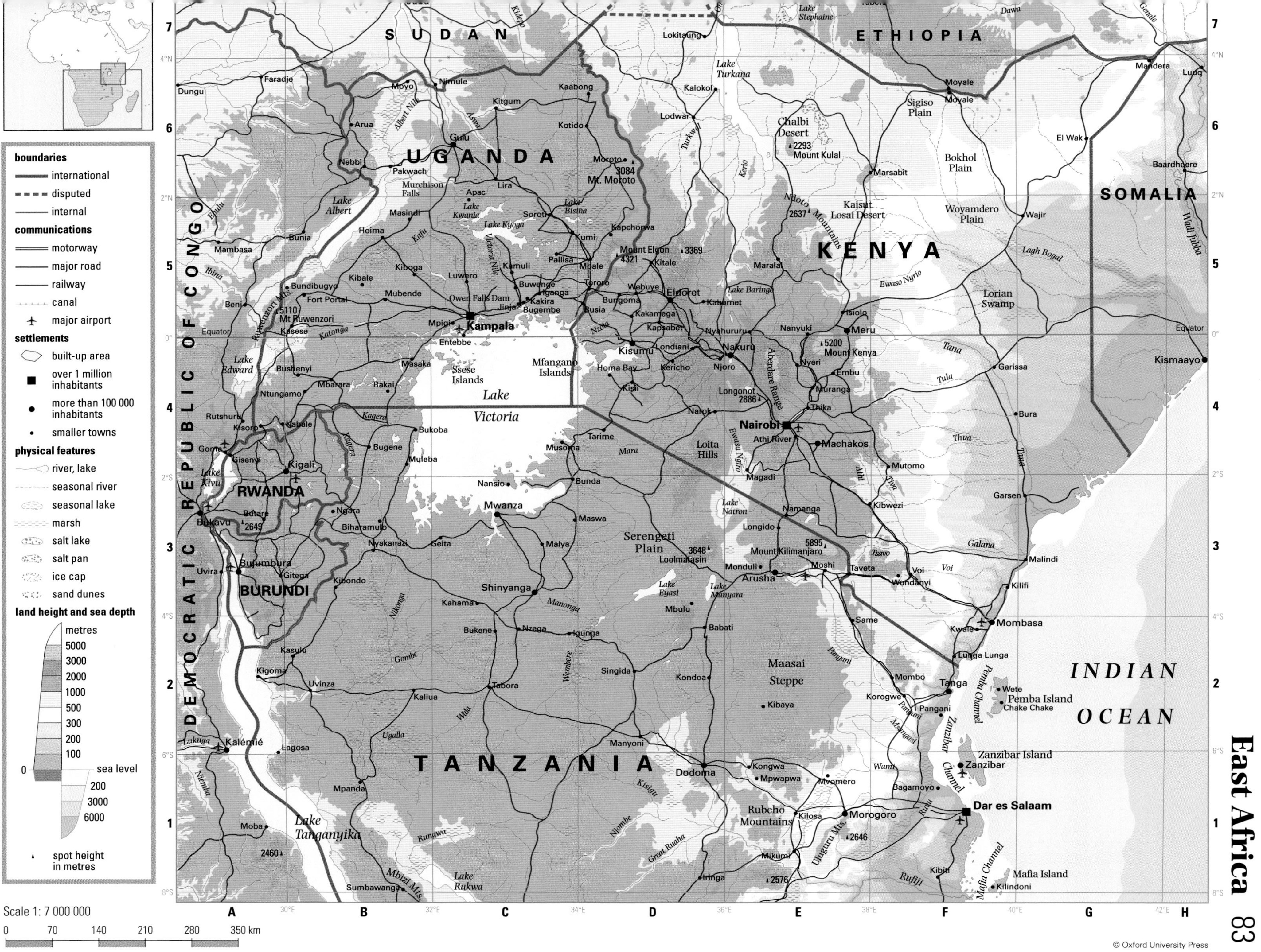

**boundaries**
—— international
- - - disputed
—— internal

**communications**
═══ motorway
——— major road
——— railway
- - - canal
✈ major airport

**settlements**
⬭ built-up area
■ over 1 million inhabitants
● more than 100 000 inhabitants
• smaller towns

**physical features**
river, lake
seasonal river
seasonal lake
marsh
salt lake
salt pan
ice cap
sand dunes

**land height and sea depth**

metres
5000
3000
2000
1000
500
300
200
100
sea level
200
3000
6000

▲ spot height in metres

Scale 1 : 7 000 000

0   70   140   210   280   350 km

INDIAN OCEAN

SOMALIA

ETHIOPIA

KENYA

UGANDA

SUDAN

DEMOCRATIC REPUBLIC OF CONGO

RWANDA

BURUNDI

TANZANIA

Lake Victoria

Lake Tanganyika

Maasai Steppe

Serengeti Plain

Nairobi

Kampala

Dodoma

Dar es Salaam

Mombasa

Zanzibar

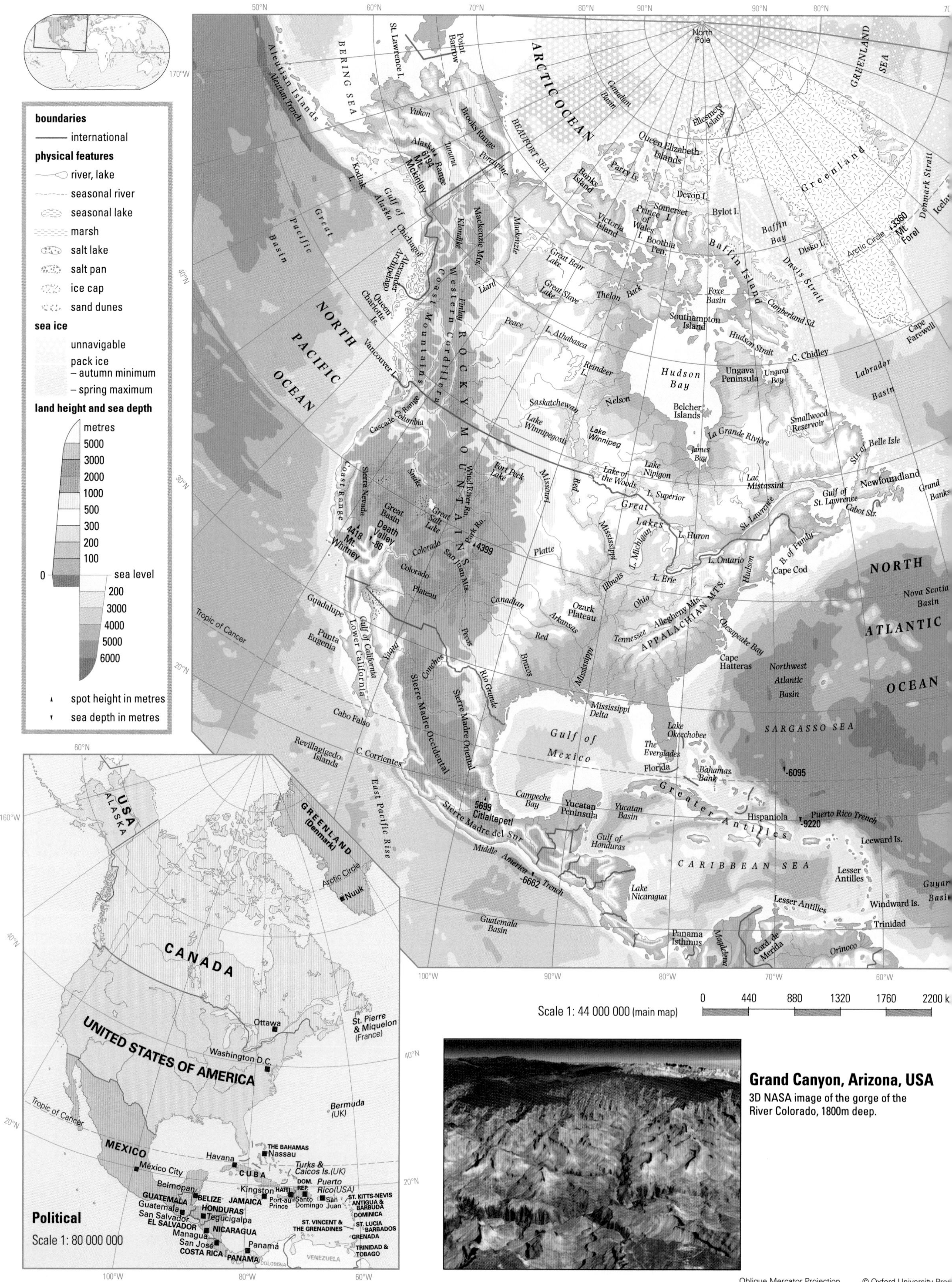

## boundaries
— international

## physical features
river, lake
seasonal river
seasonal lake
marsh
salt lake
salt pan
ice cap
sand dunes

## sea ice
unnavigable
pack ice
– autumn minimum
– spring maximum

## land height and sea depth
metres
5000
3000
2000
1000
500
300
200
100
0 — sea level
200
3000
4000
5000
6000

▲ spot height in metres
▼ sea depth in metres

**Political**
Scale 1: 80 000 000

Scale 1: 44 000 000 (main map)

0   440   880   1320   1760   2200 k

**Grand Canyon, Arizona, USA**
3D NASA image of the gorge of the
River Colorado, 1800m deep.

Oblique Mercator Projection

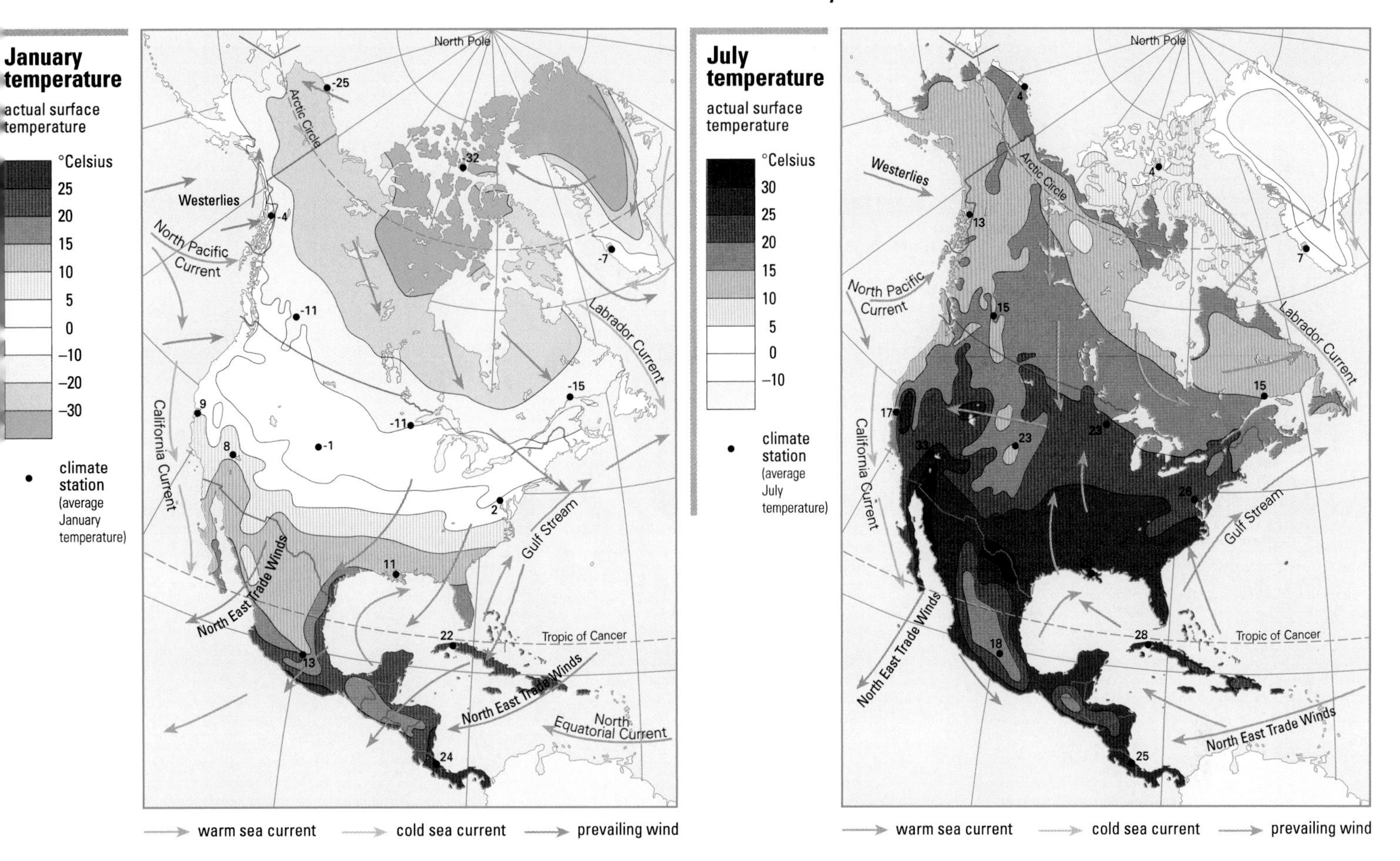

**January temperature**

actual surface temperature

°Celsius
- 25
- 20
- 15
- 10
- 5
- 0
- −10
- −20
- −30

● climate station (average January temperature)

**July temperature**

actual surface temperature

°Celsius
- 30
- 25
- 20
- 15
- 10
- 5
- 0
- −10

● climate station (average July temperature)

→ warm sea current  →→ cold sea current  → prevailing wind

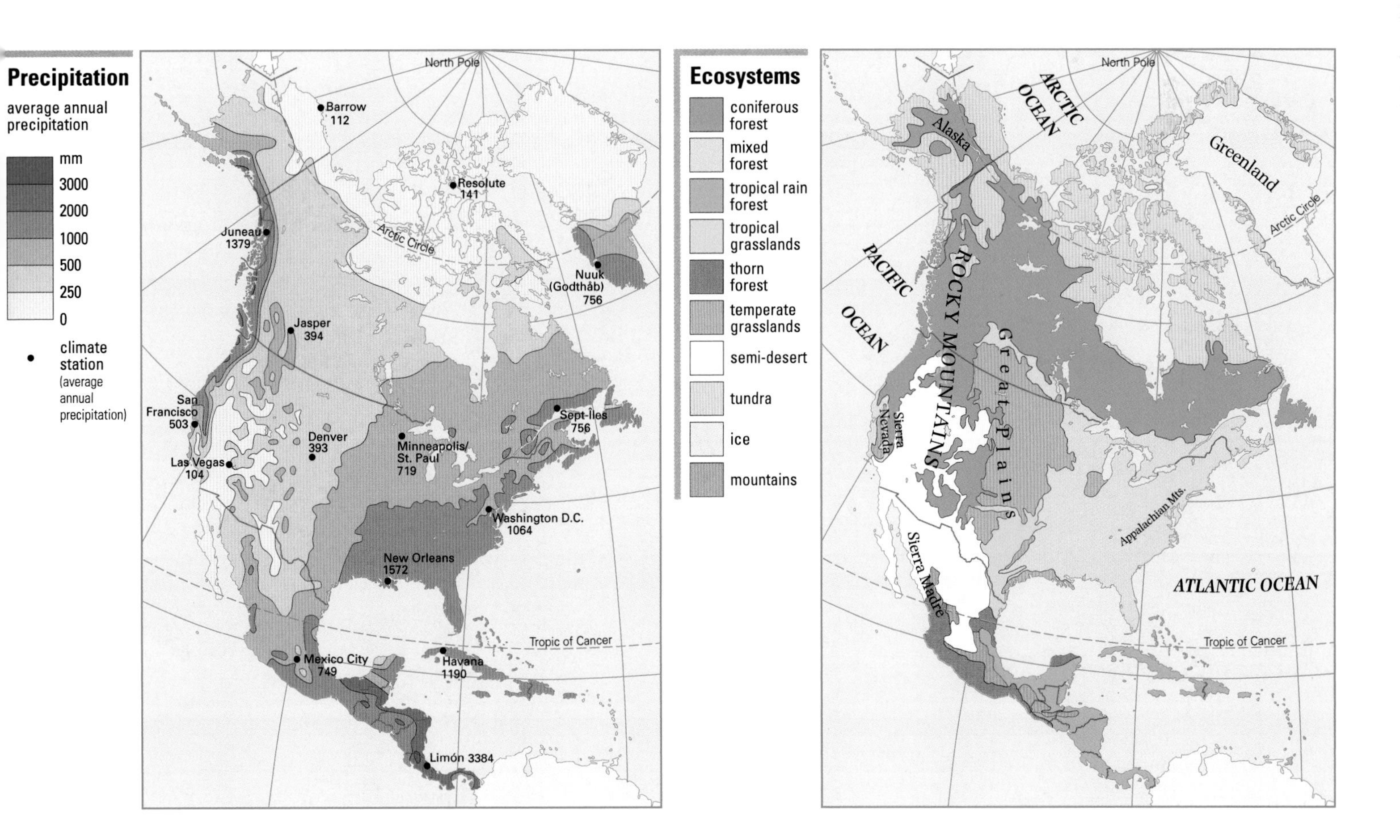

**Precipitation**

average annual precipitation

mm
- 3000
- 2000
- 1000
- 500
- 250
- 0

● climate station (average annual precipitation)

Barrow 112
Resolute 141
Nuuk (Godthåb) 756
Juneau 1379
Jasper 394
Sept-Îles 756
San Francisco 503
Denver 393
Minneapolis/ St. Paul 719
Las Vegas 104
Washington D.C. 1064
New Orleans 1572
Mexico City 749
Havana 1190
Limón 3384

**Ecosystems**

- coniferous forest
- mixed forest
- tropical rain forest
- tropical grasslands
- thorn forest
- temperate grasslands
- semi-desert
- tundra
- ice
- mountains

ARCTIC OCEAN
Alaska
Greenland
Arctic Circle
PACIFIC OCEAN
ROCKY MOUNTAINS
Great Plains
Sierra Nevada
Sierra Madre
Appalachian Mts.
ATLANTIC OCEAN
Tropic of Cancer

Scale 1: 55 000 000

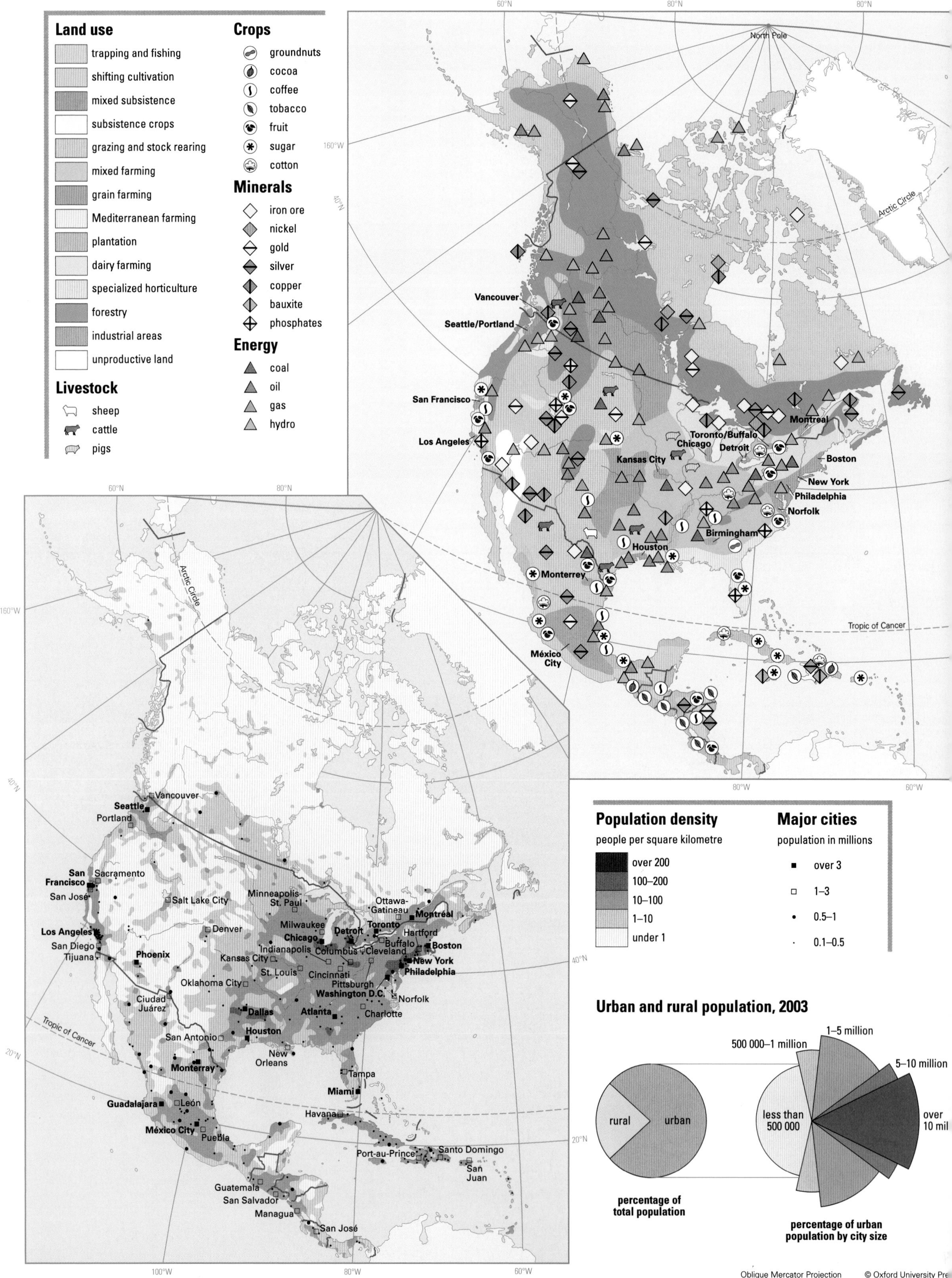

## Land use

- trapping and fishing
- shifting cultivation
- mixed subsistence
- subsistence crops
- grazing and stock rearing
- mixed farming
- grain farming
- Mediterranean farming
- plantation
- dairy farming
- specialized horticulture
- forestry
- industrial areas
- unproductive land

## Livestock

- sheep
- cattle
- pigs

## Crops

- groundnuts
- cocoa
- coffee
- tobacco
- fruit
- sugar
- cotton

## Minerals

- iron ore
- nickel
- gold
- silver
- copper
- bauxite
- phosphates

## Energy

- coal
- oil
- gas
- hydro

## Population density

people per square kilometre

- over 200
- 100–200
- 10–100
- 1–10
- under 1

## Major cities

population in millions

- over 3
- 1–3
- 0.5–1
- 0.1–0.5

## Urban and rural population, 2003

rural    urban

percentage of
total population

1–5 million

500 000–1 million

5–10 million

less than
500 000

over
10 mil

percentage of urban
population by city size

Oblique Mercator Projection    © Oxford University Pre

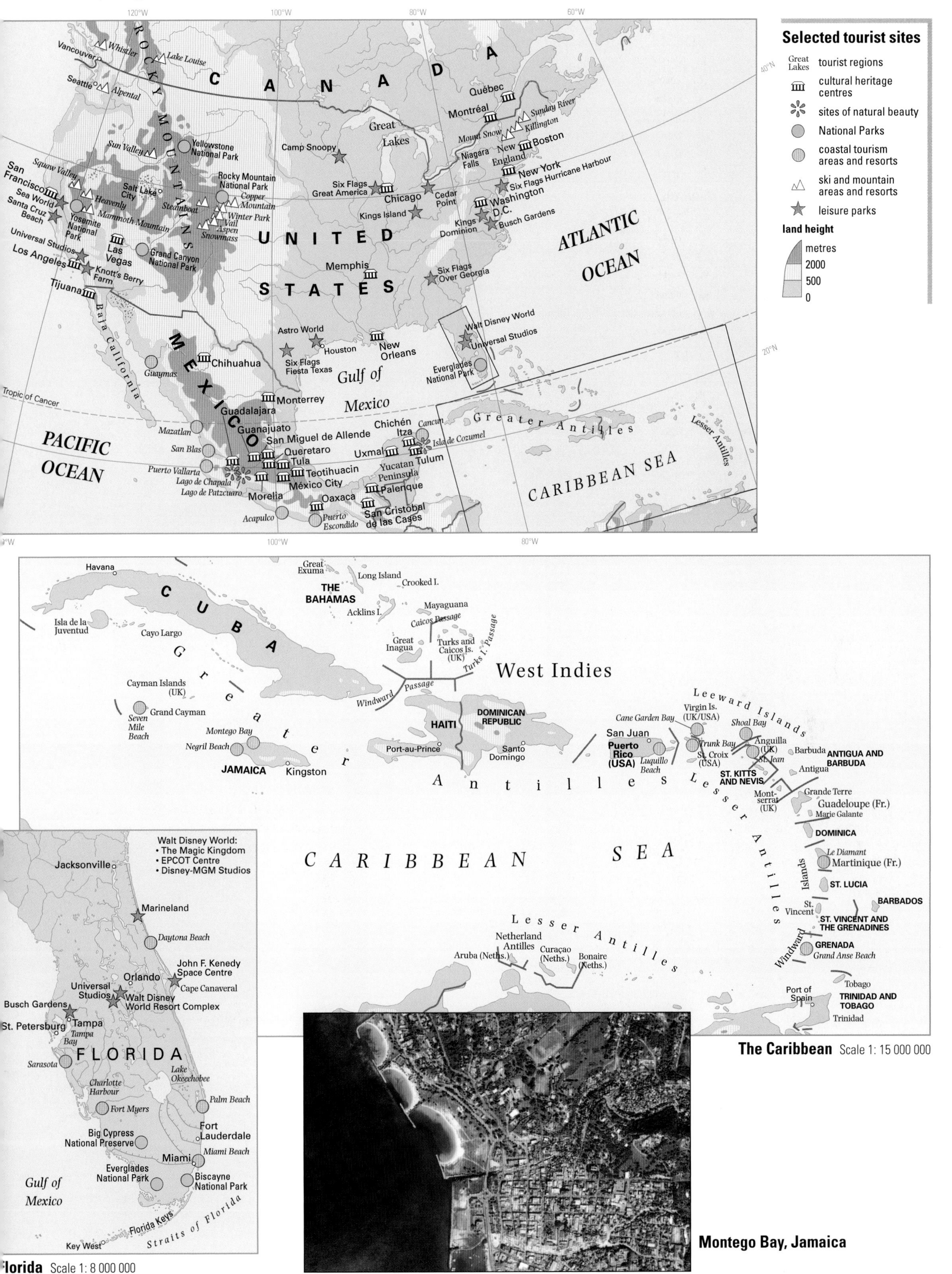

Scale 1: 40 000 000 (main map)

**Selected tourist sites**

Great Lakes — tourist regions

cultural heritage centres

sites of natural beauty

National Parks

coastal tourism areas and resorts

ski and mountain areas and resorts

leisure parks

land height
metres
2000
500
0

**The Caribbean** Scale 1: 15 000 000

**Montego Bay, Jamaica**

Florida Scale 1: 8 000 000

Walt Disney World:
• The Magic Kingdom
• EPCOT Centre
• Disney-MGM Studios

Oblique Mercator Projection   © Oxford University Press

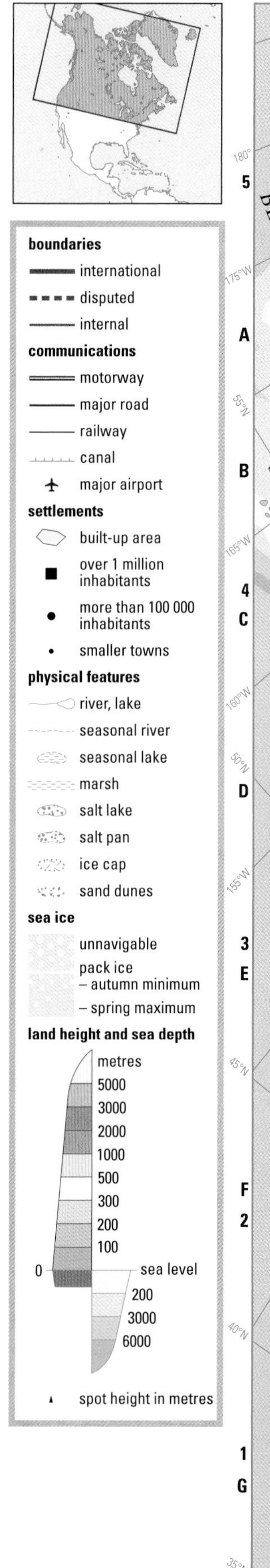

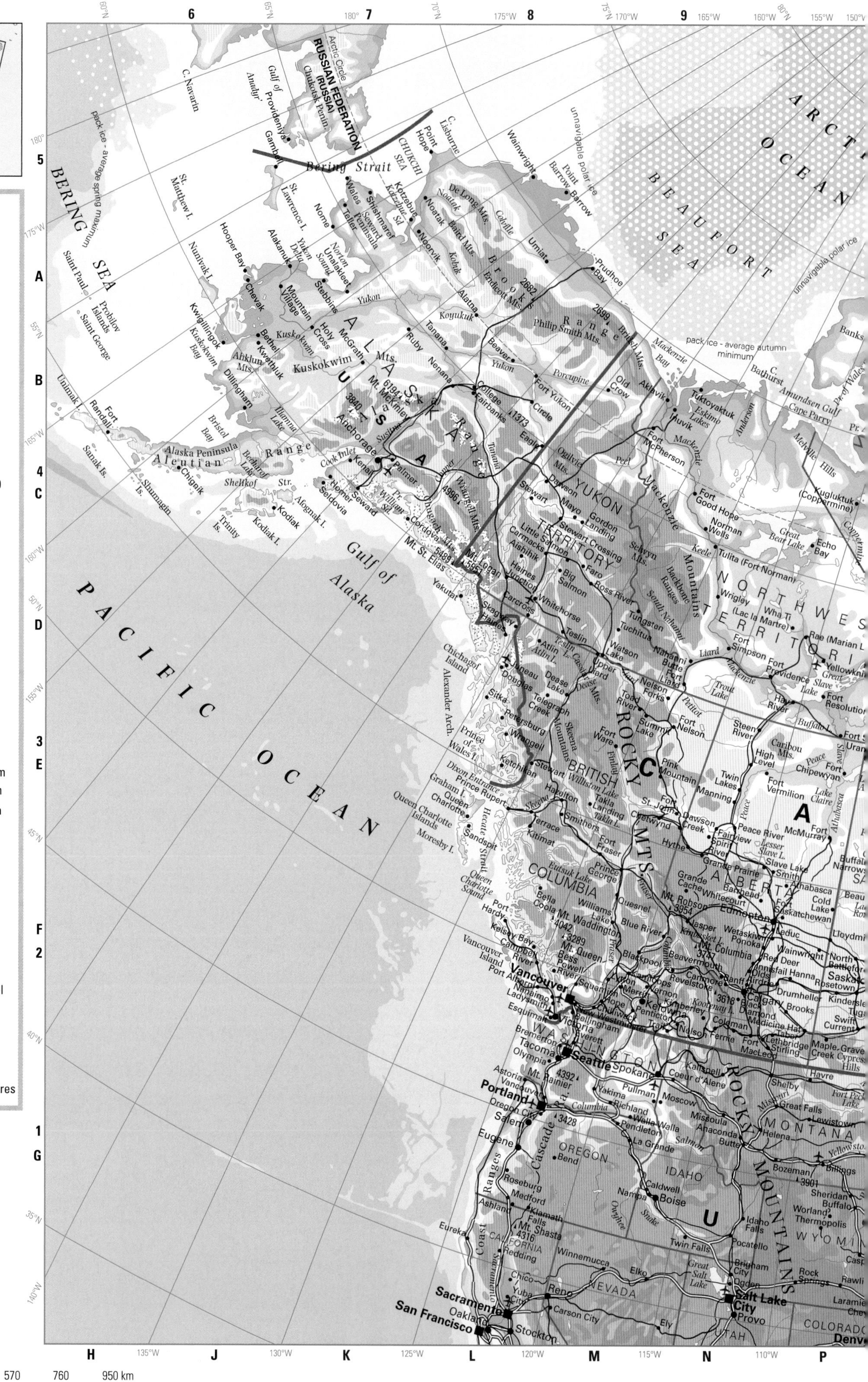

Scale 1: 19 000 000

0 190 380 570 760 950 km

Zenithal Equidistant Projection

© Oxford University Press

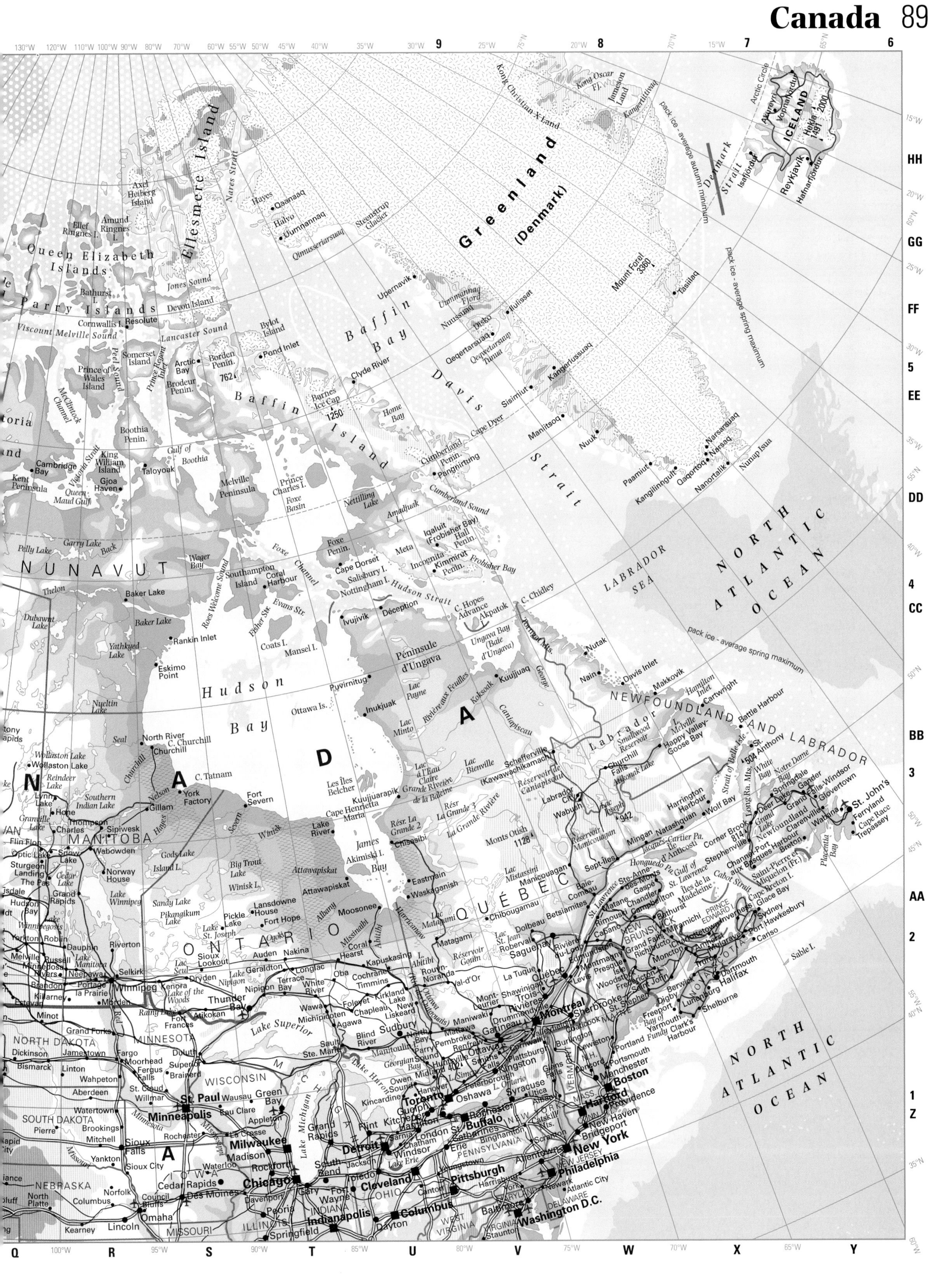
Greenland (Denmark)
Baffin Bay
Davis Strait
Baffin Island
Hudson Bay
Ellesmere Island
Queen Elizabeth Islands
Parry Islands
NUNAVUT
MANITOBA
ONTARIO
QUÉBEC
NEWFOUNDLAND AND LABRADOR
LABRADOR SEA
NORTH ATLANTIC OCEAN
ICELAND
Lake Superior
Lake Huron
Lake Michigan
Montreal
Toronto
Ottawa
Québec
Boston
New York
Philadelphia
Washington D.C.
Chicago
Detroit
Minneapolis
St. Paul
Winnipeg
Halifax
St. John's
Reykjavík

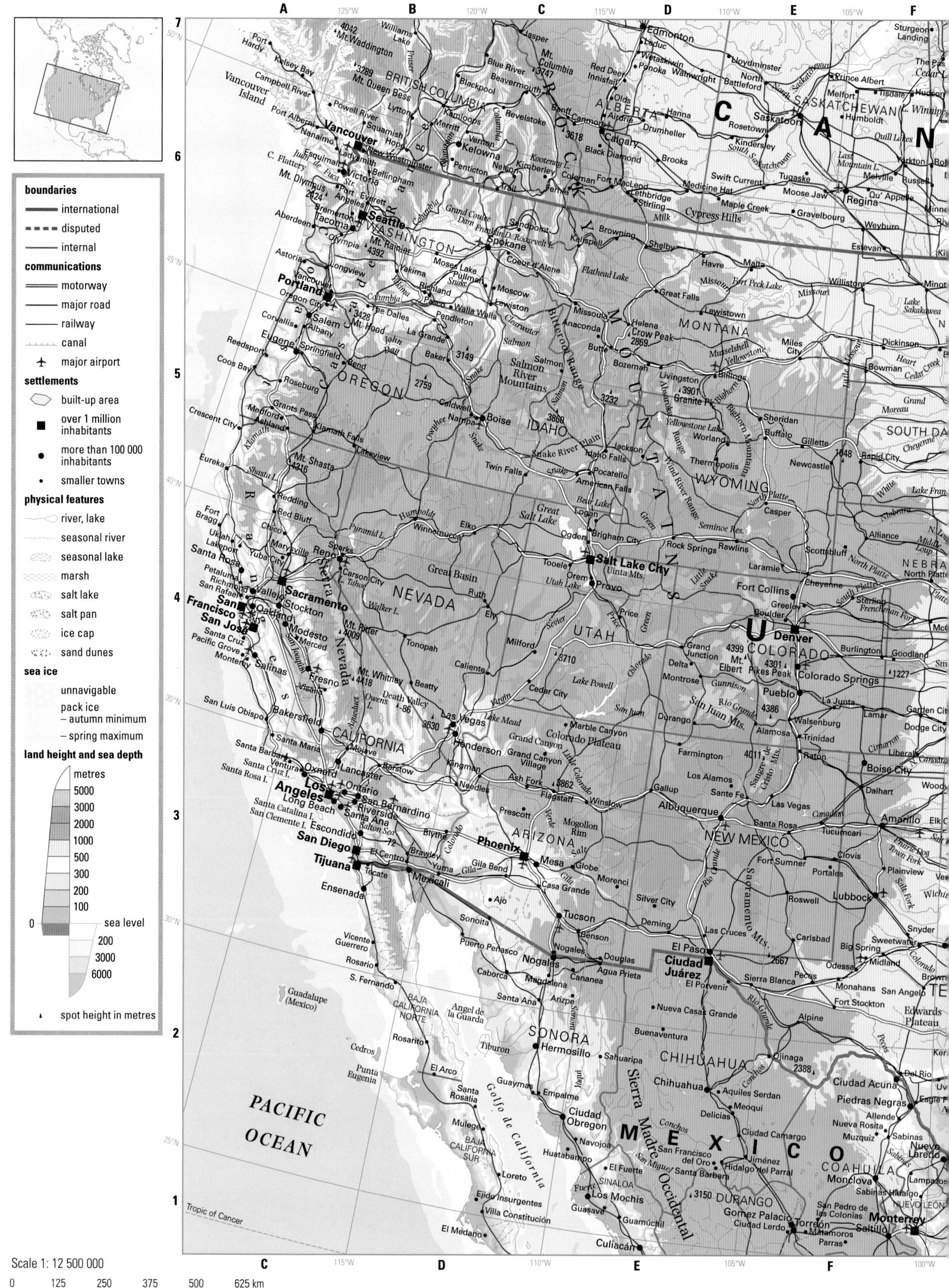

**boundaries**
——— international
- - - disputed
——— internal

**communications**
═══ motorway
——— major road
——— railway
⊥⊥⊥ canal
✈ major airport

**settlements**
⬡ built-up area
■ over 1 million inhabitants
● more than 100 000 inhabitants
• smaller towns

**physical features**
river, lake
seasonal river
seasonal lake
marsh
salt lake
salt pan
ice cap
sand dunes

**sea ice**
unnavigable
pack ice
– autumn minimum
– spring maximum

**land height and sea depth**
metres
5000
3000
2000
1000
500
300
200
100
0 sea level
200
3000
6000

▲ spot height in metres

Scale 1: 12 500 000

0 125 250 375 500 625 km

PACIFIC OCEAN

Conical Orthomorphic Projection

© Oxford University Press

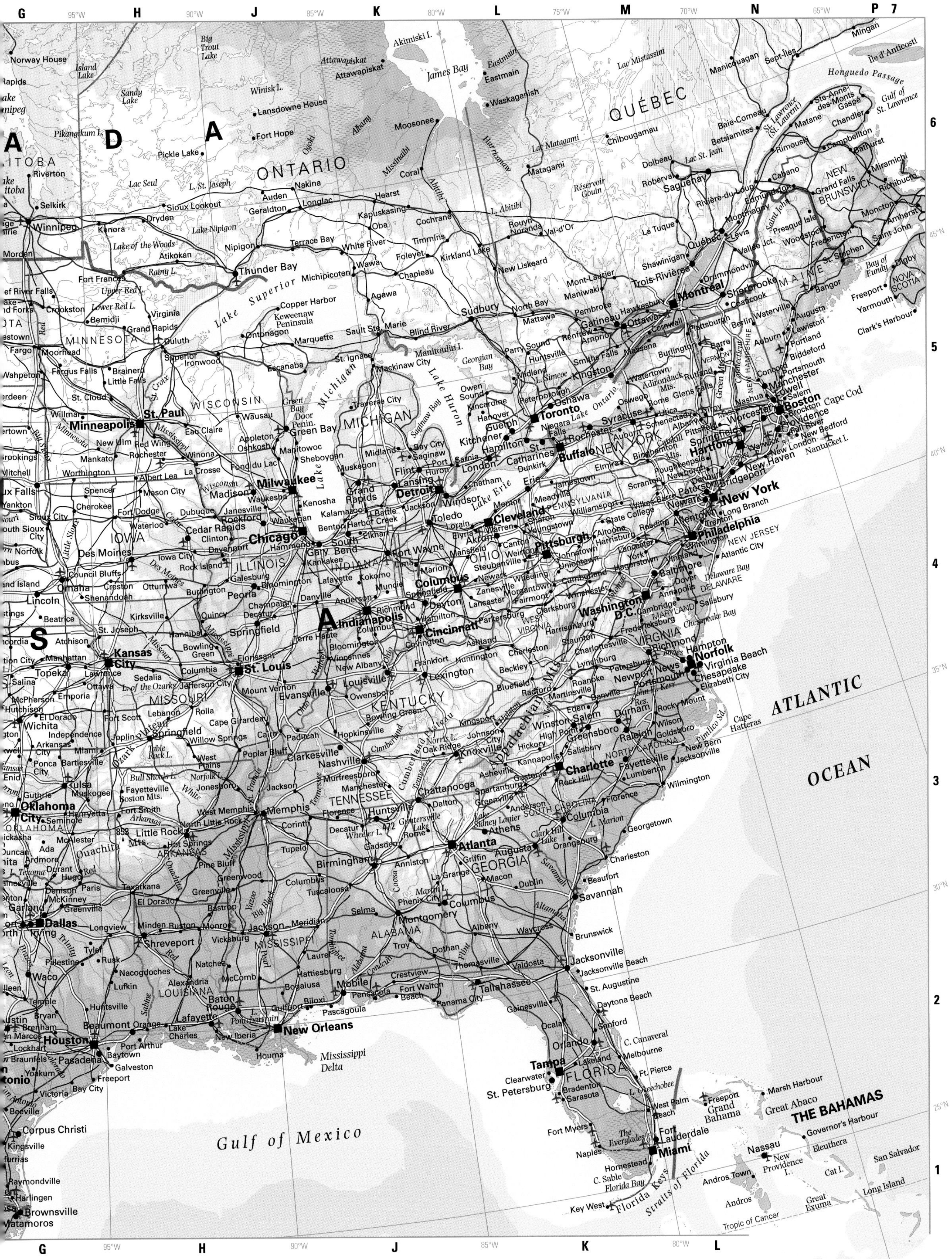

G
H
J
K
L
M
N
P 7
95°W
90°W
85°W
80°W
75°W
70°W
65°W
6
5
4
3
2
1
45°N
40°N
35°N
30°N
25°N
ONTARIO
QUÉBEC
NEW BRUNSWICK
NOVA SCOTIA
MAINE
VERMONT
NEW HAMPSHIRE
NEW YORK
PENNSYLVANIA
NEW JERSEY
DELAWARE
MARYLAND
VIRGINIA
WEST VIRGINIA
NORTH CAROLINA
SOUTH CAROLINA
GEORGIA
FLORIDA
ALABAMA
MISSISSIPPI
LOUISIANA
TENNESSEE
KENTUCKY
OHIO
INDIANA
ILLINOIS
MICHIGAN
WISCONSIN
MINNESOTA
IOWA
MISSOURI
ARKANSAS
OKLAHOMA
THE BAHAMAS
ATLANTIC
OCEAN
Gulf of Mexico
James Bay
Lake Superior
Lake Michigan
Lake Huron
Lake Erie
Lake Ontario
Gulf of St. Lawrence
Straits of Florida
Mississippi Delta
Appalachian Mts.
Ozark Plateau
Cumberland Plateau
Tropic of Cancer
Winnipeg
Thunder Bay
Duluth
Minneapolis
St. Paul
Milwaukee
Chicago
Detroit
Toronto
Montréal
Ottawa
Québec
Buffalo
Cleveland
Pittsburgh
Boston
New York
Philadelphia
Baltimore
Washington D.C.
Richmond
Norfolk
Columbus
Cincinnati
Indianapolis
Louisville
St. Louis
Kansas City
Oklahoma City
Tulsa
Dallas
Houston
New Orleans
Memphis
Nashville
Atlanta
Charlotte
Jacksonville
Tampa
Orlando
Miami
Nassau
Key West
Brownsville
Corpus Christi
Savannah
Charleston
Birmingham
Montgomery
Mobile
Tallahassee
Little Rock
Shreveport
Jackson
Knoxville
Chattanooga
Raleigh
Hartford
Providence
Halifax
Sudbury
Cape Cod
Cape Hatteras

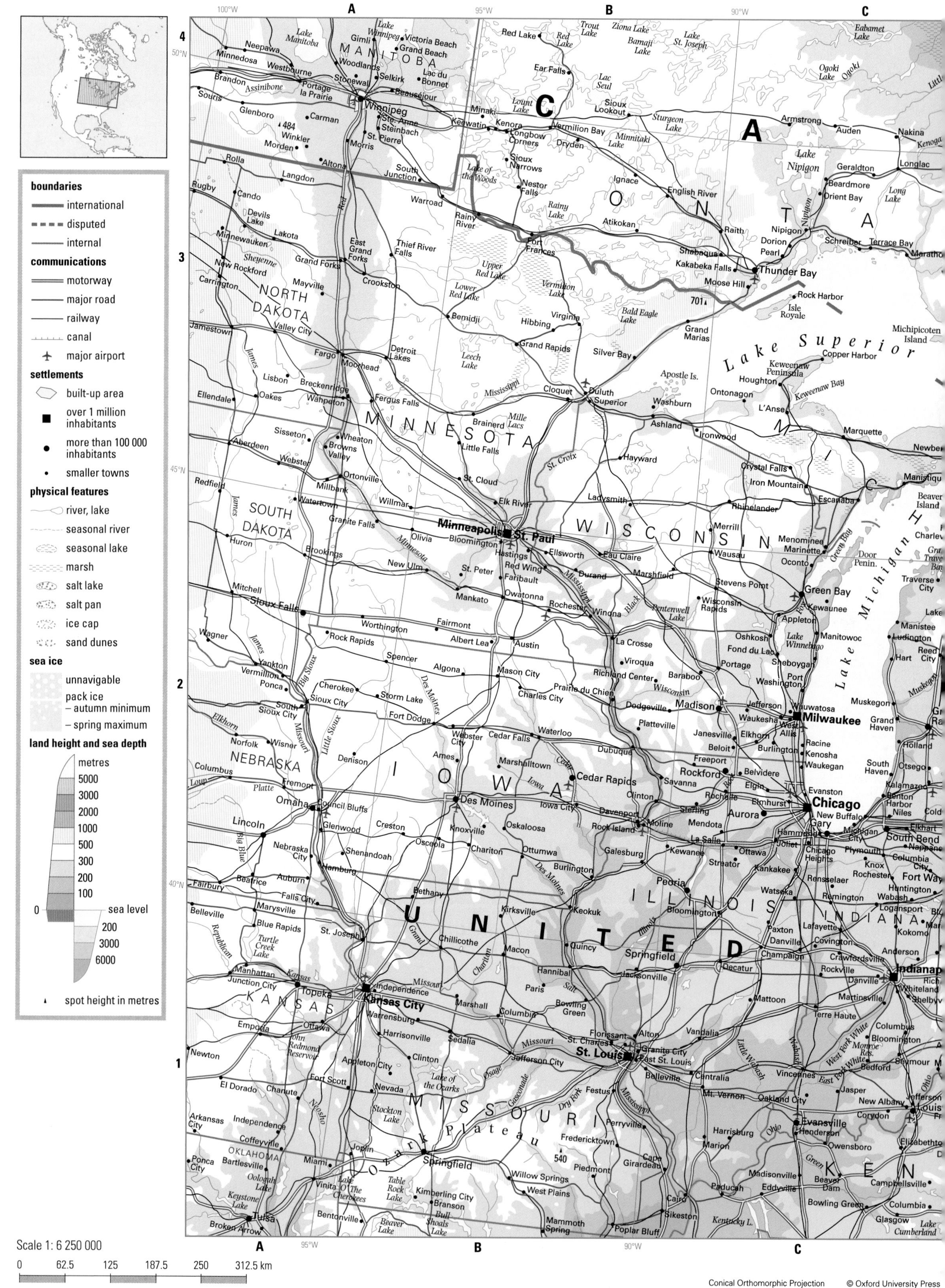
boundaries
international
disputed
internal
communications
motorway
major road
railway
canal
major airport
settlements
built-up area
over 1 million inhabitants
more than 100 000 inhabitants
smaller towns
physical features
river, lake
seasonal river
seasonal lake
marsh
salt lake
salt pan
ice cap
sand dunes
sea ice
unnavigable
pack ice
– autumn minimum
– spring maximum
land height and sea depth
metres
5000
3000
2000
1000
500
300
200
100
0
sea level
200
3000
6000
spot height in metres
A
B
C
100°W
95°W
90°W
50°N
45°N
40°N
1
2
3
4
C A N A D A
MANITOBA
O N T A R I O
NORTH DAKOTA
SOUTH DAKOTA
MINNESOTA
WISCONSIN
MICHIGAN
IOWA
NEBRASKA
KANSAS
MISSOURI
ILLINOIS
INDIANA
OKLAHOMA
KENTUCKY
U N I T E D
Lake Superior
Lake Michigan
Lake Winnipeg
Lake Manitoba
Lake Nipigon
Lake of the Woods
Upper Red Lake
Lower Red Lake
Rainy Lake
Lac Seul
Leech Lake
Mille Lacs
Lake Winnebago
Green Bay
Keweenaw Bay
Isle Royale
Apostle Is.
Ozark Plateau
Lake of the Ozarks
Mississippi
Missouri
Ohio
Wabash
Illinois
Des Moines
Red
Winnipeg
Minneapolis
St. Paul
Duluth
Superior
Thunder Bay
Fargo
Moorhead
Grand Forks
Sioux Falls
Sioux City
Omaha
Council Bluffs
Lincoln
Des Moines
Cedar Rapids
Davenport
Madison
Milwaukee
Green Bay
Chicago
Gary
Rockford
Aurora
Peoria
Springfield
Decatur
Kansas City
Topeka
St. Joseph
Independence
St. Louis
East St. Louis
Springfield
Tulsa
Evansville
Indianapolis
South Bend
Fort Wayne
Rochester
St. Cloud
Marquette
La Crosse
Dubuque
Quincy
Columbia
Jefferson City
Joplin
Scale 1: 6 250 000
0
62.5
125
187.5
250
312.5 km
Conical Orthomorphic Projection
© Oxford University Press

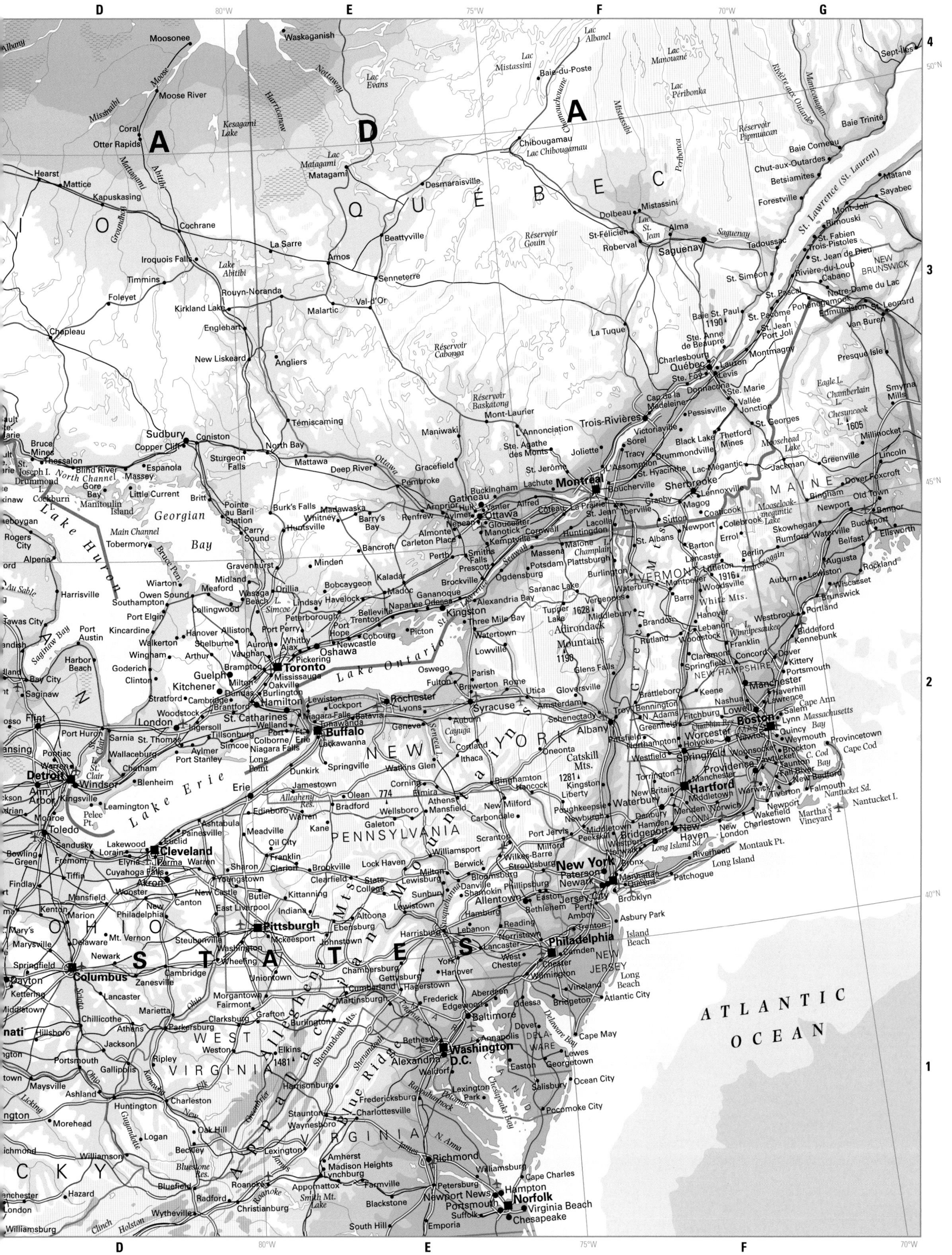
D
E
F
G
80°W
75°W
70°W
50°N
45°N
40°N
4
3
2
1
C A N A D A
Q U É B E C
O
ATLANTIC OCEAN
U N I T E D S T A T E S
NEW YORK
PENNSYLVANIA
OHIO
WEST VIRGINIA
VIRGINIA
NEW HAMPSHIRE
VERMONT
MAINE
MASS.
CONN.
NEW JERSEY
DELAWARE
NEW BRUNSWICK
Lake Huron
Georgian Bay
Lake Ontario
Lake Erie
Lake Abitibi
Lake Simcoe
Lake Champlain
Moosehead Lake
Lac Mistassini
Lac St. Jean
Réservoir Gouin
Réservoir Cabonga
Réservoir Baskatong
Réservoir Pipmuacan
Saguenay
St. Lawrence (St. Laurent)
Ottawa
Appalachian Mountains
Adirondack Mountains
White Mts.
Catskill Mts.
Blue Ridge
Moosonee
Waskaganish
Moose River
Chibougamau
Baie-Comeau
Sept-Îles
Hearst
Kapuskasing
Cochrane
Timmins
Rouyn-Noranda
Val-d'Or
Amos
Sudbury
North Bay
Trois-Rivières
Québec
Montréal
Ottawa
Gatineau
Sherbrooke
Toronto
Hamilton
Kingston
London
Windsor
Detroit
Toledo
Cleveland
Buffalo
Rochester
Syracuse
Albany
Burlington
Portland
Augusta
Bangor
Manchester
Boston
Worcester
Providence
Hartford
New York
Newark
Philadelphia
Pittsburgh
Columbus
Baltimore
Washington D.C.
Richmond
Norfolk
Virginia Beach
Chesapeake
Charleston
Huntington
Harrisburg
Erie
Cape Cod
Nantucket I.
Martha's Vineyard
Long Island
Cape May
Atlantic City

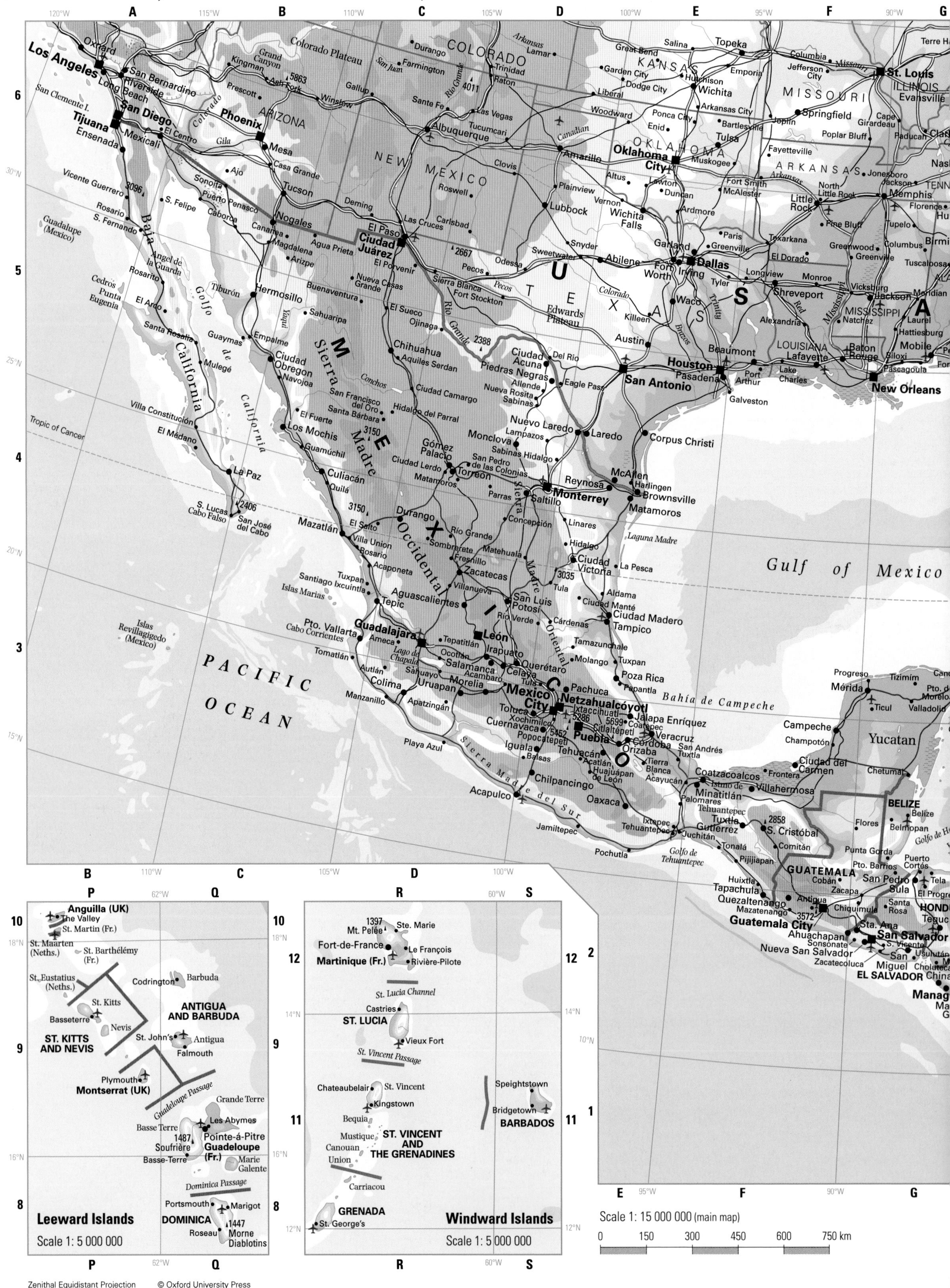

**Leeward Islands**

Scale 1 : 5 000 000

**Windward Islands**

Scale 1 : 5 000 000

Scale 1 : 15 000 000 (main map)

0    150    300    450    600    750 km

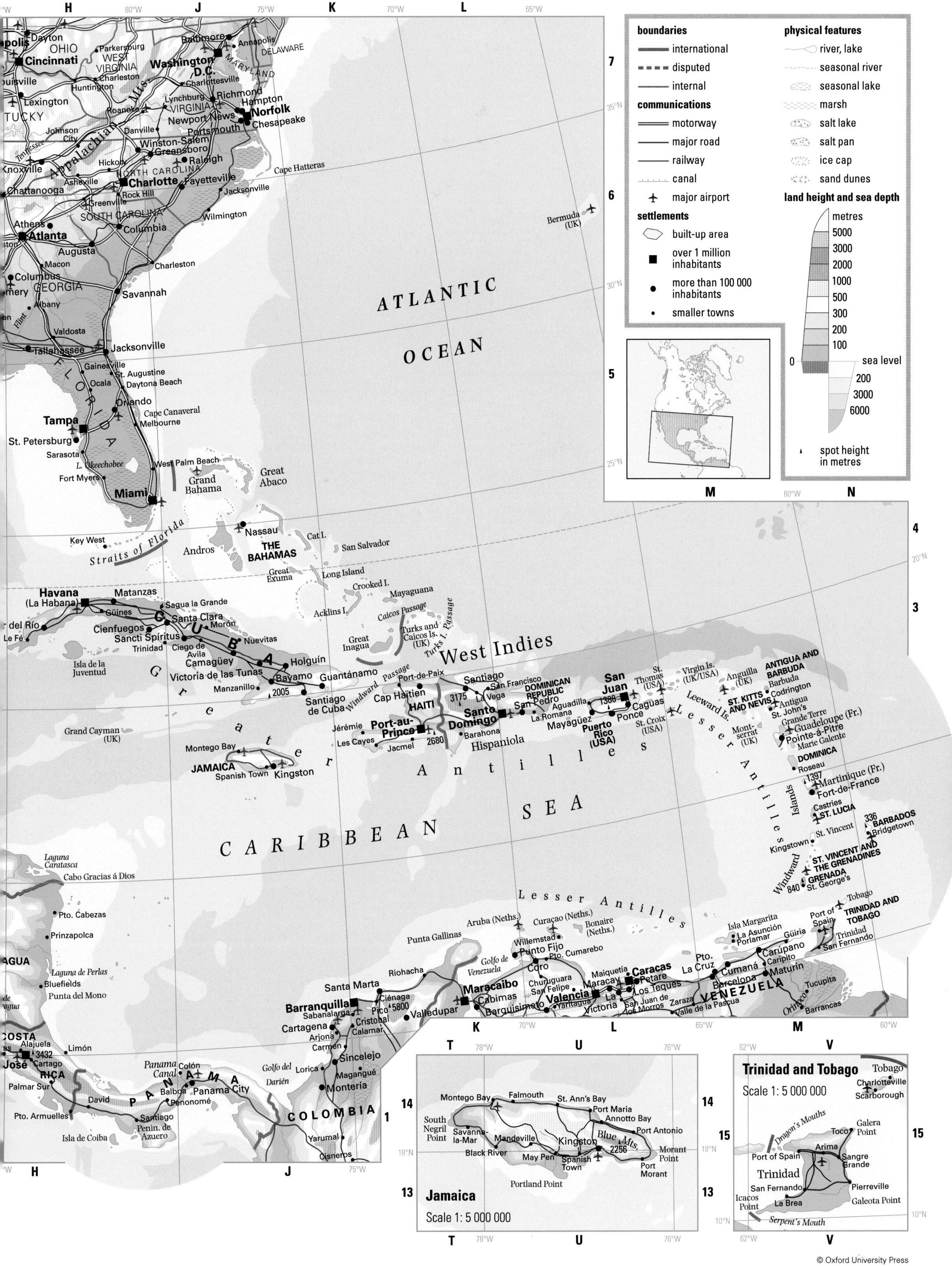
boundaries
international
disputed
internal
communications
motorway
major road
railway
canal
major airport
settlements
built-up area
over 1 million inhabitants
more than 100 000 inhabitants
smaller towns
physical features
river, lake
seasonal river
seasonal lake
marsh
salt lake
salt pan
ice cap
sand dunes
land height and sea depth
metres
5000
3000
2000
1000
500
300
200
100
0
sea level
200
3000
6000
spot height in metres
ATLANTIC OCEAN
CARIBBEAN SEA
West Indies
Greater Antilles
Lesser Antilles
Leeward Is.
Windward Islands
Windward Passage
Caicos Passage
Turks I. Passage
Straits of Florida
Bermuda (UK)
Cape Hatteras
OHIO
Dayton
Cincinnati
Parkersburg
WEST VIRGINIA
Charleston
Huntington
Washington D.C.
Baltimore
Annapolis
DELAWARE
MARYLAND
Charlottesville
Lexington
Lynchburg
Richmond
Roanoke
VIRGINIA
Hampton
Norfolk
Newport News
Chesapeake
Portsmouth
Johnson City
Danville
Winston-Salem
Greensboro
Raleigh
Appalachian Mts.
Tennessee
Knoxville
Hickory
Asheville
Charlotte
NORTH CAROLINA
Fayetteville
Jacksonville
Chattanooga
Greenville
Rock Hill
SOUTH CAROLINA
Wilmington
Columbia
Athens
Atlanta
Augusta
Macon
Charleston
Columbus
GEORGIA
Savannah
Albany
Flint
Valdosta
Tallahassee
Jacksonville
Gainesville
St. Augustine
Ocala
Daytona Beach
FLORIDA
Orlando
Cape Canaveral
Melbourne
Tampa
St. Petersburg
Sarasota
L. Okeechobee
West Palm Beach
Fort Myers
Miami
Grand Bahama
Great Abaco
Key West
Nassau
Andros
THE BAHAMAS
Cat I.
San Salvador
Great Exuma
Long Island
Crooked I.
Mayaguana
Acklins I.
Great Inagua
Turks and Caicos Is. (UK)
Havana (La Habana)
Matanzas
Güines
Sagua la Grande
Santa Clara
Cienfuegos
Morón
Sancti Spíritus
Trinidad
Ciego de Avila
Nuevitas
CUBA
Camagüey
Holguín
Isla de la Juventud
Victoria de las Tunas
Bayamo
Guantánamo
Manzanillo
2005
Santiago de Cuba
Grand Cayman (UK)
Montego Bay
JAMAICA
Spanish Town
Kingston
Port-de-Paix
Cap Haïtien
HAITI
Santiago
San Francisco
La Vega
DOMINICAN REPUBLIC
3175
Santo Domingo
San Pedro
La Romana
Jérémie
Port-au-Prince
Les Cayes
Jacmel
2680
Barahona
Hispaniola
San Juan
Aguadilla
1338
Mayagüez
Puerto Rico (USA)
Ponce
Caguas
St. Thomas (USA)
St. Croix (USA)
Virgin Is. (UK/USA)
Anguilla (UK)
ANTIGUA AND BARBUDA
Barbuda
Codrington
ST. KITTS AND NEVIS
Antigua
St. John's
Montserrat (UK)
Grande Terre
Guadeloupe (Fr.)
Pointe-à-Pitre
Marie Galente
DOMINICA
Roseau
1397
Martinique (Fr.)
Fort-de-France
Castries
ST. LUCIA
336
BARBADOS
Bridgetown
St. Vincent
Kingstown
ST. VINCENT AND THE GRENADINES
GRENADA
840
St. George's
Tobago
TRINIDAD AND TOBAGO
Port of Spain
Trinidad
San Fernando
Laguna Caratasca
Cabo Gracias á Dios
Pto. Cabezas
Prinzapolca
Laguna de Perlas
Bluefields
Punta del Mono
Alajuela
3432
San José
Cartago
COSTA RICA
Limón
Palmar Sur
Pto. Armuelles
David
Isla de Coiba
Panama Canal
Colón
Balboa
Panama City
Penonomé
PANAMA
Santiago
Penin. de Azuero
Golfo del Darién
Aruba (Neths.)
Curaçao (Neths.)
Bonaire (Neths.)
Punta Gallinas
Willemstad
Punto Fijo
Pto. Cumarebo
Golfo de Venezuela
Coro
Riohacha
Santa Marta
Ciénaga
Pico 5800
Barranquilla
Sabanalarga
Cristobal
Calamar
Cartagena
Arjona
Carmen
Valledupar
Maracaibo
Cabimas
Barquisimeto
Churuguara
San Felipe
Yaritagua
Valencia
Maracay
Maiquetía
Caracas
Petare
Los Teques
La Victoria
San Juan de los Morros
Zaraza
Valle de la Pascua
VENEZUELA
La Cruz
Pto.
Barcelona
Cumaná
Isla Margarita
La Asunción
Porlamar
Güiria
Carúpano
Caripito
Maturín
Tucupita
Orinoco
Barrancas
Sincelejo
Lorica
Magangué
Montería
COLOMBIA
Yarumal
Cisneros
Jamaica
Scale 1: 5 000 000
Montego Bay
Falmouth
St. Ann's Bay
Port Maria
Annotto Bay
Port Antonio
South Negril Point
Savanna-la-Mar
Mandeville
Black River
May Pen
Kingston
Spanish Town
Blue Mts.
2256
Morant Point
Port Morant
Portland Point
Trinidad and Tobago
Scale 1: 5 000 000
Tobago
Charlotteville
Scarborough
Dragon's Mouths
Toco
Galera Point
Port of Spain
Arima
Sangre Grande
Trinidad
San Fernando
Pierreville
Icacos Point
La Brea
Galeota Point
Serpent's Mouth

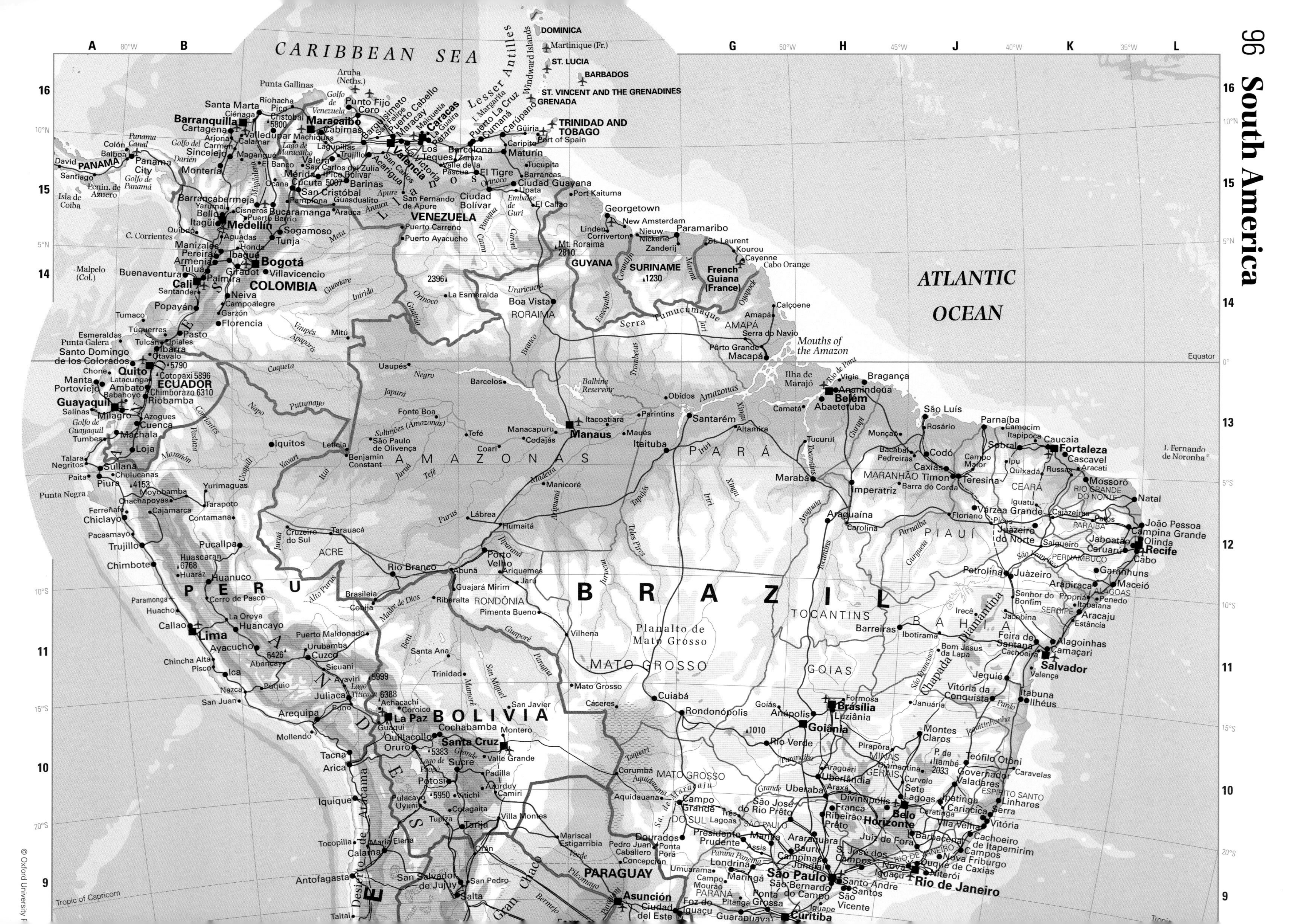

ATLANTIC OCEAN

CARIBBEAN SEA

*Lesser Antilles*

DOMINICA
ST. LUCIA
BARBADOS
ST. VINCENT AND THE GRENADINES
GRENADA
TRINIDAD AND TOBAGO
Port of Spain

Windward Islands
Martinique (Fr.)

Aruba (Neths.)

PANAMA
Panama City
Panama Canal
Colón
Balboa

COLOMBIA
Bogotá
Medellín
Cali
Barranquilla
Cartagena
Santa Marta

VENEZUELA
Caracas
Valencia
Maracaibo
Ciudad Bolívar

GUYANA
Georgetown

SURINAME
Paramaribo

French Guiana (France)
Cayenne

ECUADOR
Quito
Guayaquil

PERU
Lima
Callao
Cuzco
Arequipa
Iquitos

BRAZIL
Manaus
Belém
Fortaleza
Recife
Salvador
Brasília
Goiânia
Belo Horizonte
Rio de Janeiro
São Paulo
Curitiba
São Luís
Teresina
Natal
João Pessoa
Maceió

BOLIVIA
La Paz
Santa Cruz
Cochabamba
Sucre
Oruro
Potosí

PARAGUAY
Asunción
Ciudad del Este

A M A Z O N A S

M A T O G R O S S O

Planalto de Mato Grosso

Mouths of the Amazon

Tropic of Capricorn

© Oxford University P

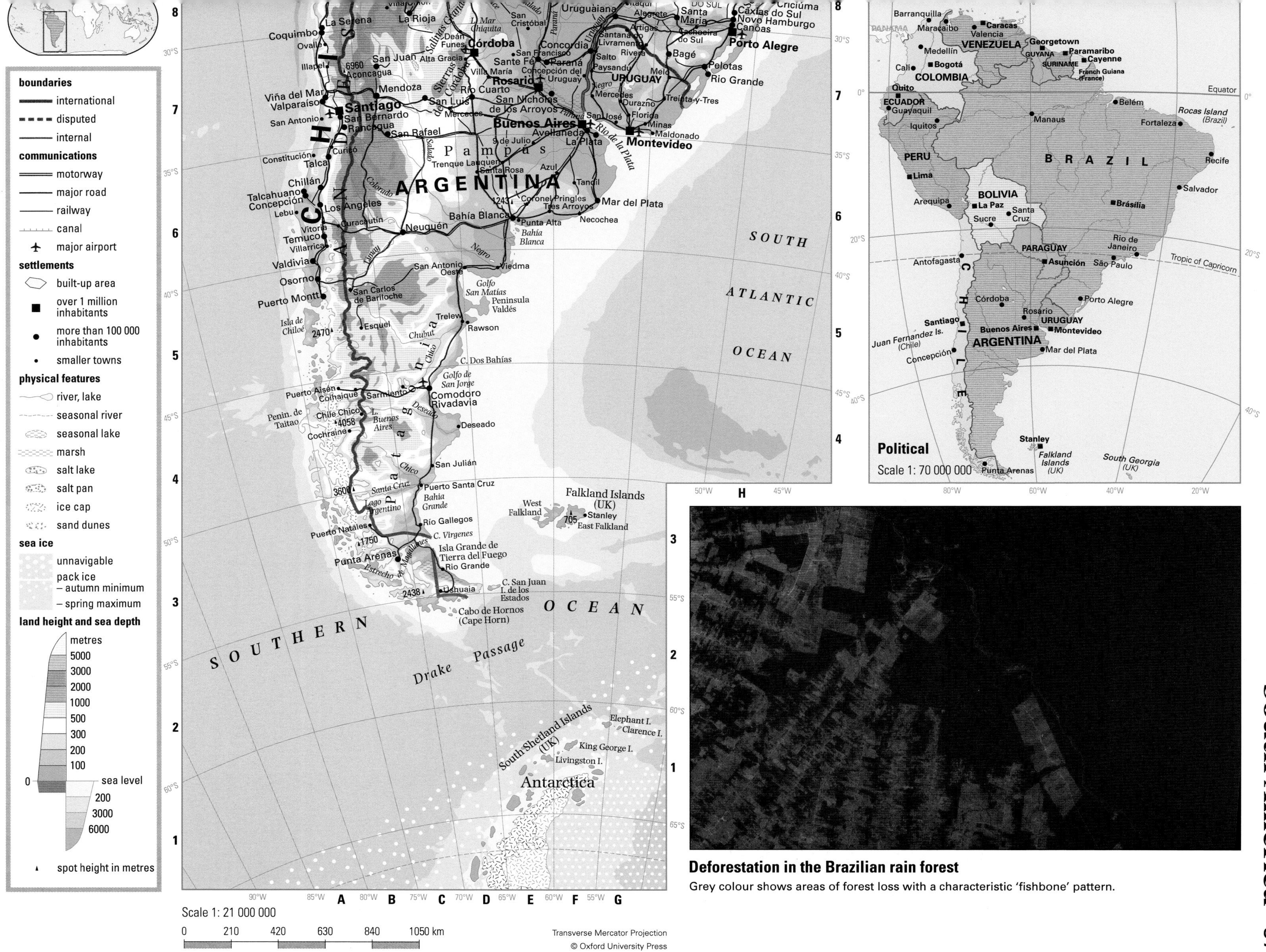

**Deforestation in the Brazilian rain forest**
Grey colour shows areas of forest loss with a characteristic 'fishbone' pattern.

Political
Scale 1: 70 000 000

Scale 1: 21 000 000

Transverse Mercator Projection
© Oxford University Press

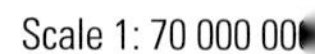

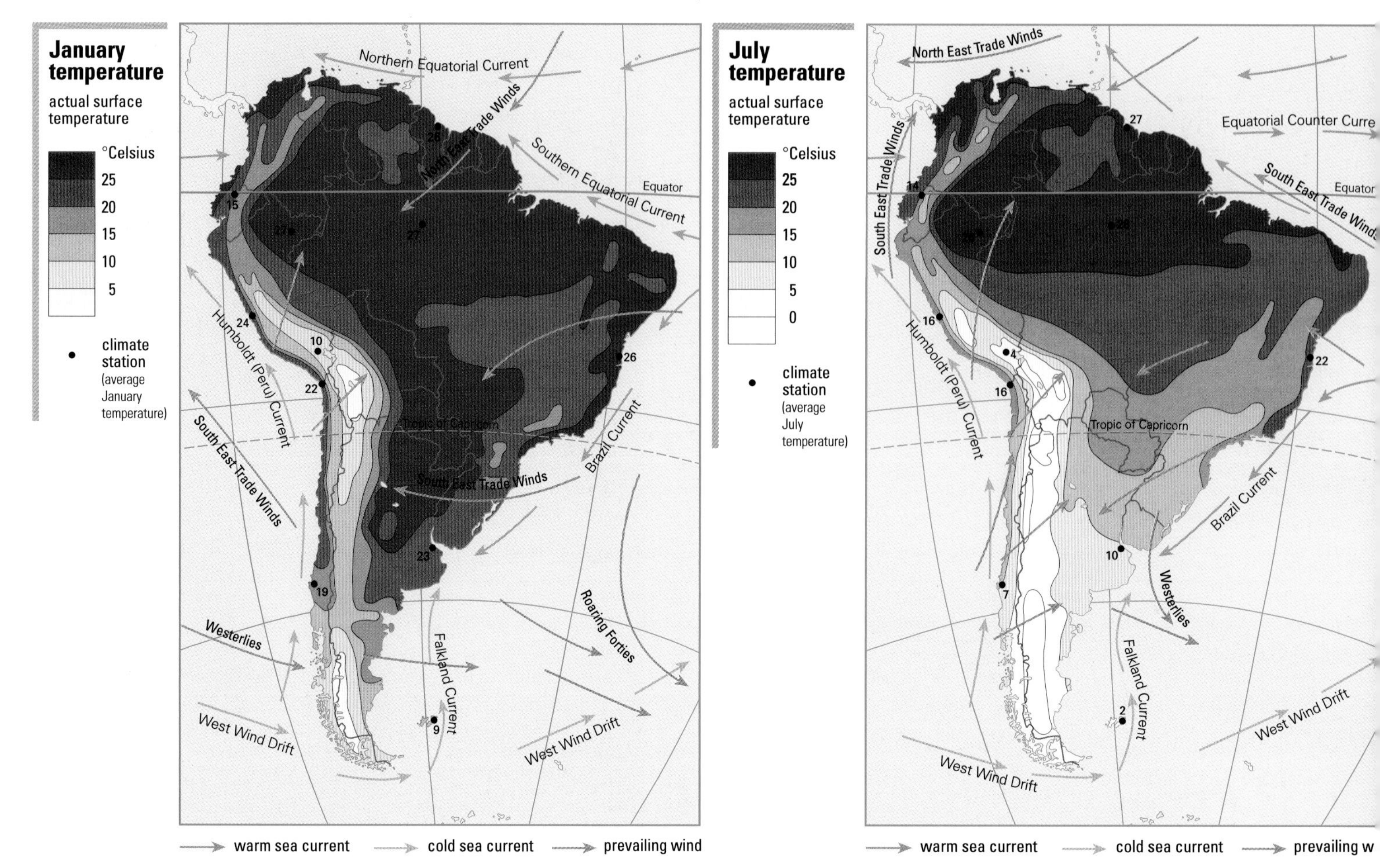

## January temperature

actual surface temperature

°Celsius
25
20
15
10
5

• climate station (average January temperature)

Northern Equatorial Current
North East Trade Winds
Southern Equatorial Current
Equator
Humboldt (Peru) Current
South East Trade Winds
Tropic of Capricorn
South East Trade Winds
Brazil Current
Westerlies
Roaring Forties
West Wind Drift
Falkland Current
West Wind Drift

⟶ warm sea current ⟶ cold sea current ⟶ prevailing wind

## July temperature

actual surface temperature

°Celsius
25
20
15
10
5
0

• climate station (average July temperature)

North East Trade Winds
Equatorial Counter Curre
South East Trade Winds
Equator
South East Trade Winds
Humboldt (Peru) Current
Tropic of Capricorn
Brazil Current
Westerlies
Falkland Current
West Wind Drift
West Wind Drift

⟶ warm sea current ⟶ cold sea current ⟶ prevailing w

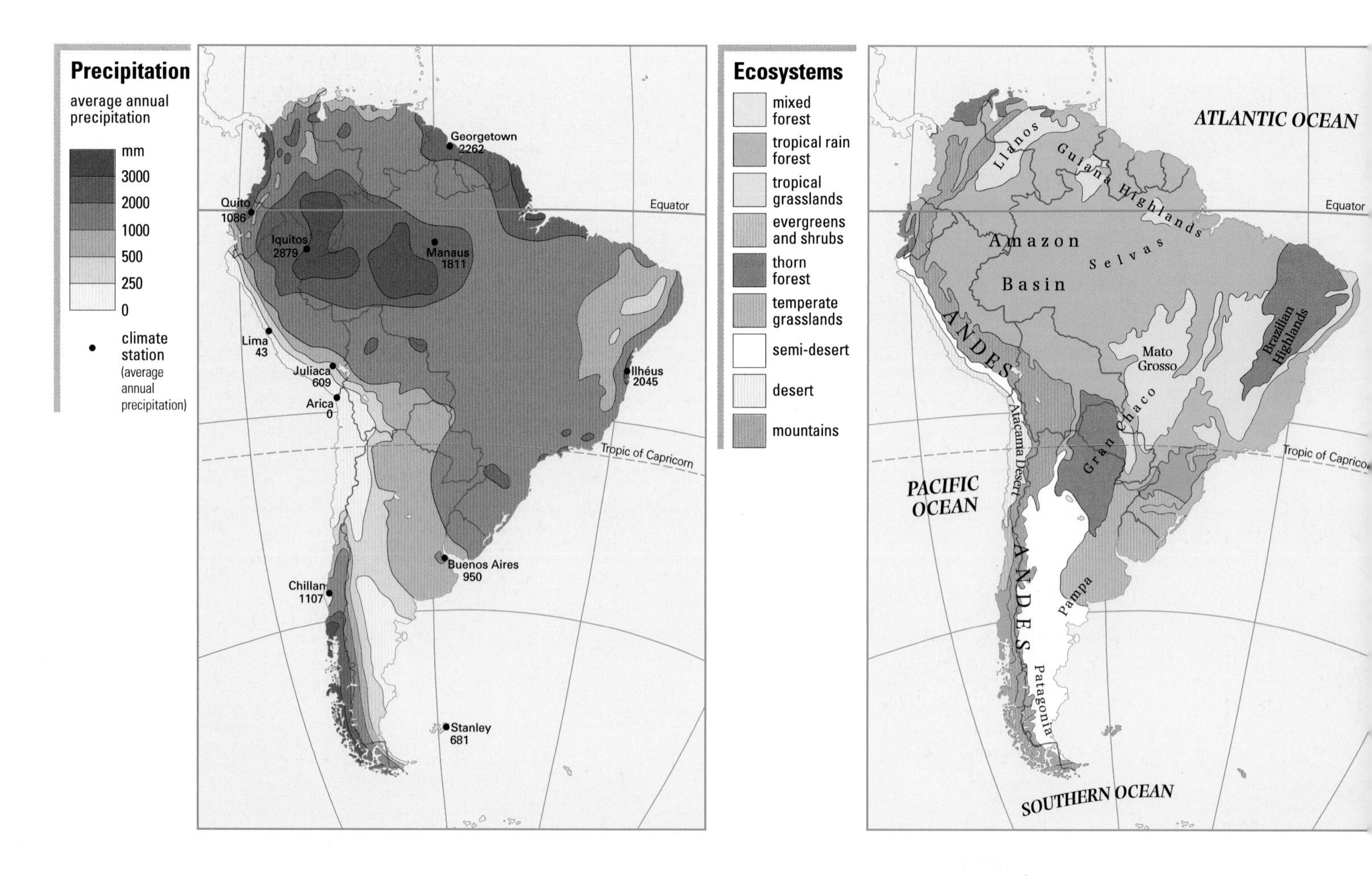

## Precipitation

average annual precipitation

mm
3000
2000
1000
500
250
0

• climate station (average annual precipitation)

Georgetown 2262
Equator
Quito 1086
Iquitos 2879
Manaus 1811
Lima 43
Juliaca 609
Arica 0
Ilhéus 2045
Tropic of Capricorn
Buenos Aires 950
Chillan 1107
Stanley 681

## Ecosystems

mixed forest
tropical rain forest
tropical grasslands
evergreens and shrubs
thorn forest
temperate grasslands
semi-desert
desert
mountains

ATLANTIC OCEAN
Llanos
Guiana Highlands
Equator
Amazon Basin
Selvas
ANDES
Mato Grosso
Brazilian Highlands
Atacama Desert
Gran Chaco
Tropic of Capricorn
PACIFIC OCEAN
Pampa
ANDES
Patagonia
SOUTHERN OCEAN

Oblique Mercator Projection   © Oxford University

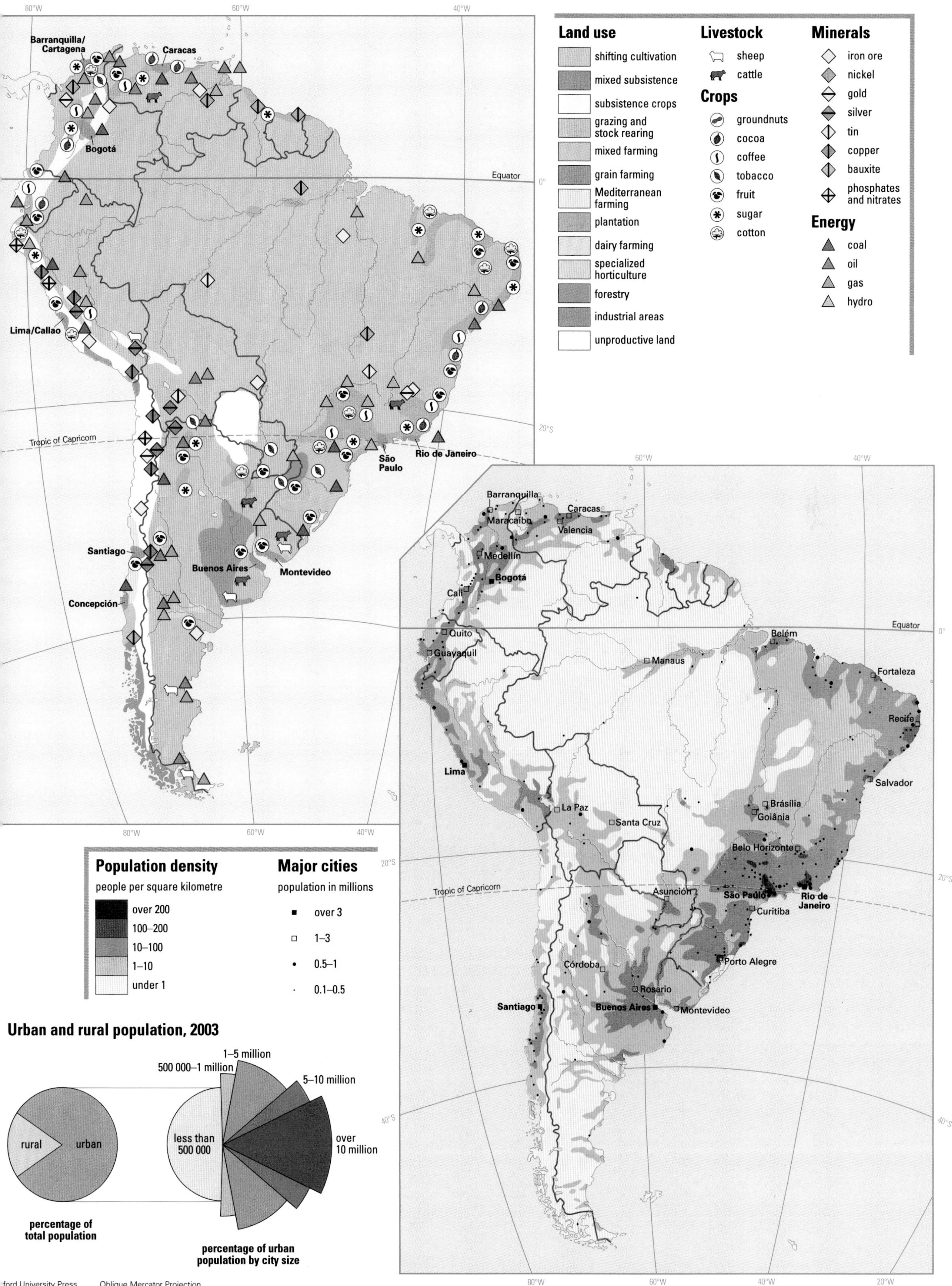

**Land use**

shifting cultivation

mixed subsistence

subsistence crops

grazing and
stock rearing

mixed farming

grain farming

Mediterranean
farming

plantation

dairy farming

specialized
horticulture

forestry

industrial areas

unproductive land

**Livestock**

sheep

cattle

**Crops**

groundnuts

cocoa

coffee

tobacco

fruit

sugar

cotton

**Minerals**

iron ore

nickel

gold

silver

tin

copper

bauxite

phosphates
and nitrates

**Energy**

coal

oil

gas

hydro

**Population density**

people per square kilometre

over 200

100–200

10–100

1–10

under 1

**Major cities**

population in millions

over 3

1–3

0.5–1

0.1–0.5

**Urban and rural population, 2003**

rural  urban

percentage of
total population

less than
500 000

500 000–1 million

1–5 million

5–10 million

over
10 million

percentage of urban
population by city size

# Pacific Ocean

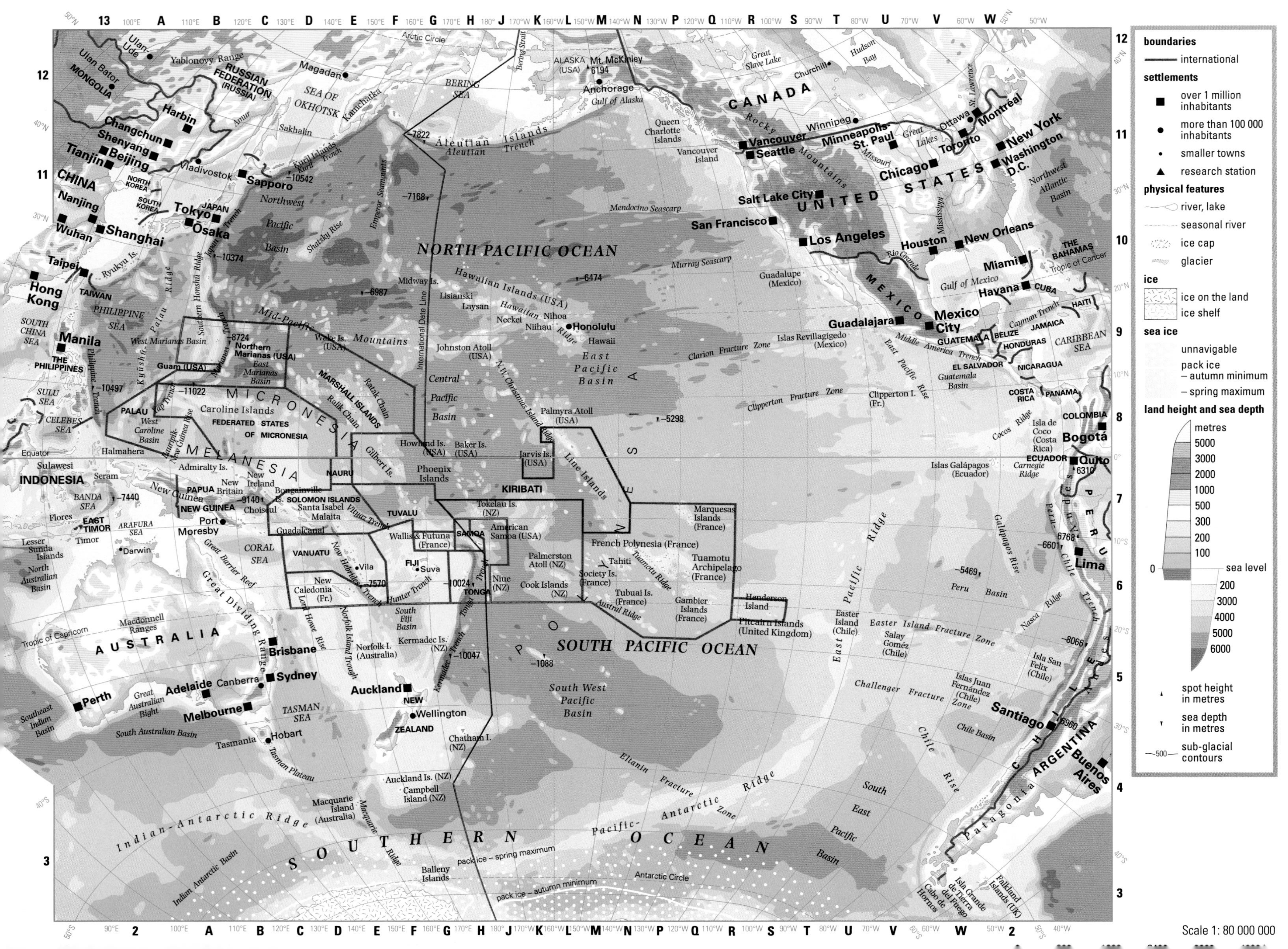

boundaries
international
settlements
over 1 million inhabitants
more than 100 000 inhabitants
smaller towns
research station
physical features
river, lake
seasonal river
ice cap
glacier
ice
ice on the land
ice shelf
sea ice
unnavigable
pack ice – autumn minimum – spring maximum
land height and sea depth
metres
5000
3000
2000
1000
500
300
200
100
0
sea level
200
3000
4000
5000
6000
spot height in metres
sea depth in metres
sub-glacial contours
Scale 1: 80 000 000
NORTH PACIFIC OCEAN
SOUTH PACIFIC OCEAN
SOUTHERN OCEAN
CANADA
UNITED STATES
MEXICO
PERU
ARGENTINA
CHILE
COLOMBIA
ECUADOR
AUSTRALIA
NEW ZEALAND
PAPUA NEW GUINEA
INDONESIA
THE PHILIPPINES
CHINA
JAPAN
RUSSIAN FEDERATION (RUSSIA)
MONGOLIA
MICRONESIA
MELANESIA
POLYNESIA
KIRIBATI
TUVALU
FIJI
VANUATU
TONGA
SAMOA
NAURU
PALAU
MARSHALL ISLANDS
FEDERATED STATES OF MICRONESIA
SOLOMON ISLANDS
Hawaiian Islands (USA)
Honolulu
Line Islands
French Polynesia (France)
Pitcairn Islands (United Kingdom)
Easter Island (Chile)
Islas Galápagos (Ecuador)
East Pacific Rise
Pacific-Antarctic Ridge
Peru-Chile Trench
Aleutian Trench
International Date Line
Equator
Tropic of Capricorn
Antarctic Circle
Arctic Circle
Los Angeles
San Francisco
Vancouver
Seattle
Anchorage
Mexico City
Lima
Santiago
Buenos Aires
Bogotá
Quito
Sydney
Melbourne
Brisbane
Auckland
Wellington
Tokyo
Ōsaka
Shanghai
Beijing
Manila
Hong Kong
Taipei

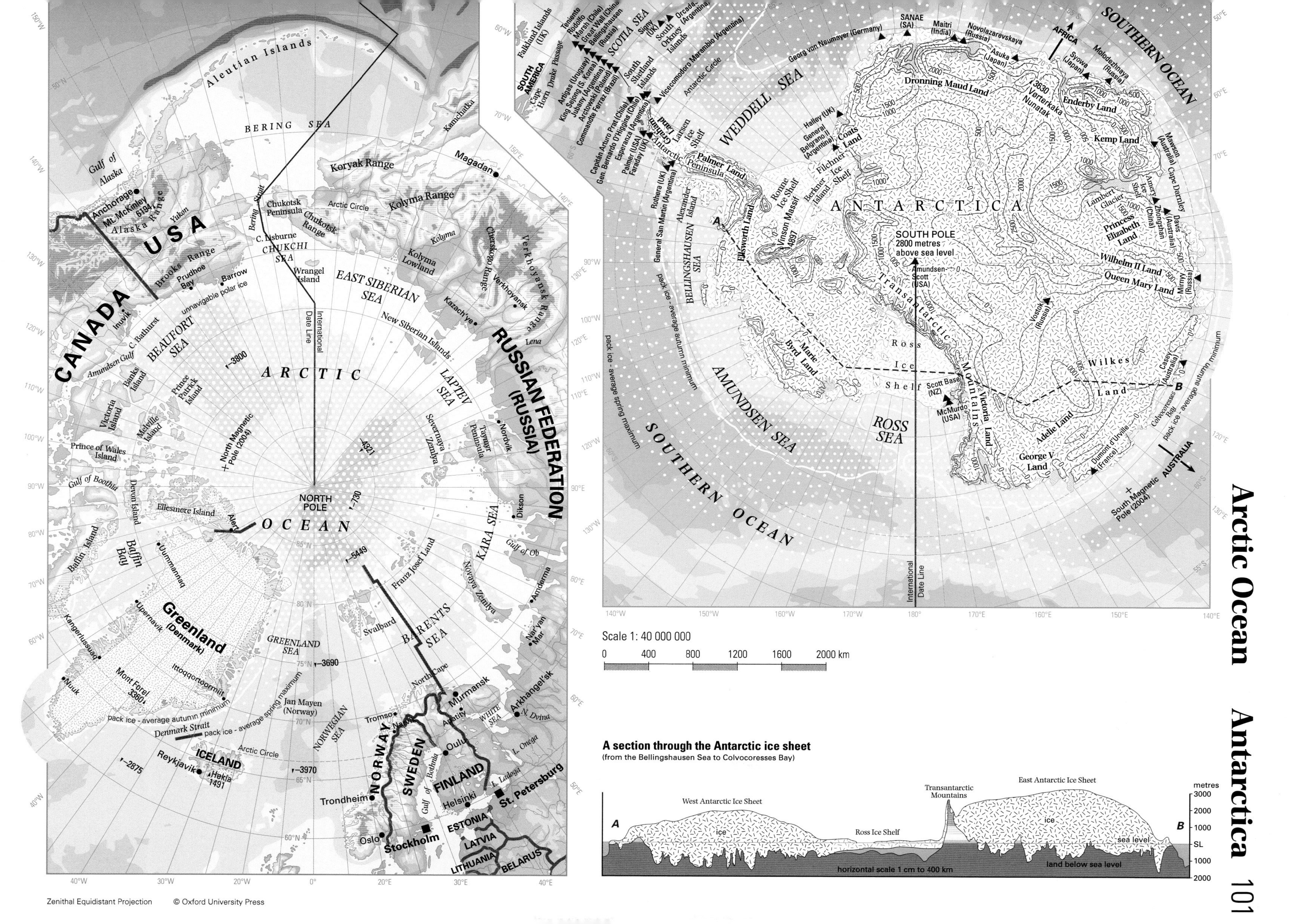

**A section through the Antarctic ice sheet**
(from the Bellingshausen Sea to Colvocoresses Bay)

Scale 1 : 40 000 000

Zenithal Equidistant Projection

© Oxford University Press

Equatorial scale 1: 95 000 000 (main map)

international boundary
• capital city

**10** 180° A 160°W B 140°W C 120°W D 100°W E 80°W F 60°W G 40°W H 20°W

80°N

**9** Arctic Circle

Greenland (Denmark)

USA

• Nuuk       Reykjavik • ICELAND

60°N

C A N A D A

**8**

Ottawa •

40°N

UNITED STATES OF AMERICA

• Washington D.C.

*N O R T H*

REPUBLIC OF IRELAND
Dublin •

Azores (Portugal)      PORTUGAL
Lisbon •

Madeira (Portugal)      MORO

**7**

*A T L A N T I C*

Tropic of Cancer

20°N

Bermuda (UK)

*O C E A N*

Canary Islands (Spain)

Laayoune
WESTERN SAHARA

Hawaiian Islands (USA) •

MEXICO

THE BAHAMAS

MAURITANIA

• Havana      CUBA
• Mexico City      JAMAICA      HAITI      DOMINICAN REPUBLIC

• Nouakchott

**6**

BELIZE      • Kingston      Puerto Rico (USA)      ANTIGUA AND BARBUDA
Belmopan      ST. KITTS AND NEVIS      DOMINICA

CAPE VERDE      Dakar •      SENEGAL
THE GAMBIA      • Bamako

GUATEMALA      HONDURAS      ST. LUCIA
Guatemala City      • Tegucigalpa      ST. VINCENT AND      BARBADOS
San Salvador •      THE GRENADINES      GRENADA
EL SALVADOR      NICARAGUA

GUINEA-BISSAU      GUINEA
Conakry •      CO
SIERRA LEONE      D'IV
Freetown •      Yamous
Monrovia •
LIBERIA

Managua •
COSTA      • San José      Caracas •      TRINIDAD AND TOBAGO
RICA      • Panama City
PANAMA      VENEZUELA

Georgetown •
COLOMBIA      GUYANA      SURINAME
• Bogotá      Paramaribo •      • Cayenne
French Guiana (France)

Galapagos Islands (Ecuador) •      Quito •
ECUADOR

0° Equator

*P A C I F I C*

KIRIBATI

**5**

American Samoa

SAMOA

*O C E A N*

PERU

• Lima

B R A Z I L

Ascension Islan

French Polynesia (France)

• La Paz      • Brasília

St. Helena

Cook Islands (New Zealand)

BOLIVIA

20°S

TONGA
Tropic of Capricorn

Pitcairn Island (UK)

PARAGUAY

*S O U T H*

**4**

Easter Island (Chile)

• Asunción

CHILE

*A T L A N T I*

URUGUAY
• Santiago      Buenos Aires •      • Montevideo

*O C E A N*

40°S

Chatham Islands (NZ) •

**3**

ARGENTINA

Tristan da Cunha (UK) •

Falkland Islands (UK) •

South Georgia (UK) •

A      B      C      D      E      F      G      H
160°W   140°W   120°W   100°W   80°W   60°W   40°W   20°W

Antarctic Circle

A N T A      A

A 160°W B 140°W C 120°W D 100°W E 80°W F 60°W G 40°W H 20°W

40°W      20°W      undefined

N O R W A Y

UNITED KINGDOM
ARGENTINA

Antarctic Circle

60°W

CHILE

Prime Meridian

AUSTRALIA

80°W

80°E

A N T A R C T I C A

100°W      100°E

120°W      120°E

FRANCE

AUSTRALIA

NEW ZEALAND

140°W      160°W      180°      160°E      140°E

Europe      Asia      North America

Africa

Oceania      South America

Antarctica

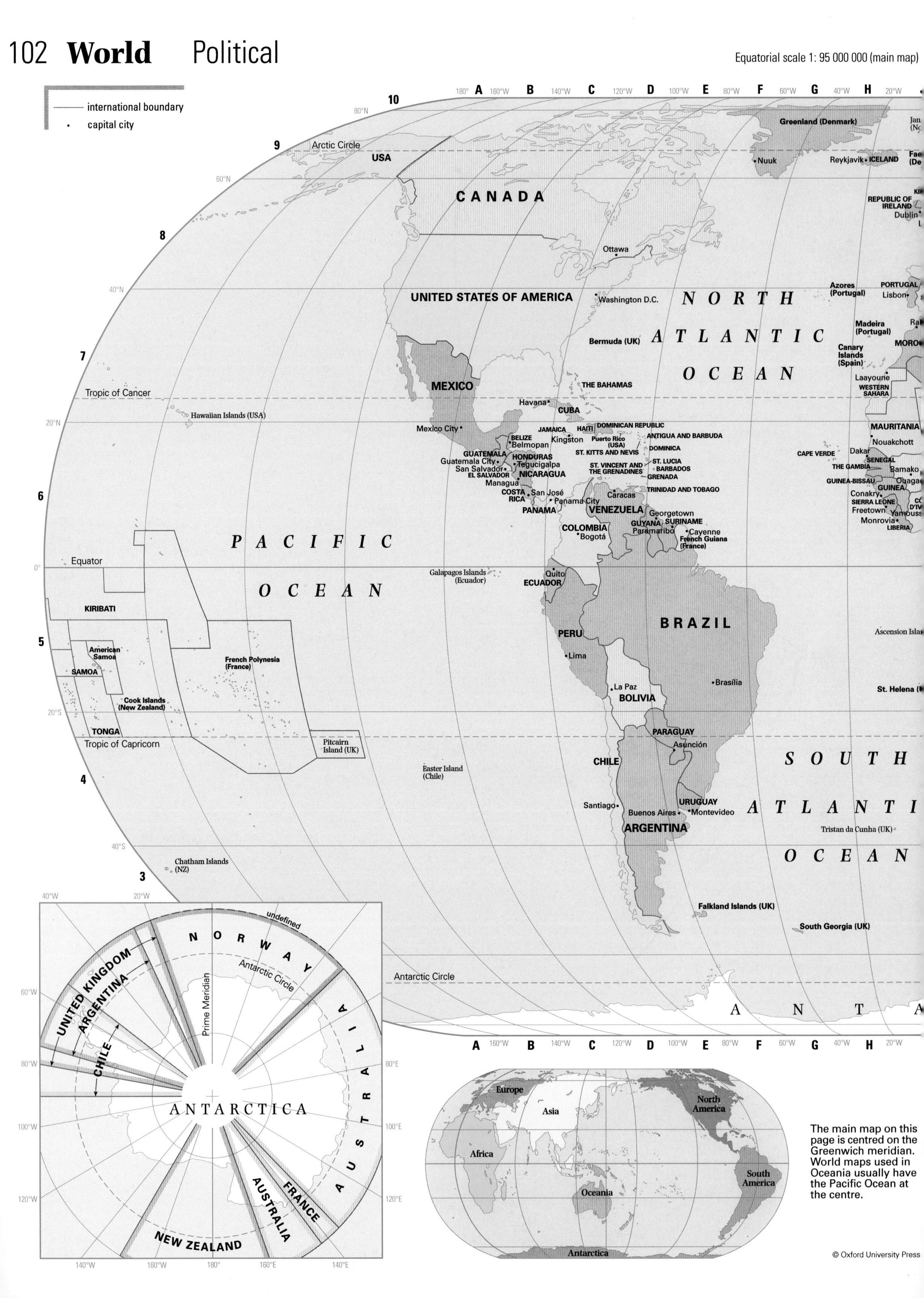

The main map on this page is centred on the Greenwich meridian. World maps used in Oceania usually have the Pacific Ocean at the centre.

© Oxford University Press

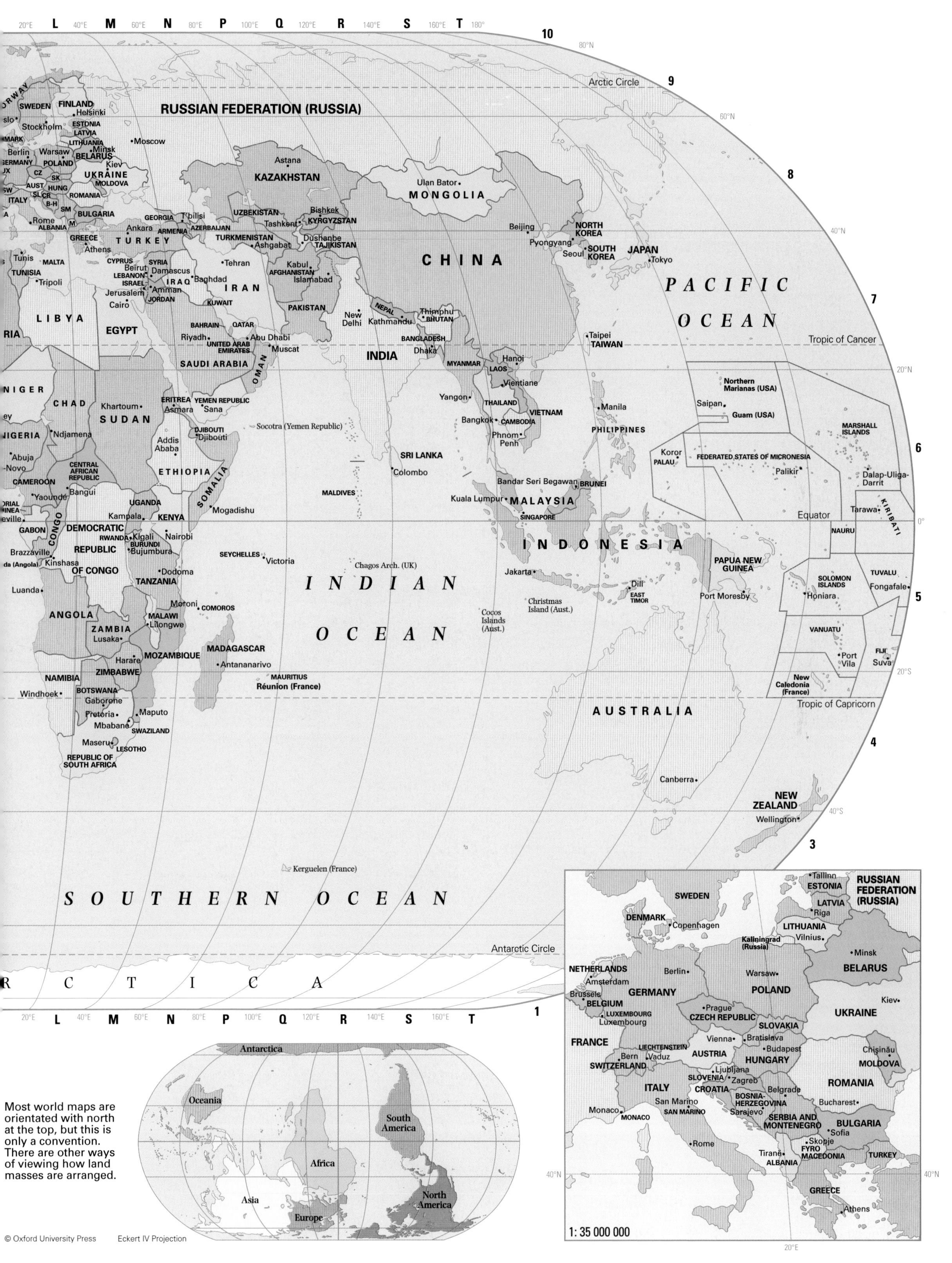

Most world maps are orientated with north at the top, but this is only a convention. There are other ways of viewing how land masses are arranged.

 Eckert IV Projection

Equatorial scale 1: 95 000 00

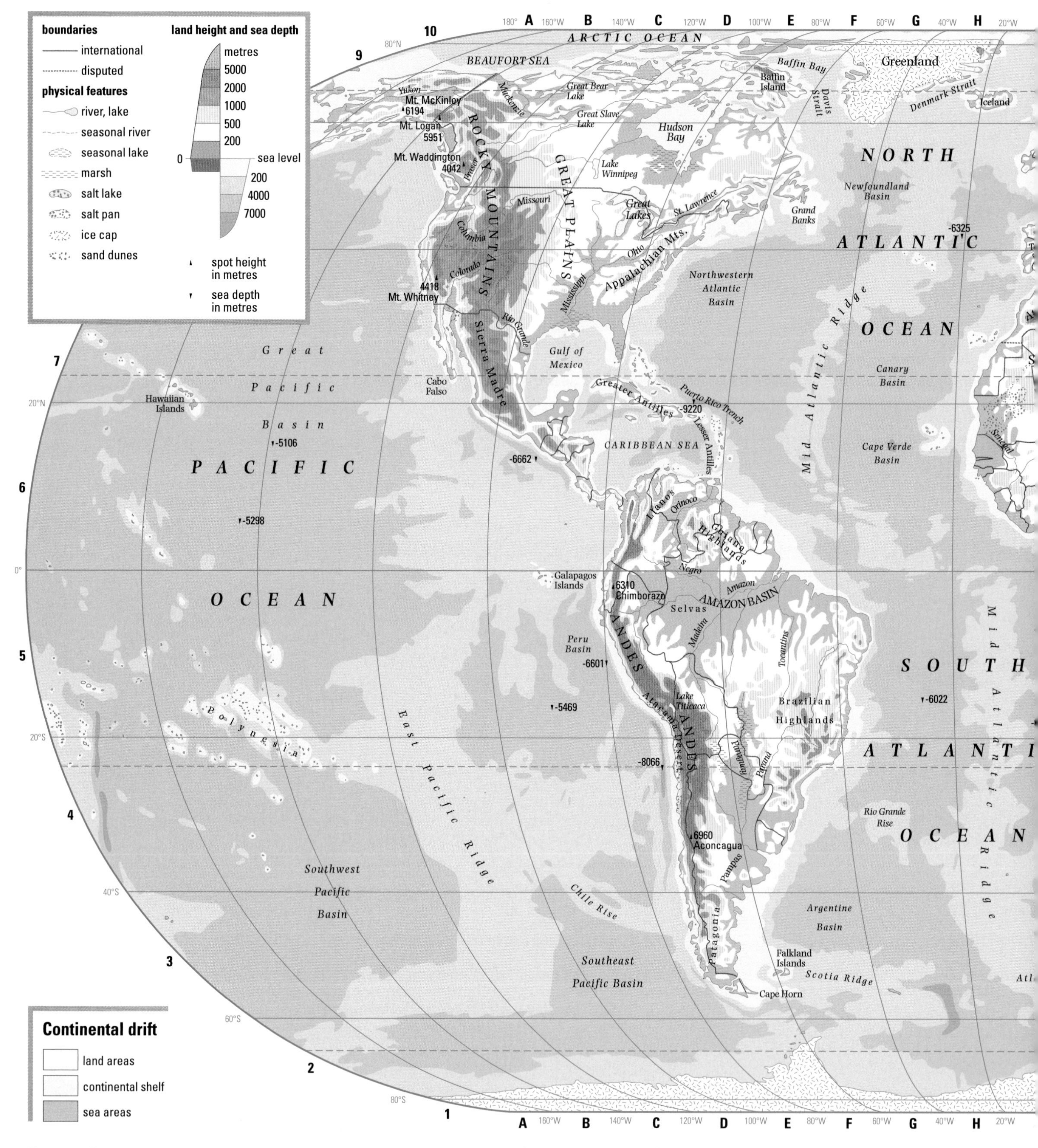

**boundaries**
— international
······ disputed

**physical features**
river, lake
seasonal river
seasonal lake
marsh
salt lake
salt pan
ice cap
sand dunes

**land height and sea depth**

metres
5000
2000
1000
500
200
0        sea level
         200
         4000
         7000

▲ spot height in metres
▼ sea depth in metres

**Continental drift**

land areas
continental shelf
sea areas

**Present day**

**100 million years ago (Cretaceous period)**

**200 million years ago (Triassic period)**

© Oxford University Press

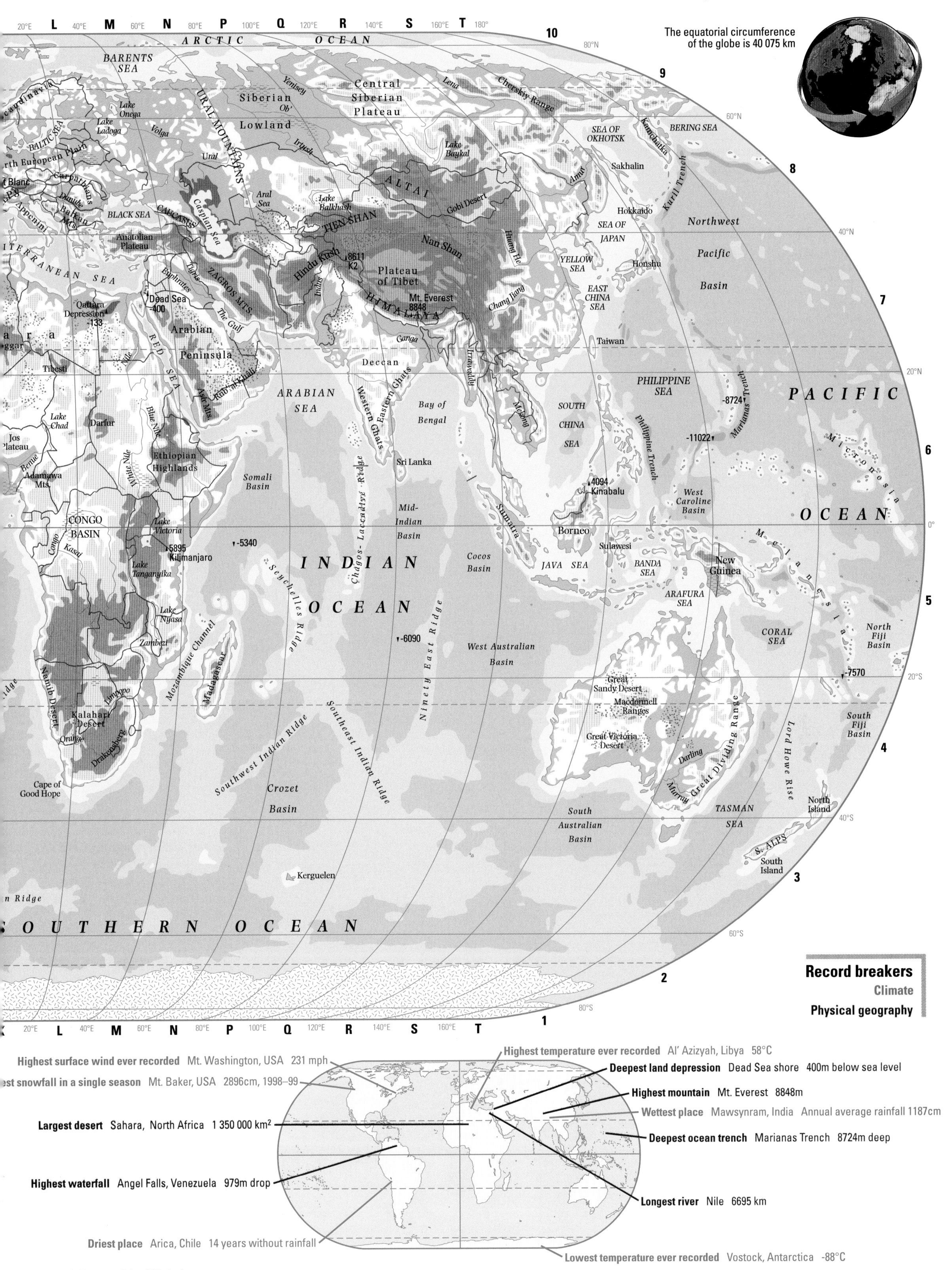

The equatorial circumference of the globe is 40 075 km

**L** 20°E **M** 40°E **N** 60°E **P** 80°E **Q** 100°E **R** 120°E **S** 140°E **T** 160°E **T** 180° **10**

ARCTIC OCEAN

BARENTS SEA

Scandinavia

Lake Onega
Lake Ladoga

Baltic Sea
North European Plain

Volga

Ural

URAL MOUNTAINS

Yenisey

Ob'

Irtysh

Siberian Lowland

Central Siberian Plateau

Lena

Cherskiy Range

Kamchatka

Kuril Trench

SEA OF OKHOTSK

BERING SEA

60°N

9

8

80°N

Mt Blanc
Carpathians
Danube
Balkan Mts
Apennini

BLACK SEA

CAUCASUS

Caspian Sea

Aral Sea

Lake Balkhash

TIEN SHAN

ALTAI

Gobi Desert

Nan Shan

Amur

Sakhalin

Hokkaido

Northwest Pacific Basin

40°N

MEDITERRANEAN SEA

Anatolian Plateau

Qattara Depression -133

Sahara

Tibesti

Hoggar

Euphrates

Tigris

ZAGROS MTS

Dead Sea -400

The Gulf

Arabian Peninsula

Hindu Kush

K2 8611

Plateau of Tibet

HIMALAYA

Mt. Everest 8848

Chang Jiang

Huang He

YELLOW SEA

EAST CHINA SEA

SEA OF JAPAN

Honshu

Taiwan

PACIFIC OCEAN

7

20°N

RED SEA

Nile

Asir Mts

Rub' al Khali

Ganga

Deccan

ARABIAN SEA

Western Ghats

Eastern Ghats

Bay of Bengal

Irrawaddy

Mekong

SOUTH CHINA SEA

PHILIPPINE SEA

-8724

Marianas Trench

-11022

Philippine Trench

Micronesia

6

Lake Chad
Darfur
Jos Plateau
Benue
Adamawa Mts

Blue Nile
White Nile
Mbe
Ethiopian Highlands

Somali Basin

Sri Lanka

Mid-Indian Basin

Sumatra

4094 Kinabalu

Borneo

Sulawesi

West Caroline Basin

OCEAN

0°

CONGO BASIN

Congo
Kasai

Lake Victoria

5895 Kilimanjaro

Lake Tanganyika

-5340

Chagos-Laccadive Ridge

INDIAN OCEAN

Seychelles Ridge

Cocos Basin

JAVA SEA

BANDA SEA

New Guinea

Melanesia

5

Namib Desert

Lake Nyasa

Zambezi

Mozambique Channel

Madagascar

-6090

Ninety East Ridge

West Australian Basin

ARAFURA SEA

CORAL SEA

North Fiji Basin

-7570

20°S

Ridge

Kalahari Desert

Orange

Drakensberg

Limpopo

Southwest Indian Ridge

Southeast Indian Ridge

Great Sandy Desert

Macdonnell Ranges

Great Victoria Desert

Darling

Great Dividing Range

Murray

South Fiji Basin

4

Cape of Good Hope

Crozet Basin

South Australian Basin

TASMAN SEA

Lord Howe Rise

S. ALPS

North Island

South Island

40°S

Kerguelen

3

n Ridge

SOUTHERN OCEAN

2

1

60°S

80°S

**K** 20°E **L** 40°E **M** 60°E **N** 80°E **P** 100°E **Q** 120°E **R** 140°E **S** 160°E **T**

**Record breakers**
Climate
**Physical geography**

**Highest surface wind ever recorded**  Mt. Washington, USA  231 mph

**...est snowfall in a single season**  Mt. Baker, USA  2896cm, 1998–99

**Largest desert**  Sahara, North Africa  1 350 000 km²

**Highest waterfall**  Angel Falls, Venezuela  979m drop

**Driest place**  Arica, Chile  14 years without rainfall

**Highest temperature ever recorded**  Al' Azizyah, Libya  58°C

**Deepest land depression**  Dead Sea shore  400m below sea level

**Highest mountain**  Mt. Everest  8848m

**Wettest place**  Mawsynram, India  Annual average rainfall 1187cm

**Deepest ocean trench**  Marianas Trench  8724m deep

**Longest river**  Nile  6695 km

**Lowest temperature ever recorded**  Vostock, Antarctica  -88°C

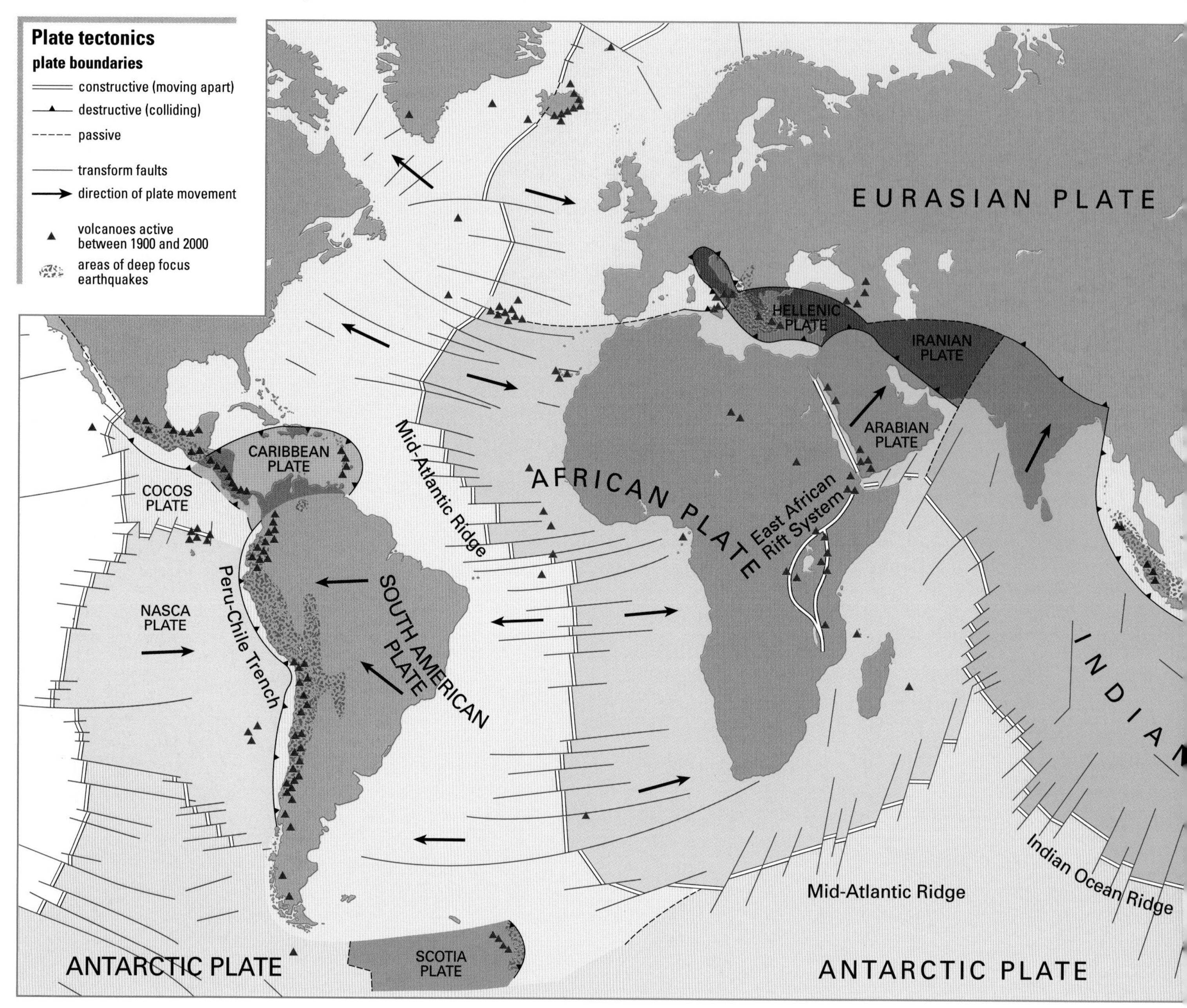

**Plate tectonics**

**plate boundaries**
- constructive (moving apart)
- ▲ destructive (colliding)
- passive
- transform faults
- → direction of plate movement
- ▲ volcanoes active between 1900 and 2000
- areas of deep focus earthquakes

EURASIAN PLATE

HELLENIC PLATE
IRANIAN PLATE
ARABIAN PLATE

CARIBBEAN PLATE
COCOS PLATE
NASCA PLATE

AFRICAN PLATE
East African Rift System

Mid-Atlantic Ridge
Peru-Chile Trench

SOUTH AMERICAN PLATE

INDIAN

Indian Ocean Ridge
Mid-Atlantic Ridge

ANTARCTIC PLATE
SCOTIA PLATE
ANTARCTIC PLATE

## Structure of the Earth

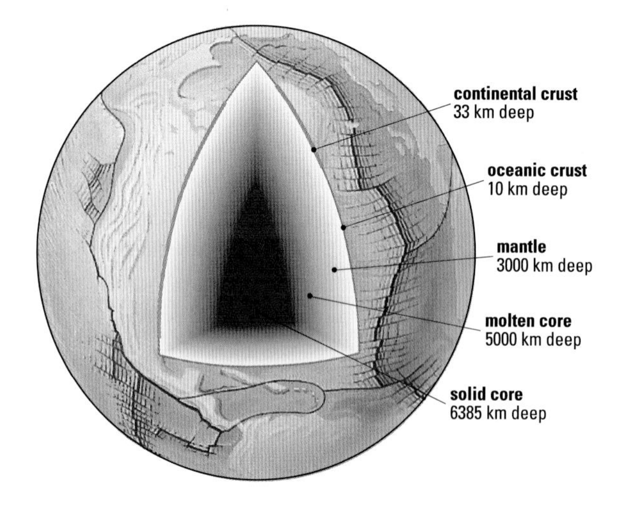

- **continental crust** 33 km deep
- **oceanic crust** 10 km deep
- **mantle** 3000 km deep
- **molten core** 5000 km deep
- **solid core** 6385 km deep

## Mt. St. Helens

A digital elevation model (DEM) of the stratovolcano which erupted on 18 May, 1980 in Washington State, USA.

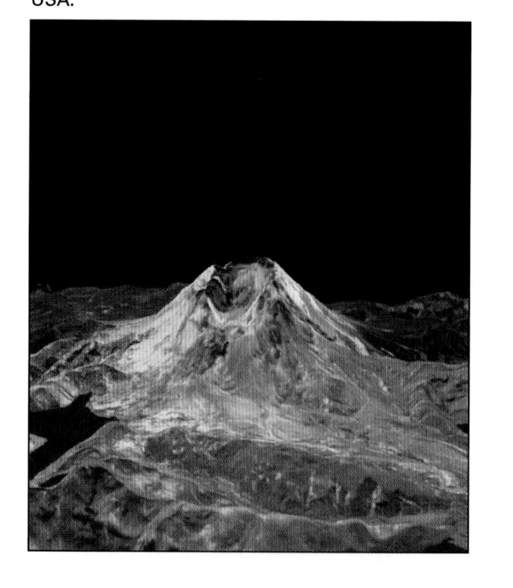

### Deadliest earthquakes, 1990–2003
force measured on the Richter scale

| Year | Place | Force | Deaths |
|------|-------|-------|--------|
| 1990 | Northwestern Iran | 7.7 | 37 000 |
| 1990 | Luzon, Philippines | 7.7 | 1660 |
| 1991 | Afghanistan/Pakistan | 6.8 | 1000 |
| 1991 | Uttar Pradesh, India | 6.1 | 1500 |
| 1992 | Erzincan, Turkey | 6.7 | 2000 |
| 1992 | Flores Island, Indonesia | 7.5 | 2500 |
| 1993 | Maharashtra, India | 6.3 | 9800 |
| 1994 | Cauca, Colombia | 6.8 | 1000 |
| 1995 | Kobe, Japan | 7.2 | 5500 |
| 1995 | Sakhalin Island, Russia | 7.6 | 2000 |
| 1997 | Ardabil, Iran | unknown | >1000 |
| 1997 | Khorash, Iran | 7.1 | >1600 |
| 1998 | Takhar, Afghanistan | 6.1 | >3800 |
| 1998 | Northeastern Afghanistan | 7.1 | >3000 |
| 1999 | Western Colombia | 6.0 | 1124 |
| 1999 | Izmit, Turkey | 7.4 | >17 000 |
| 1999 | Central Taiwan | 7.6 | 2295 |
| 1999 | Ducze, Turkey | 7.2 | >700 |
| 2001 | Gujarat, India | 6.9 | >20 000 |
| 2002 | Baghlan, Afghanistan | 6.0 | >2000 |
| 2003 | Northern Algeria | 6.8 | 2266 |
| 2003 | Southeastern Iran | 6.6 | 31 000 |

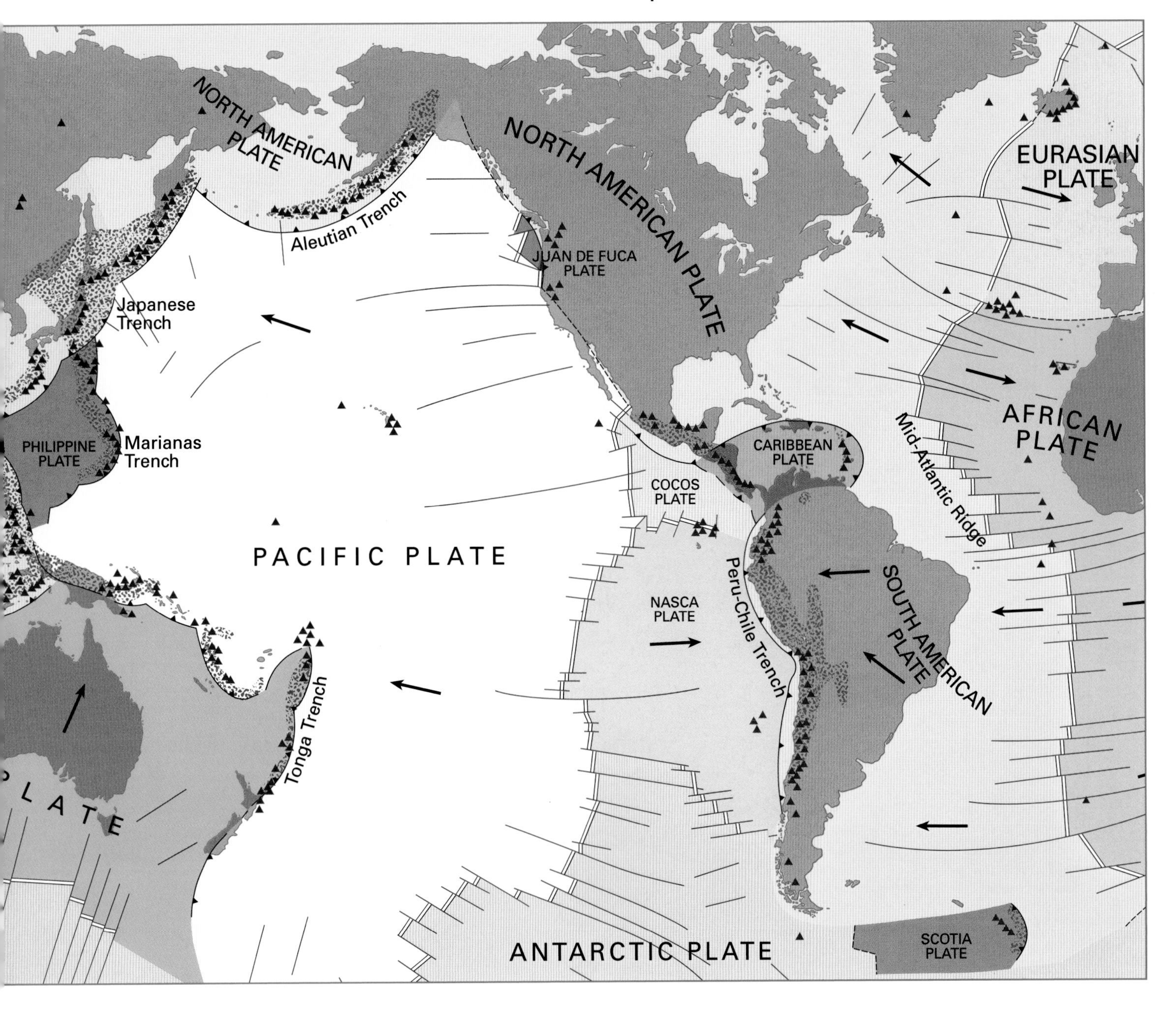

NORTH AMERICAN PLATE

NORTH AMERICAN PLATE

Aleutian Trench

Japanese Trench

PHILIPPINE PLATE

Marianas Trench

PACIFIC PLATE

JUAN DE FUCA PLATE

COCOS PLATE

CARIBBEAN PLATE

NASCA PLATE

Peru-Chile Trench

SOUTH AMERICAN PLATE

EURASIAN PLATE

AFRICAN PLATE

Mid-Atlantic Ridge

Tonga Trench

PLATE

ANTARCTIC PLATE

SCOTIA PLATE

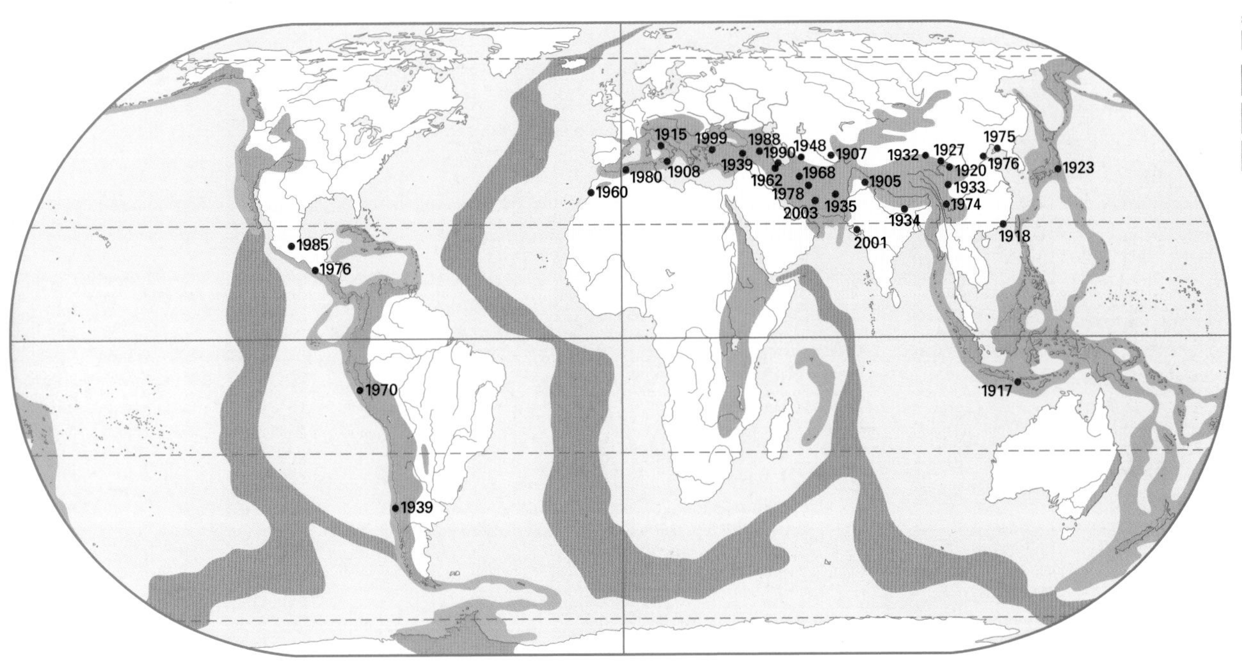

**Earthquakes**

mobile areas (on land)

mobile areas (under sea)

mid-oceanic ridges

• earthquakes causing more than 10 000 deaths, 1900–2003

1915 1999 1988 1948
1980 1908 1939 1990 1907 1932 1927 1975
1960 1962 1968 1920 1976 1923
1978 1905 1933
2003 1935 1974
2001 1934 1918

1985
1976

1917

1970

1939

Eckert IV Projection

## January temperature

actual surface temperature

°Celsius

32
24
16
8
0
−8
−16
−24
−32
−40

→ warm sea current
→ cold sea current

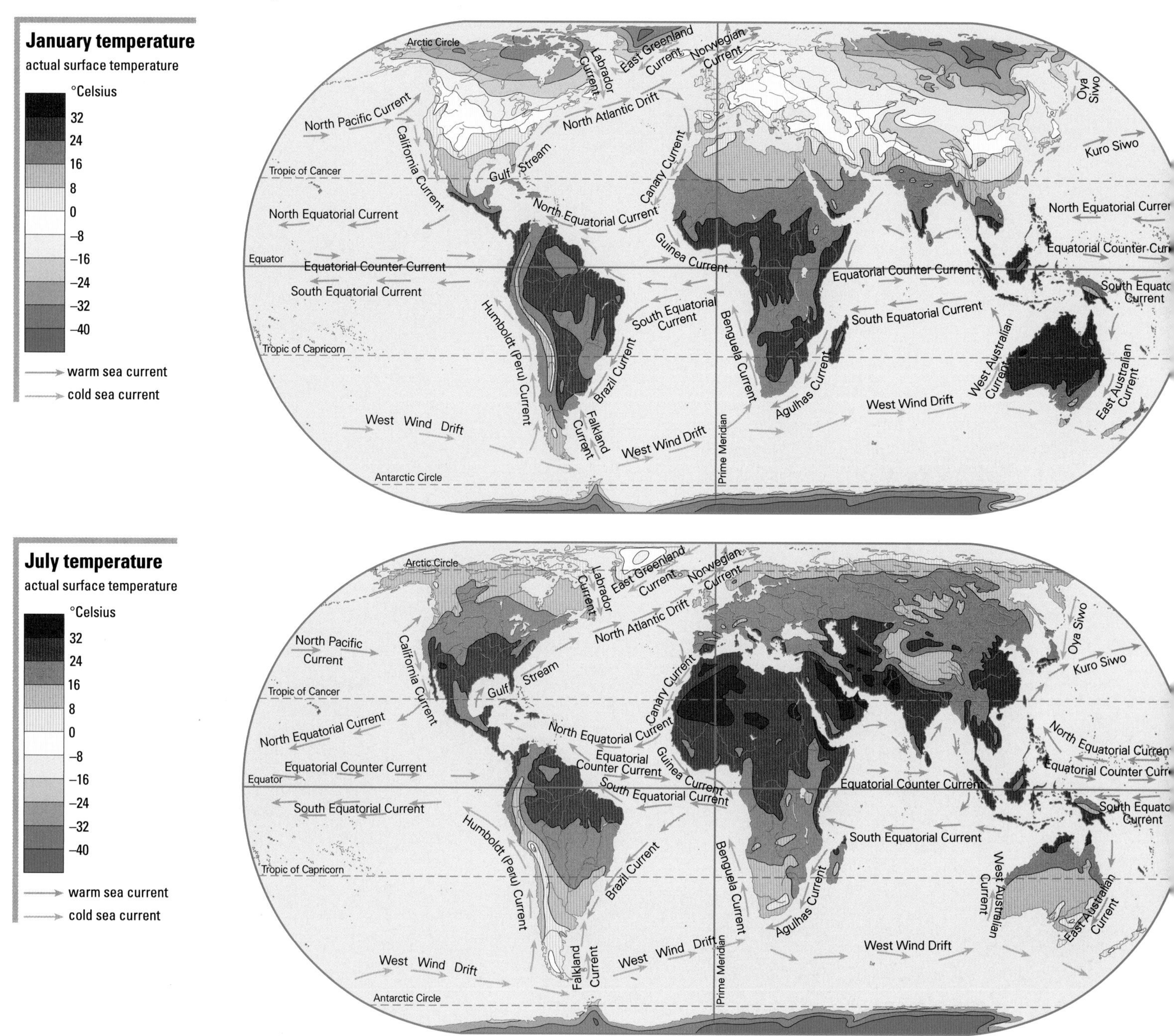

## July temperature

actual surface temperature

°Celsius

32
24
16
8
0
−8
−16
−24
−32
−40

→ warm sea current
→ cold sea current

## Antarctic ozone 'hole'

Three dimensional image of ozone depletion over Antarctica in September, 1998. The lowest ozone concentration is shown in blue.

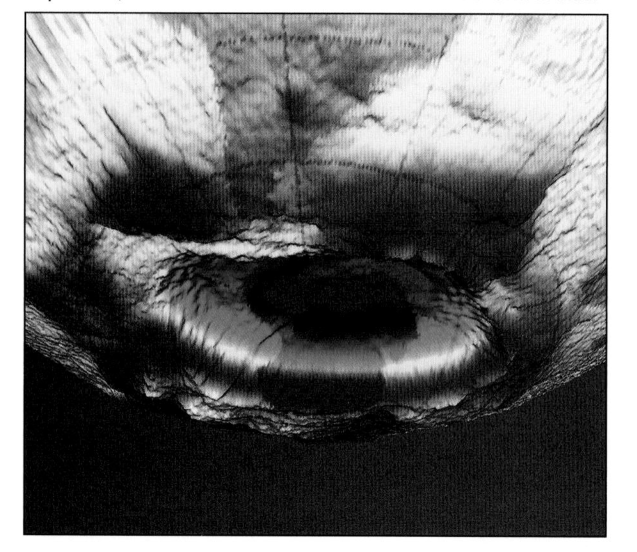

## Global warming

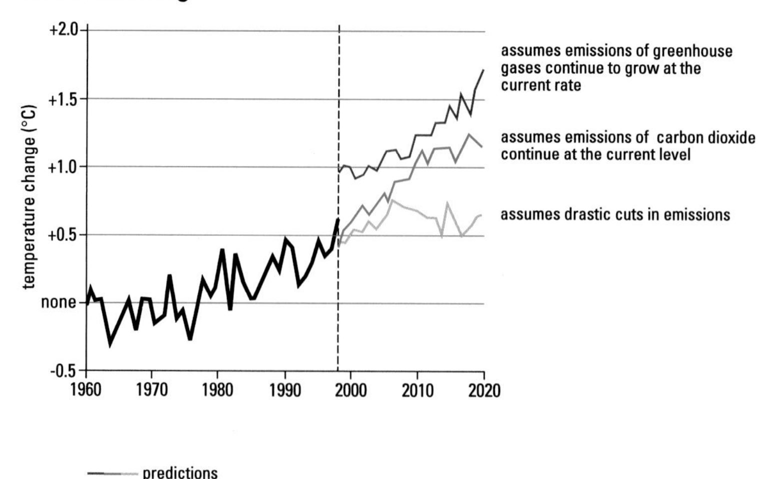

assumes emissions of greenhouse gases continue to grow at the current rate

assumes emissions of carbon dioxide continue at the current level

assumes drastic cuts in emissions

temperature change (°C)

+2.0
+1.5
+1.0
+0.5
none
−0.5

1960  1970  1980  1990  2000  2010  2020

— predictions
— actual temperature change

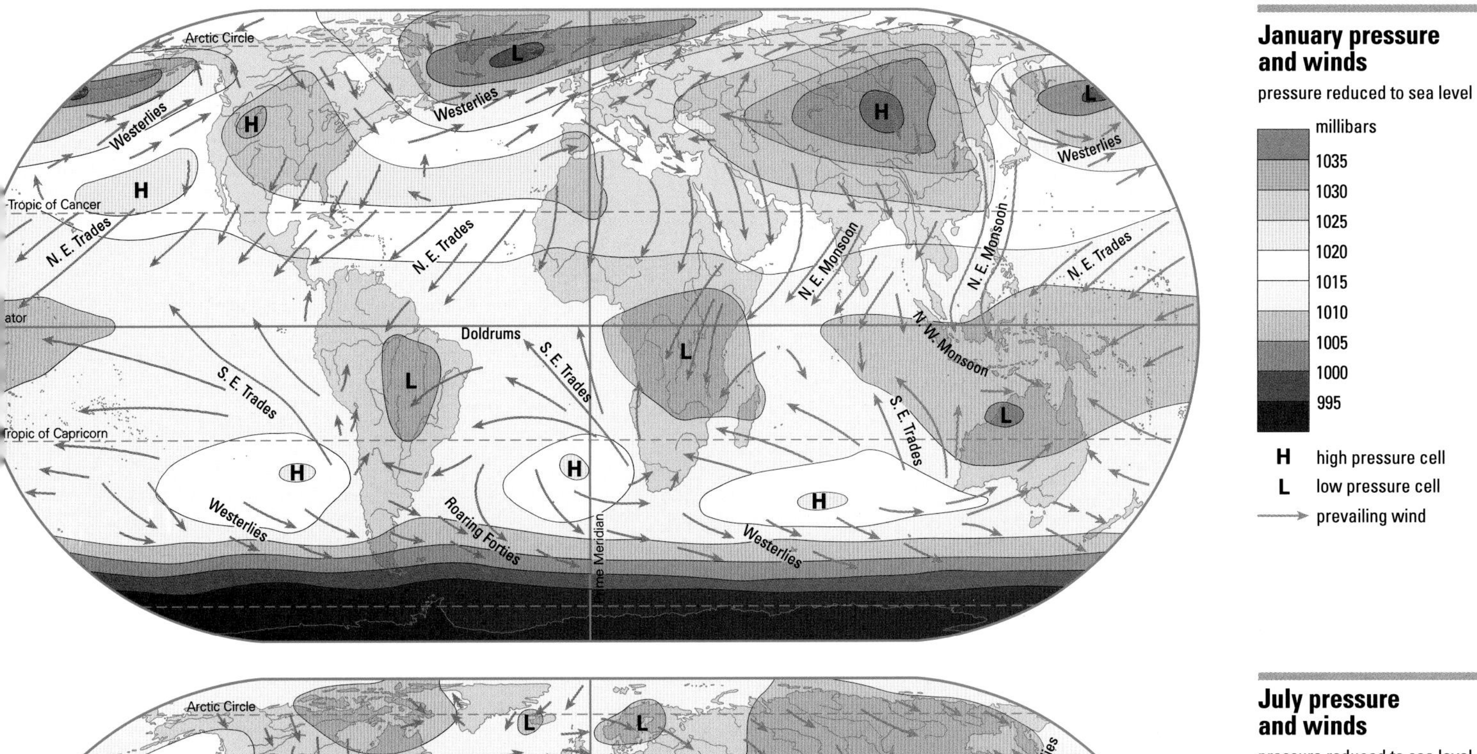

**January pressure and winds**

pressure reduced to sea level

millibars
1035
1030
1025
1020
1015
1010
1005
1000
995

H  high pressure cell
L  low pressure cell
prevailing wind

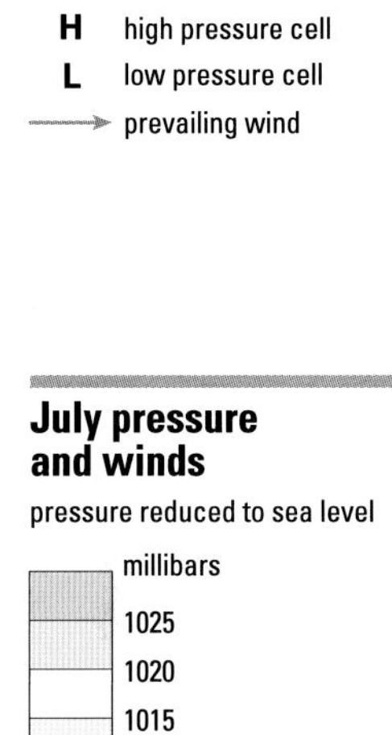

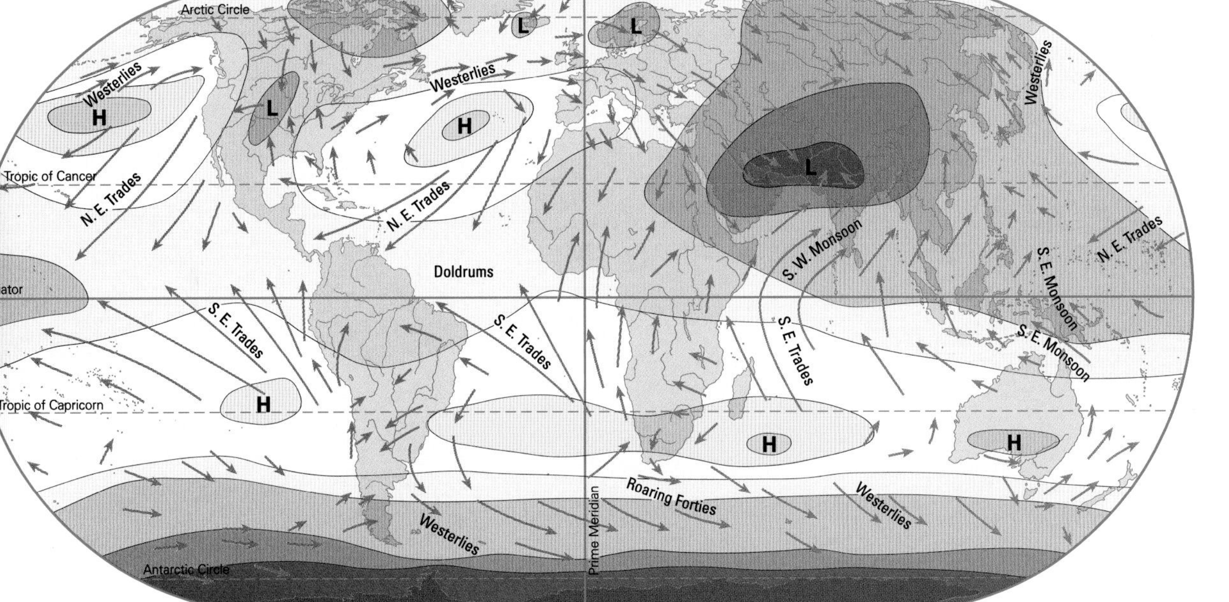

**July pressure and winds**

pressure reduced to sea level

millibars
1025
1020
1015
1010
1005
1000
995

H  high pressure cell
L  low pressure cell
prevailing wind

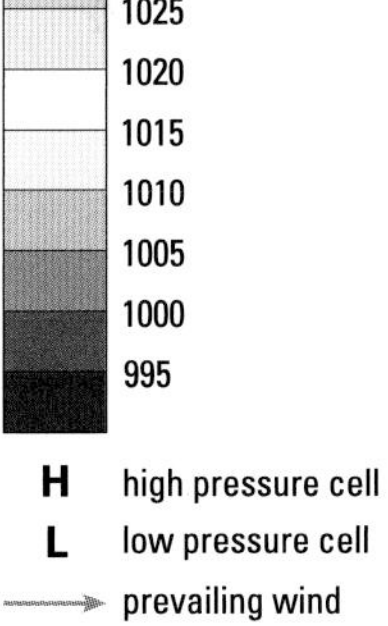

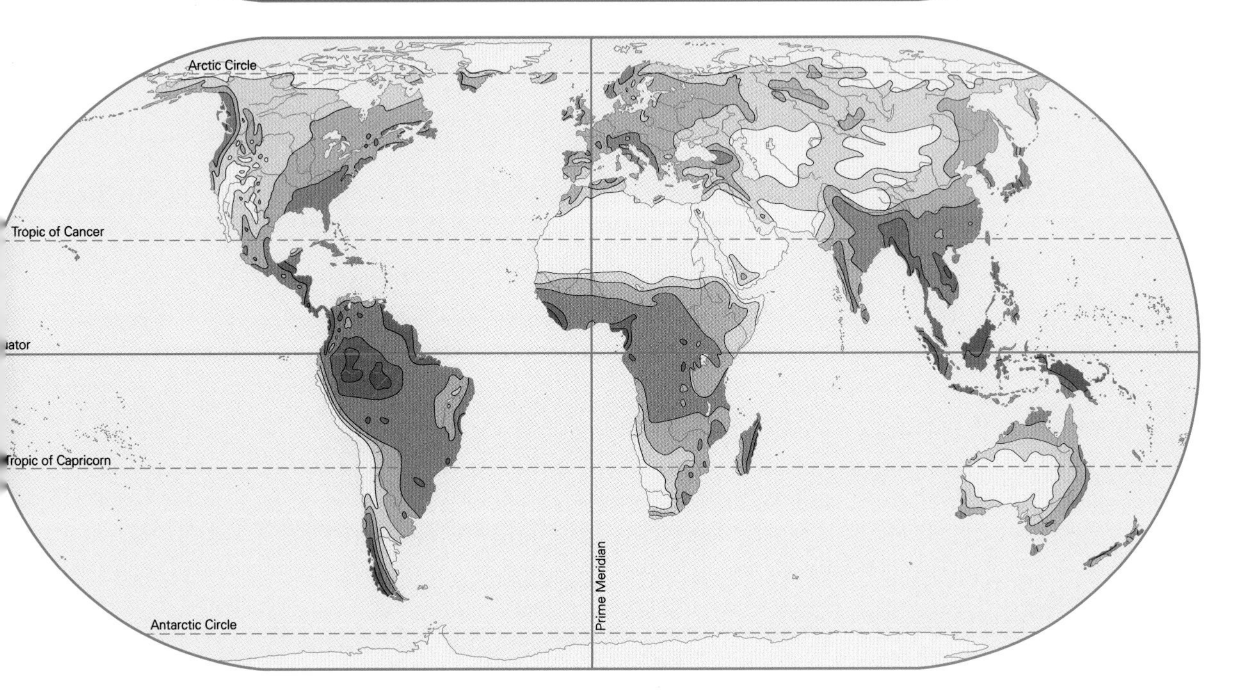

**Precipitation**

average annual precipitation

mm
3000
2000
1000
500
250

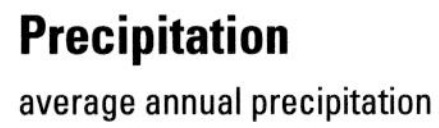

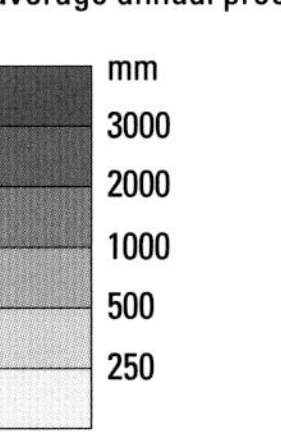

Equatorial scale 1: 95 000 0C

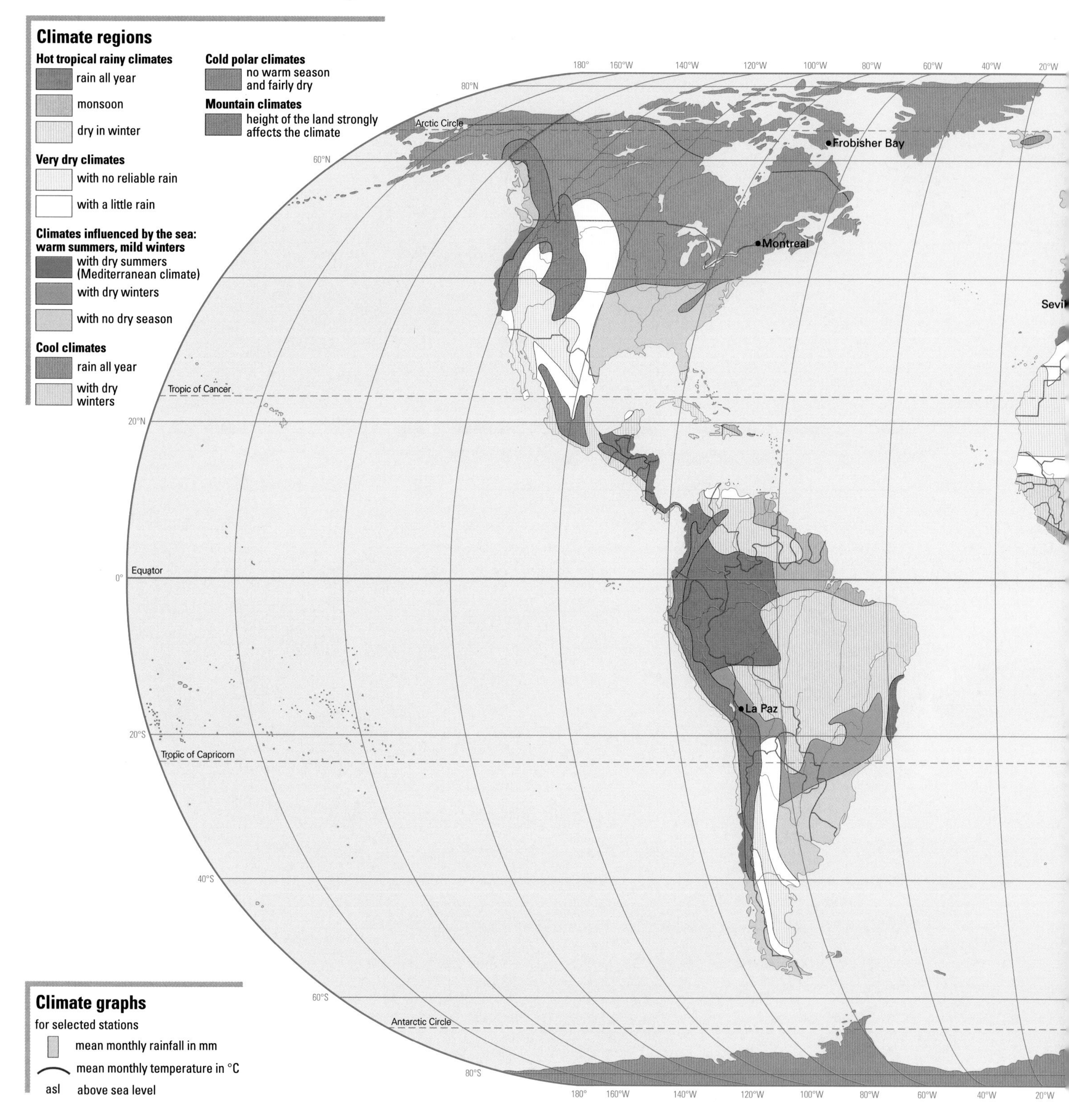

## Climate regions

### Hot tropical rainy climates
- rain all year
- monsoon
- dry in winter

### Very dry climates
- with no reliable rain
- with a little rain

### Climates influenced by the sea: warm summers, mild winters
- with dry summers (Mediterranean climate)
- with dry winters
- with no dry season

### Cool climates
- rain all year
- with dry winters

### Cold polar climates
- no warm season and fairly dry

### Mountain climates
- height of the land strongly affects the climate

## Climate graphs

for selected stations

- mean monthly rainfall in mm
- mean monthly temperature in °C

asl    above sea level

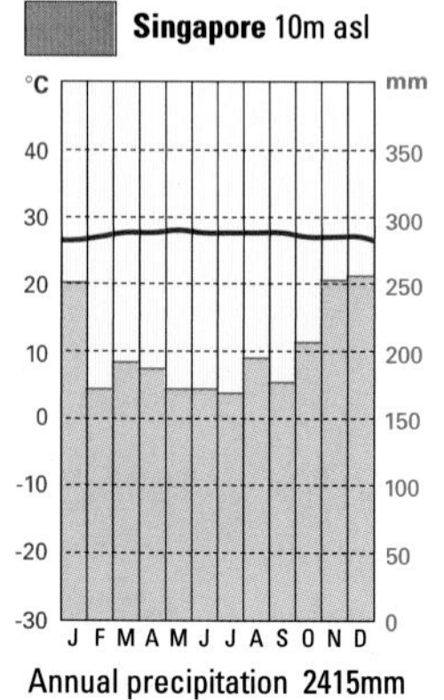

**Singapore** 10m asl

Annual precipitation 2415mm

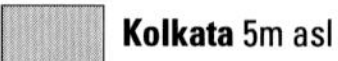

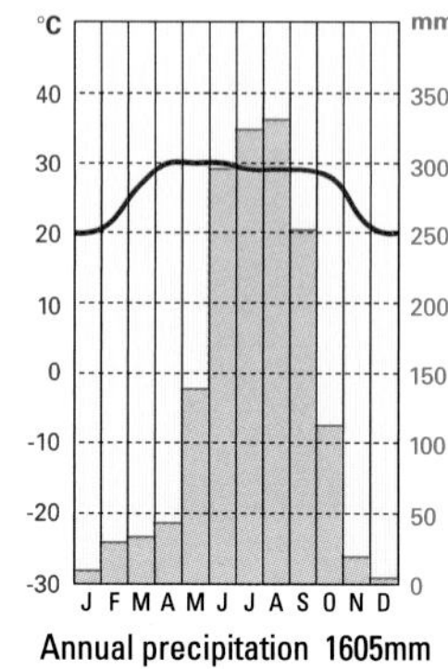

**Kolkata** 5m asl

Annual precipitation 1605mm

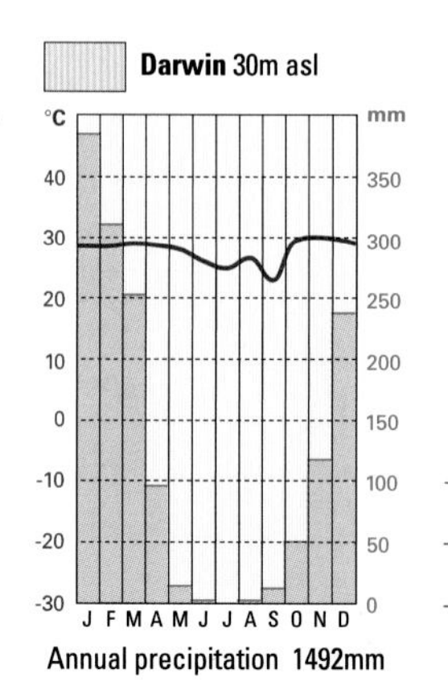

**Darwin** 30m asl

Annual precipitation 1492mm

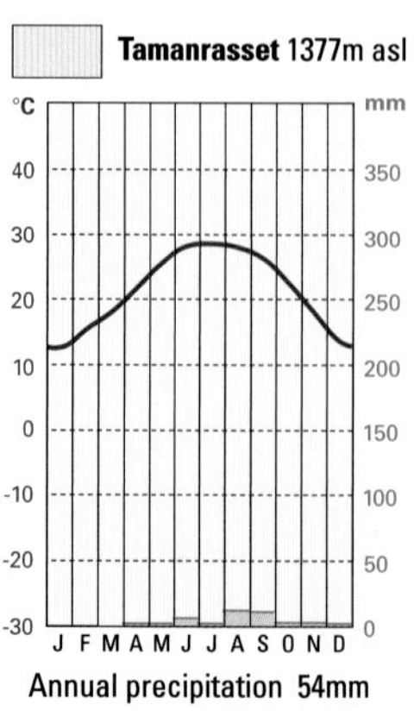

**Tamanrasset** 1377m asl

Annual precipitation 54mm

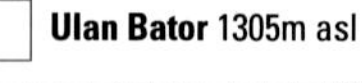

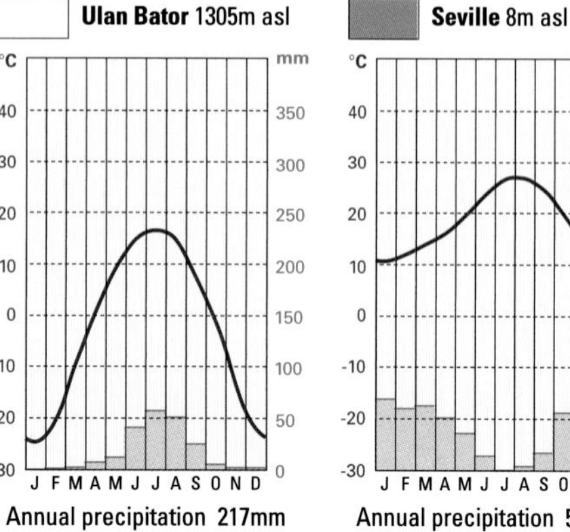

**Ulan Bator** 1305m asl

Annual precipitation 217mm

**Seville** 8m asl

Annual precipitation 534mm

Eckert IV Projection          © Oxford University Press

Ulan Bator
Vladivostock
Tokyo
Tamanrasset
Kolkata
Singapore
Darwin
Johannesburg

80°N
Arctic Circle
60°N
40°N
Tropic of Cancer
20°N
Equator  0°
Tropic of Capricorn
20°S
40°S
60°S
Antarctic Circle
80°S

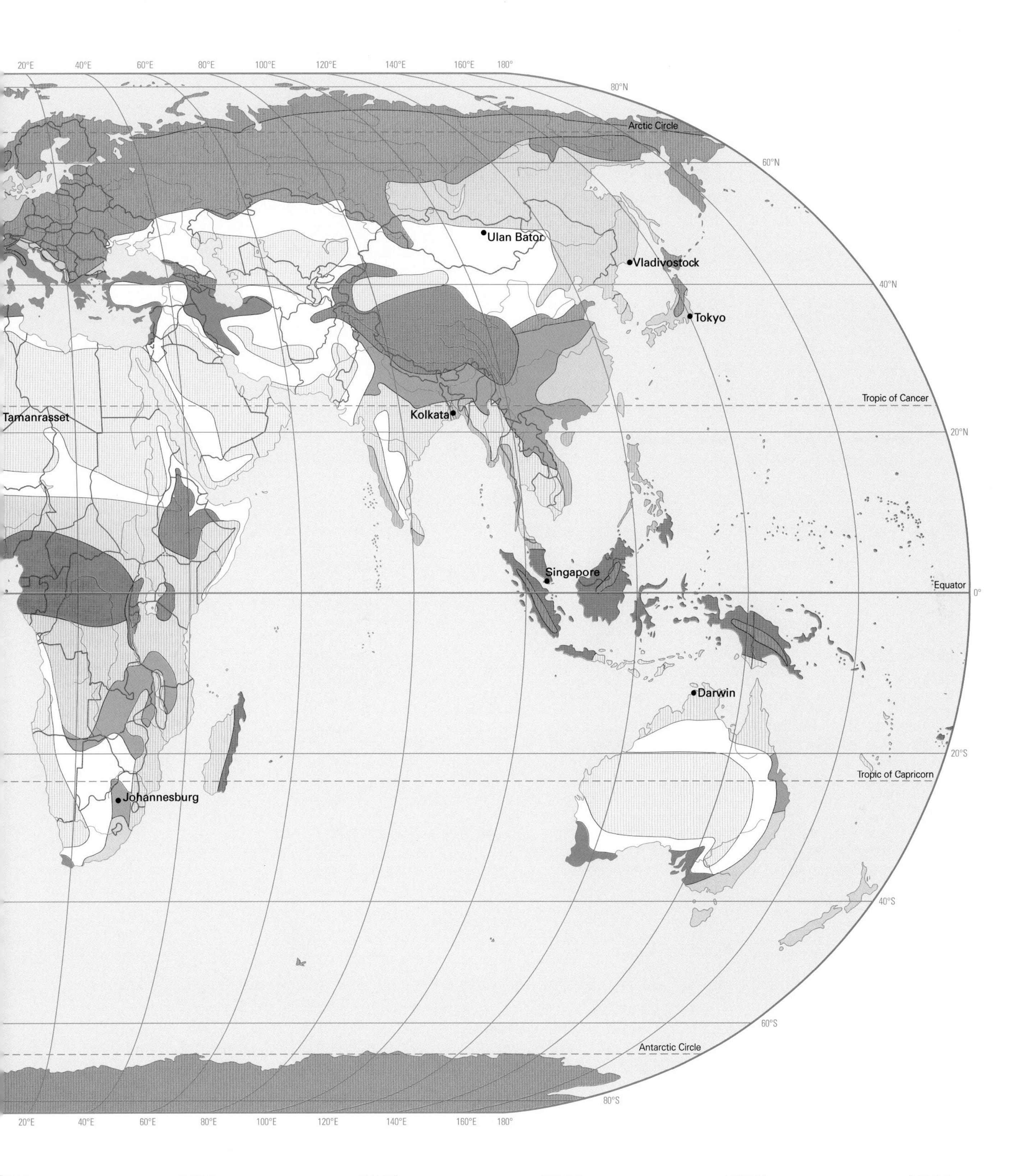

20°E  40°E  60°E  80°E  100°E  120°E  140°E  160°E  180°

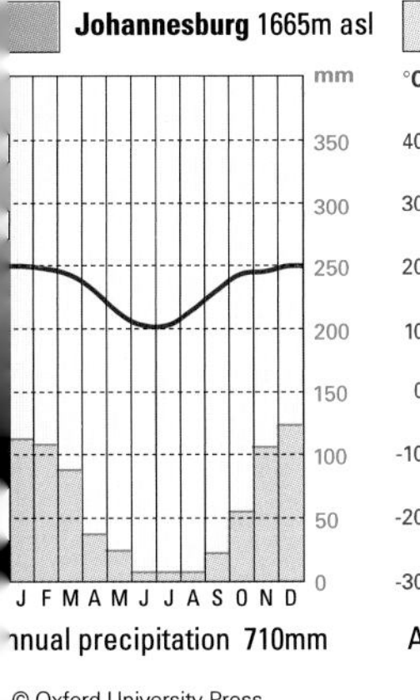

| | | | | | |
|---|---|---|---|---|---|
| **Johannesburg** 1665m asl | **Tokyo** 6m asl | **Montreal** 57m asl | **Vladivostock** 29m asl | **Frobisher Bay** 21m asl | **La Paz** 3632m asl |
| Annual precipitation  710mm | Annual precipitation  1565mm | Annual precipitation  1047mm | Annual precipitation  600mm | Annual precipitation  427mm | Annual precipitation  610mm |

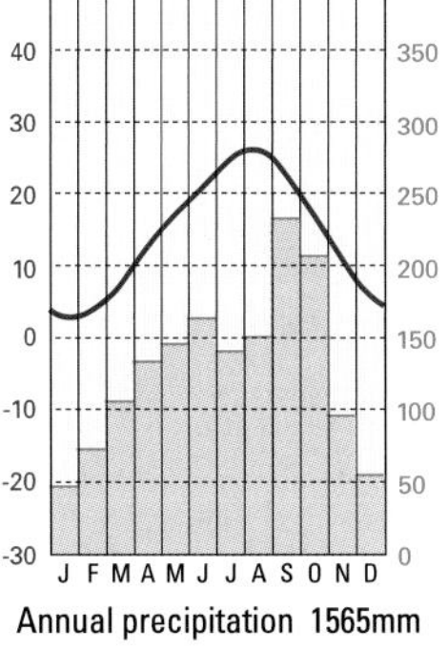

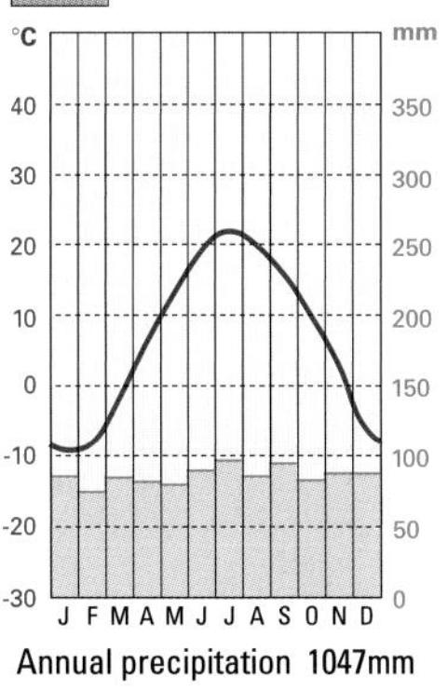

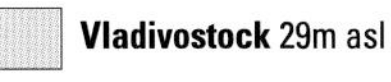

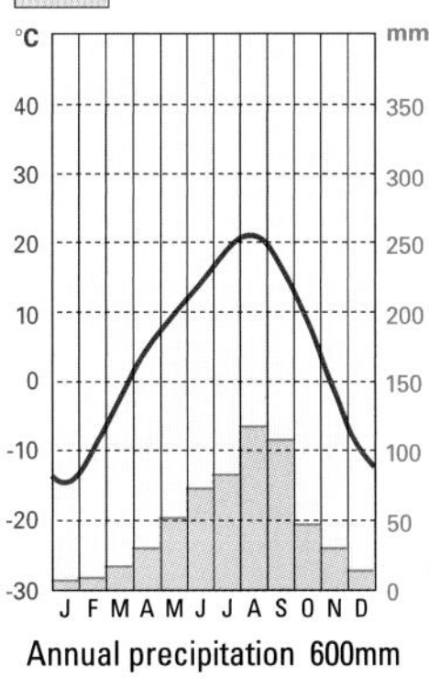

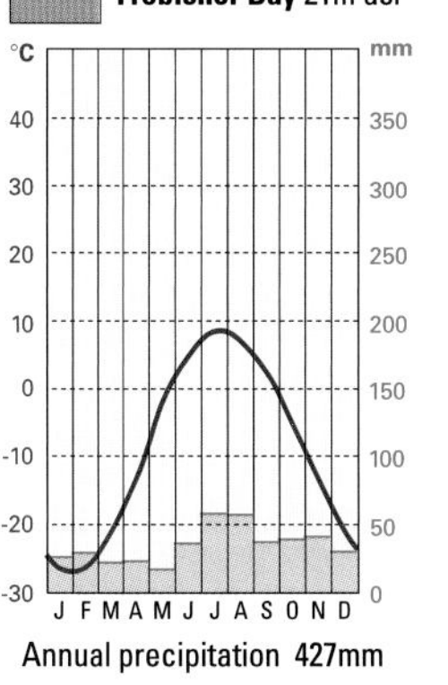

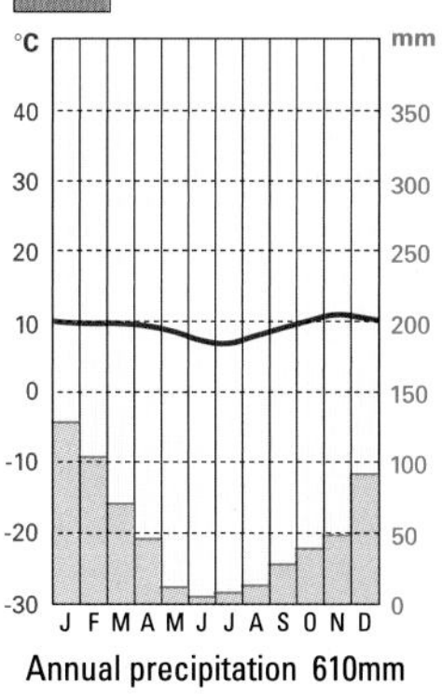

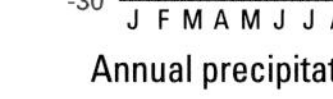

## Climate data

Averages are for 1961–1990

| | | |
|---|---|---|
| **Denver** 1626m | | climate station and its height above sea level |
| Temperature (°C) | high | average daily maximum temperature |
| | **mean** | average monthly temperature |
| | low | average daily minimum temperature |
| Rainfall (mm) | | average monthly precipitation |

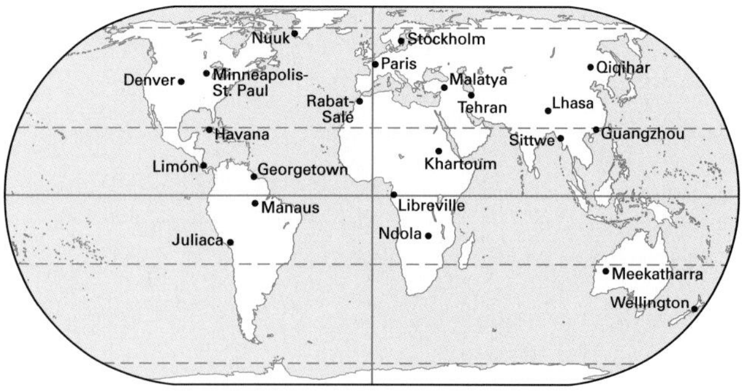


| | | Jan | Feb | Mar | Apr | May | Jun | Jul | Aug | Sep | Oct | Nov | Dec | YEAR |
|---|---|---|---|---|---|---|---|---|---|---|---|---|---|---|
| **Denver** 1626m | | | | | | | | | | | | | | |
| Temperature (°C) | high | 6.2 | 8.1 | 11.2 | 16.6 | 21.6 | 27.4 | 31.2 | 29.9 | 24.9 | 19.1 | 11.4 | 6.9 | 17.9 |
| | **mean** | -1.3 | 0.8 | 3.9 | 9.0 | 14.0 | 19.4 | 23.1 | 21.9 | 16.8 | 10.8 | 3.9 | -0.6 | 10.1 |
| | low | -8.8 | -6.6 | -3.4 | 1.4 | 6.4 | 11.3 | 14.8 | 13.8 | 8.7 | 2.4 | -3.7 | -8.1 | 2.4 |
| Rainfall (mm) | | 13 | 15 | 33 | 43 | 61 | 46 | 49 | 38 | 32 | 25 | 22 | 16 | 393 |
| **Georgetown** 2m | | | | | | | | | | | | | | |
| Temperature (°C) | high | 28.6 | 28.9 | 29.2 | 29.5 | 29.4 | 29.2 | 29.6 | 30.2 | 30.8 | 30.8 | 30.2 | 29.1 | 29.6 |
| | **mean** | 26.1 | 26.4 | 26.7 | 27.0 | 26.8 | 26.5 | 26.6 | 27.0 | 27.5 | 27.6 | 27.2 | 26.4 | 26.8 |
| | low | 23.6 | 23.9 | 24.2 | 24.4 | 24.3 | 23.8 | 23.5 | 23.8 | 24.2 | 24.4 | 24.2 | 23.8 | 24.0 |
| Rainfall (mm) | | 185 | 89 | 111 | 141 | 286 | 328 | 268 | 201 | 98 | 107 | 186 | 262 | 2262 |
| **Guangzhou** 42m | | | | | | | | | | | | | | |
| Temperature (°C) | high | 18.3 | 18.4 | 21.6 | 25.5 | 29.4 | 31.3 | 32.7 | 32.6 | 31.4 | 28.6 | 24.4 | 20.5 | 26.2 |
| | **mean** | 13.3 | 14.3 | 17.7 | 21.9 | 25.6 | 27.3 | 28.5 | 28.3 | 27.1 | 24.0 | 19.4 | 15.0 | 21.9 |
| | low | 5.0 | 6.6 | 10.7 | 16.1 | 20.7 | 23.5 | 25.7 | 25.2 | 22.6 | 17.6 | 11.9 | 6.5 | 16.0 |
| Rainfall (mm) | | 43 | 65 | 85 | 182 | 284 | 258 | 228 | 221 | 172 | 79 | 42 | 24 | 1683 |
| **Havana** 50m | | | | | | | | | | | | | | |
| Temperature (°C) | high | 25.8 | 26.1 | 27.6 | 28.6 | 29.8 | 30.5 | 31.3 | 31.6 | 31.0 | 29.2 | 27.7 | 26.5 | 28.8 |
| | **mean** | 22.2 | 22.4 | 23.7 | 24.8 | 26.1 | 26.9 | 27.6 | 27.8 | 27.4 | 26.2 | 24.5 | 23.0 | 25.2 |
| | low | 18.6 | 18.6 | 19.7 | 20.9 | 22.4 | 23.4 | 23.8 | 24.1 | 23.8 | 23.0 | 21.3 | 19.5 | 21.6 |
| Rainfall (mm) | | 64 | 69 | 46 | 54 | 98 | 182 | 106 | 100 | 144 | 181 | 88 | 58 | 1190 |
| **Juliaca** 3827m | | | | | | | | | | | | | | |
| Temperature (°C) | high | 16.7 | 16.7 | 16.5 | 16.8 | 16.6 | 16.0 | 16.0 | 17.0 | 17.6 | 18.6 | 18.8 | 17.7 | 17.1 |
| | **mean** | 10.2 | 10.1 | 9.9 | 8.7 | 6.4 | 4.5 | 4.3 | 5.8 | 8.1 | 9.5 | 10.2 | 10.4 | 8.2 |
| | low | 3.6 | 3.5 | 3.2 | 0.6 | -3.8 | -7.0 | -7.5 | -5.4 | -1.4 | 0.3 | 1.5 | 3.0 | -0.8 |
| Rainfall (mm) | | 133 | 109 | 99 | 43 | 10 | 3 | 2 | 6 | 22 | 41 | 55 | 86 | 609 |
| **Khartoum** 380m | | | | | | | | | | | | | | |
| Temperature (°C) | high | 30.8 | 33.0 | 36.8 | 40.1 | 41.9 | 41.3 | 38.4 | 37.3 | 39.1 | 39.3 | 35.2 | 31.8 | 37.1 |
| | **mean** | 23.2 | 25.0 | 28.7 | 31.9 | 34.5 | 34.3 | 32.1 | 31.5 | 32.5 | 32.4 | 28.1 | 24.5 | 29.9 |
| | low | 15.6 | 17.0 | 20.5 | 23.6 | 27.1 | 27.3 | 25.9 | 25.3 | 26.0 | 25.5 | 21.0 | 17.1 | 22.7 |
| Rainfall (mm) | | 0 | 0 | 0 | 0.5 | 4 | 5 | 46 | 75 | 25 | 5 | 1 | 0 | 161 |
| **Lhasa** 3650m | | | | | | | | | | | | | | |
| Temperature (°C) | high | 6.9 | 9.0 | 12.1 | 15.6 | 19.3 | 22.7 | 22.1 | 21.1 | 19.7 | 16.3 | 11.2 | 7.7 | 15.3 |
| | **mean** | -2.1 | 1.1 | 4.6 | 8.1 | 11.9 | 15.5 | 15.3 | 14.5 | 12.8 | 8.1 | 2.2 | -1.7 | 7.5 |
| | low | -10.1 | -6.8 | -3.0 | 0.9 | 5.0 | 9.3 | 10.1 | 9.4 | 7.5 | 1.3 | -4.9 | -9.0 | 0.8 |
| Rainfall (mm) | | 1 | 1 | 2 | 5 | 27 | 72 | 119 | 123 | 58 | 10 | 2 | 1 | 421 |
| **Libreville** 15m | | | | | | | | | | | | | | |
| Temperature (°C) | high | 29.5 | 30.0 | 30.2 | 30.1 | 29.4 | 27.6 | 26.4 | 26.8 | 27.5 | 28.0 | 28.4 | 29.0 | 28.6 |
| | **mean** | 26.8 | 27.0 | 27.1 | 26.6 | 26.7 | 25.4 | 24.3 | 24.3 | 25.4 | 25.7 | 25.9 | 26.2 | 26.0 |
| | low | 24.1 | 24.0 | 23.9 | 23.1 | 24.0 | 23.2 | 22.1 | 21.8 | 23.4 | 23.4 | 23.4 | 23.4 | 23.3 |
| Rainfall (mm) | | 250 | 243 | 363 | 339 | 247 | 54 | 7 | 14 | 104 | 427 | 490 | 303 | 2841 |
| **Limón** 3m | | | | | | | | | | | | | | |
| Temperature (°C) | high | 27.9 | 28.6 | 29.6 | 29.6 | 28.5 | 27.5 | 27.7 | 27.7 | 27.2 | 27.0 | 27.1 | 27.7 | 28.0 |
| | **mean** | 24.0 | 24.3 | 25.0 | 25.8 | 26.1 | 25.9 | 25.2 | 25.6 | 25.7 | 25.4 | 25.1 | 24.3 | 25.2 |
| | low | 20.3 | 20.3 | 20.9 | 21.6 | 22.2 | 22.3 | 22.1 | 22.1 | 22.2 | 21.9 | 21.6 | 20.9 | 21.5 |
| Rainfall (mm) | | 319 | 201 | 193 | 287 | 281 | 276 | 408 | 289 | 163 | 198 | 367 | 402 | 3384 |
| **Malatya** 849m | | | | | | | | | | | | | | |
| Temperature (°C) | high | 2.9 | 5.3 | 11.1 | 18.2 | 23.5 | 29.2 | 33.8 | 33.4 | 28.9 | 20.9 | 11.8 | 5.7 | 18.7 |
| | **mean** | -0.4 | 1.5 | 6.9 | 13.0 | 17.8 | 22.9 | 27.0 | 26.5 | 22.0 | 14.8 | 7.6 | 2.4 | 13.5 |
| | low | -3.2 | -1.7 | 2.4 | 7.7 | 11.8 | 16.1 | 19.8 | 19.4 | 15.2 | 9.5 | 3.7 | -0.3 | 8.4 |
| Rainfall (mm) | | 42 | 36 | 60 | 61 | 50 | 22 | 3 | 2 | 6 | 40 | 47 | 42 | 411 |
| **Manaus** 84m | | | | | | | | | | | | | | |
| Temperature (°C) | high | 30.5 | 30.4 | 30.6 | 30.7 | 30.8 | 31.0 | 31.3 | 32.6 | 32.9 | 32.8 | 32.1 | 31.3 | 31.4 |
| | **mean** | 26.1 | 26.0 | 26.1 | 26.3 | 26.3 | 26.4 | 26.5 | 27.0 | 27.5 | 27.6 | 27.3 | 26.7 | 26.7 |
| | low | 23.1 | 23.1 | 23.2 | 23.3 | 23.3 | 23.0 | 22.7 | 23.0 | 23.5 | 23.7 | 23.7 | 23.5 | 23.3 |
| Rainfall (mm) | | 260 | 288 | 314 | 300 | 256 | 114 | 88 | 58 | 83 | 126 | 183 | 217 | 2287 |
| | | **Jan** | **Feb** | **Mar** | **Apr** | **May** | **Jun** | **Jul** | **Aug** | **Sep** | **Oct** | **Nov** | **Dec** | **YEAR** |

| | | Jan | Feb | Mar | Apr | May | Jun | Jul | Aug | Sep | Oct | Nov | Dec |
|---|---|---|---|---|---|---|---|---|---|---|---|---|---|
| **Meekatharra** 518m | | | | | | | | | | | | | |
| Temperature (°C) | high | 38.1 | 36.5 | 34.5 | 29.2 | 23.6 | 19.7 | 18.9 | 21.0 | 25.4 | 29.4 | 33.1 | 36.5 |
| | **mean** | 31.2 | 30.1 | 28.0 | 23.2 | 17.8 | 14.3 | 13.2 | 14.8 | 18.4 | 22.2 | 25.9 | 29.3 |
| | low | 24.3 | 23.7 | 21.5 | 17.1 | 11.9 | 8.9 | 7.5 | 8.5 | 11.4 | 15.0 | 18.6 | 22.1 |
| Rainfall (mm) | | 26 | 30 | 22 | 17 | 27 | 36 | 25 | 12 | 6 | 7 | 14 | 11 |
| **Minneapolis-St. Paul** 255m | | | | | | | | | | | | | |
| Temperature (°C) | high | -6.3 | -3.0 | 4.0 | 13.6 | 20.8 | 26.0 | 28.9 | 27.1 | 21.5 | 14.9 | 5.0 | -3.6 |
| | **mean** | -11.2 | -7.8 | -0.6 | 8.0 | 14.7 | 20.1 | 23.1 | 21.4 | 15.8 | 9.3 | 0.7 | -7.8 |
| | low | -16.2 | -12.7 | -5.2 | 2.3 | 8.7 | 14.2 | 17.3 | 15.7 | 10.2 | 3.8 | -3.8 | -12.1 |
| Rainfall (mm) | | 24 | 22 | 49 | 62 | 86 | 103 | 90 | 92 | 69 | 56 | 39 | 27 |
| **Ndola** 1270m | | | | | | | | | | | | | |
| Temperature (°C) | high | 26.6 | 26.9 | 27.4 | 27.5 | 26.6 | 25.1 | 25.2 | 27.5 | 30.5 | 31.5 | 29.4 | 27.0 |
| | **mean** | 20.8 | 20.8 | 21.0 | 20.5 | 18.6 | 16.5 | 16.7 | 19.2 | 22.5 | 23.7 | 22.5 | 21.0 |
| | low | 17.1 | 17.1 | 16.5 | 14.4 | 10.8 | 7.9 | 7.8 | 10.2 | 13.6 | 16.2 | 17.1 | 17.2 |
| Rainfall (mm) | | 29.3 | 249 | 170 | 46 | 4 | 1 | 0 | 0 | 3 | 32 | 130 | 306 |
| **Nuuk** 70m | | | | | | | | | | | | | |
| Temperature (°C) | high | -4.4 | -4.5 | -4.8 | -0.8 | 3.5 | 7.7 | 10.6 | 9.9 | 6.3 | 1.7 | -1.0 | -3.3 |
| | **mean** | -7.4 | -7.8 | -8.0 | -3.9 | 0.6 | 3.9 | 6.5 | 6.1 | 3.5 | -0.6 | -3.6 | -6.2 |
| | low | -10.1 | -10.6 | -10.6 | -6.1 | -1.5 | 1.3 | 3.8 | 3.8 | 1.6 | -2.5 | -5.8 | -8.7 |
| Rainfall (mm) | | 39 | 47 | 50 | 46 | 55 | 62 | 82 | 89 | 88 | 70 | 74 | 54 |
| **Paris** 65m | | | | | | | | | | | | | |
| Temperature (°C) | high | 6.0 | 7.6 | 10.8 | 14.4 | 18.2 | 21.5 | 24.0 | 23.8 | 20.8 | 16.0 | 10.1 | 6.8 |
| | **mean** | 3.4 | 4.2 | 6.6 | 9.5 | 13.2 | 16.4 | 18.4 | 18.0 | 15.3 | 11.4 | 6.7 | 4.2 |
| | low | 0.9 | 1.3 | 2.9 | 5.0 | 8.3 | 11.2 | 12.9 | 12.7 | 10.6 | 7.7 | 3.8 | 1.7 |
| Rainfall (mm) | | 54 | 46 | 54 | 47 | 63 | 58 | 84 | 52 | 54 | 56 | 56 | 56 |
| **Qiqihar** 148m | | | | | | | | | | | | | |
| Temperature (°C) | high | -12.7 | -7.8 | 2.3 | 12.9 | 21.0 | 26.2 | 27.8 | 26.1 | 20.1 | 11.1 | -1.3 | -10.4 |
| | **mean** | -19.2 | -14.8 | -4.5 | 6.1 | 14.4 | 20.3 | 22.8 | 20.9 | 14.0 | 4.8 | -7.1 | -16.2 |
| | low | -24.5 | -20.9 | -11.0 | -0.9 | 7.3 | 14.2 | 17.9 | 16.2 | 8.5 | -0.7 | -12.0 | -21.2 |
| Rainfall (mm) | | 1 | 2 | 5 | 15 | 31 | 64 | 138 | 94 | 45 | 19 | 4 | 3 |
| **Rabat Sale** 75m | | | | | | | | | | | | | |
| Temperature (°C) | high | 17.2 | 17.7 | 19.2 | 20.0 | 22.1 | 24.1 | 26.8 | 27.1 | 26.4 | 24.0 | 20.6 | 17.7 |
| | **mean** | 12.6 | 13.1 | 14.2 | 15.2 | 17.4 | 19.8 | 22.2 | 22.4 | 21.5 | 19.0 | 15.9 | 13.2 |
| | low | 8.0 | 8.6 | 9.2 | 10.4 | 12.7 | 15.4 | 17.6 | 17.7 | 16.7 | 14.1 | 11.1 | 8.7 |
| Rainfall (mm) | | 77 | 74 | 61 | 62 | 25 | 7 | 1 | 1 | 6 | 44 | 97 | 101 |
| **Sittwe** 5m | | | | | | | | | | | | | |
| Temperature (°C) | high | 28.0 | 29.4 | 31.4 | 34.1 | 31.5 | 29.5 | 28.9 | 28.9 | 30.1 | 31.1 | 30.3 | 28.5 |
| | **mean** | 21.4 | 22.7 | 24.8 | 28.9 | 28.3 | 27.1 | 26.8 | 26.7 | 27.4 | 27.6 | 25.7 | 22.6 |
| | low | 14.7 | 15.9 | 18.2 | 23.6 | 25.1 | 24.6 | 24.7 | 24.5 | 24.6 | 24.0 | 21.0 | 16.6 |
| Rainfall (mm) | | 11 | 8 | 5 | 44 | 268 | 1091 | 1155 | 1025 | 537 | 289 | 105 | 17 |
| **Stockholm** 52m | | | | | | | | | | | | | |
| Temperature (°C) | high | -0.7 | -0.6 | 3.0 | 8.6 | 15.7 | 20.7 | 21.9 | 20.4 | 15.1 | 9.9 | 4.5 | 1.1 |
| | **mean** | -2.8 | -3.0 | 0.1 | 4.6 | 10.7 | 15.6 | 17.2 | 16.2 | 11.9 | 7.5 | 2.6 | -1.0 |
| | low | -5.0 | -5.3 | -2.7 | 1.1 | 6.3 | 11.3 | 13.4 | 12.7 | 9.0 | 5.3 | 0.7 | -3.2 |
| Rainfall (mm) | | 39 | 27 | 26 | 30 | 30 | 45 | 72 | 66 | 55 | 50 | 53 | 46 |
| **Tehran** 1191m | | | | | | | | | | | | | |
| Temperature (°C) | high | 7.2 | 9.9 | 15.4 | 21.9 | 28.0 | 34.1 | 36.8 | 35.4 | 31.5 | 24.0 | 16.5 | 9.8 |
| | **mean** | 3.0 | 5.3 | 10.3 | 16.4 | 22.1 | 27.5 | 30.4 | 29.2 | 25.3 | 18.5 | 11.6 | 5.6 |
| | low | -1.1 | 0.7 | 5.2 | 10.9 | 16.1 | 20.9 | 24.0 | 23.0 | 19.2 | 12.9 | 6.7 | 1.3 |
| Rainfall (mm) | | 37 | 34 | 37 | 28 | 15 | 3 | 3 | 1 | 1 | 14 | 21 | 36 |
| **Wellington** 8m | | | | | | | | | | | | | |
| Temperature (°C) | high | 21.3 | 21.1 | 19.8 | 17.3 | 14.8 | 12.8 | 12.0 | 12.7 | 14.2 | 15.9 | 17.8 | 19.6 |
| | **mean** | 17.8 | 17.7 | 16.6 | 14.3 | 11.9 | 10.1 | 9.2 | 9.8 | 11.2 | 12.8 | 14.5 | 16.4 |
| | low | 14.4 | 14.3 | 13.5 | 11.3 | 9.1 | 7.3 | 6.4 | 6.9 | 8.3 | 9.7 | 11.3 | 13.2 |
| Rainfall (mm) | | 67 | 48 | 76 | 87 | 99 | 113 | 111 | 106 | 82 | 81 | 74 | 74 |
| | | **Jan** | **Feb** | **Mar** | **Apr** | **May** | **Jun** | **Jul** | **Aug** | **Sep** | **Oct** | **Nov** | **Dec** |

## Tropical revolving storms

temperature 27°C and over at mean sea level

**August–September**
Maximum frequency in
northern hemisphere

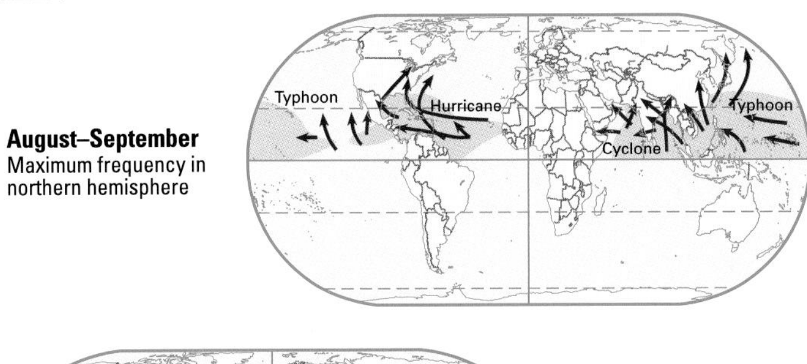

**January–March**
Maximum frequency in
southern hemisphere

## Hurricane Floyd, Florida

Winds in this hurricane reached 225km per hour and caused 40 deaths.
US NOAA satellite image, 15 September, 1999.

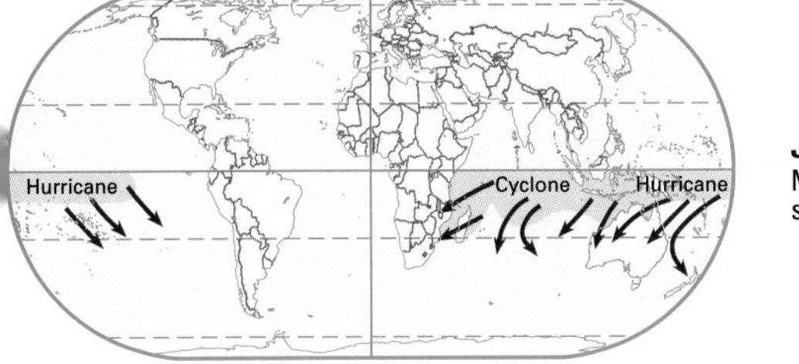

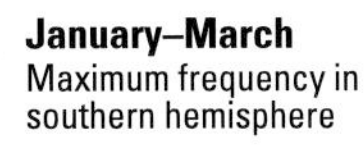

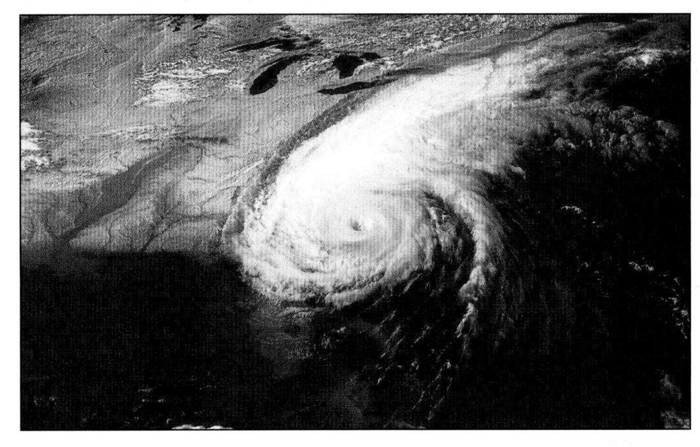

## Drought and flood

areas where severe drought may occur

major river flood plains susceptible to flooding

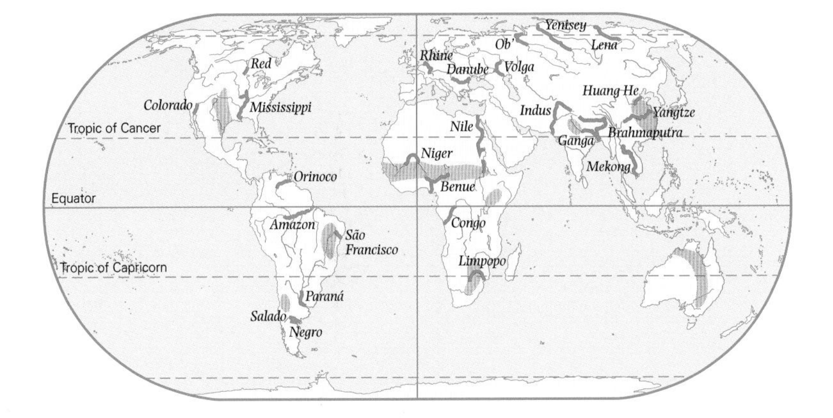

## Dust storms, South West Africa

Dust streaming from SW African coastal
deserts into the Atlantic Ocean.
NASA SeaWiFS image, 6 June, 2000.

## El Niño

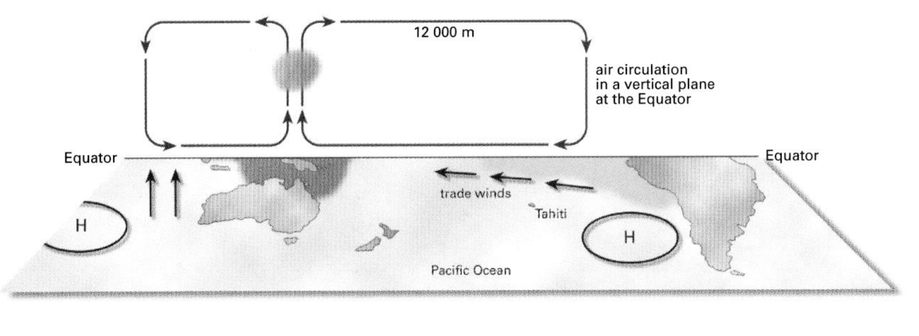

**Normal year**
The Humboldt current carries cold water north along the coast of Peru.
High temperatures in S.E. Asia draw in the S.E. Trade Winds which
push the surface waters west. Rainfall in S.E. Asia is high. Cold water
continues to flow north along the coast of S. America and this is rich in
plankton and fish.

**El Niño year**
Weaker S.E. Trade Winds allow hot water from the western Pacific to
drift eastwards. Warm waters appear in Peru at about Christmas time.
Arid coastal areas in S. America suffer torrential rains. Coastal fish
stocks move to deeper cold water out of reach of small boats.
Drought occurs in S.E. Asia.

Equatorial scale 1: 105 000 000

## Ecosystems

vegetation types are those which would occur naturally without interference by people

**coniferous forest**
cone bearing trees

**deciduous and mixed forest**
leaf shedding and coniferous trees

**tropical rain forest**
many species of lush, tall trees

**tropical grasslands (savannah)**
tall grass parkland with scattered trees

**evergreen trees and shrubs**
plants and small trees with leathery leaves

**thorn forest**
low trees and shrubs with spines or thorns

**temperate grasslands**
prairies, steppes, pampas, and veld

**semi-desert**
short grasses and drought-resistant scrub

**desert**
sand and stones, very little vegetation

**tundra**
moss and lichen, with few trees

**ice**
no vegetation

**mountains**
thin soils, steep slopes, and high altitude affects type of vegetation

**ice**
Aerial view of Jameson Land, towards Liverpool Land, Greenland

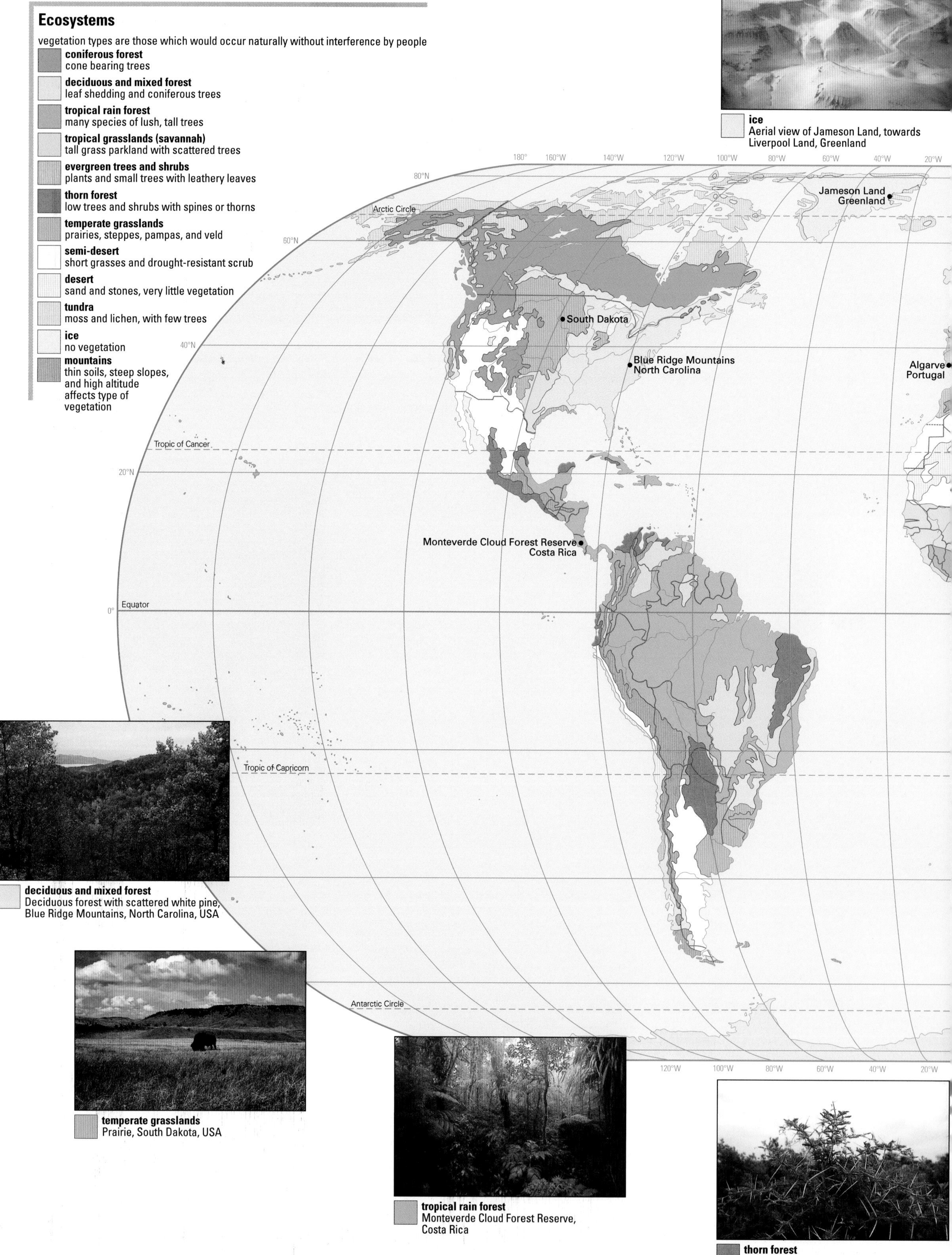

**deciduous and mixed forest**
Deciduous forest with scattered white pine, Blue Ridge Mountains, North Carolina, USA

**temperate grasslands**
Prairie, South Dakota, USA

**tropical rain forest**
Monteverde Cloud Forest Reserve, Costa Rica

**thorn forest**
Acacia thorns, Hwange, Zimbabwe

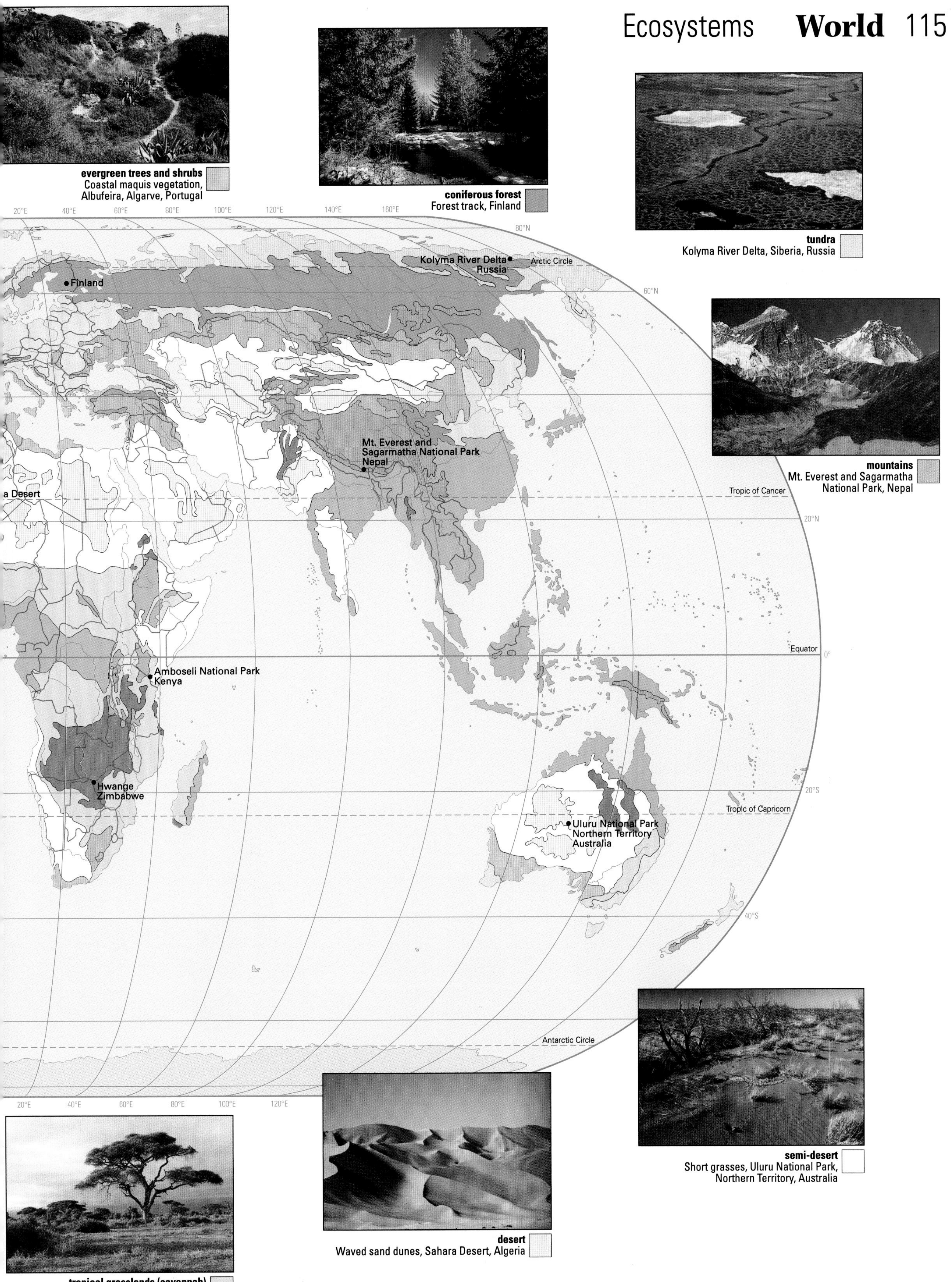

**evergreen trees and shrubs**
Coastal maquis vegetation,
Albufeira, Algarve, Portugal

**coniferous forest**
Forest track, Finland

**tundra**
Kolyma River Delta, Siberia, Russia

**mountains**
Mt. Everest and Sagarmatha
National Park, Nepal

**semi-desert**
Short grasses, Uluru National Park,
Northern Territory, Australia

**desert**
Waved sand dunes, Sahara Desert, Algeria

**tropical grasslands (savannah)**
Amboseli National Park, Kenya

Finland

Kolyma River Delta
Russia

Arctic Circle

Mt. Everest and
Sagarmatha National Park
Nepal

Tropic of Cancer

20°N

Amboseli National Park
Kenya

Equator    0°

Hwange
Zimbabwe

20°S

Tropic of Capricorn

Uluru National Park
Northern Territory
Australia

40°S

Antarctic Circle

a Desert

20°E   40°E   60°E   80°E   100°E   120°E   140°E   160°E

80°N

60°N

20°E   40°E   60°E   80°E   100°E   120°E

Eckert IV Projection    © Oxford University Press

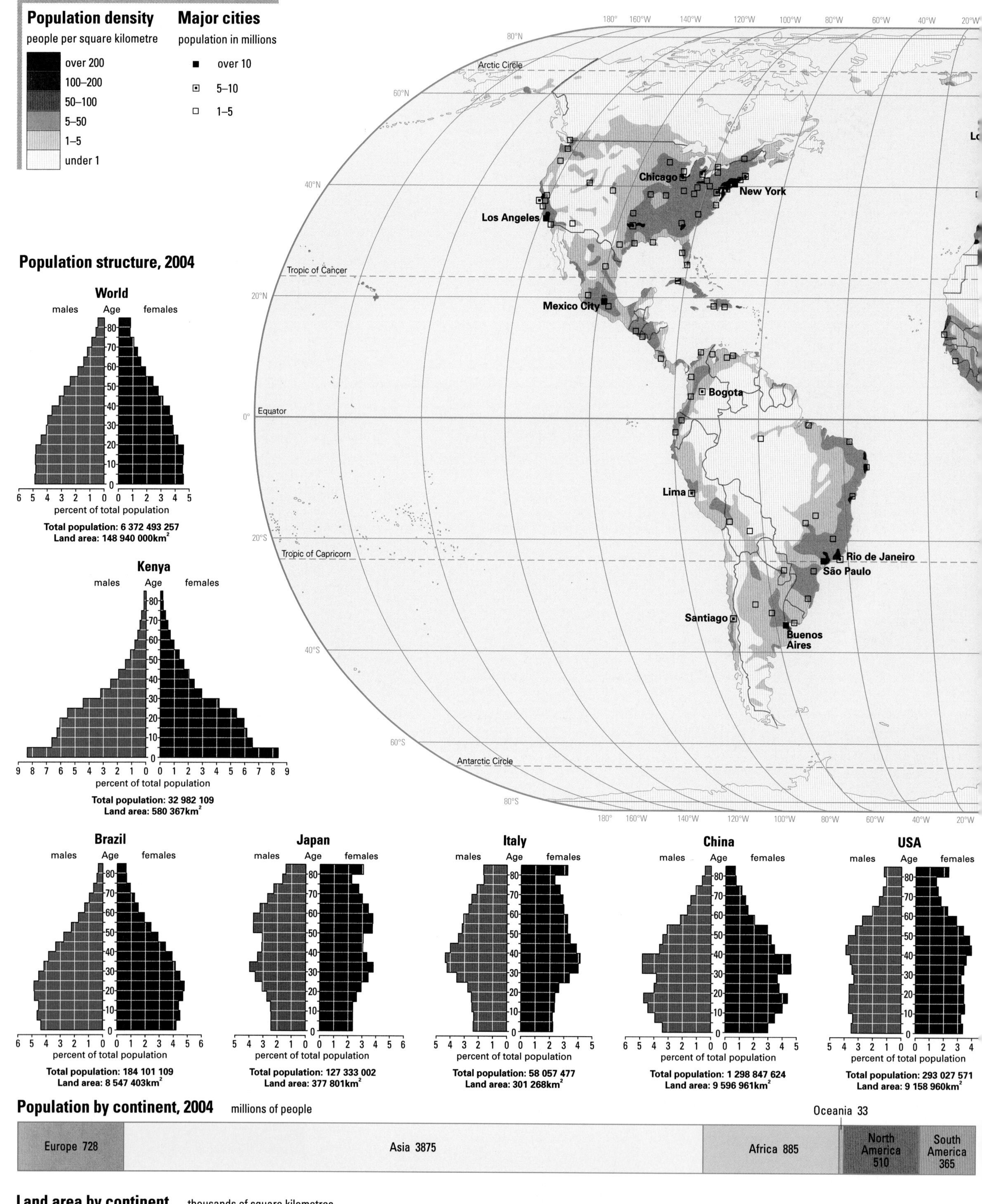

## Population density
people per square kilometre

- over 200
- 100–200
- 50–100
- 5–50
- 1–5
- under 1

## Major cities
population in millions

- ■ over 10
- ⊡ 5–10
- □ 1–5

## Population structure, 2004

### World
males  Age  females

percent of total population

Total population: 6 372 493 257
Land area: 148 940 000km²

### Kenya
males  Age  females

percent of total population

Total population: 32 982 109
Land area: 580 367km²

### Brazil
males  Age  females

percent of total population

Total population: 184 101 109
Land area: 8 547 403km²

### Japan
males  Age  females

percent of total population

Total population: 127 333 002
Land area: 377 801km²

### Italy
males  Age  females

percent of total population

Total population: 58 057 477
Land area: 301 268km²

### China
males  Age  females

percent of total population

Total population: 1 298 847 624
Land area: 9 596 961km²

### USA
males  Age  females

percent of total population

Total population: 293 027 571
Land area: 9 158 960km²

## Population by continent, 2004  millions of people

| Europe 728 | Asia 3875 | Africa 885 | Oceania 33 | North America 510 | South America 365 |
|---|---|---|---|---|---|

## Land area by continent  thousands of square kilometres

| Europe 10 498 | Asia 44 387 | Africa 30 335 | Oceania 8503 | North America 24 241 | South America 17 832 | Antarctica 13 340 |
|---|---|---|---|---|---|---|

Map labels: Chicago, New York, Los Angeles, Mexico City, Bogotá, Lima, Rio de Janeiro, São Paulo, Santiago, Buenos Aires

Arctic Circle, Tropic of Cancer, Equator, Tropic of Capricorn, Antarctic Circle

Eckert IV Projection  © Oxford University Press

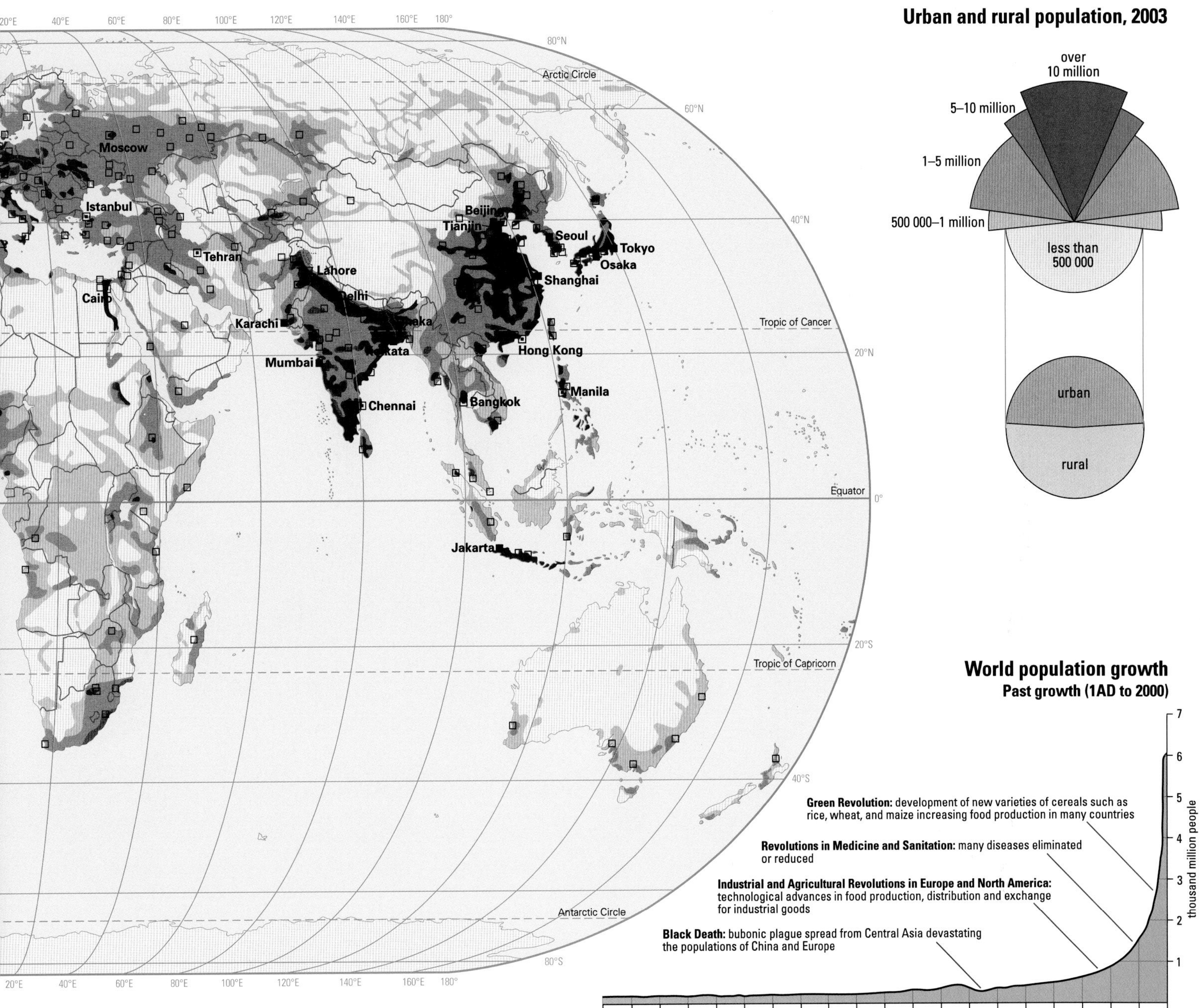

## Urban and rural population, 2003

over 10 million
5–10 million
1–5 million
500 000–1 million
less than 500 000

urban
rural

## World population growth
### Past growth (1AD to 2000)

**Green Revolution:** development of new varieties of cereals such as rice, wheat, and maize increasing food production in many countries

**Revolutions in Medicine and Sanitation:** many diseases eliminated or reduced

**Industrial and Agricultural Revolutions in Europe and North America:** technological advances in food production, distribution and exchange for industrial goods

**Black Death:** bubonic plague spread from Central Asia devastating the populations of China and Europe

thousand million people

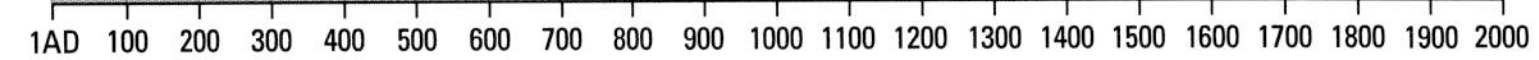

1AD 100 200 300 400 500 600 700 800 900 1000 1100 1200 1300 1400 1500 1600 1700 1800 1900 2000

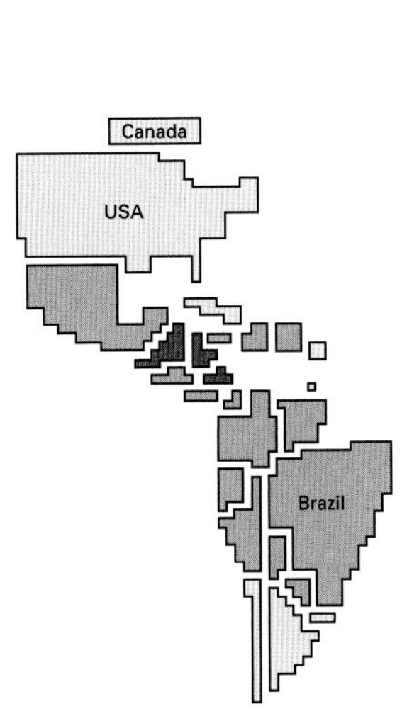

## Population cartogram, 2004

the size of each country represents the number of people living there

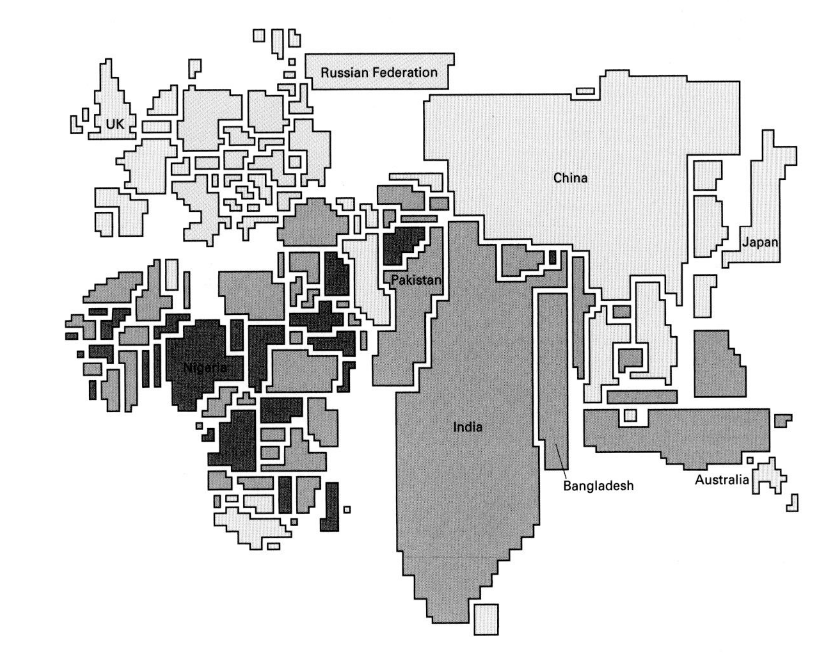

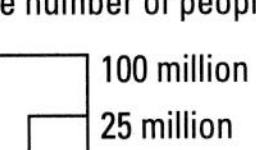

100 million
25 million
1 million

## Population change

average annual increase or decrease

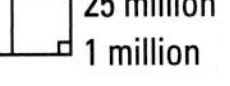

very high increase (over 2.6%)
increase above world average (1.3–2.6%)
increase below world average (0–1.3%)
decrease (by less than 1%)

## Population change, 1994–2004

percentage population gain or loss

- over 40% gain
- 30–40% gain
- 20–30% gain
- 10–20% gain
- under 10% gain
- 0–20% loss

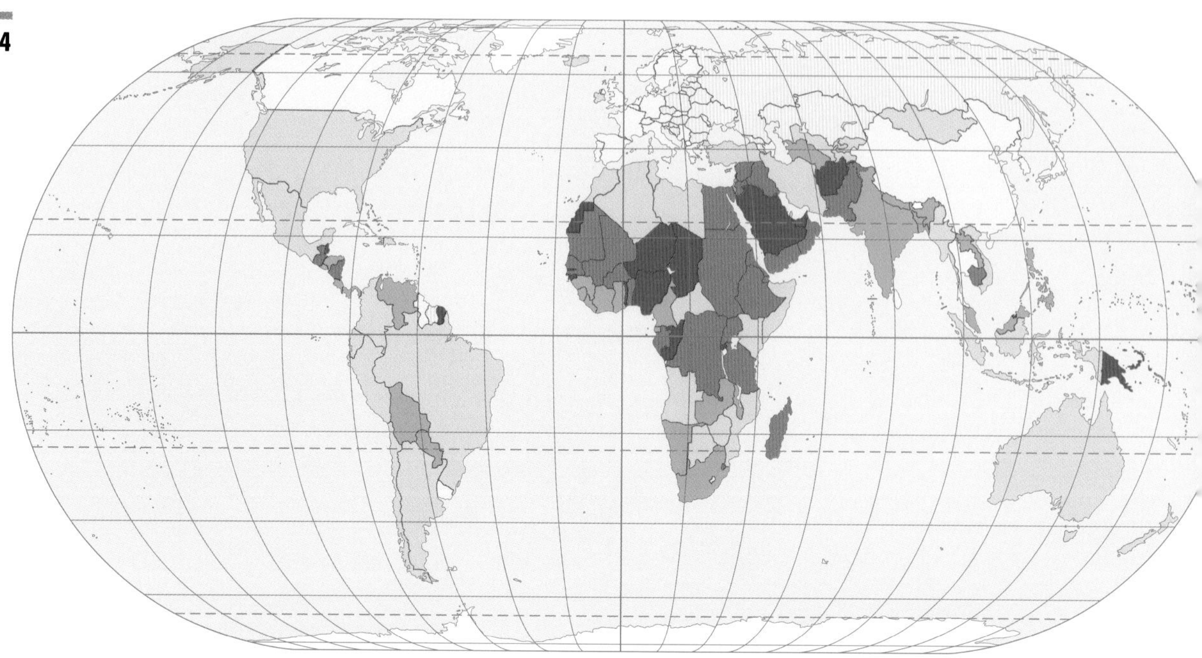

**Highest population gain**
Marshall Islands 85.2%
United Arab Emirates 83.2%
Rwanda 65.4%
São Tomé and Príncipe 60%
Afghanistan 57.7%

United Kingdom 2.2%

**Highest population loss**
Albania -9.8%
Bosnia-Herzegovina -12.5%
Estonia -13.3%
Armenia -14.6%
Georgia -17.1%

## Urban population, 2003

percentage of the population living in urban areas

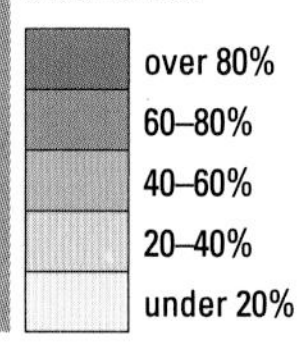

- over 80%
- 60–80%
- 40–60%
- 20–40%
- under 20%

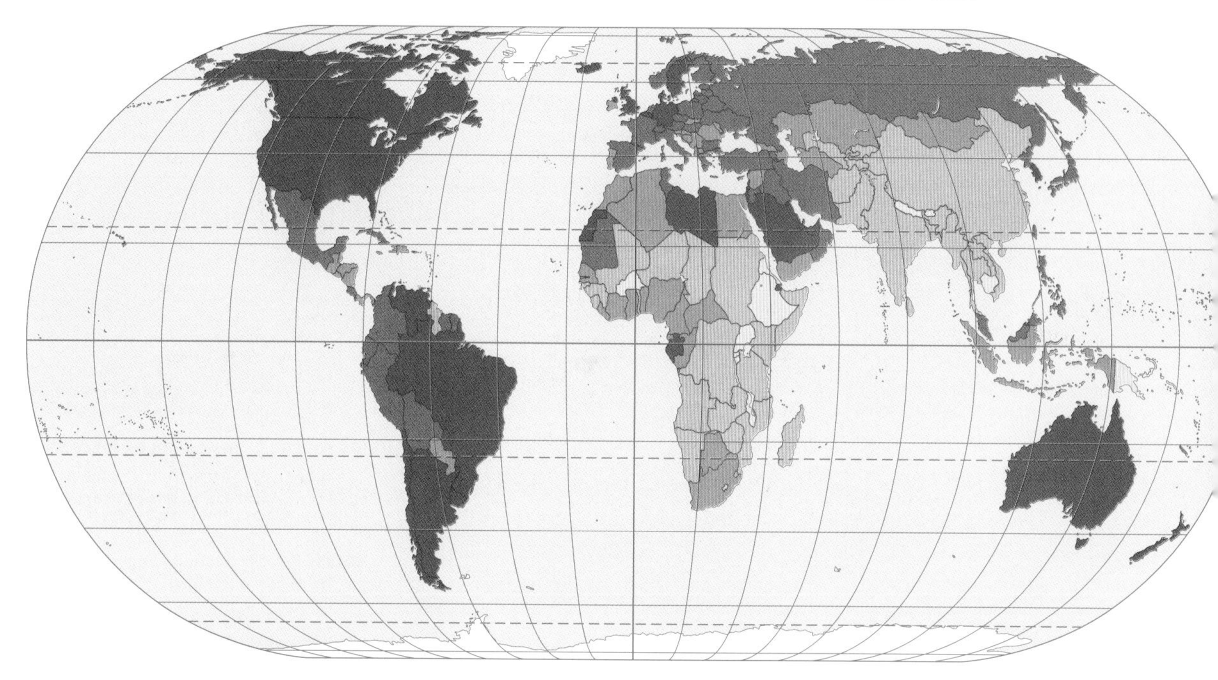

**Most urban**
Singapore 100%
Belgium 97%
Kuwait 96%
Iceland 93%
Uruguay 93%

United Kingdom 89%

**Least urban**
Papua New Guinea 13%
Uganda 12%
Burundi 10%
Bhutan 8%
East Timor 8%

## Fertility rate, 2003

average number of children born to childbearing women

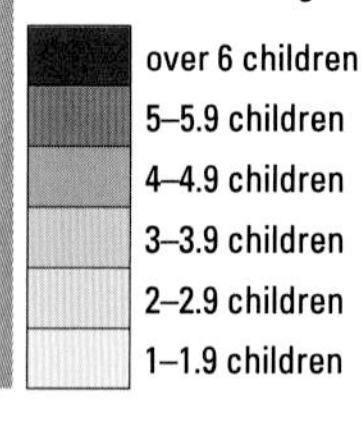

- over 6 children
- 5–5.9 children
- 4–4.9 children
- 3–3.9 children
- 2–2.9 children
- 1–1.9 children

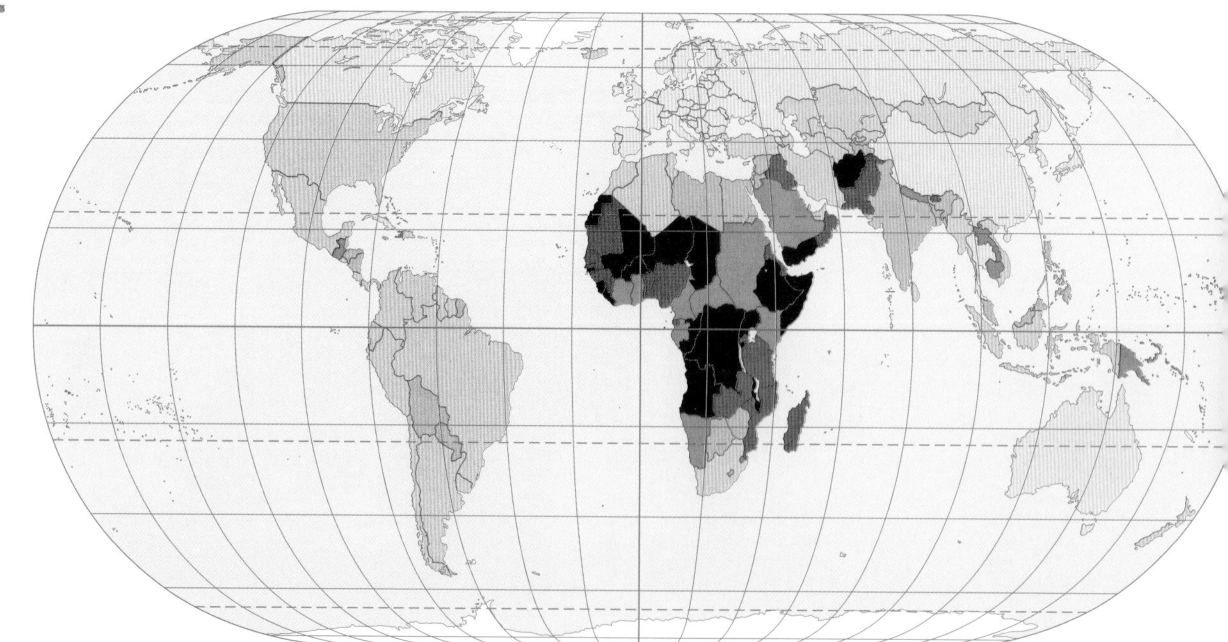

**Largest families**
Niger 8.0 children
Angola 7.2 children
Guinea-Bissau 7.1 children
Uganda 7.1 children
Yemen Republic 7.0 children

United Kingdom 1.6 children

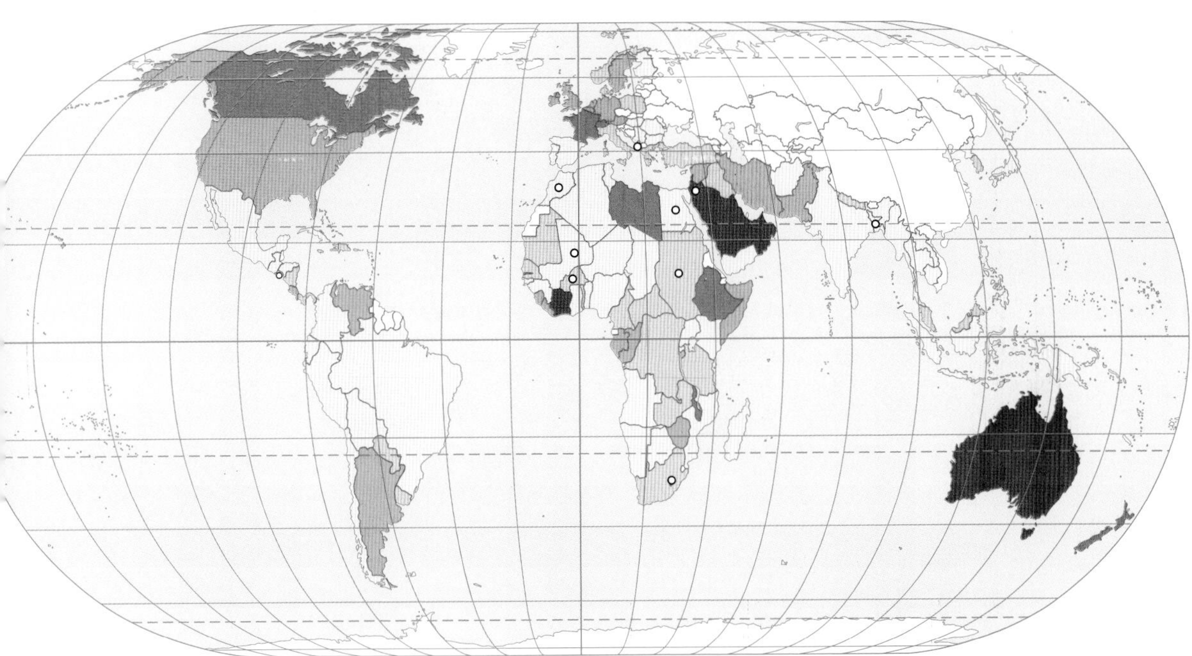

© Oxford University Press

## International migration

percentage of total population foreign born

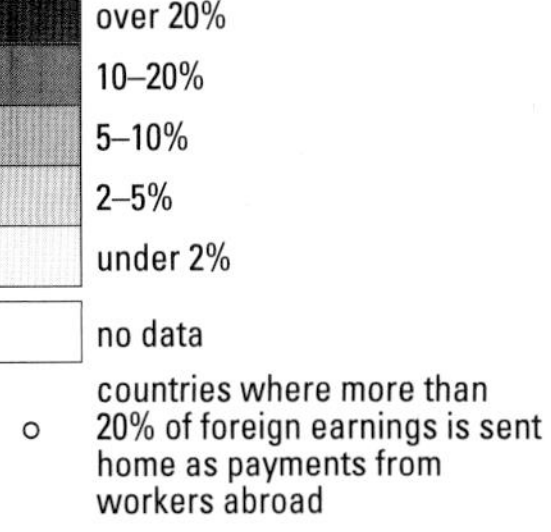

- over 20%
- 10–20%
- 5–10%
- 2–5%
- under 2%
- no data
- ○ countries where more than 20% of foreign earnings is sent home as payments from workers abroad

**Highest percentage of foreign born**
United Arab Emirates 90.2%
Oman 33.6%
Israel 30.9%
Côte d'Ivoire 29.3%
Jordan 26.4%

**United Kingdom 6.5%**

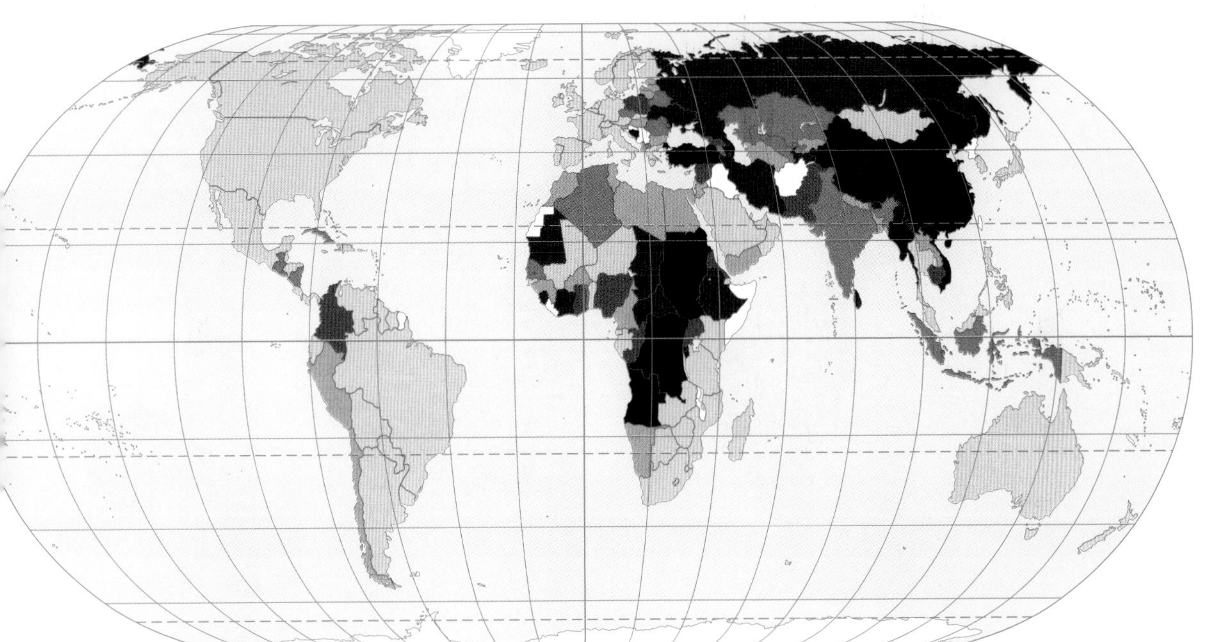

## Refugees by country of origin, 2003

number of applications for refugee status submitted during 2003

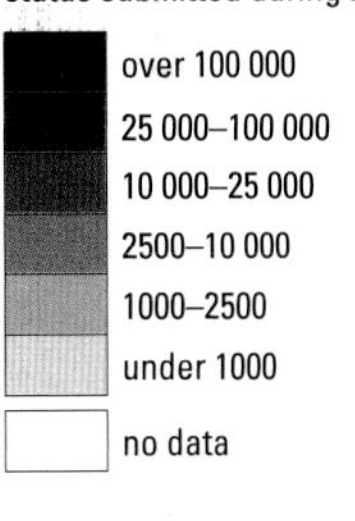

- over 100 000
- 25 000–100 000
- 10 000–25 000
- 2500–10 000
- 1000–2500
- under 1000
- no data

**Countries from which most refugees left, 2003**
Sudan 567 000
Burundi 525 000
Congo, Dem. Rep. 428 000
Vietnam 331 000
Angola 313 000
Azerbaijan 248 000
Croatia 215 000
Bosnia-Herzegovina 167 000

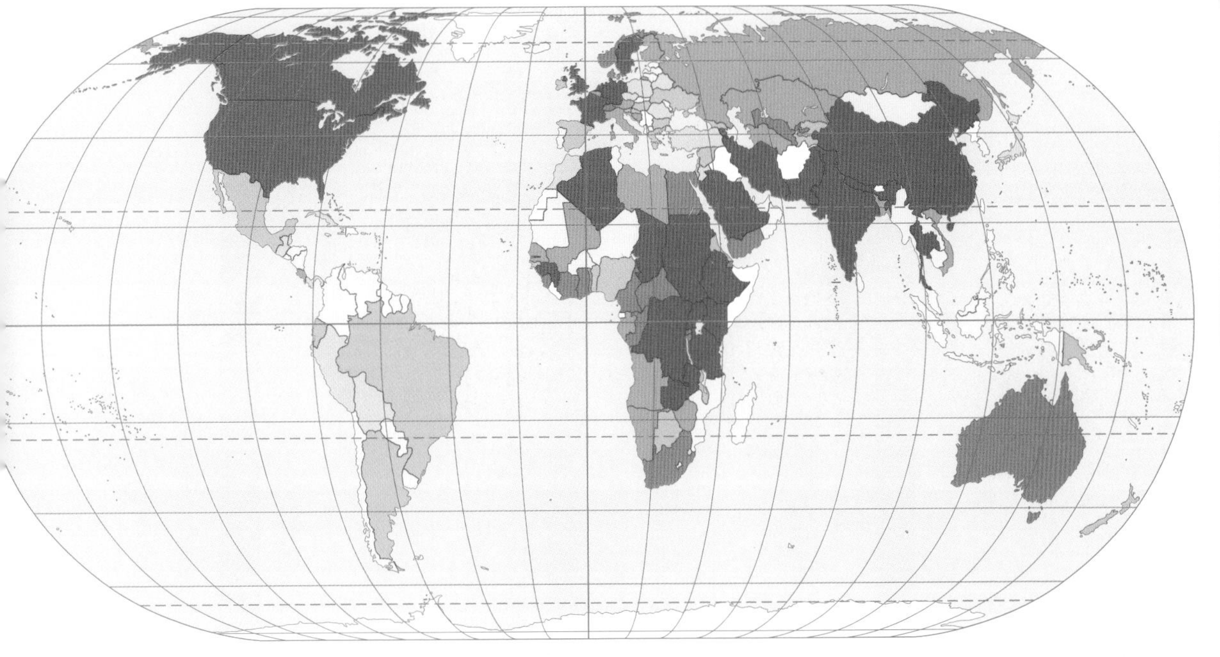

## Refugees by country of asylum, 2003

number of applications for refugee status received during 2003

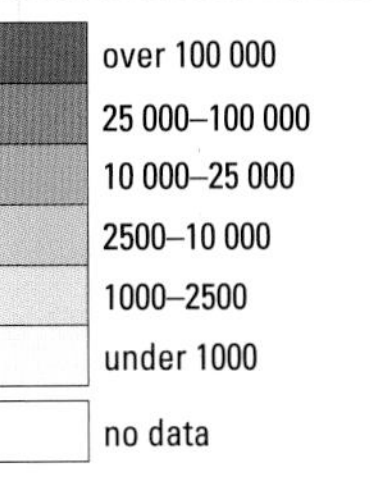

- over 100 000
- 25 000–100 000
- 10 000–25 000
- 2500–10 000
- 1000–2500
- under 1000
- no data

**Countries receiving the most refugees, 2003**
Pakistan 1 124 000
Iran 985 000
Germany 960 000
Tanzania 650 000
USA 453 000
Sudan 328 000
China 299 000
**United Kingdom 277 000**

## Purchasing power, 2002

Purchasing Power Parity (PPP) in US$
Based on Gross Domestic Product (GDP)
per person, adjusted for the local cost
of living

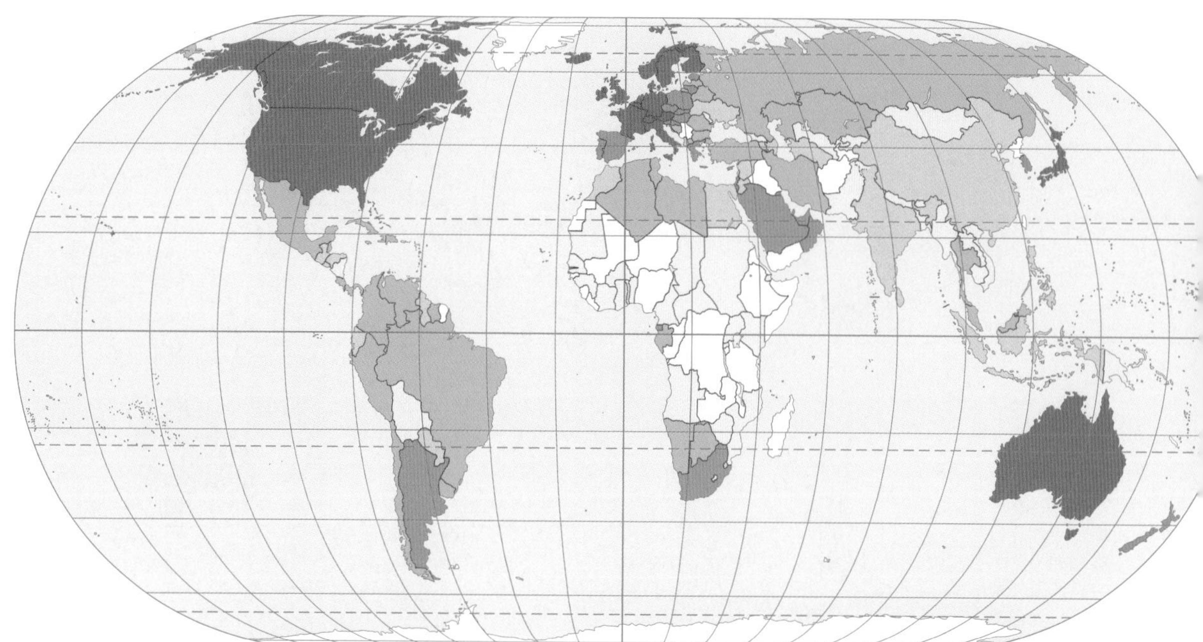

- over 25 000
- 10 000–25 000
- 5000–10 000
- 2500–5000
- 1000–2500
- under 1000
- no data

**Highest purchasing power**
Luxembourg $61 190
Norway $36 600
Ireland $36 360
United States $35 750
Denmark $30 940

**United Kingdom $26 150**

**Lowest purchasing power**
Congo, Democratic Republic $650
Burundi $630
Malawi $580
Tanzania $580
Sierra Leone $520

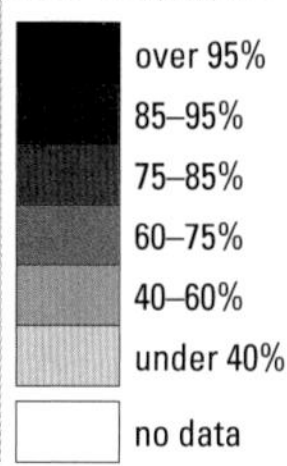

## Literacy and schooling, 2002

percentage of people aged 15 and above
who can, with understanding, both read
and write a short, simple statement on
their everyday life

- over 95%
- 85–95%
- 75–85%
- 60–75%
- 40–60%
- under 40%
- no data

**Highest literacy levels**
Georgia 100%
Estonia 99.8%
Slovenia 99.7%
Barbados 99.7%
Poland 99.7%

**United Kingdom 99%**

**Lowest literacy levels**
Sierra Leone 36%
Vanuatu 34%
Mali 19%
Niger 17.1%
Burkina 12.8%

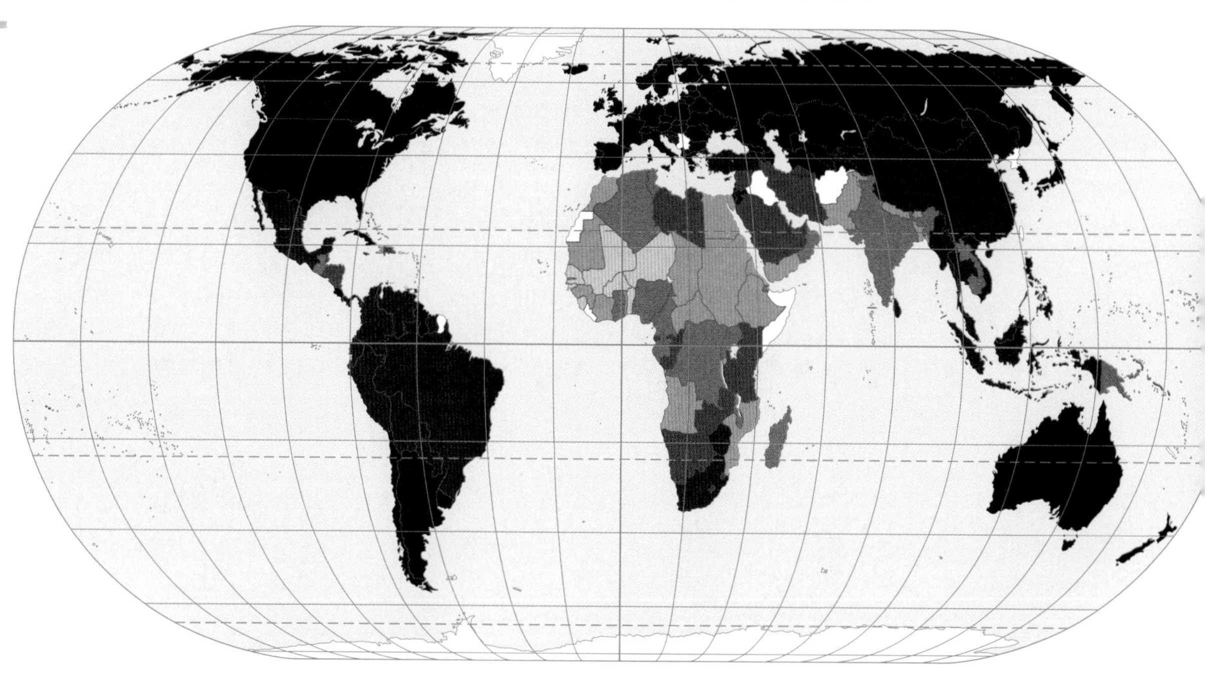

## Life expectancy, 2002

average expected lifespan of babies
born in 2002

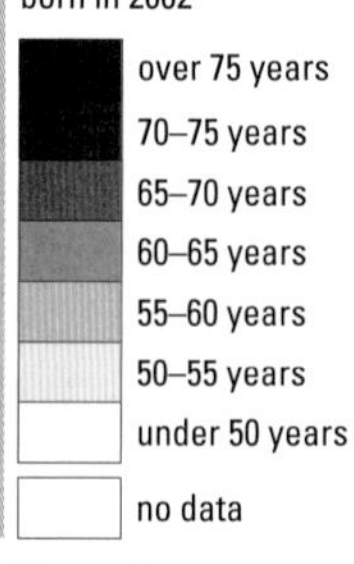

- over 75 years
- 70–75 years
- 65–70 years
- 60–65 years
- 55–60 years
- 50–55 years
- under 50 years
- no data

**Highest life expectancy**
Japan 82 years
Sweden 80 years
Iceland 80 years
Canada 79 years
Spain 79 years

**United Kingdom 78 years**

**Lowest life expectancy**
Lesotho 36 years
Swaziland 36 years
Sierra Leone 34 years
Zimbabwe 34 years
Zambia 33 years

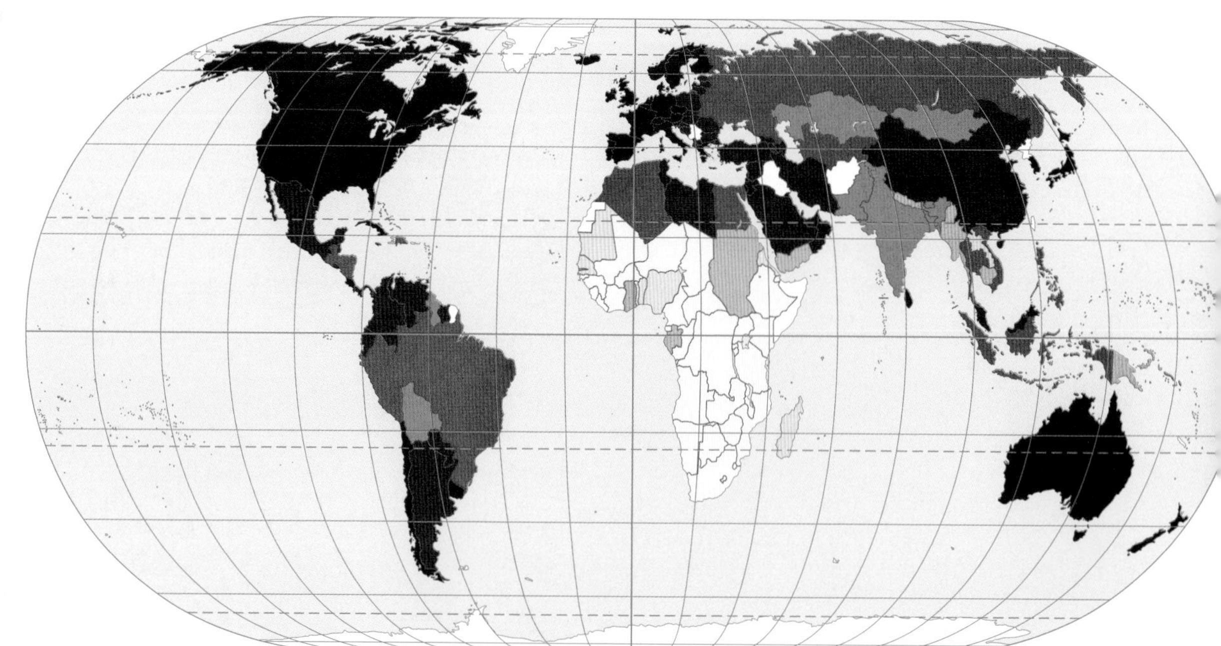

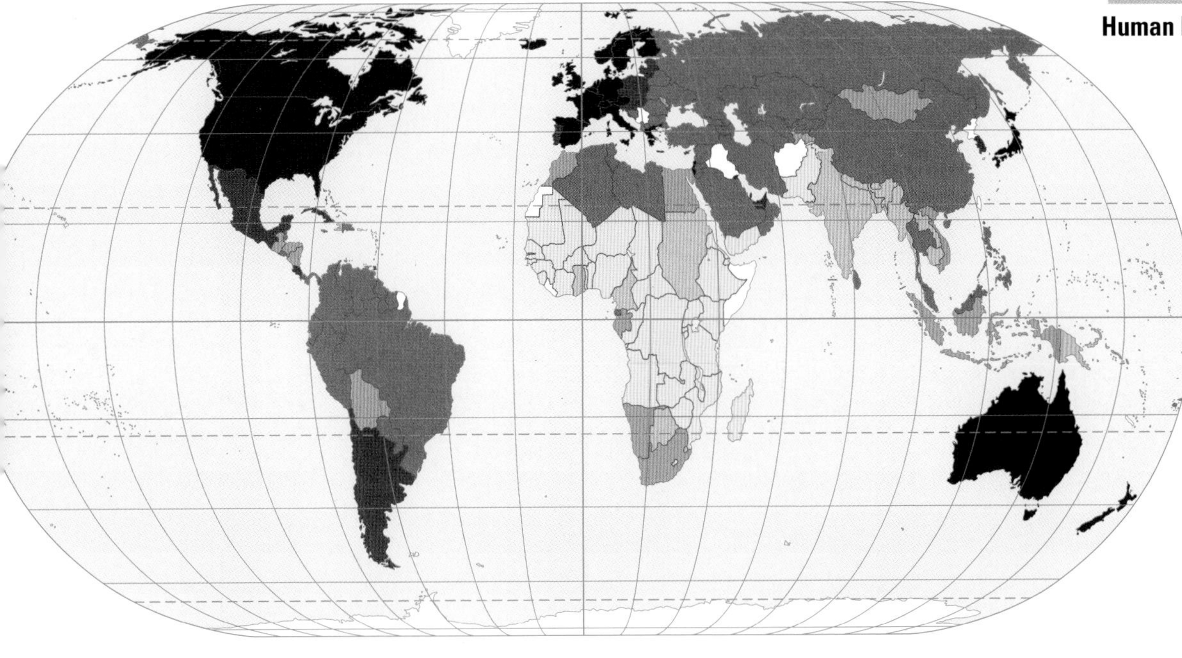

## Human Development Index (HDI), 2002

HDI measures the relative social and economic progress of a country. It combines life expectancy, adult literacy, average number of years of schooling, and purchasing power.

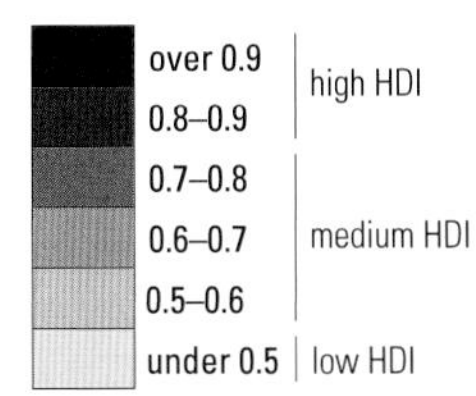

| | |
|---|---|
| over 0.9 | high HDI |
| 0.8–0.9 | |
| 0.7–0.8 | |
| 0.6–0.7 | medium HDI |
| 0.5–0.6 | |
| under 0.5 | low HDI |

**Highest HDI**
Norway 0.956
Sweden 0.946
Australia 0.946
Canada 0.943
Netherlands 0.942

**United Kingdom 0.936**

**Lowest HDI**
Burundi 0.339
Mali 0.326
Burkina 0.302
Niger 0.292
Sierra Leone 0.273

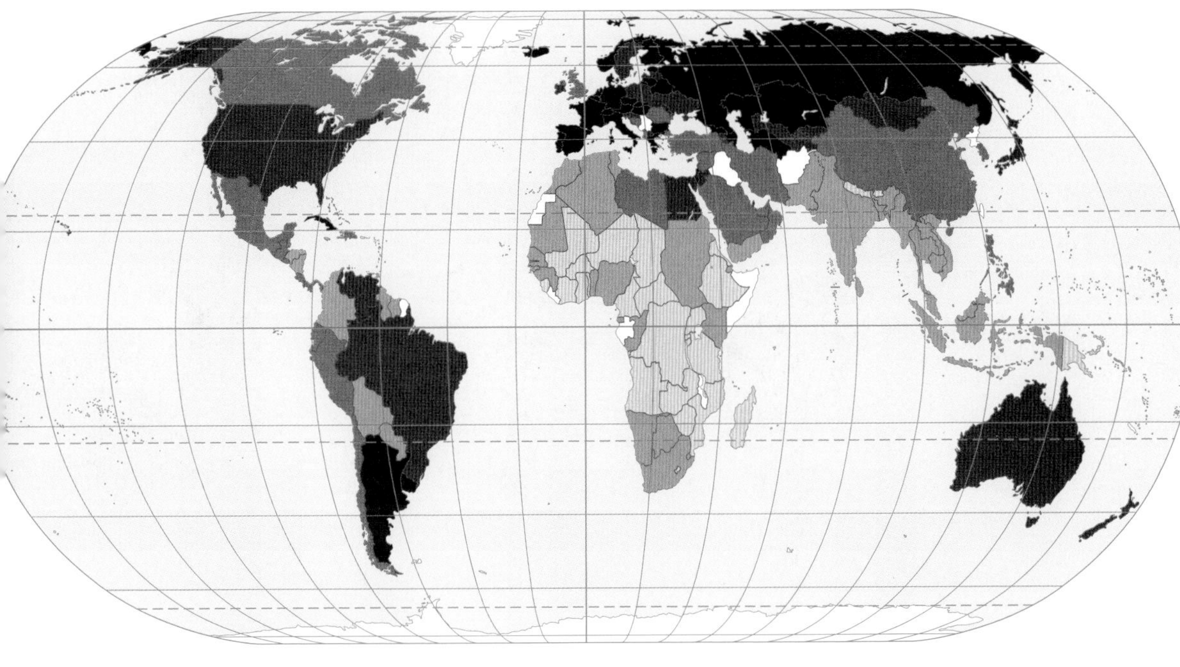

## Medical care, 1990–2003

number of doctors per 100 000 people

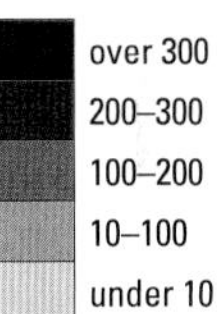

over 300
200–300
100–200
10–100
under 10

**Most doctors per 100 000 people**
Italy 607
Cuba 596
Georgia 463
Belarus 450
Greece 438

**United Kingdom 164**

**Fewest doctors per 100 000 people**
Chad 3
Ethiopia 3
Niger 3
Mozambique 2
Rwanda 2
Burundi 1

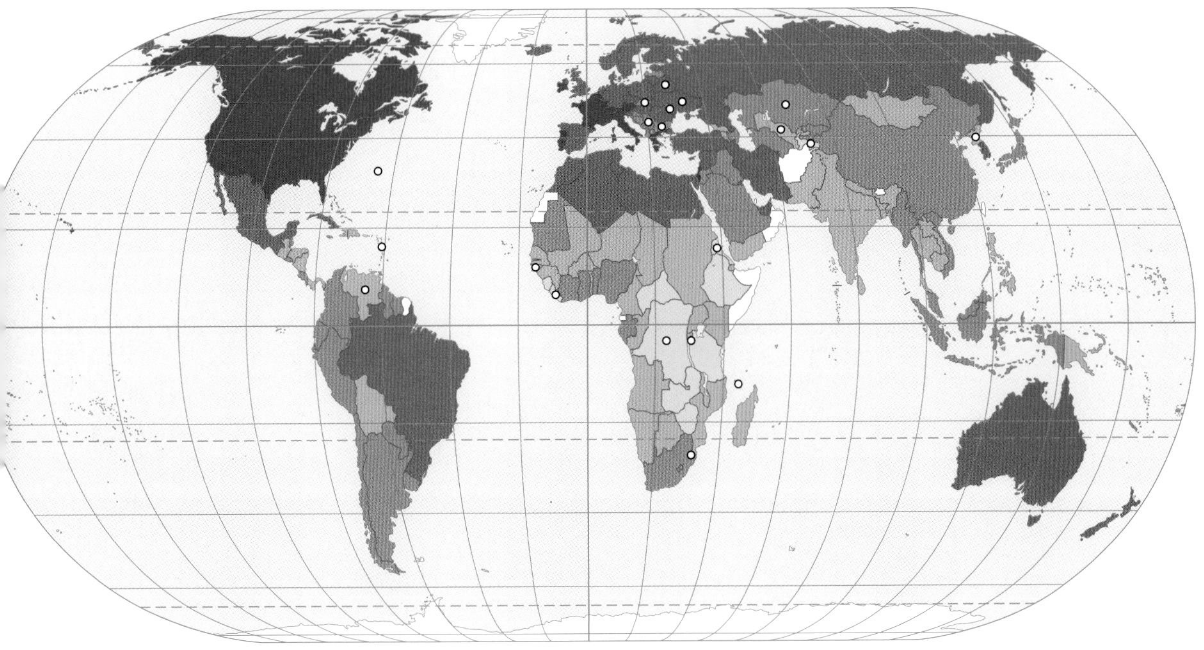

## Food consumption, 2002

average daily food intake in calories per person

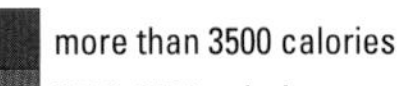

more than 3500 calories
3000–3500 calories
2500–3000 calories
2000–2500 calories
less than 2000 calories

o  average food consumption per head declining by more than 1%, 1992–2002

**Highest food consumption**
United States 3774
Portugal 3741
Greece 3721
Austria 3673
Italy 3671

**United Kingdom 3412**

**Lowest food consumption**
Tajikistan 1828
Comoros 1754
Burundi 1649
Congo, Democratic Republic 1597
Eritrea 1513

Scale 1 : 240 000 00

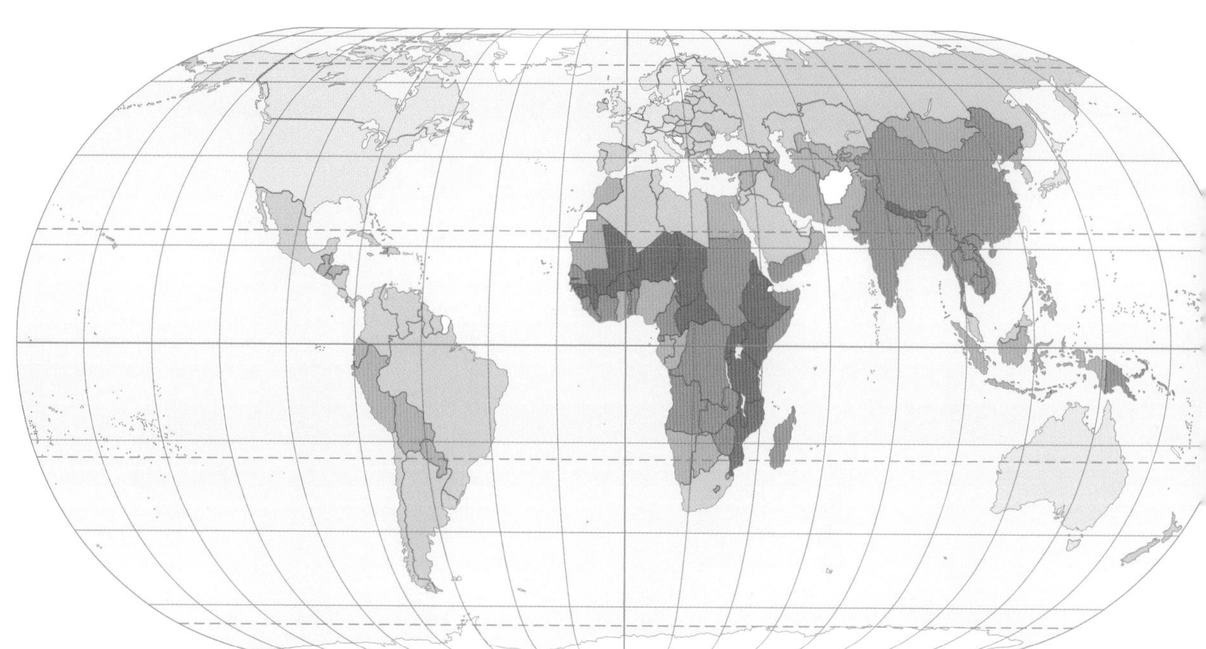

## Employment in agriculture

percentage of the labour force

- over 80%
- 60–80%
- 30–60%
- 10–30%
- under 10%
- no data

**Highest employment in agriculture**
Bhutan 94%
Nepal 94%
Burkina 92%
Burundi 92%
Rwanda 92%

**Lowest employment in agriculture**
Bahrain 2%
Brunei 2%
United Kingdom 2%
Kuwait 1%
Singapore 0%

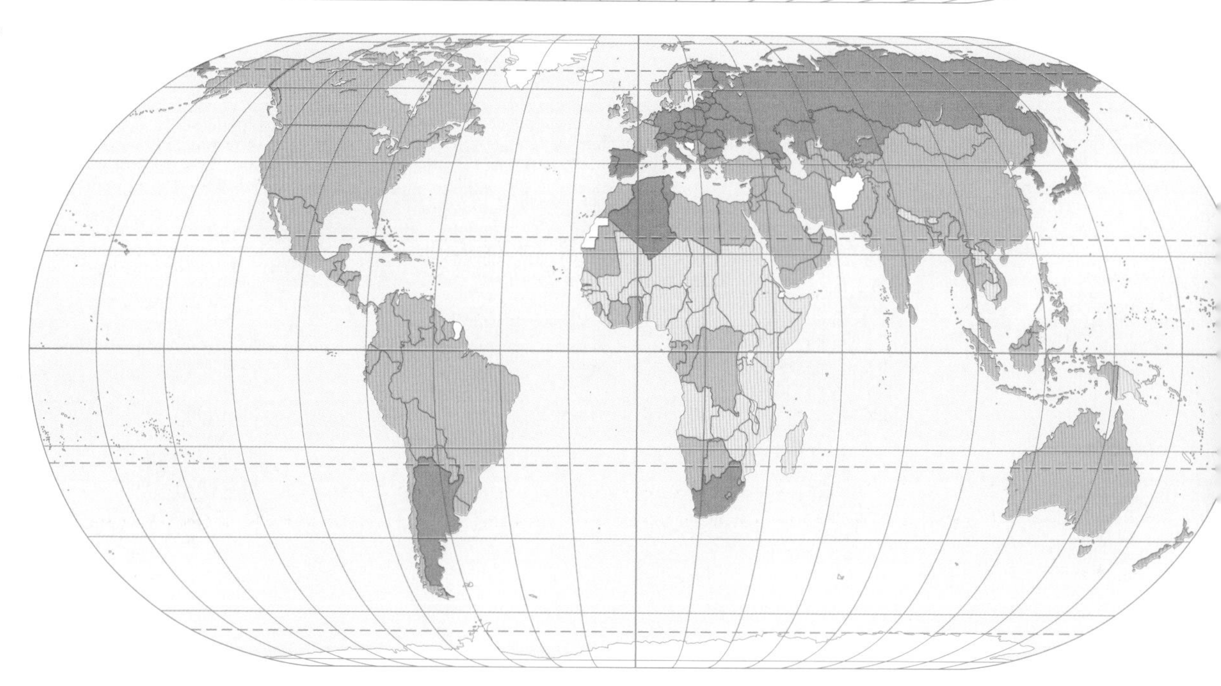

## Employment in industry

percentage of the labour force

- over 80%
- 60–80%
- 30–60%
- 10–30%
- under 10%
- no data

**Highest employment in industry**
Bulgaria 48%
Romaina 47%
Slovenia 46%
Czech Republic 45%
Armenia 43%
Mauritius 43%

United Kingdom 29%

**Lowest employment in industry**
Bhutan 2%
Burkina 2%
Ethiopia 2%
Guinea 2%
Guinea-Bissau 2%
Mali 2%
Nepal 0%

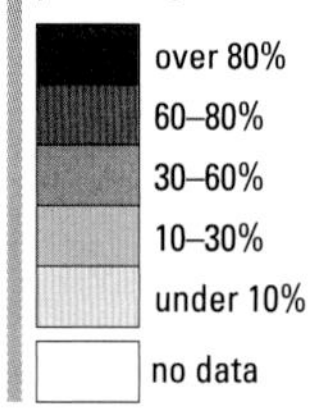

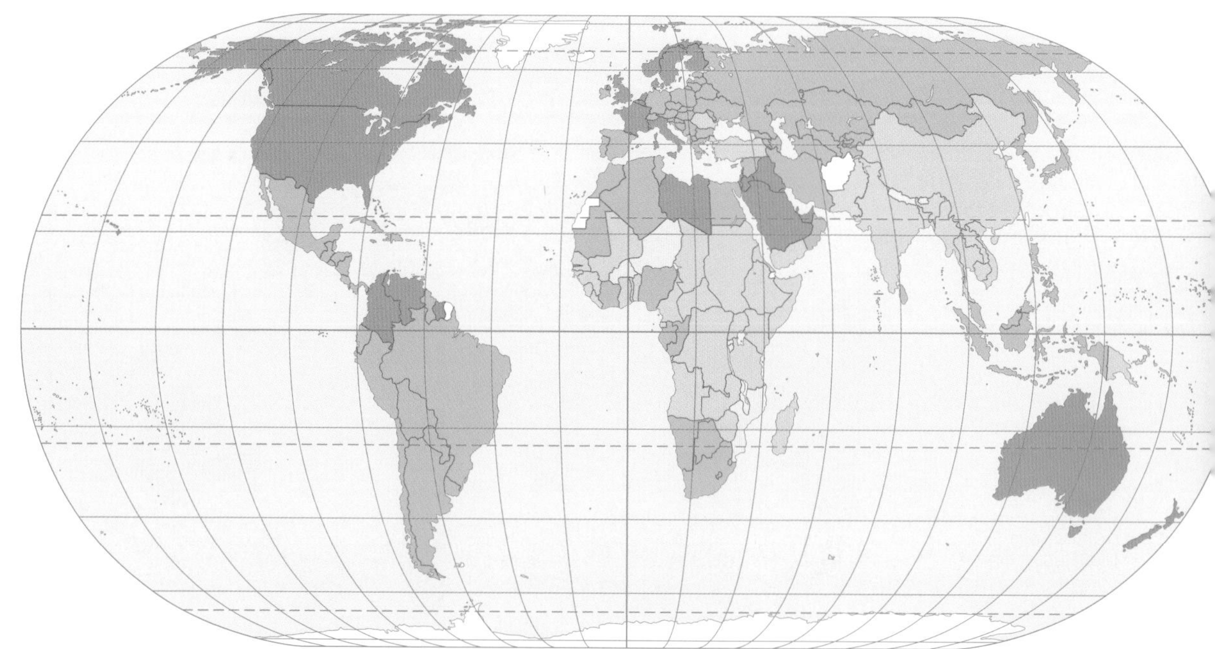

## Employment in services

percentage of the labour force

- over 80%
- 60–80%
- 30–60%
- 10–30%
- under 10%
- no data

**Highest employment in services**
Bahamas 79%
Brunei 74%
Kuwait 74%
Sweden 74%
Canada 72%

United Kingdom 69%

**Lowest employment in services**
Burkina 6%
Nepal 6%
Niger 6%
Burundi 5%
Rwanda 5%
Bhutan 4%

Eckert IV Projection        © Oxford University Press

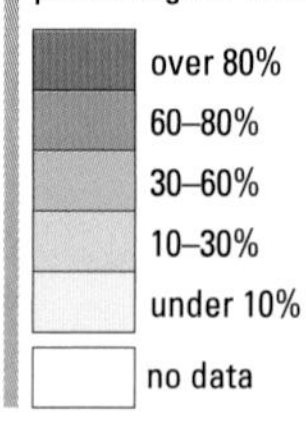

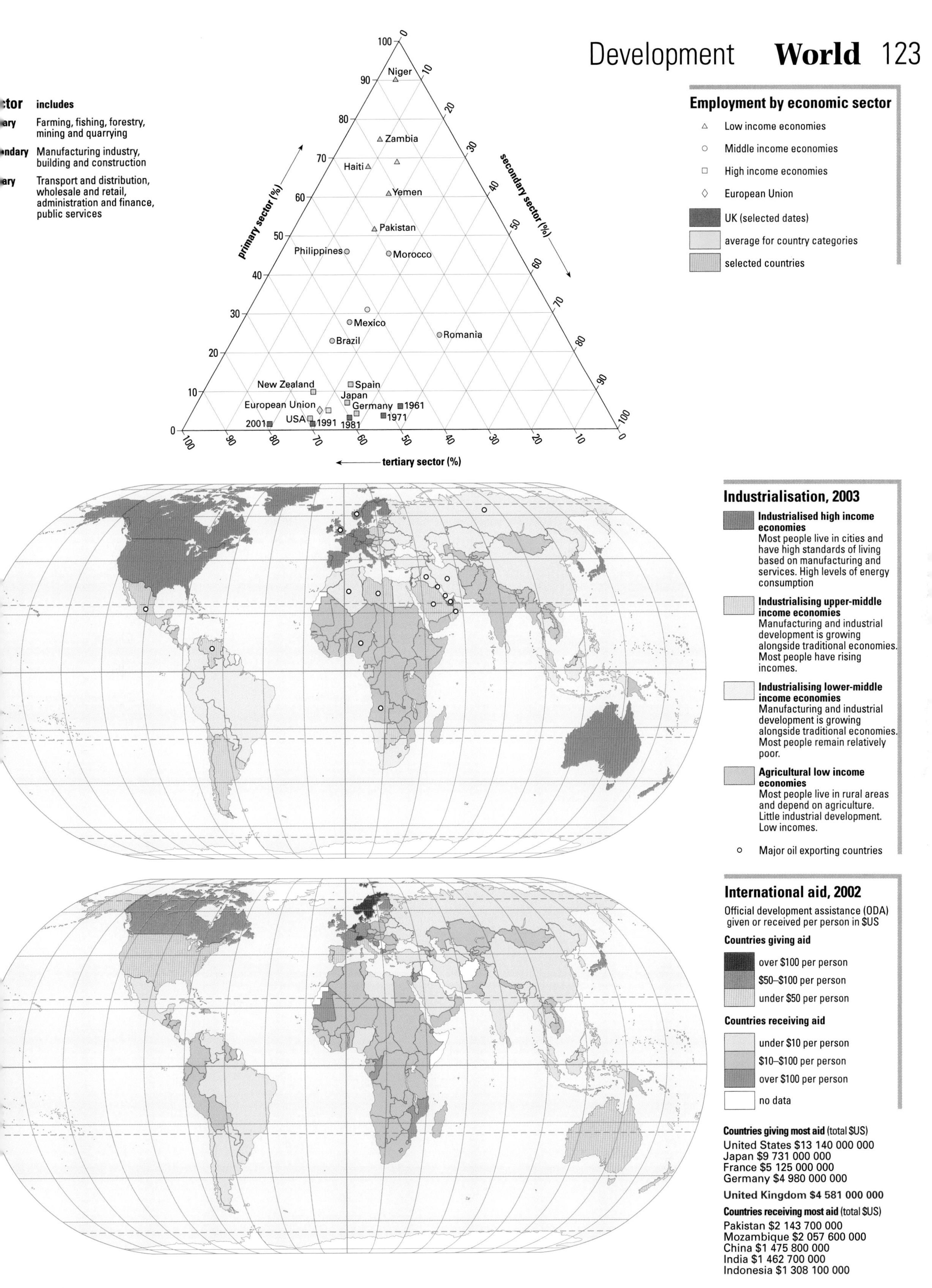

**ctor** **includes**

**ary** Farming, fishing, forestry, mining and quarrying

**ndary** Manufacturing industry, building and construction

**ary** Transport and distribution, wholesale and retail, administration and finance, public services

*primary sector (%)*

*secondary sector (%)*

*tertiary sector (%)*

Niger

Zambia

Haiti

Yemen

Pakistan

Philippines
Morocco

Mexico

Brazil
Romania

New Zealand
Spain
Japan
European Union
Germany
1961
2001 USA 1991 1981 1971

### Employment by economic sector

△ Low income economies

○ Middle income economies

□ High income economies

◇ European Union

UK (selected dates)

average for country categories

selected countries

### Industrialisation, 2003

**Industrialised high income economies**
Most people live in cities and have high standards of living based on manufacturing and services. High levels of energy consumption

**Industrialising upper-middle income economies**
Manufacturing and industrial development is growing alongside traditional economies. Most people have rising incomes.

**Industrialising lower-middle income economies**
Manufacturing and industrial development is growing alongside traditional economies. Most people remain relatively poor.

**Agricultural low income economies**
Most people live in rural areas and depend on agriculture. Little industrial development. Low incomes.

○ Major oil exporting countries

### International aid, 2002

Official development assistance (ODA) given or received per person in $US

**Countries giving aid**

over $100 per person

$50–$100 per person

under $50 per person

**Countries receiving aid**

under $10 per person

$10–$100 per person

over $100 per person

no data

**Countries giving most aid** (total $US)
United States $13 140 000 000
Japan $9 731 000 000
France $5 125 000 000
Germany $4 980 000 000

United Kingdom $4 581 000 000

**Countries receiving most aid** (total $US)
Pakistan $2 143 700 000
Mozambique $2 057 600 000
China $1 475 800 000
India $1 462 700 000
Indonesia $1 308 100 000

## Energy production, 2002

kg oil equivalent per person

- over 25 000
- 2500–25 000
- 1000–2500
- 100–1000
- under 100
- no data

**Highest energy producers**
kg oil equivalent per person

Qatar 86 470
United Arab Emirates 66 425
Brunei 59 419
Norway 58 245
Kuwait 54 378
Equatorial Guinea 23 638
Oman 22 886
Trinidad & Tobago 20 985
Saudi Arabia 20 531
Bahrain 16 301
Libya 14 492
Canada 14 261
Australia 13 559
Turkmenistan 12 721
Gabon 10 712

**United Kingdom 4535**

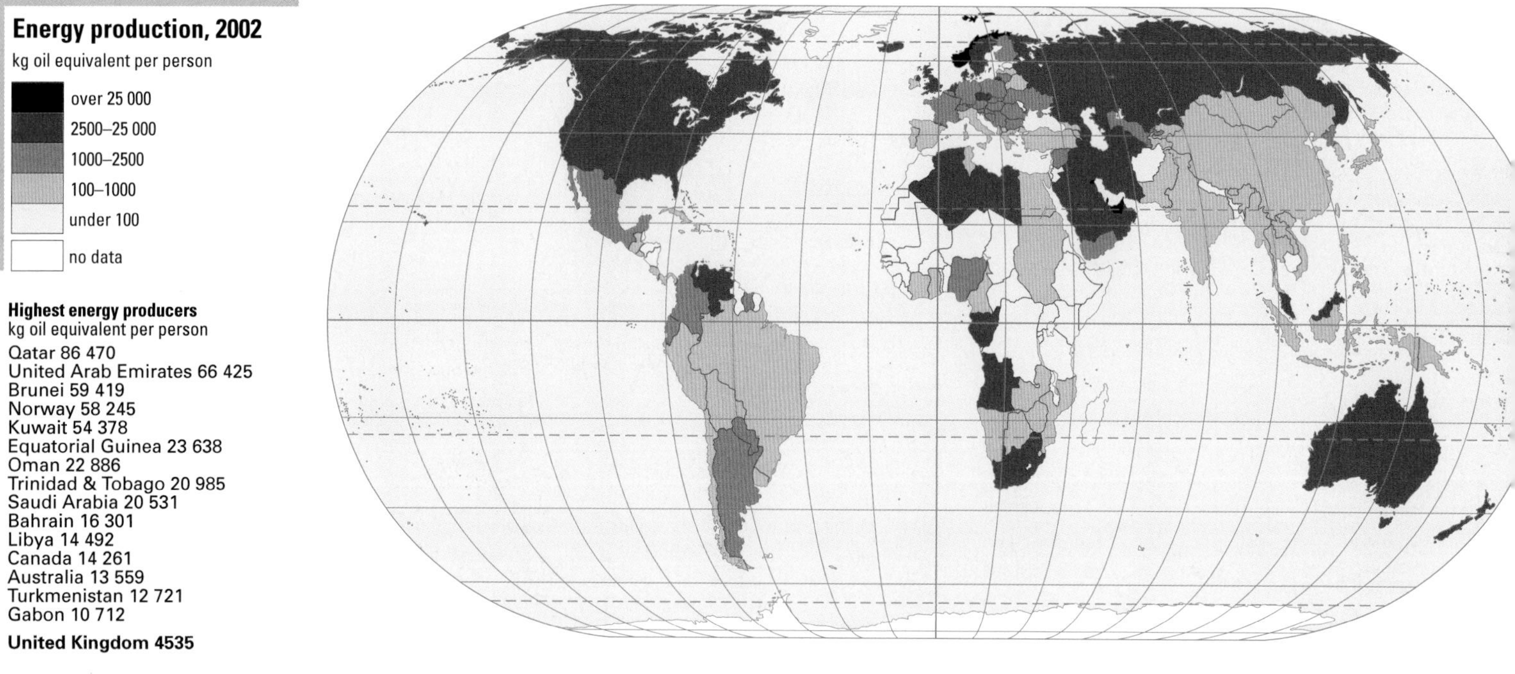

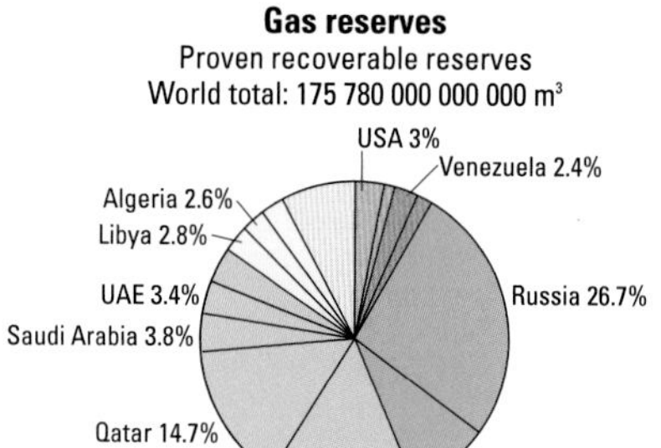

- North America
- Central and South America
- Europe and Eurasia
- Middle East
- Africa
- Asia Pacific

### Oil reserves
Proven recoverable reserves
World total: 156 700 000 000 tonnes

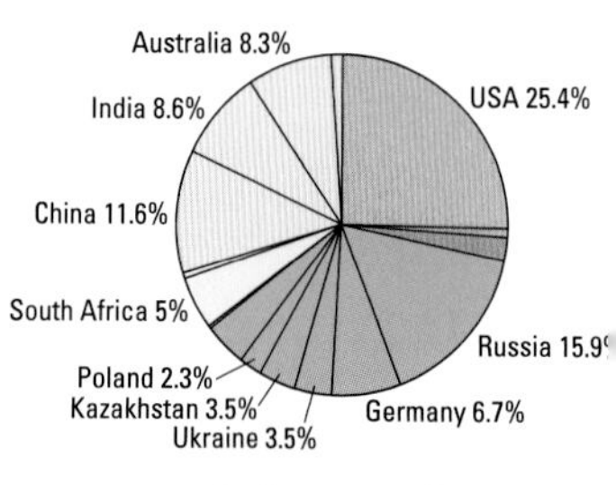

China 2.1% — USA 2.7%
Nigeria 3% — Venezuela 6.8%
Libya 3.1% — Russia 6%
Kuwait 8.4%
UAE 8.5%
Iraq 10%
Iran 11.4%
Saudi Arabia 22.9%

### Gas reserves
Proven recoverable reserves
World total: 175 780 000 000 000 m³

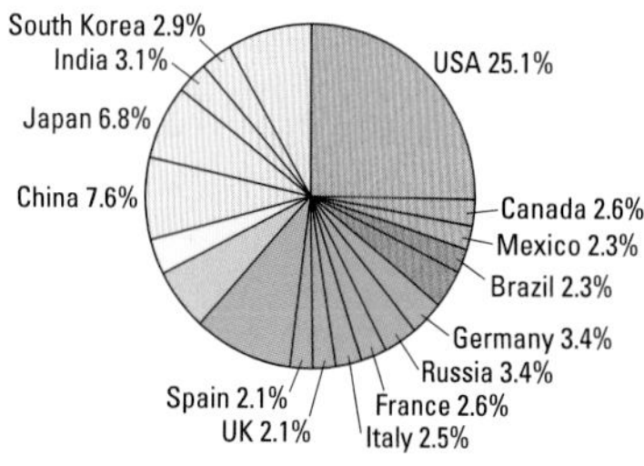

USA 3% — Venezuela 2.4%
Algeria 2.6%
Libya 2.8%
UAE 3.4% — Russia 26.7%
Saudi Arabia 3.8%
Qatar 14.7%
Iran 15.2%

### Coal reserves
Proven recoverable reserves
World total: 984 453 000 000 tonnes

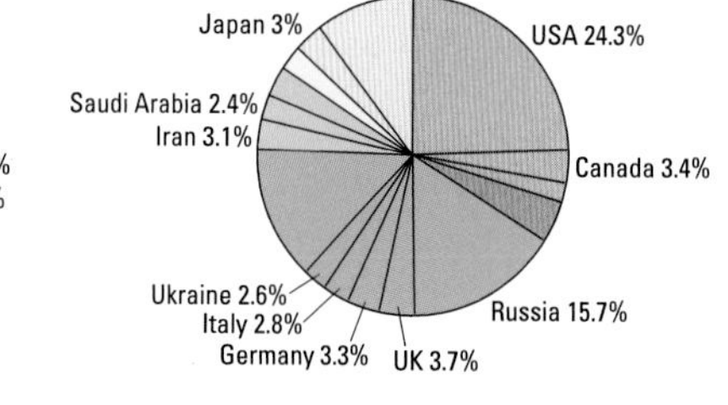

Australia 8.3%
India 8.6% — USA 25.4%
China 11.6%
South Africa 5%
Russia 15.9%
Poland 2.3%
Kazakhstan 3.5% — Germany 6.7%
Ukraine 3.5%

### Oil consumption
World total: 3 636 600 000 tonnes

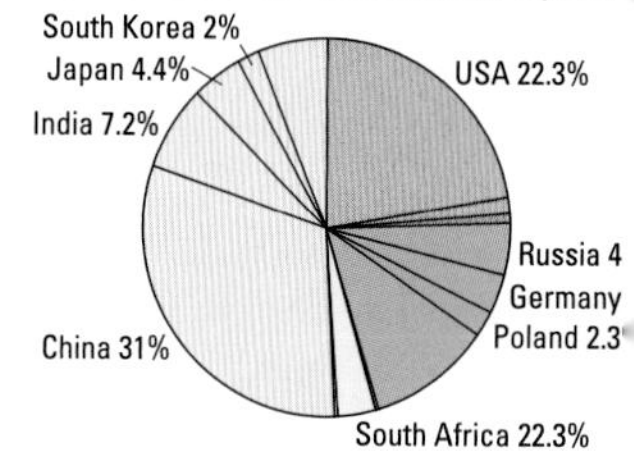

South Korea 2.9%
India 3.1% — USA 25.1%
Japan 6.8%
China 7.6%
Canada 2.6%
Mexico 2.3%
Brazil 2.3%
Germany 3.4%
Russia 3.4%
Spain 2.1% — France 2.6%
UK 2.1% — Italy 2.5%

### Gas consumption
World total: 2 331 900 000 000 m³

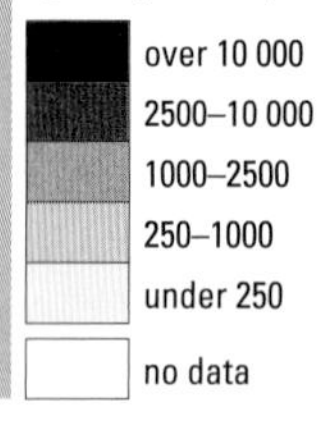

Japan 3% — USA 24.3%
Saudi Arabia 2.4%
Iran 3.1% — Canada 3.4%
Ukraine 2.6%
Italy 2.8% — Russia 15.7%
Germany 3.3% — UK 3.7%

### Coal consumption
World total: 2 578 400 000 tonnes oil equival

South Korea 2%
Japan 4.4% — USA 22.3%
India 7.2%
Russia 4
Germany
China 31% — Poland 2.3
South Africa 22.3%

## Energy consumption, 2002

kg oil equivalent per person

- over 10 000
- 2500–10 000
- 1000–2500
- 250–1000
- under 250
- no data

**Highest energy consumers**
kg oil equivalent per person

United Arab Emirates 21 648
Bahrain 15 882
Qatar 15 410
Iceland 12 065
Trinidad and Tobago 11 291

**United Kingdom 3998**

**Lowest energy consumers**
kg oil equivalent per person

Mali 29
Ethiopia 26
Afghanistan 19
Cambodia 16
Chad 8

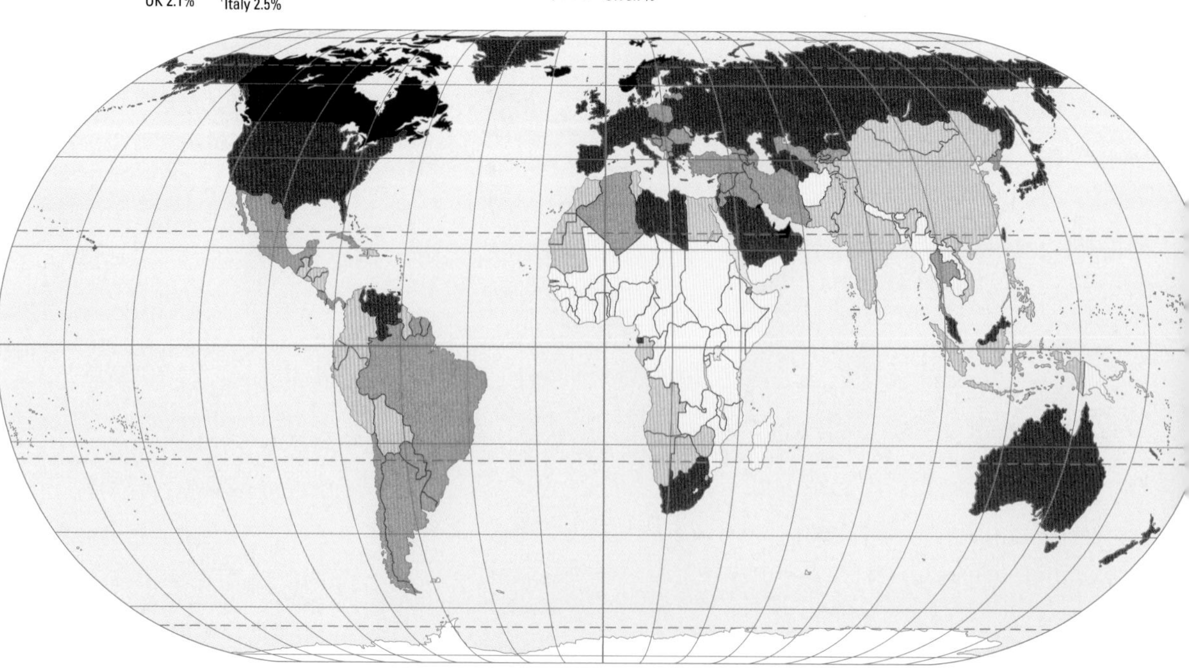

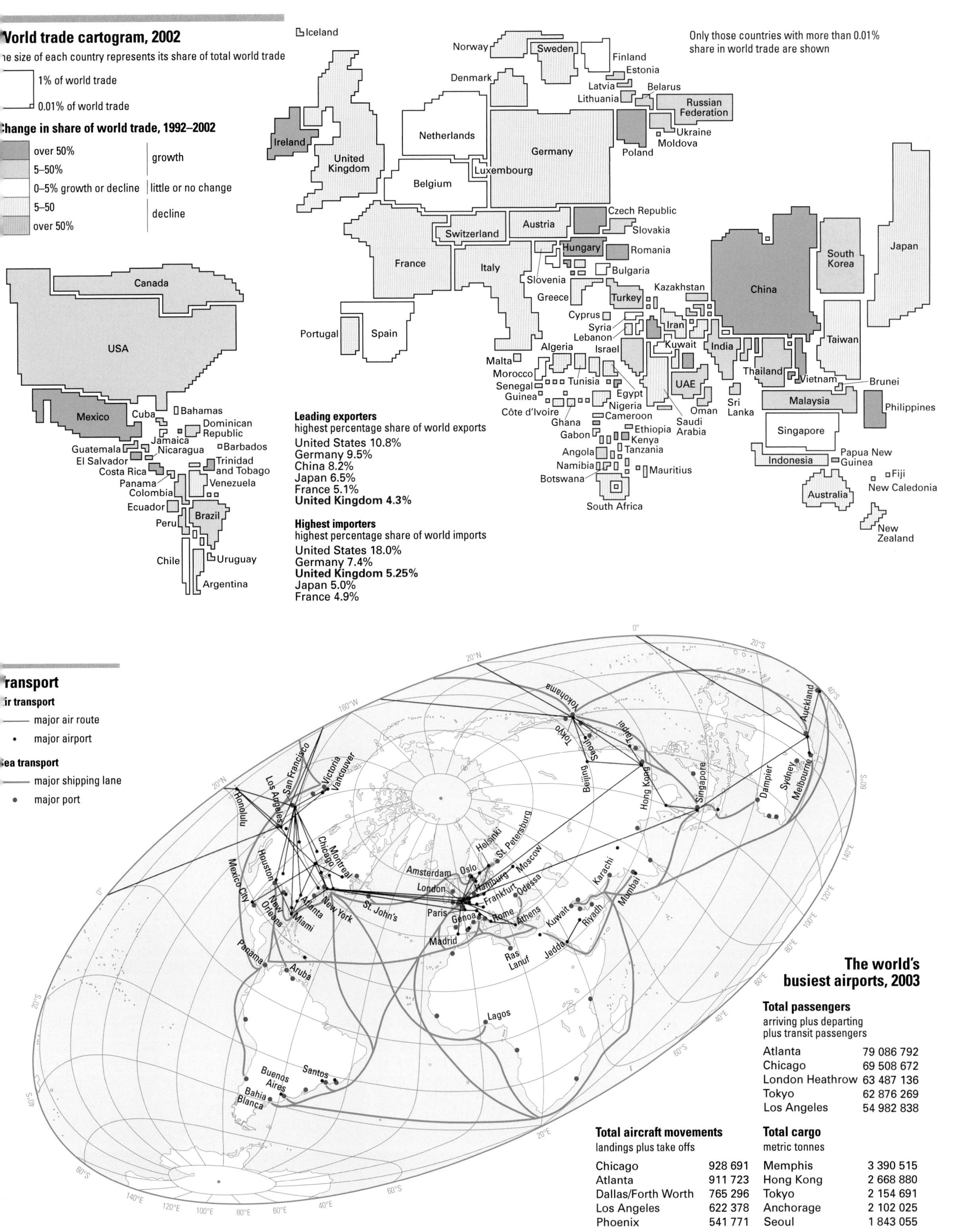

## World trade cartogram, 2002

The size of each country represents its share of total world trade

- 1% of world trade
- 0.01% of world trade

### Change in share of world trade, 1992–2002

| | |
|---|---|
| over 50% | |
| 5–50% | growth |
| 0–5% growth or decline | little or no change |
| 5–50 | |
| over 50% | decline |

Only those countries with more than 0.01% share in world trade are shown

**Leading exporters**
highest percentage share of world exports
United States 10.8%
Germany 9.5%
China 8.2%
Japan 6.5%
France 5.1%
**United Kingdom 4.3%**

**Highest importers**
highest percentage share of world imports
United States 18.0%
Germany 7.4%
**United Kingdom 5.25%**
Japan 5.0%
France 4.9%

### Transport

**Air transport**
— major air route
• major airport

**Sea transport**
— major shipping lane
• major port

### The world's busiest airports, 2003

**Total passengers**
arriving plus departing
plus transit passengers

| | |
|---|---|
| Atlanta | 79 086 792 |
| Chicago | 69 508 672 |
| London Heathrow | 63 487 136 |
| Tokyo | 62 876 269 |
| Los Angeles | 54 982 838 |

**Total aircraft movements**
landings plus take offs

| | |
|---|---|
| Chicago | 928 691 |
| Atlanta | 911 723 |
| Dallas/Forth Worth | 765 296 |
| Los Angeles | 622 378 |
| Phoenix | 541 771 |

**Total cargo**
metric tonnes

| | |
|---|---|
| Memphis | 3 390 515 |
| Hong Kong | 2 668 880 |
| Tokyo | 2 154 691 |
| Anchorage | 2 102 025 |
| Seoul | 1 843 055 |

© Oxford University Press    Oblique Aitoff Projection

## Desertification and tropical deforestation

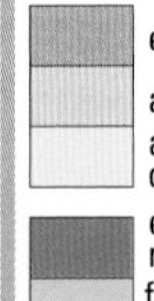

existing areas of desert

areas with a high risk of desertification

areas with a moderate risk of desertification

existing areas of tropical rain forest

former areas of tropical rain forest

**Countries losing greatest areas of forest ('000 hectares) 1990 – 2000**

| | |
|---|---|
| Brazil | 2309 |
| Indonesia | 1312 |
| Sudan | 959 |
| Zambia | 851 |
| Mexico | 631 |
| Congo, Dem. Rep. | 532 |
| Myanmar | 517 |

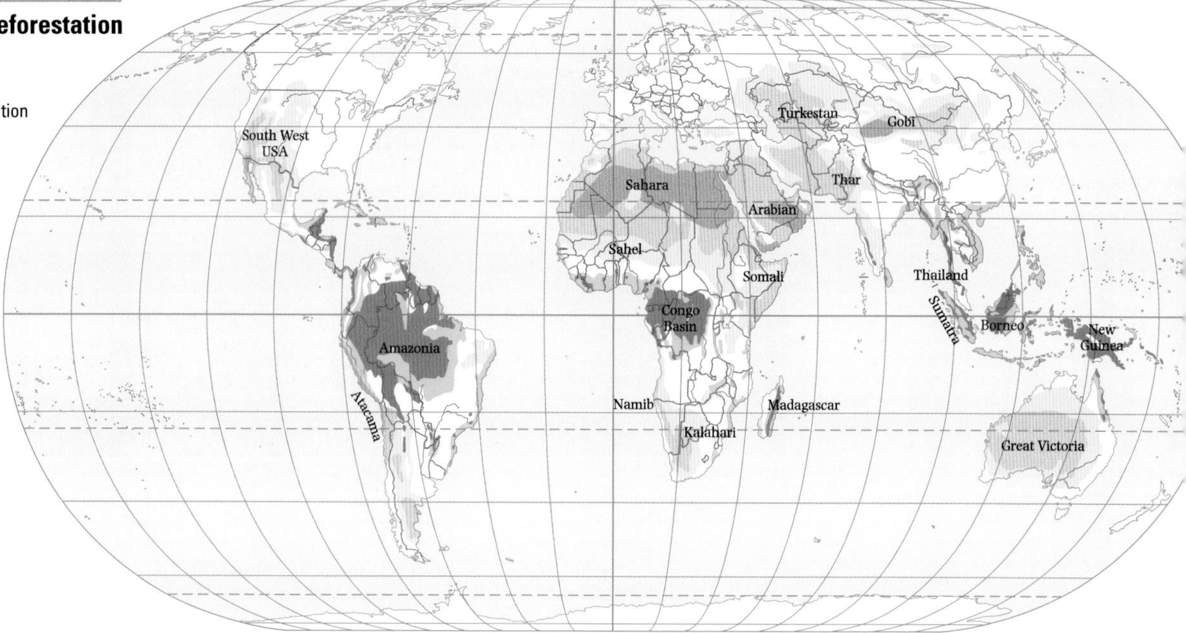

## Lake Chad, West Africa, 1973–1997

Lake Chad was once the sixth-largest lake in the world, but persistent drought since the 1960's has shrunk it to about one tenth of its former size. Wetland marsh (shown on the satellite images as red) has now largely replaced open water (shown in blue). The lake is shallow and very responsive to the high variability on rainfall in the region. People living around Lake Chad do not have secure food supplies. Farming and irrigation projects have been affected by fluctuations in the level of the lake.

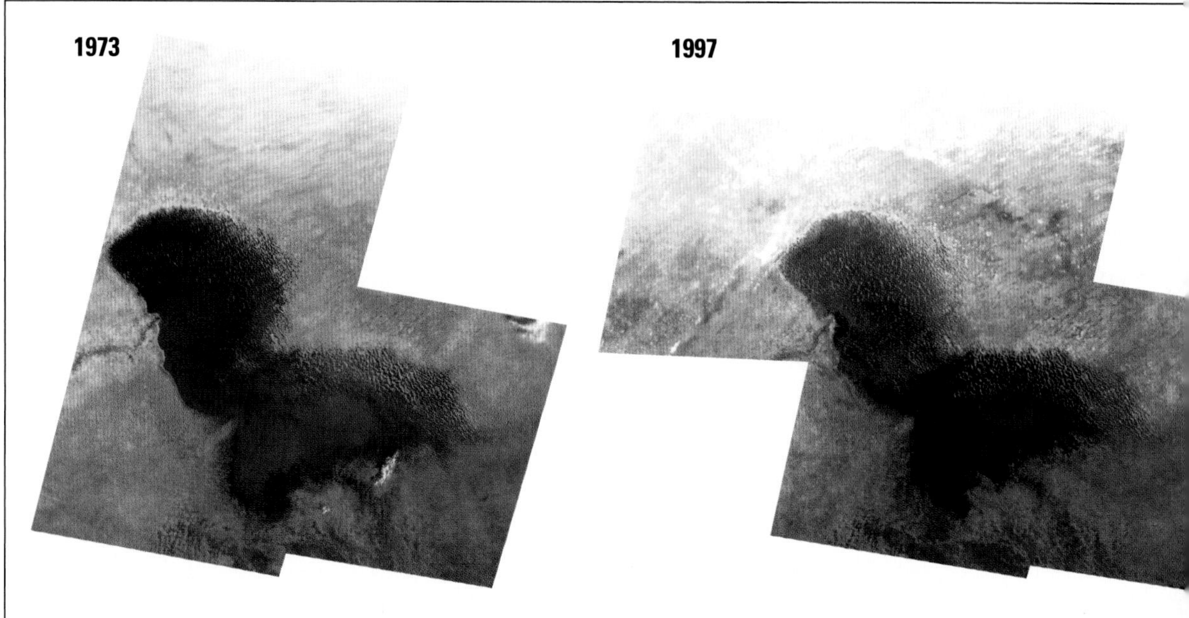

**1973**       **1997**

## Acid rain

### Sulphur and nitrogen emissions

Oxides of sulphur and nitrogen produced by burning fossil fuel react with rain to form dilute sulphuric and nitric acids

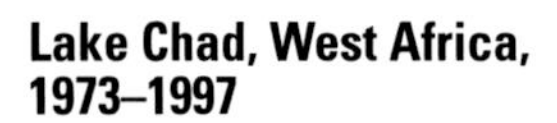

areas with high levels of fossil fuel burning

- cities where sulphur dioxide emissions are recorded and exceed World Health Organization recommended levels

### Areas of acid rain deposition

Annual mean values of pH in precipitation

pH less than 4.2 (most acidic)

pH 4.2–4.6

pH 4.6–5.0

other areas where acid rain is becoming a problem

Lower pH values are more acidic. 'Clean' rain water is slightly acidic with a pH of 5.6. The pH scale is logarithmic, so that a value of 4.6 is ten times as acidic as normal rain.

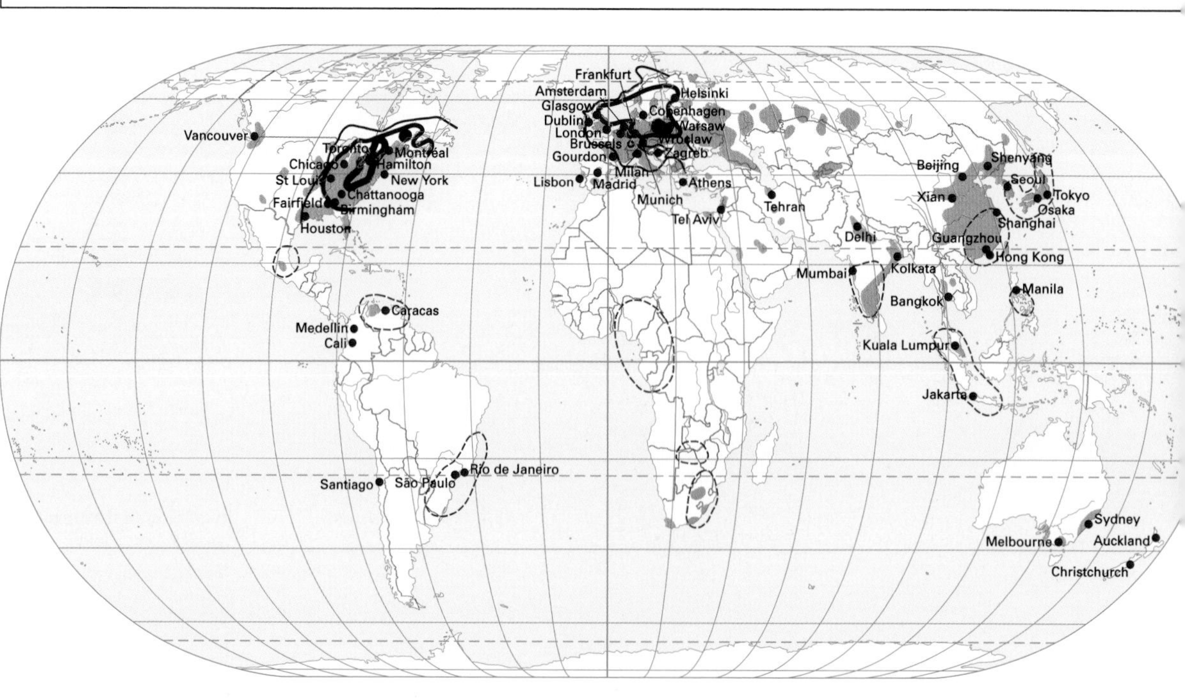

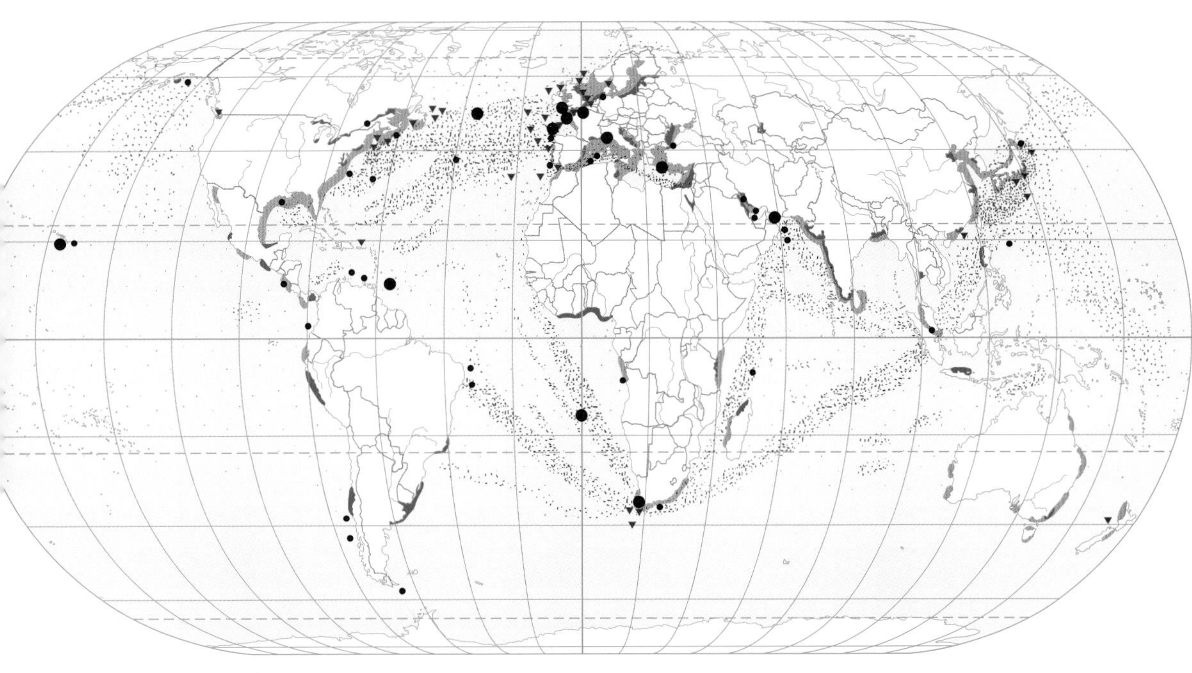

## Sea pollution

**Major oil spills**

- ● over 100 000 tonnes
- · under 100 000 tonnes
- ░░░ frequent oil slicks from shipping

**Other sea pollution**

- ▬ severe pollution
- ▬ moderate pollution
- ▼ deep sea dump sites

**Major oil spills ('000 tonnes)**

| | | |
|---|---|---|
| **1977** | *Ekofisk* well blow-out, North Sea | 270 |
| **1979** | *Ixtoc 1* well blow-out, Gulf of Mexico | 600 |
| **1979** | Collision of *Atlantic Empress* and *Aegean Captain*, off Tobago, Caribbean | 370 |
| **1983** | *Nowruz* well blow-out, The Gulf | 600 |
| **1989** | *Exxon Valdez* spills oil off the coast of Alaska | 250 |
| **1991** | Release of oil by Iraqi troops, *Sea Island* terminal, The Gulf | 799 |
| **2002** | *Prestige* oil tanker sinks off the coast of Spain | 77 |

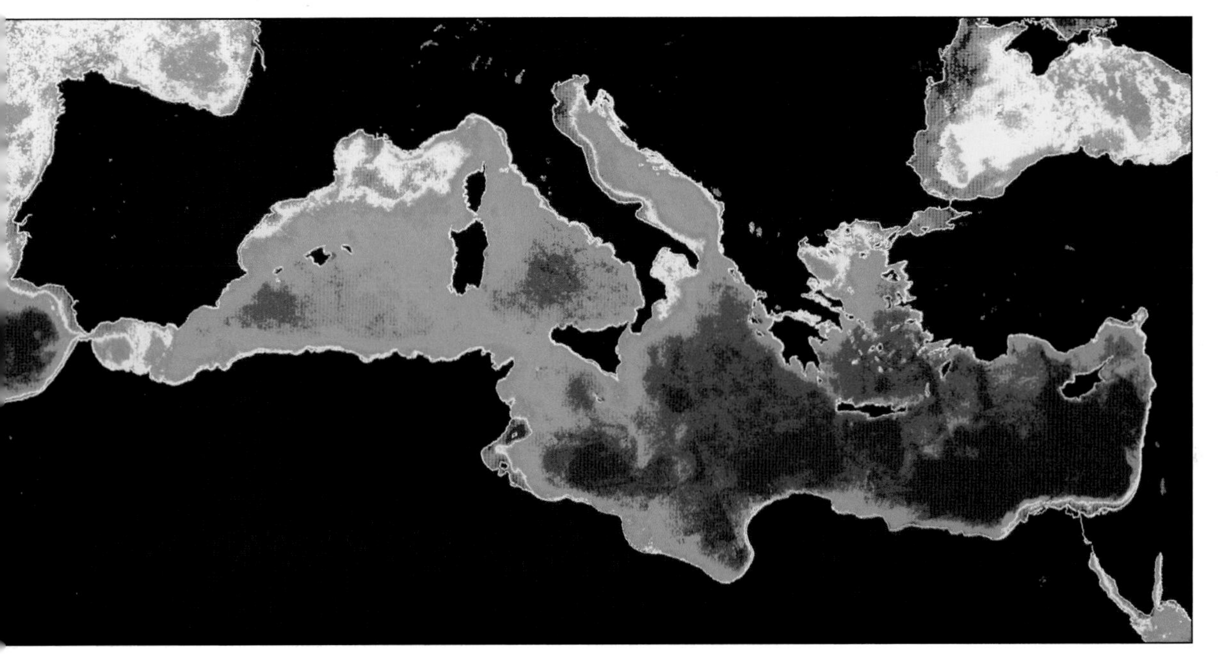

## Phytoplankton in the Mediterranean Sea

Phytoplankton are micro-organisms that thrive in shallow, polluted sea areas. In this false colour satellite image red, orange, and yellow show the highest densities of phytoplankton. Green and blue show the lowest densities.

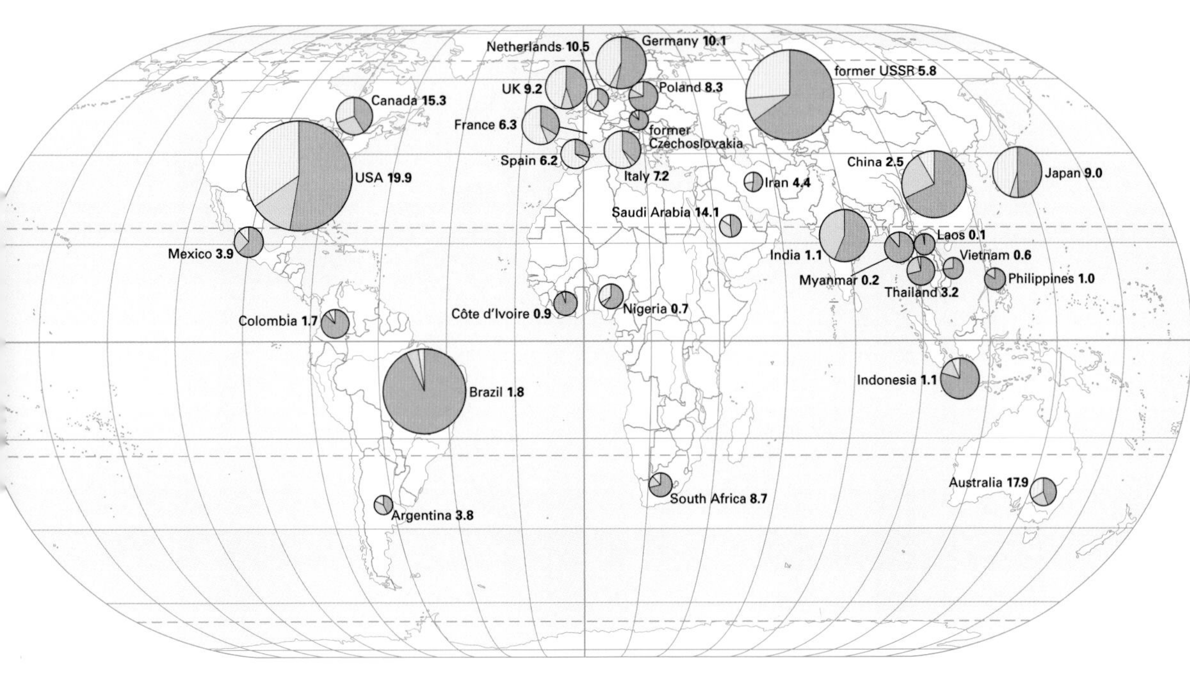

## Greenhouse gases

**Highest total emissions by country**
thousand tonnes of carbon

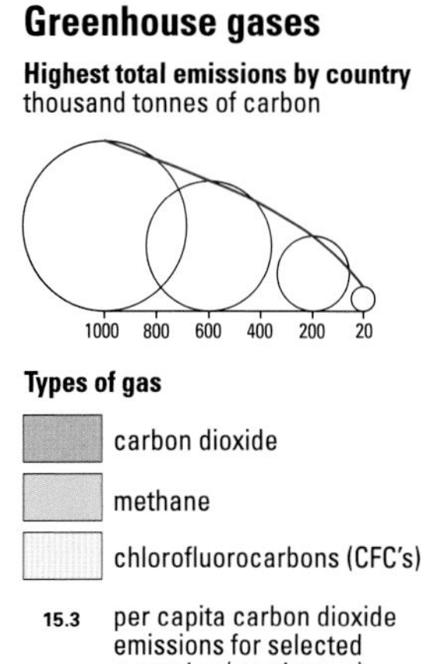

1000  800  600  400  200  20

**Types of gas**

- ▓ carbon dioxide
- ▒ methane
- ░ chlorofluorocarbons (CFC's)

**15.3** per capita carbon dioxide emissions for selected countries (metric tons)

Scale 1: 125 000 000 (main map

## Selected tourist destinations

The locations shown represent a limited selection of important tourism sites.

- 🏛 cultural/historical sites
- ✳ natural heritage sites
- ⊙ resorts
- ● tourist cities
- — main cruise routes

**land height**

| metres |
|---|
| 2000 |
| 500 |
| 0 |

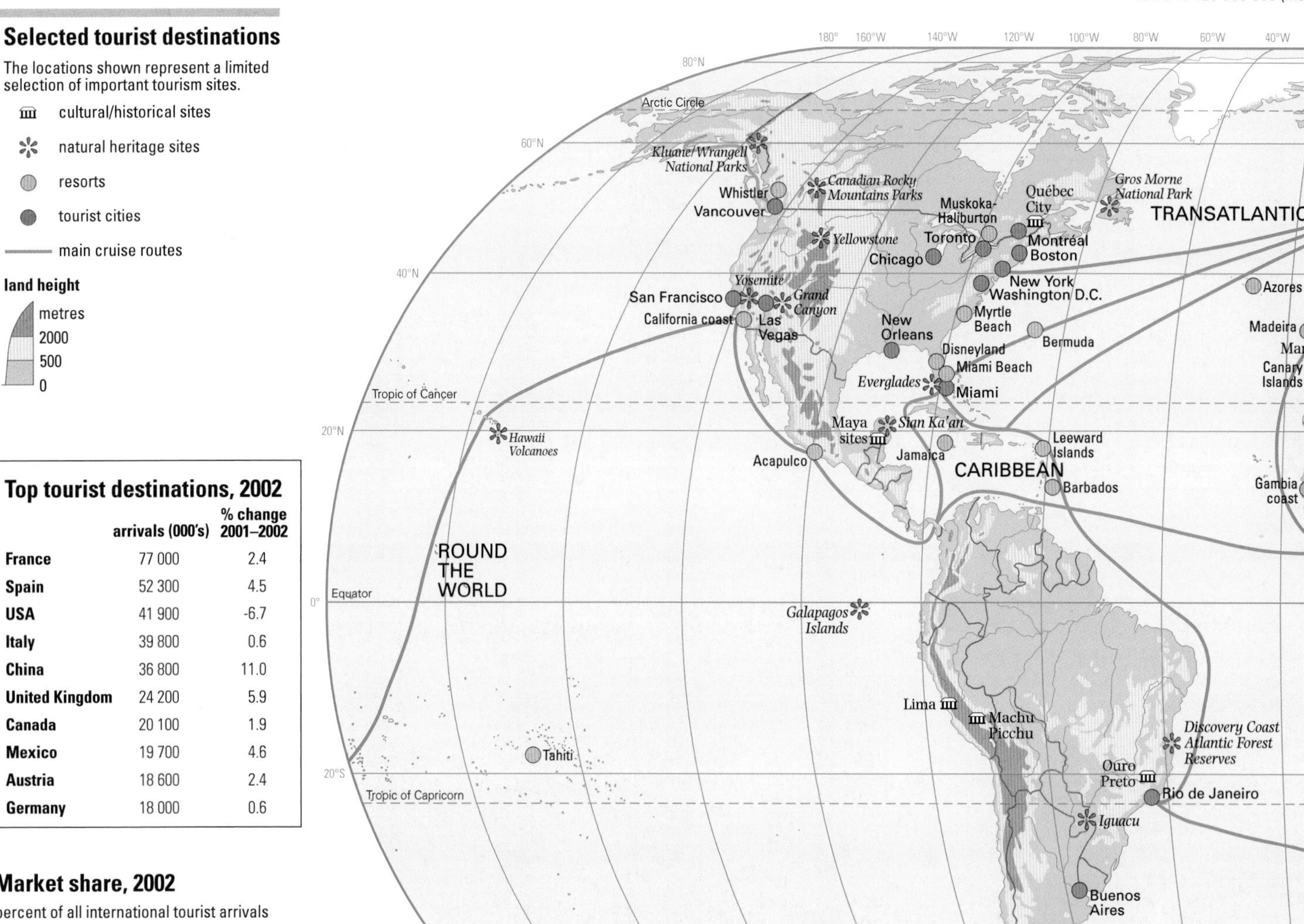

### Top tourist destinations, 2002

| | arrivals (000's) | % change 2001–2002 |
|---|---|---|
| **France** | 77 000 | 2.4 |
| **Spain** | 52 300 | 4.5 |
| **USA** | 41 900 | -6.7 |
| **Italy** | 39 800 | 0.6 |
| **China** | 36 800 | 11.0 |
| **United Kingdom** | 24 200 | 5.9 |
| **Canada** | 20 100 | 1.9 |
| **Mexico** | 19 700 | 4.6 |
| **Austria** | 18 600 | 2.4 |
| **Germany** | 18 000 | 0.6 |

## Market share, 2002

percent of all international tourist arrivals

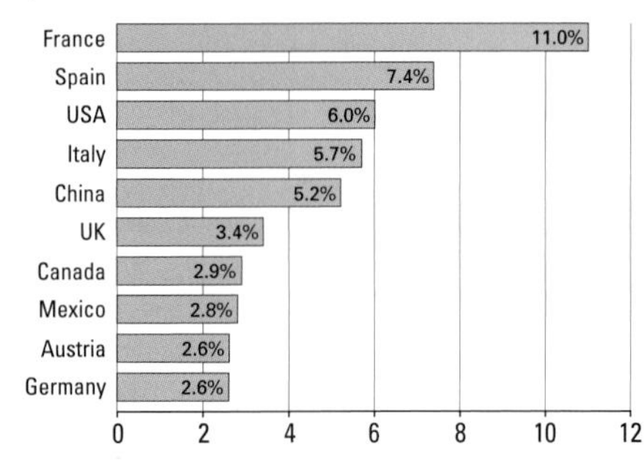

| | |
|---|---|
| France | 11.0% |
| Spain | 7.4% |
| USA | 6.0% |
| Italy | 5.7% |
| China | 5.2% |
| UK | 3.4% |
| Canada | 2.9% |
| Mexico | 2.8% |
| Austria | 2.6% |
| Germany | 2.6% |

## Earnings from tourism, 2002

tourist receipts in million $US

| | |
|---|---|
| ■ | over 5000 |
| ■ | 1000–5000 |
| ▨ | 250–1000 |
| ▨ | 100–250 |
| ▨ | under 100 |
| □ | no data |

**Highest tourist earnings (millions)**
USA $66 547
Spain $33 609
France $32 329
Italy $26 915
Germany $19 158
**United Kingdom $17 591**

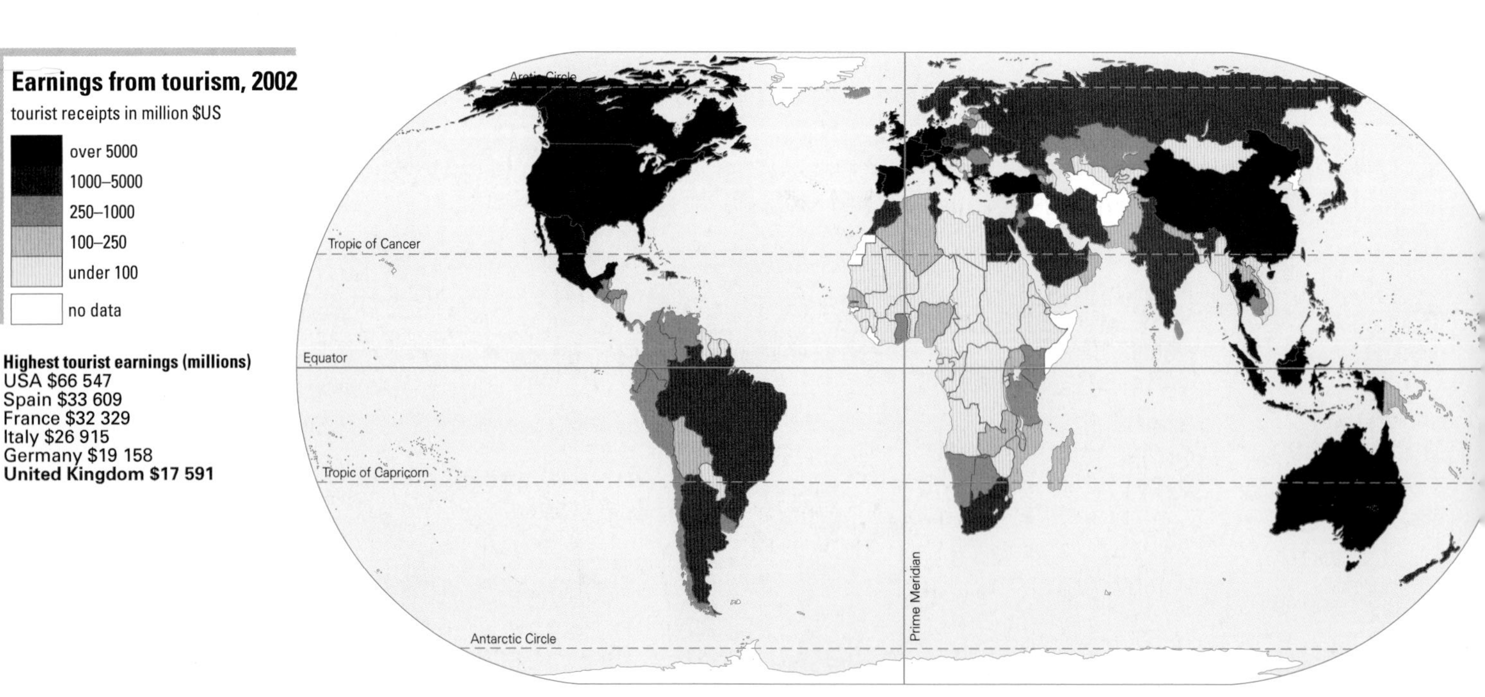

Eckert IV Projection     © Oxford University Press

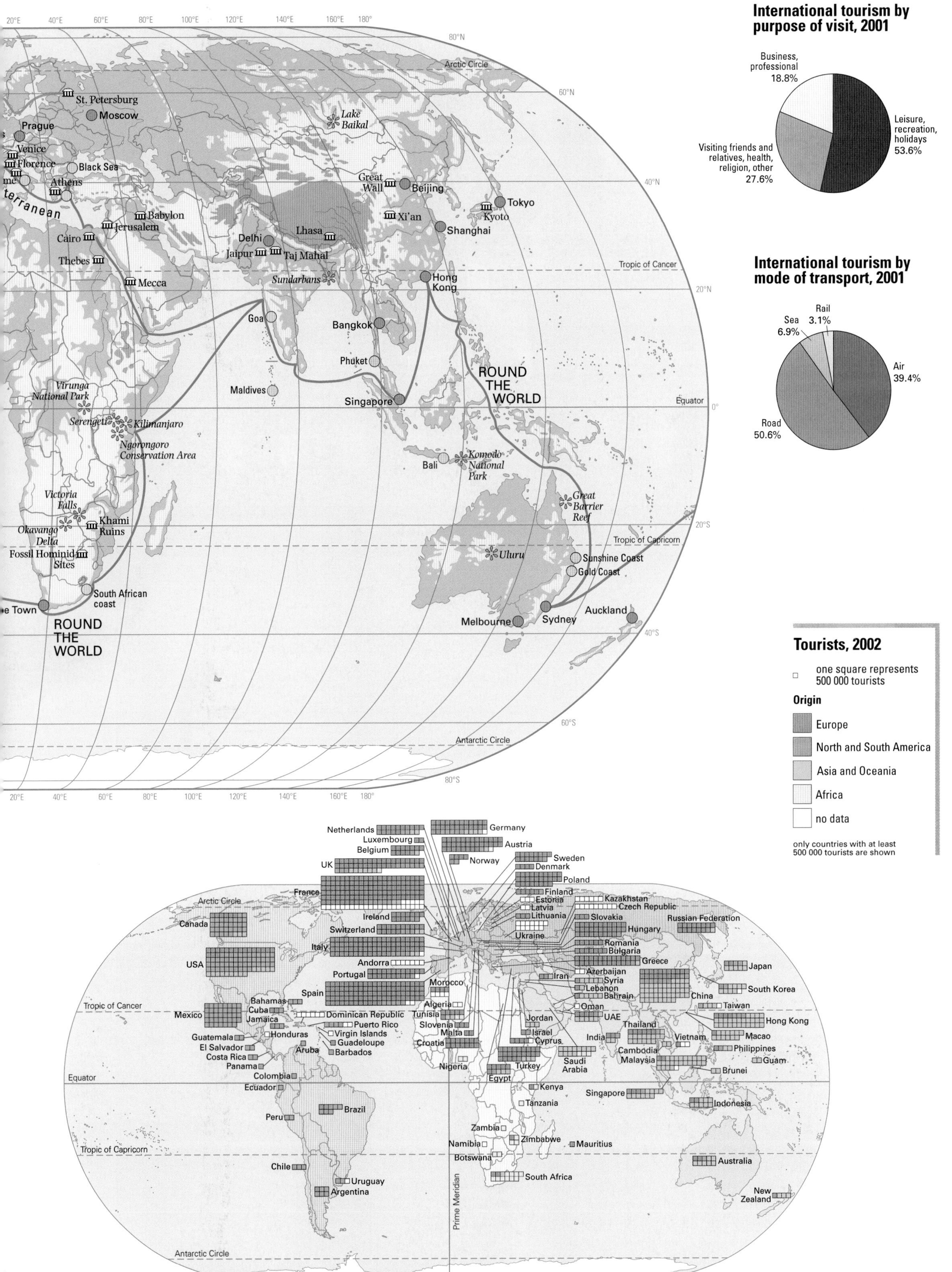

## International tourism by purpose of visit, 2001

Business, professional 18.8%

Leisure, recreation, holidays 53.6%

Visiting friends and relatives, health, religion, other 27.6%

## International tourism by mode of transport, 2001

Rail 3.1%

Sea 6.9%

Air 39.4%

Road 50.6%

St. Petersburg
Moscow
Prague
Venice
Florence
Athens
terranean
Babylon
Jerusalem
Cairo
Thebes
Mecca

Lake Baikal

Great Wall
Beijing
Xi'an
Tokyo
Kyoto
Shanghai

Delhi
Jaipur
Taj Mahal
Lhasa

Sundarbans

Goa
Bangkok
Phuket
Maldives
Singapore

Hong Kong

ROUND THE WORLD

Virunga National Park
Serengeti
Kilimanjaro
Ngorongoro Conservation Area

Bali
Komodo National Park

Great Barrier Reef

Victoria Falls
Khami Ruins
Okavango Delta
Fossil Hominid Sites

Uluru

Sunshine Coast
Gold Coast

South African coast
e Town
ROUND THE WORLD

Melbourne
Sydney
Auckland

### Tourists, 2002

□ one square represents 500 000 tourists

**Origin**

Europe

North and South America

Asia and Oceania

Africa

no data

only countries with at least 500 000 tourists are shown

Netherlands
Luxembourg
Belgium
UK
France
Ireland
Switzerland
Italy
Andorra
Portugal
Spain
Bahamas
Cuba
Jamaica
Guatemala
Honduras
El Salvador
Costa Rica
Panama
Colombia
Ecuador
Peru
Brazil
Chile
Uruguay
Argentina

Canada
USA
Mexico

Germany
Austria
Norway
Sweden
Denmark
Poland
Finland
Estonia
Latvia
Lithuania
Ukraine
Kazakhstan
Czech Republic
Slovakia
Hungary
Romania
Bulgaria
Greece
Russian Federation
Japan
South Korea
China
Taiwan
Hong Kong
Macao
Philippines
Guam
Brunei
Vietnam
Malaysia
Cambodia
Thailand
India
Singapore
Indonesia
Australia
New Zealand

Morocco
Algeria
Tunisia
Dominican Republic
Puerto Rico
Virgin Islands
Guadeloupe
Barbados
Aruba
Slovenia
Malta
Croatia
Nigeria
Egypt
Kenya
Tanzania
Zambia
Zimbabwe
Mauritius
Namibia
Botswana
South Africa

Iran
Azerbaijan
Syria
Lebanon
Bahrain
Oman
UAE
Jordan
Israel
Cyprus
Turkey
Saudi Arabia

© Oxford University Press

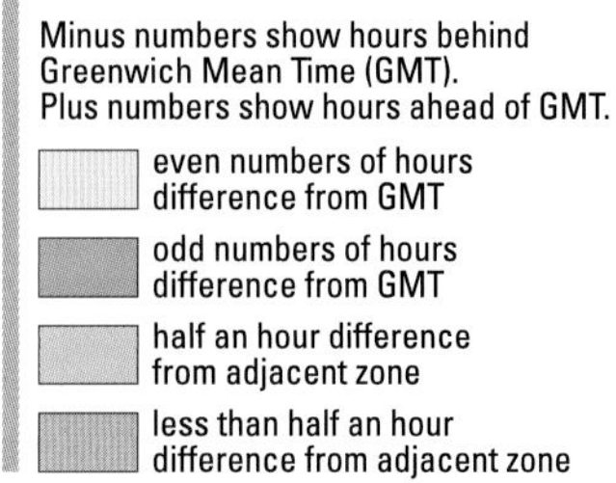

## Time zones, 2003

Minus numbers show hours behind Greenwich Mean Time (GMT). Plus numbers show hours ahead of GMT.

- even numbers of hours difference from GMT
- odd numbers of hours difference from GMT
- half an hour difference from adjacent zone
- less than half an hour difference from adjacent zone

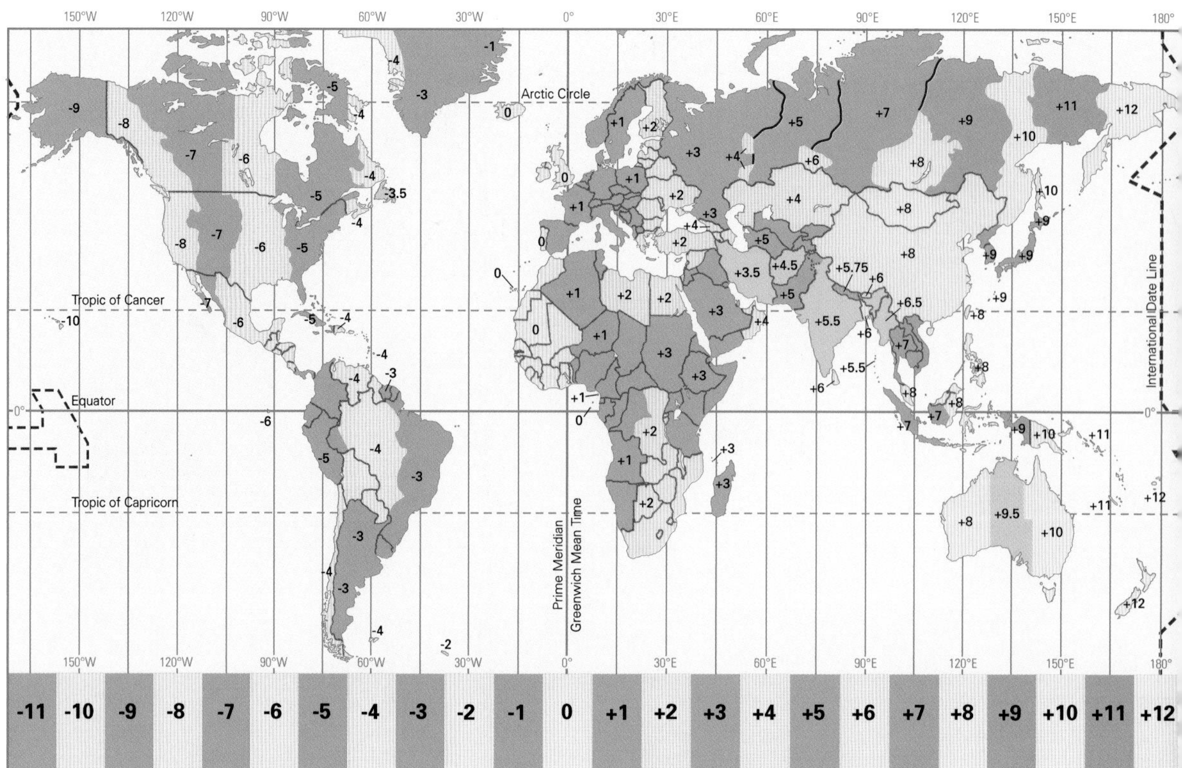

Longitude is measured from the **prime meridian** which passes through Greenwich. There are 24 standard time zones, each of 15° of longitude. The edges of these time zones usually follow international boundaries.

The **international date line** marks the point where one calendar day ends and another begins. A traveller crossing from east to west moves forward one day. Crossing from west to east the calendar goes back one day.

| -11 | -10 | -9 | -8 | -7 | -6 | -5 | -4 | -3 | -2 | -1 | 0 | +1 | +2 | +3 | +4 | +5 | +6 | +7 | +8 | +9 | +10 | +11 | +12 |

## Distance

Flight distance between cities in kilometres to convert kilometres to miles multiply by 0.62

| | | | | | | | | | | | |
|---|---|---|---|---|---|---|---|---|---|---|---|
| **Beijing** | | | | | | | | | | | |
| 19 307 | **Buenos Aires** | | | | | | | | | | |
| 1983 | 18 484 | **Hong Kong** | | | | | | | | | |
| 11 710 | 8088 | 10 732 | **Johannesburg** | | | | | | | | |
| 8145 | 11 161 | 9645 | 9071 | **London** | | | | | | | |
| 10 081 | 9871 | 11 678 | 16 676 | 8774 | **Los Angeles** | | | | | | |
| 12 468 | 7468 | 14 162 | 14 585 | 8936 | 2484 | **Mexico City** | | | | | |
| 4774 | 14 952 | 4306 | 8274 | 7193 | 14 033 | 15 678 | **Mumbai** | | | | |
| 11 000 | 8548 | 12 984 | 12 841 | 5580 | 3951 | 3371 | 12 565 | **New York** | | | |
| 8226 | 11 097 | 9613 | 8732 | 338 | 9032 | 9210 | 7032 | 5839 | **Paris** | | |
| 4468 | 15 904 | 2661 | 8860 | 10 871 | 14 146 | 16 630 | 3919 | 15 533 | 10 758 | **Singapore** | |
| 8949 | 11 800 | 7374 | 11 040 | 16 992 | 12 073 | 12 969 | 9839 | 15 989 | 16 962 | 6300 | **Sydney** |
| 2113 | 18 388 | 2903 | 13 547 | 9581 | 8823 | 11 355 | 6758 | 10 871 | 9726 | 5322 | 7823 | **Tokyo** |

## The Earth rotates from west to east

The Earth rotates on its axis once in every 24 hours.

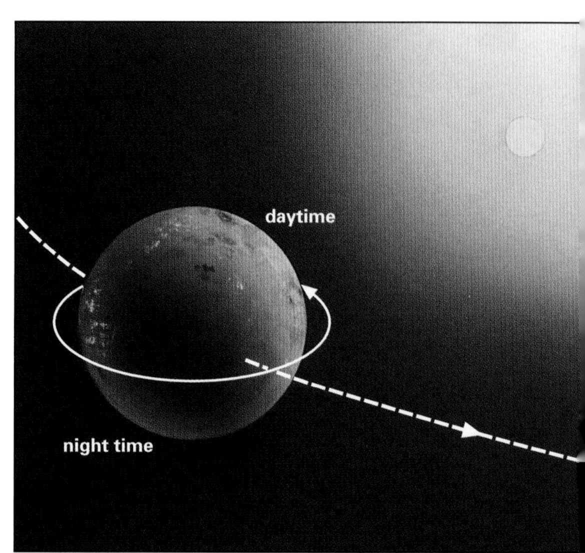

## Flying time

Typical flight times by air between cities in hours and minutes
ooo means there is no direct flight available, early 2002

| | | | | | | | | | | |
|---|---|---|---|---|---|---|---|---|---|---|
| **Beijing** | | | | | | | | | | |
| ooo | **Buenos Aires** | | | | | | | | | |
| 3.00 | ooo | **Hong Kong** | | | | | | | | |
| ooo | ooo | 13.00 | **Johannesburg** | | | | | | | |
| 10.25 | 14.50 | 13.30 | 10.50 | **London** | | | | | | |
| 13.30 | 16.40 | 14.50 | ooo | 13.00 | **Los Angeles** | | | | | |
| ooo | 13.10 | ooo | ooo | 11.05 | 4.15 | **Mexico City** | | | | |
| ooo | ooo | 7.45 | 20.15 | 10.30 | ooo | ooo | **Mumbai** | | | |
| 25.20 | 13.25 | 19.25 | 16.25 | 7.20 | 6.00 | 6.00 | 20.05 | **New York** | | |
| 10.20 | 13.50 | 12.45 | 10.55 | 1.10 | 12.30 | 12.20 | 12.10 | 7.40 | **Paris** | |
| 6.15 | ooo | 4.05 | 10.30 | 14.40 | 18.45 | ooo | 6.30 | 22.05 | 14.15 | **Singapore** |
| 12.55 | 16.35 | 8.50 | 14.30 | 22.45 | 14.35 | ooo | 14.40 | 21.45 | 22.25 | 8.55 | **Sydney** |
| 3.35 | ooo | 4.55 | ooo | 12.40 | 10.40 | 15.50 | 12.35 | 15.55 | 12.50 | 7.05 | 10.00 | **Tokyo** |

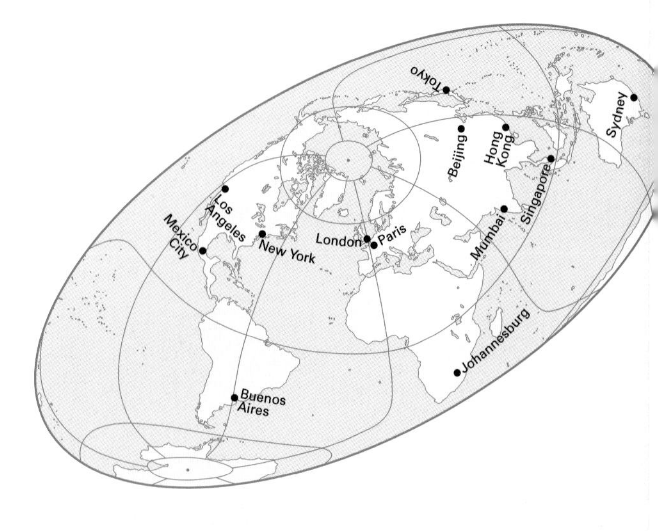

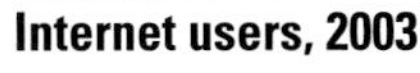

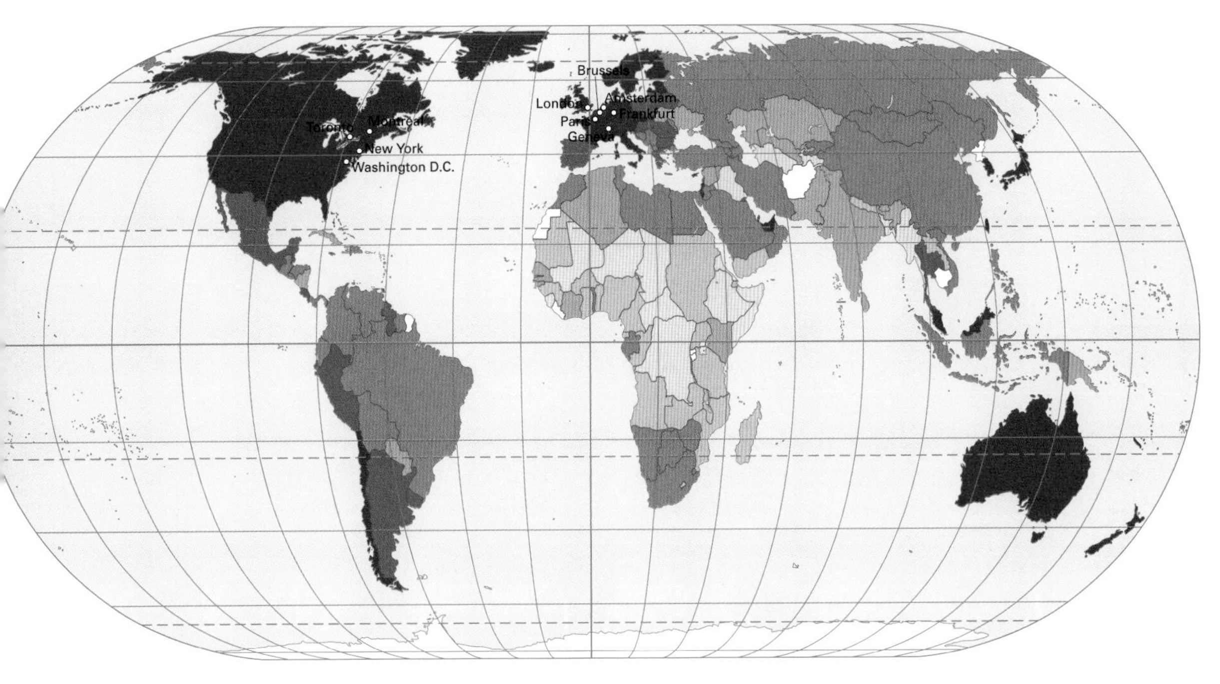

## Internet users, 2003

per 10 000 people

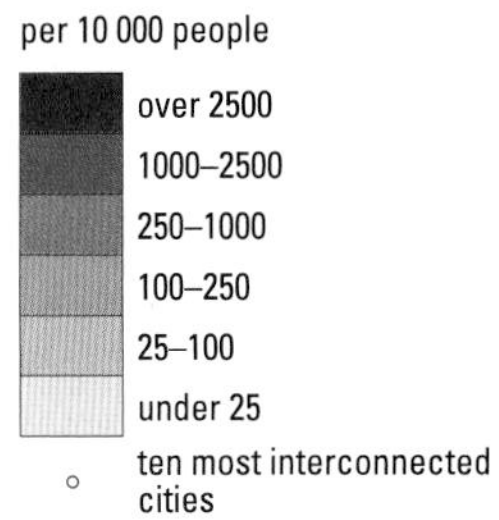

- over 2500
- 1000–2500
- 250–1000
- 100–250
- 25–100
- under 25
- ○ ten most interconnected cities

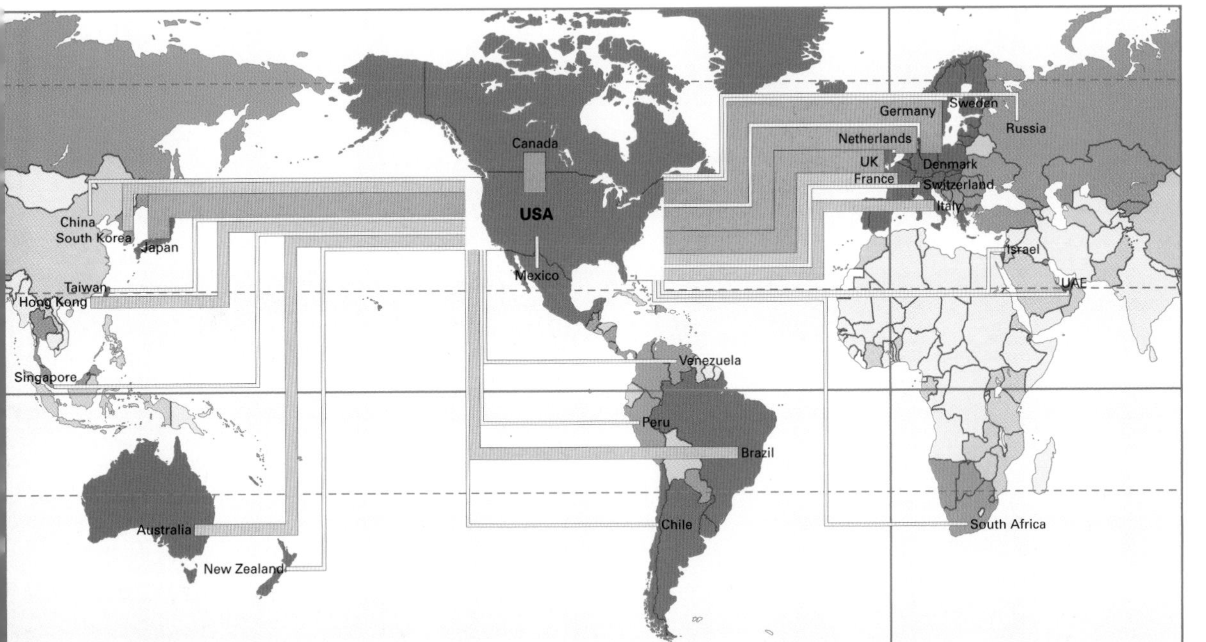

## Internet traffic, 2003

**internet providers**
per 10 000 people

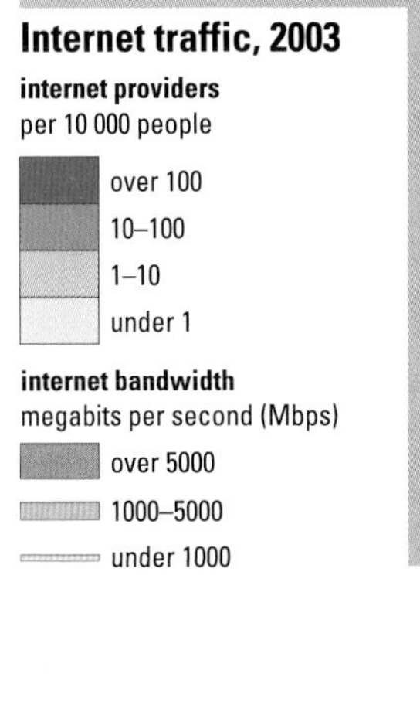

- over 100
- 10–100
- 1–10
- under 1

**internet bandwidth**
megabits per second (Mbps)

- over 5000
- 1000–5000
- under 1000

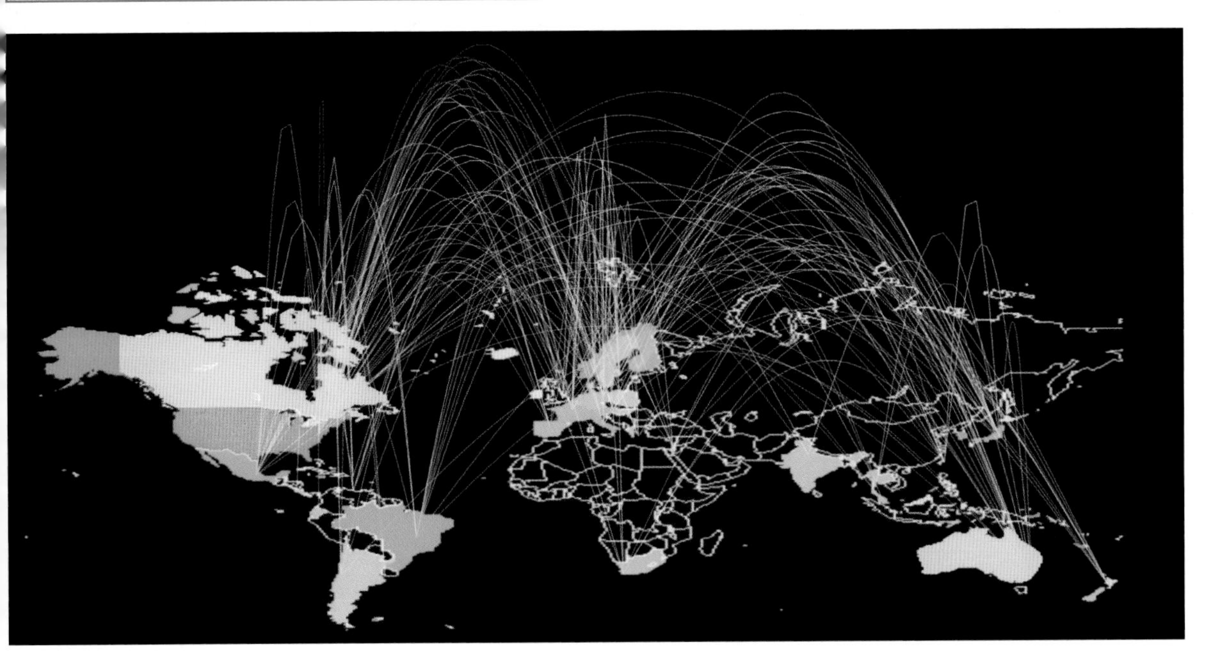

## Internet traffic flow

The 'arc map' shows internet traffic between 50 countries. Arcs are coloured to show internet traffic between countries. The height of each arc is proportional to the volume of internet traffic flowing over a link, so the highest arcs represent the greatest volume of traffic.

## Europe

Albania
Andorra
Austria
Belarus
Belgium
Bosnia-Herzegovina
Bulgaria

Greece
Hungary
Iceland
Ireland
Italy
Latvia
Liechtenstein

Norway
Poland
Portugal
Romania
Russian Federation
San Marino
Serbia and Montenegro

## Asia

Afghanistan
Armenia
Azerbaijan
Bahrain
Bangladesh
Bhutan
Brunei

Iran
Iraq
Israel
Japan
Jordan
Kazakhstan
Kuwait

Nepal
North Korea
Oman
Pakistan
Papua New Guinea
Philippines
Qatar

Tajikistan
Thailand
Turkey
Turkmenistan
United Arab Emirates
Uzbekistan
Vietnam

## Oceania

Australia
Fiji
Kiribati
Marshall Islands
Micronesia
Nauru
New Zealand

## Africa

Algeria
Angola
Benin
Botswana
Burkina
Burundi
Cameroon

Djibouti
Egypt
Equatorial Guinea
Eritrea
Ethiopia
Gabon
Gambia

Madagascar
Malawi
Mali
Mauritania
Mauritius
Morocco
Mozambique

Somalia
South Africa
Sudan
Swaziland
Tanzania
Togo
Tunisia

## North America

Antigua and Barbuda
Bahamas
Barbados
Belize
Canada
Costa Rica
Cuba

Honduras
Jamaica
Mexico
Nicaragua
Panama
St. Kitts and Nevis
St. Lucia

## S. America

Argentina
Bolivia
Brazil
Chile
Colombia
Ecuador
French Guiana

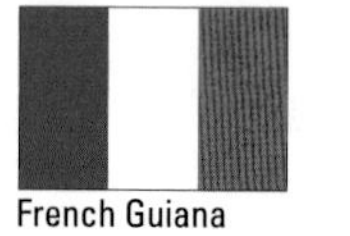

Croatia

Czech Republic

Denmark

Estonia

Finland

France

Germany

Lithuania

Luxembourg

Macedonia, FYRO

Malta

Moldova

Monaco

Netherlands

Slovakia

Slovenia

Spain

Sweden

Switzerland

Ukraine

United Kingdom

Cambodia

China

Cyprus

East Timor

Georgia

India

Indonesia

Kyrgyzstan

Laos

Lebanon

Malaysia

Maldives

Mongolia

Myanmar

Saudi Arabia

Seychelles

Singapore

South Korea

Sri Lanka

Syria

Taiwan

Yemen

Northern Marianas

Palau

Samoa

Solomon Islands

Tonga

Tuvalu

Vanuatu

Cape Verde

Central African Republic

Chad

Comoros

Congo

Congo, Dem. Rep.

Côte d'Ivoire

Ghana

Guinea

Guinea-Bissau

Kenya

Lesotho

Liberia

Libya

Namibia

Niger

Nigeria

Rwanda

Sao Tomé and Pirncipe

Senegal

Sierra Leone

Uganda

Zambia

Zimbabwe

Dominica

Dominican Republic

El Salvador

Greenland

Grenada

Guatemala

Haiti

St. Vincent & the Grenadines

Trinidad and Tobago

United States of America

Guyana

Paraguay

Peru

Suriname

Uruguay

Venezuela

The datasets below are explained on pages 14▸

| | no data |
|---|---|
| per capita | for each person |

| | Land | | Population | | | | | | | | | Employment | | |
|---|---|---|---|---|---|---|---|---|---|---|---|---|---|---|
| | Area | Arable and permanent crops | Total | Density | Change | Births | Deaths | Fertility | Infant mortality | Life expectancy | Urban | Agriculture | Industry | Servi▸ |
| | | | 2004 | 2004 | 1994–2004 | 2003 | 2003 | 2003 | 2003 | 2003 | 2003 | | | |
| | thousand km² | % of total | millions | persons per km² | % | births per 1000 | deaths per 1000 | children per mother | per 1000 live births | years | % | % | % | % |
| Afghanistan | 652 | 12.4 | 28.5 | 43.7 | 57.7 | 48 | 21 | 6.8 | 165 | 43 | 22 | ooo | ooo | oo▸ |
| Albania | 29 | 24.3 | 3.2 | 110.3 | -9.8 | 17 | 5 | 2.1 | 11 | 74 | 42 | 55 | 23 | 22 |
| Algeria | 2382 | 3.5 | 32.3 | 13.6 | 17.4 | 20 | 4 | 2.5 | 54 | 73 | 49 | 26 | 31 | 43 |
| Andorra | 0.5 | 2.2 | 0.1 | 200.0 | 53.8 | 11 | 3 | 1.3 | 4 | ooo | 92 | ooo | ooo | oo▸ |
| Angola | 1247 | 2.6 | 13.3 | 10.7 | 19.7 | 49 | 24 | 6.8 | 145 | 40 | 33 | 75 | 8 | 17 |
| Antigua and Barbuda | 0.4 | 22.7 | 0.1 | 250.0 | 51.5 | 24 | 6 | 2.7 | 17 | 71 | 37 | ooo | ooo | oo▸ |
| Argentina | 2780 | 12.6 | 37.9 | 13.6 | 10.4 | 19 | 8 | 2.4 | 16 | 74 | 89 | 12 | 32 | 56 |
| Armenia | 30 | 18.8 | 3.2 | 106.7 | -14.6 | 10 | 8 | 1.2 | 36 | 73 | 64 | 18 | 43 | 39 |
| Australia | 7741 | 6.3 | 20.1 | 2.6 | 12.6 | 13 | 7 | 1.7 | 5 | 80 | 91 | 6 | 26 | 68 |
| Austria | 84 | 17.4 | 8.1 | 96.4 | 0.9 | 9 | 9 | 1.4 | 5 | 79 | 54 | 8 | 38 | 54 |
| Azerbaijan | 87 | 23.2 | 8.3 | 95.4 | 9.3 | 14 | 6 | 1.8 | 13 | 72 | 51 | 31 | 29 | 40 |
| Bahamas, The | 14 | 0.9 | 0.3 | 21.4 | 9.5 | 18 | 5 | 2.1 | 16 | 72 | 89 | 5 | 16 | 79 |
| Bahrain | 0.7 | 8.5 | 0.7 | 937.1 | 25.4 | 20 | 3 | 2.7 | 7 | 74 | 87 | 2 | 30 | 68 |
| Bangladesh | 144 | 58.5 | 141.3 | 981.3 | 20.1 | 30 | 9 | 3.3 | 66 | 60 | 23 | 65 | 16 | 19 |
| Barbados | 0.4 | 39.5 | 0.3 | 690.0 | 13.6 | 15 | 8 | 1.7 | 13 | 72 | 50 | 14 | 30 | 56 |
| Belarus | 208 | 30.4 | 9.8 | 47.1 | -4.9 | 9 | 15 | 1.2 | 8 | 69 | 72 | 20 | 40 | 40 |
| Belgium | 33 | 25.4 | 10.4 | 315.2 | 2.9 | 11 | 10 | 1.6 | 4 | 79 | 97 | 3 | 28 | 69 |
| Belize | 23 | 4.4 | 0.3 | 11.3 | 42.2 | 28 | 5 | 3.4 | 20 | 70 | 49 | 33 | 19 | 48 |
| Benin | 113 | 25.0 | 7.3 | 64.6 | 39.3 | 41 | 14 | 5.6 | 89 | 51 | 40 | 63 | 8 | 29 |
| Bhutan | 47 | 3.5 | 1.0 | 21.3 | -44.4 | 34 | 9 | 4.7 | 61 | 66 | 21 | 94 | 2 | 4 |
| Bolivia | 1099 | 2.8 | 8.8 | 8.0 | 21.6 | 28 | 9 | 3.8 | 54 | 63 | 63 | 47 | 18 | 35 |
| Bosnia-Herzegovina | 51 | 21.3 | 3.9 | 76.5 | -12.5 | 10 | 8 | 1.2 | 9 | 74 | 43 | ooo | ooo | ooo▸ |
| Botswana | 582 | 0.7 | 1.7 | 2.9 | 19.3 | 27 | 26 | 3.5 | 62 | 36 | 54 | 46 | 20 | 34 |
| Brazil | 8547 | 7.8 | 179.1 | 21.0 | 16.5 | 20 | 7 | 2.2 | 33 | 71 | 81 | 23 | 23 | 54 |
| Brunei | 6 | 1.2 | 0.4 | 58.3 | 40.8 | 22 | 3 | 2.3 | 7 | 76 | 74 | 2 | 24 | 74 |
| Bulgaria | 111 | 32.3 | 7.8 | 70.3 | -7.6 | 9 | 14 | 1.2 | 12 | 72 | 70 | 13 | 48 | 39 |
| Burkina | 274 | 16.1 | 13.6 | 49.6 | 37.5 | 45 | 19 | 6.2 | 83 | 45 | 15 | 92 | 2 | 6 |
| Burundi | 28 | 48.5 | 6.2 | 221.4 | 5.5 | 40 | 18 | 6.2 | 74 | 43 | 8 | 92 | 3 | 5 |
| Cambodia | 181 | 21.0 | 13.1 | 72.4 | 36.9 | 32 | 10 | 4.5 | 95 | 57 | 16 | 74 | 8 | 18 |
| Cameroon | 475 | 15.1 | 16.1 | 33.9 | 24.4 | 37 | 15 | 4.9 | 77 | 48 | 48 | 70 | 9 | 21 |
| Canada | 9971 | 4.6 | 31.9 | 3.2 | 9.9 | 11 | 7 | 1.5 | 5 | 79 | 79 | 3 | 25 | 72 |
| Cape Verde | 4 | 11.2 | 0.5 | 125.0 | 34.4 | 29 | 7 | 4.0 | 31 | 69 | 53 | 30 | 30 | 40 |
| Central African Republic | 623 | 3.2 | 3.7 | 5.9 | 13.3 | 37 | 19 | 4.9 | 96 | 42 | 39 | 80 | 3 | 17 |
| Chad | 1284 | 2.8 | 9.5 | 7.4 | 52.9 | 49 | 16 | 6.6 | 103 | 49 | 24 | 83 | 4 | 13 |
| Chile | 757 | 3.0 | 16.0 | 21.1 | 14.3 | 17 | 5 | 2.4 | 8 | 76 | 87 | 19 | 25 | 56 |
| China | 9598 | 16.0 | 1300.1 | 135.5 | 7.7 | 12 | 6 | 1.7 | 32 | 71 | 41 | 72 | 15 | 13 |
| Colombia | 1139 | 3.4 | 45.3 | 39.8 | 19.7 | 23 | 6 | 2.6 | 26 | 72 | 71 | 27 | 23 | 50 |
| Comoros | 2 | 59.2 | 0.7 | 350.0 | 18.2 | 47 | 12 | 6.8 | 84 | 56 | 33 | 78 | 9 | 13 |
| Congo | 342 | 0.7 | 3.8 | 11.1 | 50.5 | 44 | 15 | 6.3 | 84 | 48 | 52 | 49 | 15 | 36 |
| Congo, Dem. Rep. | 2345 | 3.3 | 58.3 | 24.9 | 34.3 | 46 | 15 | 6.8 | 100 | 49 | 30 | 68 | 13 | 19 |
| Costa Rica | 51 | 10.3 | 4.2 | 82.4 | 28.6 | 18 | 4 | 2.1 | 10 | 79 | 59 | 26 | 27 | 47 |
| Côte d'Ivoire | 322 | 21.4 | 16.9 | 52.5 | 23.4 | 39 | 19 | 5.2 | 102 | 42 | 46 | 60 | 10 | 30 |
| Croatia | 57 | 28.1 | 4.4 | 77.2 | -5.4 | 9 | 11 | 1.3 | 7 | 75 | 56 | 16 | 34 | 50 |
| Cuba | 111 | 34.2 | 11.3 | 101.8 | 3.2 | 11 | 7 | 1.6 | 7 | 76 | 75 | 19 | 30 | 51 |
| Cyprus | 9 | 12.2 | 0.9 | 100.0 | 24.0 | 12 | 7 | 1.6 | 6 | 78 | 65 | 14 | 30 | 56 |
| Czech Republic | 79 | 41.9 | 10.2 | 129.1 | -1.3 | 9 | 11 | 1.2 | 4 | 75 | 77 | 11 | 45 | 44 |
| Denmark | 43 | 53.0 | 5.4 | 125.6 | 3.7 | 12 | 11 | 1.8 | 4 | 77 | 72 | 6 | 28 | 66 |
| Djibouti | 23 | 0.04 | 0.7 | 30.4 | 30.6 | 41 | 17 | 5.9 | 106 | 46 | 82 | ooo | ooo | ooo |
| Dominica | 0.8 | 26.7 | 0.1 | 125.0 | 35.1 | 17 | 7 | 1.9 | 16 | 74 | 71 | ooo | ooo | ooo |
| Dominican Republic | 49 | 32.8 | 8.8 | 179.6 | 13.3 | 25 | 6 | 3.0 | 31 | 69 | 64 | 25 | 29 | 46 |
| Ecuador | 284 | 10.5 | 13.4 | 47.2 | 19.4 | 25 | 4 | 3.0 | 30 | 71 | 61 | 33 | 19 | 48 |
| Egypt | 1001 | 3.4 | 73.4 | 73.3 | 30.3 | 26 | 6 | 3.2 | 38 | 68 | 43 | 40 | 22 | 38 |
| El Salvador | 21 | 43.3 | 6.7 | 319.0 | 20.7 | 26 | 6 | 3.0 | 25 | 70 | 58 | 36 | 21 | 43 |
| Equatorial Guinea | 28 | 8.2 | 0.5 | 17.9 | 28.5 | 43 | 17 | 5.9 | 105 | 49 | 45 | 66 | 11 | 23 |
| Eritrea | 118 | 4.3 | 4.4 | 37.3 | 39.3 | 39 | 13 | 5.7 | 76 | 53 | 19 | 80 | 5 | 15 |

## Wealth · Energy and trade · Quality of life

| | Purchasing power | Growth of PP | Energy consumption | Imports | Exports | Aid received (given) | Human Development Index | Health care | Food consumption | Safe water | Illiteracy male | Illiteracy female | Higher education | Cars | |
|---|---|---|---|---|---|---|---|---|---|---|---|---|---|---|---|
| 2 | 2002 | 1990–2002 | 2002 | 2003 | 2003 | 2002 | 2002 | 1990–2003 | 2002 | 2000 | 2002 | 2002 | 2002 | 2000 | |
| on $ | US$ | annual % | kg oil equivalent per capita | US$ per capita | US$ per capita | million US$ | | doctors per 100 000 people | daily calories per capita | % access | % | % | students per 100 000 people | people per car | |
| ooo | ooo | ooo | 19 | 41 | 5 | ooo | ooo | ooo | ooo | 13 | ooo | ooo | ooo | 644 | Afghanistan |
| 4.4 | 4830 | 6.0 | 684 | 588 | 143 | 317 | 0.781 | 137 | 2848 | 76 | 1 | 2 | 1165 | 36 | Albania |
| 3.8 | 5760 | 0.3 | 1027 | 409 | 774 | 361 | 0.704 | 85 | 3022 | 94 | 22 | 40 | 1459 | 34 | Algeria |
| ooo | ooo | ooo | ooo | ooo | ooo | ooo | ooo | ooo | ooo | 100 | ooo | ooo | ooo | 2 | Andorra |
| 9.2 | 2130 | -0.1 | 301 | 305 | 650 | 421 | 0.381 | 5 | 2083 | 38 | ooo | ooo | 74 | 111 | Angola |
| 0.6 | 10 920 | 2.6 | 2595 | 3797 | 316 | 14 | 0.800 | 105 | 2349 | 91 | ooo | ooo | ooo | ooo | Antigua and Barbuda |
| 4.1 | 10 880 | 1.7 | 1607 | 360 | 765 | 0 | 0.853 | 304 | 2992 | 79 | 3 | 3 | 5006 | 7 | Argentina |
| 2.4 | 3120 | 1.7 | 1336 | 415 | 222 | 294 | 0.754 | 287 | 2268 | 84 | 0 | 1 | 2504 | 1900 | Armenia |
| 6.6 | 28 260 | 2.6 | 7155 | 4479 | 3597 | (916) | 0.946 | 247 | 3054 | 100 | ooo | ooo | 5178 | 2 | Australia |
| 0.4 | 29 220 | 1.9 | 4277 | 12 162 | 11 891 | (488) | 0.934 | 323 | 3673 | 100 | ooo | ooo | 2746 | 2 | Austria |
| 5.8 | 3210 | 0.2 | 1939 | 319 | 315 | 349 | 0.746 | 359 | 2575 | ooo | ooo | ooo | 2189 | 30 | Azerbaijan |
| 4.5 | 17 280 | 0.1 | 3981 | 4991 | 2177 | 5 | 0.815 | 163 | 2755 | 96 | 5 | 4 | ooo | 4 | Bahamas, The |
| 6.2 | 17 170 | 1.5 | 15 882 | 7185 | 8938 | 71 | 0.843 | 169 | ooo | ooo | 9 | 16 | 1683 | 5 | Bahrain |
| 8.5 | 1700 | 3.1 | 105 | 69 | 50 | 913 | 0.509 | 23 | 2205 | 97 | 50 | 69 | 631 | 2808 | Bangladesh |
| 2.5 | 15 290 | 1.6 | 2082 | 4181 | 775 | 3 | 0.888 | 137 | 3091 | 100 | 0 | 0 | 2730 | 7 | Barbados |
| 3.5 | 5520 | 0.2 | 2789 | 1164 | 1008 | 39 | 0.790 | 450 | 3000 | 100 | 0 | 0 | 4485 | 9 | Belarus |
| 9.9 | 27 570 | 1.8 | 6621 | 22 745 | 24 673 | (996) | 0.942 | 419 | 3584 | ooo | ooo | ooo | 3559 | 2 | Belgium |
| 0.7 | 6080 | 1.7 | 1346 | 2131 | 792 | 22 | 0.737 | 102 | 2869 | 76 | 23 | 23 | ooo | 29 | Belize |
| 2.5 | 1070 | 2.1 | 95 | 113 | 81 | 220 | 0.421 | 10 | 2548 | 63 | 45 | 75 | 274 | 260 | Benin |
| 0.5 | 1969 | 3.6 | 227 | 229 | 137 | 74 | 0.536 | 5 | ooo | 62 | ooo | ooo | 90 | ooo | Bhutan |
| 7.9 | 2460 | 1.1 | 453 | 180 | 175 | 681 | 0.681 | 76 | 2235 | 79 | 7 | 19 | 3576 | 65 | Bolivia |
| 5.2 | 5970 | 18.0 | 1659 | 1077 | 332 | 587 | 0.781 | 145 | 2894 | ooo | 2 | 9 | ooo | 34 | Bosnia-Herzegovina |
| 5.1 | 8170 | 2.5 | 797 | 1423 | 1664 | 38 | 0.589 | 29 | 2151 | 70 | 24 | 19 | 514 | 59 | Botswana |
| 7.4 | 7770 | 1.3 | 1194 | 287 | 414 | 376 | 0.775 | 206 | 3050 | 83 | 14 | 14 | 1737 | 11 | Brazil |
| ooo | 19 210 | -0.7 | 6555 | 4798 | 12 219 | -2 | 0.867 | 99 | 2855 | ooo | 4 | 9 | 1276 | 2 | Brunei |
| 4.1 | 7130 | -1.5 | 2780 | 1392 | 963 | 381 | 0.796 | 344 | 2848 | 100 | 1 | 2 | 2981 | 5 | Bulgaria |
| 2.6 | 1100 | 1.6 | 33 | 71 | 26 | 473 | 0.302 | 4 | 2462 | ooo | 82 | 92 | 121 | 744 | Burkina |
| 0.7 | 630 | -3.9 | 34 | 22 | 5 | 172 | 0.339 | 1 | 1649 | 52 | 42 | 56 | 177 | 650 | Burundi |
| 3.5 | 2060 | 4.1 | 16 | 128 | 126 | 487 | 0.568 | 16 | 2046 | 30 | 19 | 41 | 248 | 788 | Cambodia |
| 8.7 | 2000 | -0.1 | 133 | 137 | 148 | 632 | 0.501 | 7 | 2273 | 62 | 23 | 40 | 504 | 152 | Cameroon |
| 0.5 | 29 480 | 2.2 | 10 238 | 7746 | 8623 | (2011) | 0.943 | 187 | 3589 | 100 | ooo | ooo | 3738 | 2 | Canada |
| 0.6 | 5000 | 3.4 | 122 | 649 | 26 | 92 | 0.717 | 17 | 3243 | 74 | 15 | 32 | 415 | 133 | Cape Verde |
| 1.0 | 1170 | -0.2 | 41 | 30 | 33 | 60 | 0.361 | 4 | 1980 | 60 | 35 | 67 | 175 | 422 | Central African Republic |
| 1.8 | 1020 | -0.5 | 8 | 99 | 30 | 233 | 0.379 | 3 | 2114 | 27 | 46 | 63 | 66 | 853 | Chad |
| 6.3 | 9820 | 4.4 | 1702 | 1231 | 1334 | -23 | 0.839 | 115 | 2863 | 94 | 4 | 4 | 3366 | 15 | Chile |
| 7.1 | 4580 | 8.6 | 840 | 355 | 373 | 1476 | 0.745 | 164 | 2951 | 75 | 5 | 14 | 946 | 369 | China |
| 0.1 | 6370 | 0.4 | 739 | 313 | 285 | 441 | 0.773 | 94 | 2585 | 91 | 8 | 8 | 2383 | 43 | Colombia |
| 0.2 | 1690 | -1.4 | 41 | 208 | 25 | 33 | 0.530 | 7 | 1754 | 96 | 37 | 51 | 116 | 1 | Comoros |
| 2.2 | 980 | -1.6 | 146 | 319 | 813 | 420 | 0.494 | 25 | 2162 | 51 | 11 | 23 | 418 | 119 | Congo |
| 5.0 | 650 | -8.2 | 37 | 33 | 15 | 807 | 0.365 | 7 | 1599 | 45 | 26 | 48 | 110 | 525 | Congo, Dem. Rep. |
| 6.2 | 8840 | 2.7 | 1004 | 1908 | 1523 | 5 | 0.834 | 160 | 2876 | 98 | 4 | 4 | 2065 | 22 | Costa Rica |
| 0.3 | 1520 | -0.1 | 162 | 267 | 347 | 1069 | 0.399 | 9 | 2631 | 77 | 40 | 62 | 595 | 88 | Côte d'Ivoire |
| 0.3 | 10 240 | 2.1 | 2098 | 3172 | 1383 | 167 | 0.830 | 238 | 2799 | 95 | 1 | 3 | 2511 | 5 | Croatia |
| ooo | 5259 | 3.5 | 1055 | 412 | 133 | 61 | 0.809 | 596 | 3152 | 95 | 3 | 3 | 1704 | 574 | Cuba |
| 9.4 | 18 360 | 3.2 | 3747 | 5800 | 1199 | 50 | 0.883 | 269 | 3255 | 100 | 1 | 5 | 1815 | 4 | Cyprus |
| 6.7 | 15 780 | 1.4 | 3846 | 5008 | 4777 | 393 | 0.868 | 342 | 3171 | ooo | ooo | ooo | 2774 | 3 | Czech Republic |
| 2.7 | 30 940 | 2.1 | 3870 | 10 728 | 12 507 | (1540) | 0.932 | 366 | 3439 | 100 | ooo | ooo | 3651 | 3 | Denmark |
| 0.6 | 1990 | -3.8 | 1397 | 433 | 121 | 78 | 0.454 | 13 | 2220 | 100 | 24 | 45 | 163 | 55 | Djibouti |
| 0.2 | 5640 | 1.4 | 713 | 1775 | 549 | 30 | 0.743 | 49 | 2763 | 97 | ooo | ooo | ooo | 23 | Dominica |
| 20.0 | 6640 | 4.2 | 771 | 902 | 622 | 157 | 0.738 | 190 | 2347 | 79 | 16 | 16 | ooo | 53 | Dominican Republic |
| 9.0 | 3580 | -0.3 | 706 | 501 | 464 | 216 | 0.735 | 145 | 2754 | 71 | 8 | 10 | ooo | 123 | Ecuador |
| 7.6 | 3810 | 2.5 | 801 | 161 | 91 | 1286 | 0.653 | 218 | 3338 | 95 | 33 | 56 | 3338 | 52 | Egypt |
| 3.5 | 4890 | 2.3 | 464 | 882 | 480 | 234 | 0.720 | 126 | 2584 | 74 | 18 | 23 | 1730 | 67 | El Salvador |
| 0.3 | 30 130 | 20.8 | 2509 | 2065 | 5344 | 20 | 0.703 | 25 | ooo | 43 | 7 | 24 | 201 | 143 | Equatorial Guinea |
| 0.7 | 890 | 1.5 | 58 | 153 | 13 | 230 | 0.439 | 5 | 1513 | 46 | 32 | 54 | 128 | 760 | Eritrea |

The datasets below are explained on pages 14

| | | | | | | | | | | | | | |
|---|---|---|---|---|---|---|---|---|---|---|---|---|---|
| ○○○ no data | **Land** | | **Population** | | | | | | | | | **Employment** | |
| per capita for each person | **Area** | **Arable and permanent crops** | **Total** | **Density** | **Change** | **Births** | **Deaths** | **Fertility** | **Infant mortality** | **Life expectancy** | **Urban** | **Agriculture** | **Industry** | **Servi** |
| | | | 2004 | 2004 | 1994–2004 | 2003 | 2003 | 2003 | 2003 | 2003 | 2003 | | | |
| | thousand km² | % of total | millions | persons per km² | % | births per 1000 | deaths per 1000 | children per mother | per 1000 live births | years | % | % | % | % |

| | Area | Arable and permanent crops | Total | Density | Change | Births | Deaths | Fertility | Infant mortality | Life expectancy | Urban | Agriculture | Industry | Servi |
|---|---|---|---|---|---|---|---|---|---|---|---|---|---|---|
| Estonia | 45 | 14.0 | 1.3 | 28.9 | -13.3 | 10 | 13 | 1.4 | 6 | 71 | 69 | 14 | 41 | 45 |
| Ethiopia | 1104 | 9.7 | 72.4 | 65.6 | 34.5 | 41 | 18 | 5.9 | 105 | 46 | 15 | 86 | 2 | 12 |
| Fiji | 18 | 15.6 | 0.8 | 44.4 | 2.0 | 25 | 6 | 3.3 | 22 | 67 | 39 | 46 | 15 | 39 |
| Finland | 338 | 6.5 | 5.2 | 15.4 | 2.2 | 11 | 9 | 1.8 | 3 | 79 | 62 | 8 | 31 | 61 |
| France | 552 | 35.5 | 60.0 | 108.7 | 3.6 | 13 | 9 | 1.9 | 4 | 79 | 74 | 5 | 29 | 66 |
| French Guiana | 91 | 0.1 | 0.2 | 2.0 | 49.3 | 31 | 4 | 3.9 | 12 | 75 | 75 | ○○○ | ○○○ | ○○○ |
| Gabon | 268 | 1.8 | 1.4 | 5.2 | 34.6 | 33 | 12 | 4.3 | 57 | 57 | 73 | 51 | 16 | 33 |
| Gambia, The | 11 | 22.6 | 1.5 | 136.4 | 39.3 | 41 | 13 | 5.6 | 78 | 54 | 26 | 82 | 8 | 10 |
| Georgia | 70 | 15.3 | 4.5 | 64.3 | -17.1 | 11 | 11 | 1.4 | 24 | 72 | 52 | 26 | 31 | 43 |
| Germany | 357 | 33.6 | 82.6 | 231.4 | 1.4 | 9 | 10 | 1.3 | 4 | 78 | 88 | 4 | 38 | 58 |
| Ghana | 239 | 26.5 | 21.4 | 89.5 | 26.8 | 33 | 10 | 4.4 | 64 | 58 | 44 | 59 | 13 | 28 |
| Greece | 132 | 29.1 | 11.0 | 83.3 | 5.5 | 9 | 9 | 1.3 | 6 | 78 | 60 | 23 | 27 | 50 |
| Greenland | 342 | ○○○ | 0.06 | 0.2 | 1.1 | 16 | 8 | ○○○ | ○○○ | ○○○ | ○○○ | ○○○ | ○○○ | ○○○ |
| Grenada | 0.3 | 35.3 | 0.1 | 333.3 | 8.7 | 19 | 7 | 2.1 | 17 | 71 | 39 | ○○○ | ○○○ | ○○○ |
| Guatemala | 109 | 17.5 | 12.7 | 116.5 | 30.7 | 34 | 7 | 4.4 | 39 | 66 | 39 | 52 | 17 | 31 |
| Guinea | 246 | 6.3 | 9.2 | 37.4 | 29.4 | 43 | 16 | 6.0 | 98 | 49 | 33 | 87 | 2 | 11 |
| Guinea-Bissau | 36 | 15.2 | 1.5 | 41.7 | 42.6 | 50 | 20 | 7.1 | 125 | 45 | 32 | 85 | 2 | 13 |
| Guyana | 215 | 2.4 | 0.8 | 3.7 | 8.3 | 23 | 9 | 2.4 | 53 | 63 | 36 | 22 | 25 | 53 |
| Haiti | 28 | 39.6 | 8.1 | 289.3 | 15.0 | 33 | 14 | 4.7 | 80 | 51 | 36 | 68 | 9 | 23 |
| Honduras | 112 | 12.7 | 7.0 | 62.5 | 29.1 | 33 | 5 | 4.1 | 34 | 71 | 47 | 41 | 20 | 39 |
| Hungary | 93 | 51.6 | 10.1 | 108.6 | -1.6 | 9 | 13 | 1.3 | 7 | 73 | 65 | 15 | 38 | 47 |
| Iceland | 103 | 0.07 | 0.3 | 2.7 | 12.8 | 14 | 6 | 2.0 | 2 | 81 | 94 | ○○○ | ○○○ | ○○○ |
| India | 3288 | 51.7 | 1086.6 | 330.5 | 20.2 | 25 | 8 | 3.1 | 64 | 62 | 28 | 64 | 16 | 20 |
| Indonesia | 1905 | 17.7 | 218.7 | 114.8 | 14.7 | 22 | 6 | 2.6 | 46 | 68 | 42 | 55 | 14 | 31 |
| Iran | 1633 | 10.4 | 67.4 | 41.3 | 13.6 | 18 | 6 | 2.5 | 32 | 69 | 67 | 39 | 23 | 38 |
| Iraq | 438 | 13.9 | 25.9 | 59.1 | 33.0 | 36 | 9 | 5.0 | 102 | 60 | 68 | 16 | 18 | 66 |
| Ireland | 70 | 16.0 | 4.1 | 58.6 | 14.3 | 16 | 7 | 2.0 | 5 | 77 | 60 | 14 | 29 | 57 |
| Israel | 21 | 19.2 | 6.8 | 323.8 | 25.9 | 22 | 6 | 2.9 | 5 | 79 | 92 | 4 | 29 | 67 |
| Italy | 301 | 36.7 | 57.8 | 192.0 | 1.0 | 10 | 10 | 1.3 | 5 | 80 | 90 | 9 | 31 | 60 |
| Jamaica | 11 | 25.8 | 2.6 | 236.4 | 5.1 | 20 | 7 | 2.4 | 24 | 75 | 52 | 25 | 23 | 52 |
| Japan | 378 | 12.6 | 127.6 | 337.6 | 1.9 | 9 | 8 | 1.3 | 3 | 82 | 78 | 7 | 34 | 59 |
| Jordan | 89 | 4.5 | 5.6 | 62.9 | 37.9 | 29 | 5 | 3.7 | 22 | 72 | 79 | 15 | 23 | 62 |
| Kazakhstan | 2717 | 8.0 | 15.0 | 5.5 | -8.0 | 17 | 11 | 2.0 | 52 | 64 | 57 | 22 | 32 | 46 |
| Kenya | 580 | 8.9 | 32.4 | 55.9 | 10.6 | 38 | 15 | 5.0 | 78 | 51 | 36 | 80 | 7 | 13 |
| Kiribati | 0.7 | 50.7 | 0.1 | 137.1 | 31.6 | 26 | 8 | 4.3 | 43 | 63 | 43 | ○○○ | ○○○ | ○○○ |
| Kuwait | 18 | 0.8 | 2.5 | 138.9 | 54.3 | 18 | 2 | 4.0 | 10 | 78 | 100 | 1 | 25 | 74 |
| Kyrgyzstan | 199 | 7.1 | 5.1 | 25.6 | 12.7 | 21 | 8 | 2.6 | 42 | 68 | 35 | 32 | 27 | 41 |
| Laos | 237 | 4.2 | 5.8 | 24.5 | 26.9 | 36 | 13 | 4.9 | 104 | 54 | 19 | 78 | 6 | 16 |
| Latvia | 65 | 28.8 | 2.3 | 35.4 | -9.7 | 9 | 14 | 1.3 | 9 | 72 | 68 | 16 | 40 | 44 |
| Lebanon | 10 | 30.1 | 4.5 | 450.0 | 46.3 | 23 | 7 | 3.2 | 27 | 73 | 87 | 7 | 31 | 62 |
| Lesotho | 30 | 11.0 | 1.8 | 60.0 | -1.7 | 33 | 22 | 4.4 | 90 | 38 | 17 | 40 | 28 | 32 |
| Liberia | 111 | 5.4 | 3.5 | 31.5 | 29.6 | 50 | 22 | 6.8 | 150 | 42 | 45 | ○○○ | ○○○ | ○○○ |
| Libya | 1760 | 1.2 | 5.6 | 3.2 | 14.3 | 28 | 4 | 3.6 | 28 | 76 | 86 | 11 | 23 | 66 |
| Liechtenstein | 0.2 | 25.0 | 0.03 | 150.0 | 0 | 12 | 6 | 1.5 | 4 | 80 | 21 | ○○○ | ○○○ | ○○○ |
| Lithuania | 65 | 45.8 | 3.4 | 52.3 | -8.6 | 9 | 12 | 1.3 | 7 | 72 | 67 | 18 | 41 | 41 |
| Luxembourg | 3 | ○○○ | 0.5 | 166.7 | 23.8 | 12 | 9 | 1.6 | 5 | 78 | 91 | ○○○ | ○○○ | ○○○ |
| Macedonia, FYRO | 26 | 23.8 | 2.0 | 76.9 | 7.8 | 14 | 9 | 1.7 | 12 | 73 | 59 | 21 | 40 | 39 |
| Madagascar | 587 | 6.0 | 17.5 | 29.8 | 30.6 | 43 | 12 | 5.8 | 84 | 55 | 26 | 78 | 7 | 15 |
| Malawi | 118 | 20.6 | 11.9 | 100.8 | 25.8 | 51 | 21 | 6.6 | 121 | 44 | 14 | 87 | 5 | 8 |
| Malaysia | 330 | 23.0 | 25.6 | 77.6 | 27.3 | 26 | 4 | 3.3 | 11 | 73 | 62 | 27 | 23 | 50 |
| Maldives | 0.3 | 40.0 | 0.3 | 1066.7 | 22.0 | 18 | 4 | 3.7 | 18 | 73 | 27 | 32 | 31 | 37 |
| Mali | 1240 | 3.8 | 13.4 | 10.8 | 38.4 | 50 | 17 | 7.0 | 123 | 48 | 30 | 86 | 2 | 12 |
| Malta | 0.3 | 31.3 | 0.4 | 1333.3 | 8.7 | 10 | 8 | 1.5 | 7 | 78 | 91 | ○○○ | ○○○ | ○○○ |
| Marshall Islands | 0.2 | 16.7 | 0.1 | 500.0 | 85.2 | 42 | 5 | 4.7 | 37 | 69 | 68 | ○○○ | ○○○ | ○○○ |
| Mauritania | 1026 | 0.5 | 3.0 | 2.9 | 35.7 | 42 | 15 | 5.9 | 102 | 54 | 40 | 55 | 10 | 35 |

## Wealth · Energy and trade · Quality of life

| | Purchasing power | Growth of PP | Energy consumption | Imports | Exports | Aid received (given) | Human Development Index | Health care | Food consumption | Safe water | Illiteracy male | Illiteracy female | Higher education | Cars | |
|---|---|---|---|---|---|---|---|---|---|---|---|---|---|---|---|
| 2002 | 2002 | 1990–2002 | 2002 | 2003 | 2003 | 2002 | 2002 | 1990–2003 | 2002 | 2000 | 2002 | 2002 | 2002 | 2000 | |
| billion US$ | US$ | annual % | kg oil equivalent per capita | US$ per capita | US$ per capita | million US$ | | doctors per 100 000 people | daily calories per capita | % access | % | % | students per 100 000 people | people per car | |
| 5.6 | 12 260 | 2.3 | 3180 | 5874 | 4146 | 69 | 0.853 | 313 | 3002 | ○○○ | 0 | 0 | 4459 | 3 | Estonia |
| 6.4 | 780 | 2.3 | 26 | 29 | 8 | 1307 | 0.359 | 3 | 1857 | 24 | 51 | 66 | 150 | 1433 | Ethiopia |
| 1.8 | 5440 | 1.8 | 555 | 1280 | 725 | 34 | 0.758 | 34 | 2894 | 47 | 6 | 9 | ○○○ | 17 | Fiji |
| 2.2 | 26 190 | 2.5 | 5917 | 8064 | 10 164 | (434) | 0.935 | 311 | 3100 | 100 | ○○○ | ○○○ | 5465 | 3 | Finland |
| 2.7 | 26 920 | 1.6 | 4583 | 6539 | 6475 | (5125) | 0.932 | 330 | 3654 | 100 | ○○○ | ○○○ | 3386 | 2 | France |
| ○○○ | ○○○ | ○○○ | 1920 | ○○○ | ○○○ | ○○○ | ○○○ | ○○○ | ○○○ | ○○○ | ○○○ | ○○○ | ○○○ | 7 | French Guiana |
| 4.0 | 6590 | -0.2 | 737 | 801 | 1890 | 72 | 0.648 | ○○○ | 2637 | 70 | ○○○ | ○○○ | 580 | 7 | Gabon |
| 0.4 | 1690 | -0.3 | 69 | 130 | 9 | 61 | 0.452 | 4 | 2273 | 62 | 55 | 69 | ○○○ | 186 | Gambia, The |
| 3.3 | 2260 | -3.9 | 1005 | 206 | 87 | 313 | 0.739 | 463 | 2354 | 76 | ○○○ | ○○○ | 3154 | 12 | Georgia |
| 0.4 | 27 100 | 1.3 | 432 | 7289 | 9065 | (4980) | 0.925 | 363 | 3496 | ○○○ | ○○○ | ○○○ | 2654 | 2 | Germany |
| 5.4 | 2130 | 1.8 | 174 | 159 | 122 | 653 | 0.568 | 9 | 2667 | 64 | 18 | 34 | 339 | 214 | Ghana |
| 3.9 | 18 720 | 2.2 | 3254 | 4089 | 1236 | (253) | 0.902 | 438 | 3721 | ○○○ | 1 | 4 | 4979 | 4 | Greece |
| ○○○ | ○○○ | ○○○ | 3548 | ○○○ | ○○○ | ○○○ | ○○○ | ○○○ | ○○○ | ○○○ | ○○○ | ○○○ | ○○○ | ○○○ | Greenland |
| 0.4 | 7280 | 2.7 | 841 | 2286 | 400 | 10 | 0.745 | 81 | 2932 | 94 | ○○○ | ○○○ | ○○○ | ○○○ | Grenada |
| 0.9 | 4080 | 1.3 | 314 | 546 | 214 | 249 | 0.649 | 109 | 2219 | 92 | 23 | 38 | ○○○ | 84 | Guatemala |
| 3.1 | 2100 | 1.7 | 65 | 104 | 104 | 250 | 0.425 | 13 | 2409 | 48 | ○○○ | ○○○ | ○○○ | 488 | Guinea |
| 0.2 | 710 | -2.2 | 94 | 94 | 46 | 59 | 0.350 | 17 | 2024 | 49 | 45 | 75 | 35 | 267 | Guinea-Bissau |
| 0.6 | 4260 | 4.1 | 757 | 748 | 670 | 65 | 0.719 | 26 | 2692 | 94 | 1 | 2 | ○○○ | 32 | Guyana |
| 3.7 | 1610 | -3.0 | 85 | 141 | 41 | 156 | 0.463 | 25 | 2086 | 46 | 46 | 50 | ○○○ | 252 | Haiti |
| 6.2 | 2600 | 0.3 | 384 | 470 | 191 | 435 | 0.672 | 87 | 2356 | 90 | 20 | 20 | 1388 | 165 | Honduras |
| 53.7 | 13 400 | 2.4 | 2608 | 4704 | 4203 | 472 | 0.848 | 355 | 3483 | 99 | 1 | 1 | 3515 | 4 | Hungary |
| 8.2 | 29 750 | 2.1 | 12 065 | 9752 | 8339 | 0 | 0.941 | 352 | 3249 | ○○○ | ○○○ | ○○○ | 4022 | 2 | Iceland |
| 1.5 | 2670 | 4.0 | 338 | 66 | 53 | 1463 | 0.595 | 51 | 2459 | 88 | 31 | 54 | 1023 | 218 | India |
| 9.9 | 3230 | 2.1 | 481 | 152 | 284 | 1308 | 0.692 | 16 | 2904 | 76 | 8 | 17 | 1373 | 86 | Indonesia |
| 2.1 | 6690 | 2.2 | 2190 | 385 | 546 | 116 | 0.732 | 110 | 3085 | 95 | 17 | 30 | 2341 | 44 | Iran |
| ○○○ | ○○○ | ○○○ | 1204 | 320 | 513 | ○○○ | ○○○ | ○○○ | 2619 | 85 | ○○○ | ○○○ | 1325 | 35 | Iraq |
| 2.6 | 36 360 | 6.8 | 4047 | 13 526 | 23 492 | (360) | 0.936 | 239 | 3656 | ○○○ | ○○○ | ○○○ | 4545 | 3 | Ireland |
| 4.1 | 19 530 | 1.8 | 3292 | 5425 | 4721 | 754 | 0.908 | 375 | 3666 | 99 | 3 | 7 | 4971 | 5 | Israel |
| 7.9 | 26 430 | 1.5 | 3296 | 5045 | 5066 | (2157) | 0.920 | 607 | 3671 | ○○○ | 1 | 2 | 3201 | 2 | Italy |
| 7.4 | 3980 | -0.1 | 1399 | 1375 | 453 | 24 | 0.764 | 85 | 2685 | 71 | 16 | 9 | 1694 | 23 | Jamaica |
| 5.6 | 26 940 | 1.0 | 4322 | 3010 | 3709 | (9731) | 0.938 | 202 | 2761 | 96 | ○○○ | ○○○ | 3122 | 3 | Japan |
| 9.1 | 4220 | 0.9 | 1060 | 1065 | 581 | 534 | 0.750 | 205 | 2674 | 96 | 5 | 14 | 3065 | 28 | Jordan |
| 2.3 | 5870 | -0.7 | 3461 | 559 | 865 | 188 | 0.766 | 345 | 2677 | 91 | 0 | 1 | 3414 | ○○○ | Kazakhstan |
| 1.3 | 1020 | -0.6 | 123 | 117 | 76 | 393 | 0.488 | 14 | 2090 | 49 | 10 | 22 | 314 | 174 | Kenya |
| 0.08 | ○○○ | ○○○ | 104 | 417 | 31 | ○○○ | ○○○ | ○○○ | 2859 | 47 | ○○○ | ○○○ | ○○○ | ○○○ | Kiribati |
| 35.8 | 16 240 | -1.7 | 1129 | 4531 | 8132 | 5 | 0.838 | 160 | 3010 | 100 | 15 | 19 | 1531 | 3 | Kuwait |
| 1.5 | 1620 | -3.2 | 1159 | 142 | 115 | 186 | 0.701 | 272 | 2999 | 77 | ○○○ | ○○○ | 4219 | 36 | Kyrgyzstan |
| 1.7 | 1720 | 3.8 | 173 | 93 | 67 | 278 | 0.534 | 61 | 2312 | 90 | 23 | 45 | 494 | 540 | Laos |
| 8.1 | 9210 | 0.2 | 2040 | 2259 | 1246 | 86 | 0.823 | 291 | 2938 | ○○○ | 0 | 0 | 4722 | 5 | Latvia |
| 17.7 | 4360 | 3.1 | 1543 | 1594 | 339 | 456 | 0.758 | 274 | 3196 | 100 | 8 | 19 | 3887 | 5 | Lebanon |
| 1.0 | 2420 | 2.4 | 81 | 569 | 266 | 76 | 0.493 | 7 | 2638 | 91 | 26 | 10 | 269 | 350 | Lesotho |
| ○○○ | ○○○ | ○○○ | 38 | 166 | 68 | ○○○ | ○○○ | ○○○ | 1900 | ○○○ | ○○○ | ○○○ | 1352 | 310 | Liberia |
| ○○○ | 7570 | ○○○ | 3111 | 922 | 2689 | 10 | 0.794 | 120 | 3320 | 72 | 8 | 29 | 6690 | 11 | Libya |
| ○○○ | ○○○ | ○○○ | ○○○ | ○○○ | ○○○ | ○○○ | ○○○ | ○○○ | ○○○ | ○○○ | ○○○ | ○○○ | ○○○ | 2 | Liechtenstein |
| 12.7 | 10 320 | -0.3 | 3090 | 2850 | 2094 | 147 | 0.842 | 403 | 3325 | ○○○ | 0 | 0 | 4095 | 4 | Lithuania |
| 18.6 | 61 190 | 3.7 | 9428 | 36 359 | 29 725 | (139) | 0.933 | 254 | ○○○ | ○○○ | ○○○ | ○○○ | 658 | 2 | Luxembourg |
| 3.5 | 6470 | -0.7 | 1354 | 1094 | 659 | 277 | 0.793 | 219 | 2655 | 99 | ○○○ | ○○○ | 2202 | 7 | Macedonia, FYRO* |
| 3.9 | 740 | -0.9 | 42 | 70 | 39 | 373 | 0.469 | 9 | 2005 | 47 | 26 | 39 | 198 | 288 | Madagascar |
| 1.7 | 580 | 1.1 | 48 | 64 | 42 | 377 | 0.388 | ○○○ | 2155 | 57 | 25 | 51 | 28 | 644 | Malawi |
| 36.0 | 9120 | 3.6 | 2568 | 3308 | 4011 | 86 | 0.793 | 68 | 2881 | 95 | 8 | 15 | 2458 | 6 | Malaysia |
| 0.6 | 4798 | 4.7 | 547 | 1608 | 386 | 28 | 0.752 | 78 | 2548 | 100 | 3 | 3 | ○○○ | ○○○ | Maldives |
| 2.8 | 930 | 1.7 | 29 | 107 | 80 | 472 | 0.326 | 4 | 2174 | 65 | 73 | 88 | 243 | 557 | Mali |
| 3.6 | 17 640 | 3.6 | 2351 | 8509 | 6175 | 11 | 0.875 | 291 | 3587 | 100 | 8 | 7 | 1845 | 2 | Malta |
| 0.1 | ○○○ | ○○○ | ○○○ | ○○○ | ○○○ | ○○○ | ○○○ | ○○○ | ○○○ | ○○○ | ○○○ | ○○○ | ○○○ | ○○○ | Marshall Islands |
| 1.0 | 2220 | 1.6 | 442 | 186 | 156 | 355 | 0.465 | 14 | 2772 | 37 | 49 | 69 | 289 | 280 | Mauritania |

The datasets below are explained on pages 14

| | | | ooo | no data |
| per capita | for each person |

| | **Land** | | **Population** | | | | | | | | | **Employment** | | |
|---|---|---|---|---|---|---|---|---|---|---|---|---|---|---|
| | Area | Arable and permanent crops | Total | Density | Change | Births | Deaths | Fertility | Infant mortality | Life expectancy | Urban | Agriculture | Industry | Servic |
| | | | 2004 | 2004 | 1994–2004 | 2003 | 2003 | 2003 | 2003 | 2003 | 2003 | | | |
| | thousand km² | % of total | millions | persons per km² | % | births per 1000 | deaths per 1000 | children per mother | per 1000 live births | years | % | % | % | % |
| Mauritius | 2 | 52.0 | 1.2 | 600.0 | 7.8 | 16 | 7 | 1.9 | 13 | 72 | 42 | 17 | 43 | 40 |
| Mexico | 1958 | 13.9 | 106.2 | 54.2 | 18.6 | 25 | 5 | 2.8 | 25 | 75 | 75 | 28 | 24 | 48 |
| Micronesia, Fed. States | 0.7 | 51.4 | 0.1 | 142.9 | -3.8 | 28 | 7 | 4.4 | 40 | 67 | 22 | ooo | ooo | ooo |
| Moldova | 34 | 64.3 | 4.2 | 123.5 | -3.4 | 10 | 12 | 1.2 | 18 | 68 | 45 | 33 | 30 | 37 |
| Monaco | 0.002 | ooo | 0.03 | 15 000 | -3.2 | 23 | 16 | ooo | ooo | ooo | 100 | ooo | ooo | ooo |
| Mongolia | 1567 | 0.8 | 2.5 | 1.6 | 10.4 | 18 | 6 | 2.7 | 30 | 65 | 57 | 32 | 22 | 46 |
| Morocco | 447 | 20.8 | 30.6 | 68.5 | 15.1 | 21 | 6 | 2.5 | 37 | 70 | 57 | 45 | 25 | 30 |
| Mozambique | 802 | 5.5 | 19.2 | 23.9 | 15.6 | 40 | 23 | 5.5 | 127 | 40 | 29 | 83 | 8 | 9 |
| Myanmar | 677 | 15.7 | 50.1 | 74.0 | 14.1 | 25 | 11 | 3.1 | 87 | 57 | 28 | 73 | 10 | 17 |
| Namibia | 824 | 1.0 | 1.9 | 2.3 | 22.8 | 31 | 15 | 4.2 | 38 | 47 | 33 | 49 | 15 | 36 |
| Nauru | 0.02 | ooo | 0.01 | 500.0 | 0 | 23 | 5 | 4.4 | 25 | 61 | 100 | ooo | ooo | ooo |
| Nepal | 147 | 22.4 | 24.7 | 168.0 | 24.4 | 34 | 10 | 4.1 | 64 | 59 | 14 | 94 | 0 | 6 |
| Netherlands | 41 | 22.9 | 16.3 | 397.6 | 6.0 | 12 | 9 | 1.8 | 5 | 79 | 62 | 5 | 26 | 69 |
| New Zealand | 271 | 12.5 | 4.1 | 15.1 | 13.8 | 14 | 7 | 2.0 | 6 | 78 | 78 | 10 | 25 | 65 |
| Nicaragua | 130 | 16.6 | 5.6 | 43.1 | 30.3 | 32 | 5 | 3.8 | 31 | 69 | 58 | 28 | 26 | 46 |
| Niger | 1267 | 3.6 | 12.4 | 9.8 | 40.8 | 55 | 20 | 8.0 | 123 | 45 | 21 | 90 | 4 | 6 |
| Nigeria | 924 | 35.7 | 137.3 | 148.6 | 42.3 | 42 | 13 | 5.7 | 100 | 52 | 36 | 43 | 7 | 50 |
| Northern Marianas | 0.5 | 17.4 | 0.08 | 154.0 | 6.3 | 20 | 2 | ooo | ooo | ooo | ooo | ooo | ooo | ooo |
| North Korea | 121 | 22.4 | 22.8 | 188.4 | 7.9 | 17 | 11 | 2.0 | 45 | 63 | 60 | 38 | 32 | 30 |
| Norway | 324 | 2.7 | 4.6 | 14.2 | 6.4 | 12 | 9 | 1.8 | 3 | 80 | 78 | 6 | 25 | 69 |
| Oman | 213 | 0.3 | 2.7 | 12.7 | 31.7 | 26 | 4 | 4.1 | 16 | 74 | 76 | 44 | 24 | 32 |
| Pakistan | 796 | 27.8 | 159.2 | 200.0 | 33.3 | 34 | 10 | 4.8 | 85 | 61 | 34 | 52 | 19 | 29 |
| Palau | 0.5 | 21.7 | 0.02 | 40.0 | 17.6 | 14 | 7 | 1.6 | 17 | 70 | 70 | ooo | ooo | ooo |
| Panama | 76 | 9.2 | 3.2 | 42.1 | 23.9 | 23 | 5 | 2.7 | 21 | 75 | 62 | 26 | 16 | 58 |
| Papua New Guinea | 463 | 1.9 | 5.7 | 12.3 | 42.6 | 32 | 10 | 4.1 | 60 | 57 | 15 | 79 | 7 | 14 |
| Paraguay | 407 | 7.7 | 6.0 | 14.7 | 27.7 | 30 | 5 | 3.8 | 37 | 71 | 54 | 39 | 22 | 39 |
| Peru | 1285 | 3.4 | 27.5 | 21.4 | 18.9 | 23 | 6 | 2.8 | 33 | 69 | 72 | 36 | 18 | 46 |
| Philippines | 300 | 35.7 | 83.7 | 279.0 | 22.0 | 26 | 6 | 3.5 | 29 | 70 | 48 | 46 | 15 | 39 |
| Poland | 323 | 45.5 | 38.2 | 118.3 | -0.9 | 9 | 9 | 1.2 | 8 | 75 | 62 | 27 | 36 | 37 |
| Portugal | 92 | 29.4 | 10.5 | 114.1 | 6.0 | 11 | 10 | 1.4 | 5 | 77 | 53 | 18 | 34 | 48 |
| Qatar | 11 | 1.9 | 0.7 | 63.6 | 18.0 | 20 | 4 | 4.0 | 12 | 72 | 92 | 3 | 32 | 65 |
| Romania | 238 | 41.5 | 21.7 | 91.2 | -4.5 | 10 | 12 | 1.2 | 17 | 71 | 53 | 24 | 47 | 29 |
| Russian Federation | 17 075 | 7.4 | 144.1 | 8.4 | -2.6 | 10 | 17 | 1.4 | 13 | 65 | 73 | 14 | 42 | 44 |
| Rwanda | 26 | 52.6 | 8.4 | 323.1 | 65.4 | 40 | 21 | 5.8 | 107 | 40 | 17 | 92 | 3 | 5 |
| St. Kitts and Nevis | 0.4 | 22.2 | 0.05 | 125.0 | 16.3 | 17 | 8 | 2.1 | 28 | 70 | 33 | ooo | ooo | ooo |
| St. Lucia | 0.6 | 29.0 | 0.2 | 333.3 | 39.9 | 17 | 6 | 2.2 | 14 | 72 | 30 | ooo | ooo | ooo |
| St. Vincent & the Grenadines | 0.4 | 35.9 | 0.1 | 250.0 | -9.1 | 18 | 7 | 2.1 | 19 | 72 | 44 | ooo | ooo | ooo |
| Samoa | 3.0 | 45.4 | 0.2 | 66.7 | 22.0 | 29 | 6 | 4.4 | 18 | 73 | 22 | ooo | ooo | ooo |
| San Marino | 0.06 | 16.7 | 0.03 | 500.0 | 20.0 | 10 | 7 | 1.2 | 7 | 80 | 84 | ooo | ooo | ooo |
| Sao Tome and Principe | 1.0 | 56.3 | 0.2 | 170.0 | 60.0 | 34 | 6 | 4.3 | 34 | 69 | 38 | ooo | ooo | ooo |
| Saudi Arabia | 2150 | 1.8 | 25.1 | 11.7 | 50.4 | 32 | 3 | 4.8 | 25 | 72 | 86 | 19 | 20 | 61 |
| Senegal | 197 | 12.7 | 10.9 | 55.3 | 34.1 | 37 | 11 | 5.1 | 64 | 56 | 43 | 77 | 8 | 15 |
| Serbia and Montenegro | 102 | 36.4 | 10.7 | 104.9 | 1.7 | 12 | 11 | 1.7 | 13 | 73 | 52 | ooo | ooo | ooo |
| Seychelles | 0.5 | 15.6 | 0.1 | 200.0 | 35.1 | 18 | 8 | 2.0 | 18 | 71 | 50 | ooo | ooo | ooo |
| Sierra Leone | 72 | 8.4 | 5.2 | 72.2 | 27.5 | 50 | 29 | 6.5 | 180 | 35 | 37 | 68 | 15 | 17 |
| Singapore | 1 | 1.6 | 4.2 | 4200.0 | 22.8 | 10 | 4 | 1.3 | 2 | 79 | 100 | 0 | 36 | 64 |
| Slovakia | 49 | 31.8 | 5.4 | 110.2 | 1.0 | 10 | 10 | 1.2 | 8 | 74 | 56 | 12 | 32 | 56 |
| Slovenia | 20 | 9.8 | 2.0 | 100.0 | 0.6 | 9 | 10 | 1.2 | 4 | 76 | 51 | 6 | 46 | 48 |
| Solomon Islands | 29 | 2.6 | 0.5 | 17.2 | 37.4 | 36 | 9 | 4.8 | 66 | 61 | 16 | 77 | 7 | 16 |
| Somalia | 638 | 1.7 | 8.3 | 13.0 | 14.6 | 47 | 18 | 7.1 | 124 | 47 | 33 | ooo | ooo | ooo |
| South Africa | 1221 | 12.9 | 46.9 | 38.4 | 21.4 | 24 | 13 | 2.8 | 48 | 53 | 53 | 14 | 32 | 54 |
| South Korea | 99 | 18.9 | 48.2 | 486.9 | 8.0 | 10 | 5 | 1.2 | 8 | 77 | 80 | 18 | 35 | 47 |
| Spain | 506 | 37.0 | 42.5 | 84.0 | 8.5 | 10 | 9 | 1.3 | 4 | 79 | 76 | 12 | 33 | 55 |
| Sri Lanka | 66 | 29.2 | 19.6 | 297.0 | 9.6 | 19 | 6 | 2.0 | 10 | 72 | 30 | 48 | 21 | 31 |
| Sudan | 2506 | 6.6 | 39.1 | 15.6 | 35.1 | 38 | 10 | 5.4 | 69 | 57 | 31 | 70 | 8 | 22 |

## Wealth   Energy and trade   Quality of life

| | Purchasing power | Growth of PP | Energy consumption | Imports | Exports | Aid received (given) | Human Development Index | Health care | Food consumption | Safe water | Illiteracy male | Illiteracy female | Higher education | Cars | |
|---|---|---|---|---|---|---|---|---|---|---|---|---|---|---|---|
| 2 | 2002 | 1990–2002 | 2002 | 2003 | 2003 | 2002 | 2002 | 1990–2003 | 2002 | 2000 | 2002 | 2002 | 2002 | 2000 | |
| on $S | US$ | annual % | kg oil equivalent per capita | US$ per capita | US$ per capita | million US$ | | doctors per 100 000 people | daily calories per capita | % access | % | % | students per 100 000 people | people per car | |
| 4.6 | 10 810 | 4.0 | 1187 | ooo | ooo | 24 | 0.785 | 85 | 2955 | 100 | 12 | 20 | 1050 | 26 | Mauritius |
| 6.7 | 8970 | 1.4 | 1616 | 1745 | 1617 | 136 | 0.802 | 156 | 3145 | 86 | 7 | 11 | 2095 | 11 | Mexico |
| 0.3 | ooo | ooo | ooo | ooo | ooo | ooo | ooo | ooo | ooo | ooo | ooo | ooo | ooo | ooo | Micronesia, Fed. States |
| 1.7 | 1470 | -6.9 | 925 | 330 | 187 | 142 | 0.681 | 271 | 2806 | 100 | 1 | 1 | 2429 | 25 | Moldova |
| ooo | ooo | ooo | ooo | ooo | ooo | ooo | ooo | ooo | ooo | 100 | ooo | ooo | ooo | 2 | Monaco |
| 1.1 | 1710 | 0.2 | 785 | 317 | 208 | 209 | 0.668 | 278 | 2249 | 60 | 2 | 3 | 3376 | 104 | Mongolia |
| 5.4 | 3810 | 0.8 | 369 | 1944 | 290 | 636 | 0.620 | 49 | 3052 | 82 | 37 | 62 | 1012 | 29 | Morocco |
| 3.9 | 1050 | 4.5 | 135 | 73 | 47 | 2058 | 0.354 | 2 | 2079 | 60 | 38 | 69 | 50 | 233 | Mozambique |
| ooo | 1027 | 5.7 | 100 | 53 | 53 | 121 | 0.551 | 30 | 2937 | 68 | 11 | 19 | 1313 | 1274 | Myanmar |
| 3.3 | 6210 | 0.9 | 621 | 980 | 640 | 135 | 0.607 | 29 | 2278 | 77 | 16 | 17 | 676 | 31 | Namibia |
| ooo | ooo | ooo | 4055 | ooo | ooo | ooo | ooo | ooo | ooo | ooo | ooo | ooo | ooo | ooo | Nauru |
| 5.6 | 1370 | 2.3 | 58 | 71 | 27 | 366 | 0.504 | 5 | 2453 | 81 | 38 | 74 | 463 | ooo | Nepal |
| 6.8 | 29 100 | 2.2 | 6074 | 16 208 | 18 135 | (3068) | 0.942 | 328 | 3362 | 100 | ooo | ooo | 3205 | 3 | Netherlands |
| 3.1 | 21 740 | 2.1 | 5617 | 4628 | 4115 | (110) | 0.926 | 219 | 3219 | ooo | ooo | ooo | 4729 | 2 | New Zealand |
| 2.1 | 2470 | 1.5 | 301 | 344 | 110 | 518 | 0.667 | 62 | 2298 | 79 | 23 | 23 | ooo | 98 | Nicaragua |
| 2.0 | 800 | -0.8 | 40 | 47 | 29 | 299 | 0.292 | 3 | 2130 | 59 | 75 | 91 | 129 | 590 | Niger |
| 8.7 | 860 | -0.3 | 196 | 80 | 149 | 314 | 0.466 | 27 | 2726 | 57 | 26 | 41 | ooo | 151 | Nigeria |
| ooo | ooo | ooo | ooo | ooo | ooo | ooo | ooo | ooo | ooo | ooo | ooo | ooo | ooo | ooo | Northern Marianas |
| ooo | ooo | ooo | 1229 | ooo | ooo | ooo | ooo | ooo | 2142 | 100 | ooo | ooo | ooo | ooo | North Korea |
| 1.8 | 36 600 | 3.0 | 10 985 | 8659 | 14 798 | (1517) | 0.956 | 367 | 3484 | 100 | ooo | ooo | 4345 | 3 | Norway |
| ooo | 13 340 | 0.9 | 3317 | 2529 | 4490 | 41 | 0.770 | 137 | ooo | 39 | 18 | 35 | 732 | 13 | Oman |
| 9.2 | 1940 | 1.1 | 298 | 88 | 80 | 2144 | 0.497 | 68 | 2419 | 88 | 47 | 72 | ooo | 161 | Pakistan |
| 0.1 | ooo | ooo | ooo | ooo | ooo | ooo | ooo | ooo | ooo | 79 | ooo | ooo | ooo | ooo | Palau |
| 1.8 | 6170 | 2.5 | 1764 | 1022 | 290 | 35 | 0.791 | 121 | 2272 | 87 | 7 | 8 | 3060 | 15 | Panama |
| 2.8 | 2270 | 0.5 | 222 | 236 | 395 | 203 | 0.542 | 6 | 2193 | 42 | 29 | 42 | 191 | 136 | Papua New Guinea |
| 6.4 | 4610 | -0.5 | 1670 | 368 | 228 | 57 | 0.751 | 49 | 2565 | 79 | 7 | 10 | 1642 | 66 | Paraguay |
| 4.7 | 5010 | 2.2 | 532 | 312 | 330 | 491 | 0.752 | 103 | 2571 | 77 | 9 | 20 | 3078 | 48 | Peru |
| 1.5 | 4170 | 1.1 | 354 | 485 | 448 | 560 | 0.753 | 115 | 2379 | 87 | 8 | 7 | 2973 | 110 | Philippines |
| 6.6 | 10 560 | 4.2 | 2166 | 1780 | 1402 | 1160 | 0.850 | 220 | 3375 | ooo | 0 | 0 | 4935 | 5 | Poland |
| 8.7 | 18 280 | 2.5 | 2588 | 4424 | 3078 | (293) | 0.897 | 318 | 3741 | 82 | 5 | 10 | 3801 | 3 | Portugal |
| ooo | 19 844 | ooo | 15 410 | 9508 | 20 213 | 2 | 0.833 | 220 | ooo | ooo | 15 | 18 | 987 | 4 | Qatar |
| 1.3 | 6560 | 0.1 | 1911 | 1081 | 794 | 701 | 0.778 | 189 | 3455 | 58 | 2 | 4 | 2599 | 10 | Romania |
| 7.9 | 8230 | -2.4 | 4739 | 518 | 937 | 1301 | 0.795 | 420 | 3072 | 99 | 0 | 1 | 5523 | 9 | Russian Federation |
| 1.9 | 1270 | 0.3 | 45 | 29 | 7 | 356 | 0.431 | 2 | 2084 | 41 | 25 | 37 | 173 | 889 | Rwanda |
| 0.3 | 12 420 | 3.5 | 645 | 5106 | 702 | 29 | 0.844 | 117 | 2609 | 98 | ooo | ooo | ooo | 8 | St. Kitts and Nevis |
| 0.6 | 5300 | 0.2 | 781 | 2112 | 248 | 34 | 0.777 | 58 | 2988 | 98 | ooo | ooo | 138 | 21 | St. Lucia |
| 0.3 | 5460 | 1.1 | 644 | 1835 | 349 | 5 | 0.751 | 88 | 2599 | 93 | ooo | ooo | ooo | 16 | St. Vincent & the Grenadines |
| 0.3 | 5600 | 3.2 | 420 | 770 | 84 | 38 | 0.769 | 34 | 2945 | 99 | 1 | 2 | 660 | 56 | Samoa |
| ooo | ooo | ooo | ooo | ooo | ooo | ooo | ooo | ooo | ooo | ooo | ooo | ooo | ooo | 1 | San Marino |
| .04 | 1317 | -0.4 | 293 | 268 | 45 | 26 | 0.645 | 47 | 2460 | ooo | ooo | ooo | 107 | 3 | Sao Tome and Principe |
| 9.9 | 12 650 | -0.6 | 5249 | 1609 | 3928 | 27 | 0.768 | 153 | 2845 | 95 | 16 | 31 | 1815 | 11 | Saudi Arabia |
| 4.7 | 1580 | 1.2 | 155 | 235 | 132 | 449 | 0.437 | 10 | 2280 | 78 | 51 | 70 | 284 | 104 | Senegal |
| 1.6 | ooo | ooo | 1597 | 927 | 313 | ooo | ooo | ooo | 2678 | ooo | ooo | ooo | 1927 | 5 | Serbia and Montenegro |
| 0.6 | 18 232 | 2.6 | 2497 | 5119 | 3310 | 8 | 0.853 | 132 | 2465 | ooo | 9 | 8 | ooo | 12 | Seychelles |
| 0.7 | 520 | -5.9 | 63 | 57 | 17 | 353 | 0.273 | 9 | 1936 | 28 | ooo | ooo | 162 | 131 | Sierra Leone |
| 6.1 | 24 040 | 3.8 | 9797 | 30 102 | 33 912 | 7 | 0.902 | 140 | ooo | 100 | 3 | 11 | ooo | 10 | Singapore |
| 1.4 | 12 840 | 2.1 | 3882 | 4178 | 4081 | 189 | 0.842 | 326 | 2889 | 100 | 0 | 0 | 2813 | 4 | Slovakia |
| 9.6 | 18 540 | 4.2 | 3753 | 7052 | 6501 | 171 | 0.895 | 219 | 3001 | 100 | 0 | 0 | 4932 | 3 | Slovenia |
| 0.3 | 1590 | -2.4 | 152 | ooo | ooo | 26 | 0.624 | 13 | 2265 | 71 | ooo | ooo | ooo | ooo | Solomon Islands |
| ooo | ooo | ooo | 32 | 19 | 21 | ooo | ooo | ooo | ooo | ooo | ooo | ooo | ooo | 876 | Somalia |
| 3.5 | 10 070 | -0.2 | 2557 | 907 | 805 | 657 | 0.666 | 25 | 2956 | 86 | 13 | 15 | 1482 | 11 | South Africa |
| 3.0 | 16 950 | 4.7 | 4388 | 3732 | 4045 | -82 | 0.888 | 180 | 3058 | 92 | 1 | 3 | 6712 | 6 | South Korea |
| 4.1 | 21 460 | 2.3 | 3654 | 4890 | 3690 | (1559) | 0.922 | 329 | 3371 | ooo | 1 | 3 | 4564 | 2 | Spain |
| 5.9 | 3570 | 3.4 | 244 | 348 | 267 | 344 | 0.740 | 43 | 2385 | 83 | 5 | 10 | ooo | 87 | Sri Lanka |
| 0.3 | 1820 | 3.1 | 100 | 80 | 70 | 351 | 0.505 | 16 | 2228 | 75 | 29 | 51 | 541 | 398 | Sudan |

| | ooo | no data |
|---|---|---|
| | per capita | for each person |

## Land

## Population

## Employment

| | Area | Arable and permanent crops | Total | Density | Change | Births | Deaths | Fertility | Infant mortality | Life expectancy | Urban | Agriculture | Industry | Servic |
|---|---|---|---|---|---|---|---|---|---|---|---|---|---|---|
| | | | 2004 | 2004 | 1994–2004 | 2003 | 2003 | 2003 | 2003 | 2003 | 2003 | | | |
| | thousand km² | % of total | millions | persons per km² | % | births per 1000 | deaths per 1000 | children per mother | per 1000 live births | years | % | % | % | % |
| Suriname | 163 | 0.4 | 0.4 | 2.5 | -1.2 | 23 | 7 | 2.5 | 27 | 70 | 69 | 21 | 18 | 61 |
| Swaziland | 17 | 10.9 | 1.2 | 70.6 | 36.5 | 36 | 16 | 4.5 | 78 | 43 | 25 | 40 | 22 | 38 |
| Sweden | 450 | 6.0 | 9.0 | 20.0 | 2.5 | 11 | 10 | 1.7 | 3 | 80 | 84 | ooo | ooo | ooo |
| Switzerland | 41 | 10.5 | 7.4 | 180.5 | 5.8 | 10 | 9 | 1.4 | 4 | 80 | 68 | 6 | 35 | 59 |
| Syria | 185 | 29.3 | 18.0 | 97.3 | 30.0 | 28 | 5 | 3.8 | 18 | 70 | 50 | 33 | 24 | 43 |
| Taiwan | 36 | ooo | 22.6 | 627.8 | 9.4 | 10 | 6 | 1.2 | 6 | 76 | 78 | ooo | ooo | ooo |
| Tajikistan | 143 | 7.4 | 6.6 | 46.2 | 14.9 | 25 | 6 | 3.1 | 50 | 68 | 27 | 41 | 23 | 36 |
| Tanzania | 945 | 5.4 | 36.1 | 38.2 | 31.3 | 40 | 17 | 5.3 | 105 | 45 | 22 | 84 | 5 | 11 |
| Thailand | 513 | 37.7 | 63.8 | 124.4 | 8.7 | 14 | 7 | 1.7 | 20 | 71 | 31 | 64 | 14 | 22 |
| Togo | 57 | 46.3 | 5.6 | 98.2 | 42.6 | 38 | 11 | 5.5 | 72 | 54 | 33 | 66 | 10 | 24 |
| Tonga | 0.8 | 64.0 | 0.1 | 125.0 | 3.1 | 25 | 7 | 3.4 | 13 | 71 | 32 | ooo | ooo | ooo |
| Trinidad and Tobago | 5 | 23.8 | 1.3 | 260.0 | 4.0 | 13 | 7 | 1.6 | 19 | 71 | 74 | 11 | 31 | 58 |
| Tunisia | 164 | 30.0 | 10.0 | 61.0 | 13.4 | 17 | 6 | 2.0 | 22 | 73 | 63 | 28 | 33 | 39 |
| Turkey | 775 | 36.8 | 71.3 | 92.0 | 19.4 | 21 | 7 | 2.5 | 39 | 69 | 59 | 53 | 18 | 29 |
| Turkmenistan | 488 | 3.9 | 5.7 | 11.7 | 29.4 | 25 | 9 | 2.9 | 74 | 67 | 47 | 37 | 23 | 40 |
| Tuvalu | 0.02 | ooo | 0.01 | 500.0 | 11.1 | 27 | 10 | 3.8 | 34 | ooo | 47 | ooo | ooo | ooo |
| Uganda | 241 | 29.9 | 26.1 | 108.3 | 33.7 | 47 | 17 | 6.9 | 88 | 45 | 12 | 85 | 5 | 10 |
| Ukraine | 604 | 55.7 | 47.4 | 78.5 | -9.0 | 9 | 16 | 1.2 | 10 | 68 | 68 | 20 | 40 | 40 |
| United Arab Emirates | 84 | 3.2 | 4.2 | 50.0 | 83.2 | 16 | 2 | 2.5 | 8 | 74 | 78 | 8 | 27 | 65 |
| United Kingdom | 245 | 23.9 | 59.7 | 243.7 | 2.2 | 12 | 10 | 1.7 | 5 | 78 | 89 | 2 | 29 | 69 |
| United States of America | 9364 | 18.5 | 293.6 | 31.4 | 12.7 | 14 | 8 | 2.0 | 7 | 77 | 79 | 3 | 28 | 69 |
| Uruguay | 177 | 7.6 | 3.4 | 19.2 | 6.4 | 16 | 9 | 2.2 | 14 | 75 | 93 | 14 | 27 | 59 |
| Uzbekistan | 447 | 10.8 | 26.4 | 59.1 | 18.5 | 24 | 8 | 2.9 | 62 | 70 | 37 | 34 | 25 | 41 |
| Vanuatu | 12 | 9.8 | 0.2 | 16.7 | 19.8 | 28 | 6 | 4.8 | 27 | 67 | 21 | ooo | ooo | ooo |
| Venezuela | 912 | 3.7 | 26.2 | 28.7 | 22.6 | 24 | 5 | 2.8 | 20 | 73 | 87 | 12 | 27 | 61 |
| Vietnam | 332 | 26.8 | 81.5 | 245.5 | 12.4 | 18 | 6 | 2.1 | 21 | 72 | 25 | 71 | 14 | 15 |
| Western Sahara | 252 | 0.008 | 0.3 | 1.2 | 46.3 | 29 | 8 | 4.1 | 59 | 62 | ooo | ooo | ooo | ooo |
| Yemen | 528 | 3.2 | 20.0 | 37.9 | 34.6 | 43 | 10 | 7.0 | 75 | 60 | 26 | 61 | 17 | 22 |
| Zambia | 753 | 7.0 | 10.9 | 14.5 | 24.4 | 42 | 24 | 5.6 | 95 | 35 | 35 | 75 | 8 | 17 |
| Zimbabwe | 391 | 8.6 | 12.7 | 32.5 | 13.9 | 32 | 20 | 4.0 | 65 | 41 | 32 | 68 | 8 | 24 |

## Explanation of datasets

### Land

| | |
|---|---|
| **Area** | does not include areas of lakes and seas |
| **Arable and permanent crops** | percentage of total land area used for arable and permanent crops |

### Population

| | |
|---|---|
| **Total** | estimate for mid 2004 |
| **Density** | the total population of a country divided by its land area |
| **Change** | percentage change in population between 1994 and 2004. Negative numbers indicate a decrease |
| **Births** | number of births per one thousand people in one year |
| **Deaths** | number of deaths per one thousand people in one year |
| **Fertility** | average number of children born to child bearing women |
| **Infant mortality** | number of deaths of children under one year per 1000 live births |
| **Life expectancy** | number of years a baby born now can expect to live |
| **Urban** | percentage of the population living in towns and cities |

### Employment

| | |
|---|---|
| **Agriculture** | percentage of the labour force employ in agriculture |
| **Industry** | percentage of the labour force employ in industry |
| **Services** | percentage of the labour force employ in services |

Macedonia, FYRO*    Former Yugoslav Republic of Macedonia

## Health

## Energy and trade

## Quality of life

| GNI | Purchasing power | Growth of PP | Energy consumption | Imports | Exports | Aid received (given) | Human Development Index | Health care | Food consumption | Safe water | Illiteracy male | Illiteracy female | Higher education | Cars | |
|---|---|---|---|---|---|---|---|---|---|---|---|---|---|---|---|
| 2002 | 2002 | 1990–2002 | 2002 | 2003 | 2003 | 2002 | 2002 | 1990–2003 | 2002 | 2000 | 2002 | 2002 | 2002 | 2000 | |
| million US$ | US$ | annual % | kg oil equivalent per capita | US$ per capita | US$ per capita | million US$ | | doctors per 100 000 people | daily calories per capita | % access | % | % | students per 100 000 people | people per car | |
| 0.7 | 6590 | 0.5 | 2248 | 1598 | 993 | 12 | 0.780 | 50 | 2652 | 95 | ooo | ooo | 1196 | 8 | Suriname |
| 1.4 | 4550 | 0.1 | 326 | 931 | 818 | 25 | 0.519 | 15 | 2322 | ooo | 18 | 20 | 452 | 38 | Swaziland |
| 1.5 | 26 050 | 2.0 | 6212 | 9233 | 11 305 | (1848) | 0.946 | 287 | 3185 | 100 | ooo | ooo | 4276 | 2 | Sweden |
| 4.2 | 30 010 | 0.4 | 4316 | 12 964 | 13 533 | (863) | 0.936 | 350 | 3526 | 100 | ooo | ooo | 2310 | 2 | Switzerland |
| 9.2 | 3620 | 1.8 | 1249 | 285 | 315 | 81 | 0.710 | 142 | 3038 | 80 | 9 | 26 | 549 | 107 | Syria |
| ooo | ooo | ooo | 4567 | 5624 | 6636 | ooo | ooo | ooo | ooo | ooo | ooo | ooo | ooo | 5 | Taiwan |
| 1.1 | 980 | -8.1 | 852 | 140 | 127 | 168 | 0.671 | 212 | 1828 | 69 | 0 | 1 | 1268 | ooo | Tajikistan |
| 9.6 | 580 | 0.7 | 51 | 61 | 34 | 1233 | 0.407 | 4 | 1975 | 54 | 15 | 31 | 63 | 735 | Tanzania |
| 2.2 | 7010 | 2.9 | 1208 | 1222 | 1298 | 296 | 0.768 | 30 | 2467 | 80 | 5 | 10 | 3386 | 37 | Thailand |
| 1.3 | 1480 | -0.7 | 85 | 117 | 91 | 51 | 0.495 | 6 | 2345 | 54 | 26 | 55 | 286 | 174 | Togo |
| 0.2 | 6850 | 2.2 | 471 | ooo | ooo | 22 | 0.787 | 35 | ooo | 100 | 1 | 1 | 358 | 19 | Tonga |
| 7.2 | 9430 | 2.9 | 11 291 | 2780 | 3477 | -7 | 0.801 | 75 | 2732 | 86 | 1 | 2 | 888 | 6 | Trinidad and Tobago |
| 9.6 | 6760 | 3.1 | 865 | 1103 | 811 | 475 | 0.745 | 70 | 3238 | 99 | 17 | 37 | 2314 | 24 | Tunisia |
| 4.0 | 6390 | 1.3 | 1150 | 981 | 659 | 636 | 0.751 | 123 | 3357 | 83 | 6 | 22 | 2493 | 17 | Turkey |
| 6.7 | 4300 | -3.2 | 3146 | 517 | 744 | 41 | 0.752 | 300 | 2742 | 58 | 1 | 2 | ooo | ooo | Turkmenistan |
| ooo | ooo | ooo | ooo | ooo | ooo | ooo | ooo | ooo | ooo | 100 | ooo | ooo | ooo | ooo | Tuvalu |
| 5.9 | 1390 | 3.9 | 36 | 50 | 22 | 638 | 0.493 | 5 | 2410 | 50 | 21 | 41 | 289 | 923 | Uganda |
| 7.7 | 4870 | -6.0 | 3382 | 476 | 477 | 484 | 0.777 | 299 | 3054 | 55 | 0 | 1 | 4411 | 10 | Ukraine |
| ooo | 22 420 | -1.6 | 21 678 | 8909 | 16 292 | 4 | 0.824 | 177 | 3225 | ooo | 24 | 19 | 859 | 7 | United Arab Emirates |
| 6.2 | 26 150 | 2.4 | 3998 | 6592 | 5138 | (4581) | 0.936 | 164 | 3412 | 100 | ooo | ooo | 3740 | 2 | United Kingdom |
| 0.1 | 35 750 | 2.0 | 8486 | 4477 | 2487 | (13 140) | 0.939 | 279 | 3774 | 100 | ooo | ooo | 5537 | 2 | United States of America |
| 4.8 | 7830 | 1.4 | 1122 | 648 | 650 | 13 | 0.833 | 387 | 2828 | 98 | 3 | 2 | 2953 | 11 | Uruguay |
| 1.5 | 1670 | -0.9 | 2074 | 100 | 127 | 189 | 0.709 | 293 | 2241 | 85 | 0 | 1 | 890 | ooo | Uzbekistan |
| 0.2 | 2890 | -0.1 | 127 | 500 | 129 | 28 | 0.570 | 12 | 2587 | 88 | ooo | ooo | 344 | 50 | Vanuatu |
| 2.6 | 5380 | -1.0 | 2990 | 364 | 926 | 57 | 0.778 | 200 | 2336 | 84 | 7 | 7 | 2676 | 13 | Venezuela |
| 4.9 | 2300 | 5.9 | 269 | 306 | 248 | 1277 | 0.691 | 54 | 2566 | 56 | 6 | 13 | 970 | 562 | Vietnam |
| ooo | ooo | ooo | 390 | ooo | ooo | ooo | ooo | ooo | ooo | ooo | ooo | ooo | ooo | ooo | Western Sahara |
| 9.4 | 870 | 2.5 | 201 | 177 | 198 | 584 | 0.482 | 22 | 2038 | 69 | 31 | 72 | 926 | 101 | Yemen |
| 3.5 | 840 | -1.2 | 248 | 144 | 90 | 641 | 0.389 | 7 | 1927 | 64 | 14 | 26 | 232 | 107 | Zambia |
| 6.2 | 2400 | -0.8 | 423 | 224 | 89 | 201 | 0.491 | 6 | 1943 | 85 | 6 | 14 | 483 | 61 | Zimbabwe |

## Explanation of datasets

## Health

**GNI** Gross National Income (GNI) is the total value of goods and services produced in a country plus income from abroad.

**Purchasing power** Gross Domestic Product (GDP) is the total value of goods and services produced in a country. Purchasing power parity (PPP) is GDP per person, adjusted for the local cost of living

**Growth of PP** average annual growth (or decline, shown as a negative value in the table) in purchasing power. This figure shows whether people are becoming better or worse off

## Energy and trade

**Energy consumption** consumption of commercial energy per person shown as the equivalent in kilograms of oil

**Imports** total value of imports per person shown in US dollars

**Exports** total value of exports per person shown in US dollars

**Aid received (given)** amount of economic aid a country has received. Negative values indicate that the repayment of loans exceeds the amount of aid received. Figures in brackets show aid given

## Quality of life

**HDI** Human Development Index (HDI) measures the relative social and economic progress of a country. It combines life expectancy, adult literacy, average number of years of schooling, and purchasing power. Economically more developed countries have an HDI approaching 1.0. Economically less developed countries have an HDI approaching 0.

**Health care** number of doctors in each country per 100 000 people

**Food consumption** average number of calories consumed by each person each day

**Safe water** percentage of the population with access to safe drinking water

**Illiteracy** percentage of men and women who are unable to read and write

**Higher education** number of students in higher education per 100 000 people

**Cars** the number of people for every car

## How to use the index

To find a place on an atlas map use either the grid code or latitude and longitude.

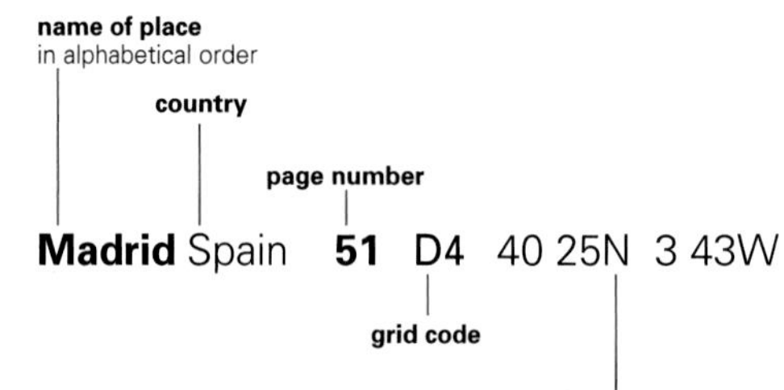

name of place
in alphabetical order

country

page number

**Madrid** Spain **51** D4 40 25N 3 43W

grid code

latitude and longitude

## Grid code

**Madrid** Spain **51** D4 40 25N 3 43W

Madrid is in grid square D4

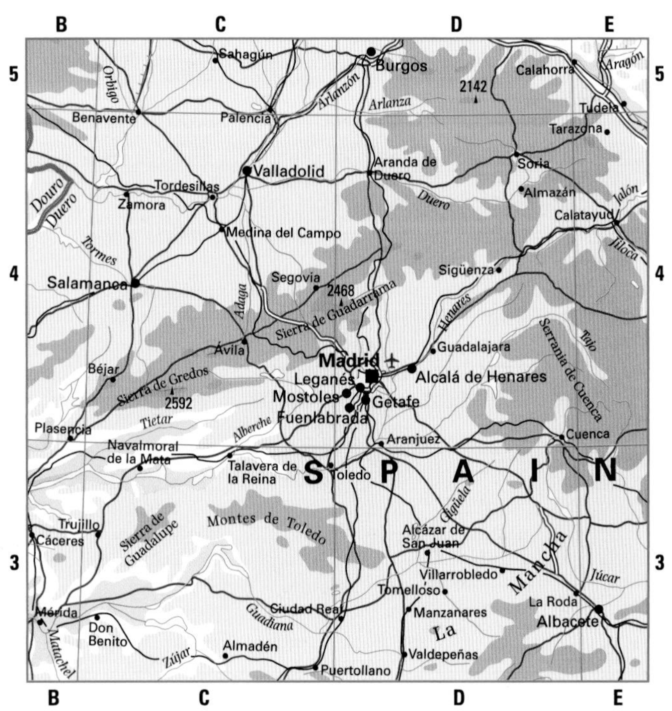

## Latitude and longitude

**Madrid** Spain **51** D4 40 25N 3 43W

Madrid is at latitude 40 degrees, 25 minutes north and 3 degrees, 43 minutes west

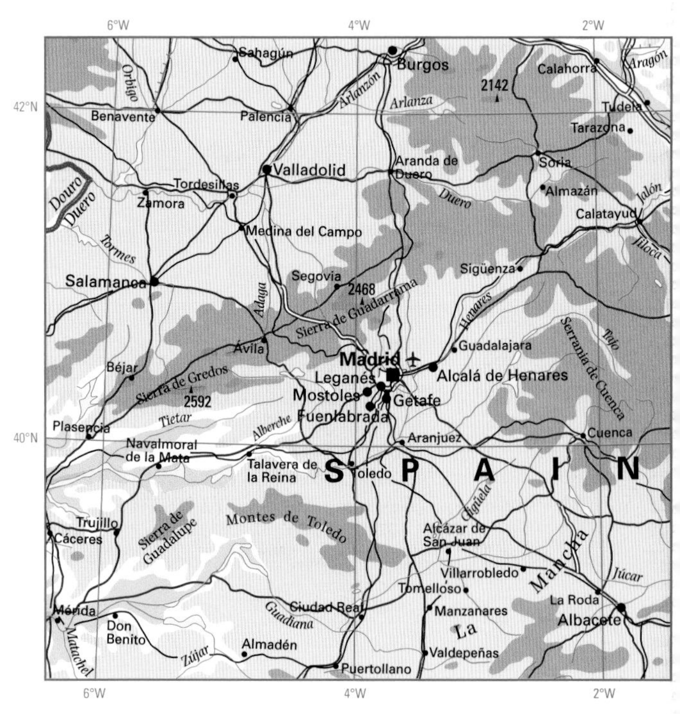

## Geographical abbreviations

| | |
|---|---|
| *admin* | administrative area |
| *Arch.* | Archipelago |
| *b.* | bay or harbour |
| *c.* | cape, point, or headland |
| *can.* | canal |
| *co.* | county |
| *d.* | desert |
| *fj.* | fjord |
| *G.* | Gunung; Gebel |
| *g.* | gulf |
| *geog. reg.* | geographical region |
| *i.* | island |
| *is.* | islands |
| *Kep.* | Kepulauan |
| *l.* | lake, lakes, lagoon |
| *mt.* | mountain, peak, or spot height |
| *mts.* | mountains |
| *NP* | National Park |
| *P.* | Pulau |
| *p; pen* | peninsula |
| *Peg* | Pegunungan |
| *plat.* | plateau |
| *prov.* | province |
| *Pt.* | Point |
| *Pta.* | Punta |
| *Pte.* | Pointe |
| *Pto.* | Porto; Puerto |
| *r.* | river |
| *Ra.* | Range |
| *res.* | reservoir |
| *salt l.* | salt lake |
| *sd.* | sound, strait, or channel |
| *St.* | Saint |

| | |
|---|---|
| *Ste.* | Sainte |
| *Str.* | Strait |
| *sum.* | summit |
| *tn.* | town or other populated place |
| *v.* | valley |
| *vol.* | volcano |

## Political abbreviations

| | |
|---|---|
| Aust. | Australia |
| Bahamas | The Bahamas |
| CAR | Central African Republic |
| Col. | Columbia |
| CDR | Congo Democratic Republic |
| Czech Rep. | Czech Republic |
| Dom. Rep. | Dominican Republic |
| Eq. Guinea | Equatorial Guinea |
| Fr. | France |
| Med. Sea | Mediterranean Sea |
| Neths | Netherlands |
| NI | Northern Ireland |
| NZ | New Zealand |
| Philippines | The Philippines |
| PNG | Papua New Guinea |
| Port. | Portugal |
| RoI | Republic of Ireland |
| RSA | Republic of South Africa |
| Sp. | Spain |
| Switz. | Switzerland |
| UAE | United Arab Emirates |
| UK | United Kingom |
| USA | United States of America |
| W. Indies | West Indies |
| Yemen | Yemen Republic |

# A

**Aachen** Germany 52 C5 50 46N 6 06E
**Aalten** Belgium 48 B4 51 56N 6 35E
**Ābādān** Iran 65 C5 30 20N 48 15E
**Abbeville** France 50 E6 50 06N 1 51E
**Abbey** RoI 21 C3 53 06N 8 24W
**Abbeyfeale** RoI 21 B2 52 24N 9 18W
**Abbeyleix** RoI 21 D2 52 55N 7 20W
**Abbeytown** England 12 C3 54 51N 3 17W
**Abbotsbury** England 19 E2 50 40N 2 36W
**Aberaeron** Wales 14 B2 52 49N 44 43W
**Abercarn** Wales 14 C1 51 39N 3 08W
**Aberchirder** Scotland 9 G2 57 33N 2 38W
**Aberdare** Wales 14 C1 51 43N 3 27W
**Aberdaron** Wales 14 B2 52 49N 44 43W
**Aberdeen** Scotland 9 G2 57 10N 2 04W
**Aberdeen** USA 91 G6 45 28N 98 30W
**Aberdeen City** *u.a.* Scotland 9 G2 57 10N 2 00W
**Aberdeenshire** *u.a.* Scotland 9 G2 57 10N 2 50W
**Aberdyfi** Wales 14 B2 52 33N 4 02W
**Aberfeldy** Scotland 9 F1 56 37N 3 54W
**Aberffraw** Wales 14 B3 53 12N 4 28W
**Aberfoyle** Scotland 11 E3 56 11N 4 23W
**Abergavenny** Wales 14 C1 51 50N 3 00W
**Abergele** Wales 14 C3 53 17N 3 34W
**Aberlady** Scotland 11 G3 56 01N 2 51W
**Aberporth** Wales 14 B2 52 07N 4 34W
**Abersoch** Wales 14 B2 52 50N 4 31W
**Abertillery** Wales 14 C1 51 45N 3 09W
**Aberystwyth** Wales 14 B2 52 25N 4 05W
**Abidjan** Côte d'Ivoire 80 D3 5 19N 4 01W
**Abilene** USA 90 G3 32 27N 99 45W
**Abingdon** England 16 B2 51 41N 1 17W
**Abington** Scotland 11 F2 55 29N 3 42W
**Aboyne** Scotland 9 G2 57 05N 2 50W
**Abu Dhabi** UAE 65 F3 24 28N 54 25E
**Abuja** Nigeria 80 F3 9 10N 7 11E
**Acapulco** Mexico 94 E3 16 51N 99 56W
**Accra** Ghana 80 D3 5 33N 0 15W
**Accrington** England 12 D2 53 46N 2 21W
**Achill Head** RoI 20 A3 53 59N 10 13W
**Achill Island** RoI 20 A3 53 55N 10 05W
**Achill Sound** *tn.* RoI 20 B3 53 56N 9 54W
**Achnacroish** Scotland 8 D1 56 32N 5 30W
**Achnasheen** Scotland 9 D2 57 35N 5 06W
**Ackworth Moor Top** England 13 E2 53 39N 1 20W
**Aclare** RoI 20 C4 54 02N 8 54W
**Acle** England 17 E3 52 38N 1 33E
**A Coruña** Spain 51 A5 43 22N 8 24W
**Adana** Turkey 57 E2 37 00N 35 19E
**Adare** RoI 21 C2 52 34N 8 48W
**Ad Dammām** Saudi Arabia 65 F4 26 25N 50 06E
**Addis Ababa** Ethiopia 81 N3 9 03N 38 42E
**Addlestone** England 16 C2 51 22N 0 31W
**Adelaide** Australia 73 G4 34 56S 138 36E
**Aden** Yemen 64 E1 12 50N 45 03E
**Aden, Gulf of** Indian Ocean 65 E1 12 30N 47 30E
**Adriatic Sea** Med. Sea 54 F5 43 00N 15 00E
**Adrigole** RoI 21 B1 51 44N 9 42W
**Adur** *r.* England 16 C1 50 55N 0 20W
**Adwick le Street** England 13 E2 53 34N 1 11W
**Aegean Sea** Med. Sea 55 L3 39 00N 24 00E
**AFGHANISTAN** 66 A6/B6
**Agen** France 50 E3 44 12N 0 38E
**Aghaville** RoI 21 B1 51 39N 9 21W
**Agra** India 66 D5 27 09N 78 00E
**Ağri Daği** *mt.* Turkey 57 F2 39 44N 44 15E
**Agropoli** Italy 54 F4 40 21N 14 59E
**Ahascragh** RoI 20 C3 53 24N 8 20W
**Ahmadabad** India 66 C4 23 03N 72 40E
**Ahoghill** NI 10 C1 54 51N 6 22W
**Ailsa Craig** *i.* Scotland 10 D2 55 16N 5 07W
**Airdrie** Scotland 11 F2 55 52N 3 59W
**Aire** *r.* England 13 E2 53 40N 1 00W
**Aire-sur-l'Adour** France 50 D2 43 42N 0 15W
**Aix-en-Provence** France 50 G2 43 31N 5 27E
**Aix-les-Bains** France 50 G3 45 41N 5 55E
**Ajaccio** France 50 J1 41 55N 8 43E
**Ajmer** India 66 C5 26 29N 74 40E
**Akron** USA 91 K5 41 04N 81 31W
**Alabama** *state* USA 91 J3 32 00N 87 00W
**Åland** *is.* Finland 49 E3 60 15N 20 00E
**Alaska** *state* USA 88 D6/F6 63 10N 157 30W
**Alaska, Gulf of** USA 88 F5/G5 58 00N 147 00W
**Alaska Range** USA 88 E6/G6 62 30N 152 30W
**Albacete** Spain 51 E3 39 00N 1 52W
**ALBANIA** 55 H4/J4
**Al Başrah** Iraq 65 E5 30 30N 47 50E
**Albert** France 50 F6 50 00N 2 40E
**Alberta** *prov.* Canada 88 M4/N4 54 00N 117 30W
**Albertville** France 50 H3 45 40N 6 24E
**Albi** France 50 F2 43 56N 2 08E
**Ålborg** Denmark 49 C2 57 05N 9 50E
**Albrighton** England 15 D2 52 38N 2 16W
**Albuquerque** USA 90 E4 35 05N 106 38W
**Alcalá de Henares** Spain 51 D4 40 28N 3 22W
**Alcantarilla** Spain 51 E2 37 59N 1 12W
**Alcázar de San Juan** Spain 51 D3 39 24N 3 12W
**Alcester** England 15 E2 52 13N 1 52W
**Alcira** Spain 51 E3 39 10N 0 27W
**Aldbourne** England 16 B2 51 30N 1 37W
**Aldbrough** England 13 F2 53 50N 0 06W
**Alde** *r.* England 17 E3 52 10N 1 30E
**Aldeburgh** England 17 E3 52 09N 1 35E
**Alderley Edge** *tn.* England 12 D2 53 18N 2 15W
**Alderney** *i.* Channel Islands 19 E1 49 43N 2 12W
**Aldershot** England 16 C2 51 15N 0 47W
**Aldridge** England 15 E2 52 36N 1 55W
**Aldsworth** England 15 E1 51 48N 1 46W
**Alençon** France 50 E5 48 25N 0 05E
**Aleppo** Syria 64 C6 36 14N 37 10E
**Alès** France 50 G3 44 08N 4 05E
**Alessándria** Italy 54 C6 44 55N 8 37E
**Ålesund** Norway 49 C3 62 28N 6 11E
**Aleutian Range** USA 88 D5 56 30N 159 00W
**Alexandria** Egypt 81 L8 31 13N 29 55E
**Alexandria** Scotland 11 E2 55 59N 4 36W
**Alexandroúpoli** Greece 55 L4 40 51N 25 53E
**Alford** England 13 G2 53 17N 0 11E
**Alford** Scotland 9 G2 57 13N 2 42W
**Alfreton** England 13 E2 53 06N 1 23W
**Algeciras** Spain 51 C2 36 08N 5 27W
**ALGERIA** 80 D7/F7
**Algiers** Algeria 80 E9 36 50N 3 00E
**Alicante** Spain 51 E3 38 21N 0 29W
**Alice** USA 91 G2 27 45N 98 06W
**Alice Springs** *tn.* Australia 72 F6 23 42S 133 52E
**Alkmaar** Neths 48 C5 52 38N 4 44E
**Allahabad** India 66 E5 25 27N 81 50E
**Allegheny Mountains** USA 93 E1/2 40 00N 79 00W
**Allen** *r.* England 12 D3 54 50N 2 20W
**Allendale Town** England 12 D3 54 54N 2 15W
**Allentown** USA 91 L5 40 37N 75 30W
**Alloa** Scotland 11 F3 56 07N 3 49W
**Alma** USA 93 D2 43 23N 84 40W
**Almadén** Spain 51 C3 38 47N 4 50W
**Almansa** Spain 51 E3 38 52N 1 06W
**Almaty** Kazakhstan 62 K5 43 19N 76 55E
**Almería** Spain 51 D2 36 50N 2 26W
**Älmhult** Sweden 49 D2 56 32N 14 10E
**Aln** *r.* England 11 H2 55 25N 1 45W
**Alness** Scotland 9 E2 57 41N 4 15W
**Alnwick** England 11 H2 55 25N 1 42W
**Alpes Maritimes** *mts.* France/Italy 50 H2 44 15N 6 45E
**Alpi Dolomitiche** *mts.* Italy 54 D7 46 00N 12 00E
**Alpi Pennine** *mts.* Italy/Switz. 52 C3 45 55N 7 30E
**Alps** *mts.* Europe 52 D3 46 00N 7 30E
**Alsager** England 12 D2 53 06N 2 19W
**Alston** England 12 D3 54 49N 2 26W
**Altai** *mts.* Mongolia 68 F7 47 00N 92 30E
**Altamira** Brazil 96 G13 3 13S 52 15W
**Altnaharra** Scotland 9 E3 58 16N 4 27W
**Alton** England 16 C2 51 09N 0 59W
**Altrincham** England 12 D2 53 24N 2 21W
**Alyth** Scotland 9 F1 56 37N 3 13W
**Amarillo** USA 90 F4 35 14N 101 50W
**Amazonas** *r.* Brazil 96 G13 2 00S 53 00W
**Amble** England 11 H2 55 20N 1 34W
**Ambleside** England 12 D3 54 26N 2 58W
**Ambon** Indonesia 71 H3 3 41S 128 10E
**Amersfoort** Neths 48 D5 52 09N 5 23E
**Amersham** England 16 C2 51 40N 0 38W
**Amesbury** England 16 B2 51 10N 1 47W
**Amfípolí** Greece 55 K4 40 48N 23 52E
**Amiens** France 50 F5 49 54N 2 18E
**Amlwch** Wales 14 B3 53 25N 4 20W
**Amman** Jordan 64 C5 31 04N 46 17E
**Ammanford** Wales 14 B1 51 48N 3 58W
**Ampthill** England 16 C3 52 02N 0 30W
**Amritsar** India 66 C6 31 35N 74 56E
**Amsterdam** Neths 48 C5 52 22N 4 54E
**Amstetten** Austria 52 G4 48 08N 14 52E
**Amur** *r.* Asia 69 N8 52 30N 126 30E
**Anascaul** RoI 21 A2 52 09N 10 04W
**Anchorage** USA 88 F6 61 10N 150 00W
**Ancona** Italy 54 E5 43 37N 13 31E
**Ancroft** England 11 G2 55 42N 2 00W
**Andalsnes** Norway 49 C3 62 33N 7 43E
**Andaman Islands** India 67 G2 12 00N 94 00E
**Anderlecht** Belgium 48 C3 50 50N 4 20E
**Andermatt** Switz. 52 D3 46 38N 8 36E
**Andes** *mts.* S. America 96/97 B14/C6 10 00S 77 00W
**ANDORRA** 50 E2
**Andorra la Vella** Andorra 50 E2 42 30N 1 30E
**Andover** England 16 B2 51 13N 1 28W
**Andreas** Isle of Man 12 B3 54 22N 4 26W
**Ándros** *i.* Greece 55 L2 37 49N 24 54E
**Andújar** Spain 51 C3 38 02N 4 03W
**Angara** *r.* Russia 63 M8 58 00N 97 30E
**Angers** France 50 D4 47 29N 0 32W
**Angle** Wales 14 A1 51 41N 5 06W
**Anglesey** *i.* Wales 14 B3 53 18N 4 25W
**Anglesey, Isle of** *u.a.* Wales 14 B3 53 18N 4 25W
**ANGOLA** 82 C5
**Angoulême** France 50 E3 45 40N 0 10E
**Anguilla** *i.* Leeward Islands 94 P10 18 14N 63 05W
**Angus** *u.a.* Scotland 9 G1 56 45N 3 00W
**Ankara** Turkey 57 D2 39 56N 32 50E
**'Annaba** Algeria 80 F9 36 55N 7 47E
**Annalong** NI 10 D1 54 06N 5 55W
**Annan** *r.* Scotland 11 F2 55 05N 3 20W
**Annan** Scotland 11 F1 54 59N 3 16W
**Annapolis** USA 91 L4 38 59N 76 30W
**Annapurna** *mt.* Nepal 66 E5 28 34N 83 50E
**Annbank** Scotland 11 E2 55 28N 4 30W
**Annecy** France 50 H3 45 54N 6 07E
**Annfield Plain** England 13 E3 54 52N 1 45W
**Ansbach** Germany 52 E4 49 18N 10 36E
**Anshan** China 69 M6 41 05N 122 58E
**Anstey** England 15 E2 52 40N 1 10W
**Anston** England 13 E2 53 22N 1 13W
**Anstruther** Scotland 11 G3 56 14N 2 42W
**Antalya** Turkey 57 D2 36 53N 30 42E
**Antananarivo** Madagascar 82 J4 18 52S 47 30E
**Antarctica** 101
**Antequera** Spain 51 C2 37 01N 4 34W
**Antibes** France 50 H2 43 35N 7 07E
**ANTIGUA & BARBUDA** 94 Q9
**Antofagasta** Chile 96 C9 23 40S 70 23W
**Antrim** NI 10 C1 54 43N 6 13W
**Antrim** *district* NI 10 C1 54 40N 6 10W
**Antrim Mountains** NI 10 C2 55 00N 6 10W
**Antwerpen** Belgium 48 C4 51 13N 4 25W
**Aosta** Italy 54 B6 45 43N 7 19E
**Apeldoorn** Neths 48 D5 52 13N 5 57E
**Appalachian Mountains** USA 91 K4 37 00N 82 00W
**Appennini** *mts.* Italy 54 C6/F4 43 00N 12 30E
**Appleby-in-Westmorland** England 12 D3 53 36N 2 29W
**Appledore** England 18 C3 51 03N 4 12W
**Appleton** USA 91 J5 44 17N 88 24W
**'Aqaba, Gulf of** Middle East 64 D4 28 40N 34 40E
**Arabian Sea** Indian Ocean 65 H3 22 00N 63 00E
**Araguaína** Brazil 96 H12 7 16S 48 18W
**Aral Sea** Asia 62 H5/J6 45 00N 60 00E
**Aranda de Duero** Spain 51 D4 41 40N 3 41W
**Aran Fawddwy** *mt.* Wales 14 C2 52 47N 3 41W
**Aran Island** RoI 20 C4 55 00N 8 40W
**Aran Islands** RoI 21 B3 53 10N 9 50W
**Aranjuez** Spain 51 D4 40 02N 3 37W
**Arbil** Iraq 64 D6 36 12N 44 01E
**Arbroath** Scotland 9 G1 56 34N 2 35W
**Arcachon** France 50 D3 44 40N 1 11W
**Arctic Ocean** 101
**Ardara** RoI 20 C4 54 46N 8 25W
**Ardbeg** Scotland 10 C2 55 39N 6 05W
**Ardee** RoI 20 E3 53 52N 6 33W
**Ardennes** *mts.* Belgium 48 D3 50 10N 5 45E
**Ardfert** RoI 21 B2 52 20N 9 47W
**Ardglass** NI 10 D1 54 16N 5 37W
**Ardle** *r.* Scotland 9 F1 56 40N 3 30W
**Ardminish** Scotland 10 D2 55 42N 5 44W
**Ardmore** RoI 21 D1 51 57N 7 43W
**Ardmore Point** Scotland 10 C2 55 41N 6 01W
**Ardnamurchan, Point of** Scotland 8 C1 56 44N 6 14W
**Ardrossan** Scotland 11 E2 55 39N 4 49W
**Ards** *district* NI 10 D1 54 35N 5 35W
**Ards Peninsula** NI 10 D1 54 25N 5 30W
**Ardvasar** Scotland 8 D2 57 03N 5 55W
**Arendal** Norway 49 C2 58 27N 8 56E
**Arequipa** Peru 96 C10 16 25S 71 32W
**Arezzo** Italy 54 D5 43 28N 11 53E
**Argentan** France 50 D5 48 45N 0 01W
**ARGENTINA** 97 D5
**Argyll and Bute** *u.a.* Scotland 10 D3 56 10N 5 00W
**Århus** Denmark 49 D2 56 15N 10 10E
**Arica** Chile 96 C10 18 30S 70 20W
**Arinagour** Scotland 8 C1 56 37N 6 31W
**Ariquemes** Brazil 96 E12 9 55S 63 06W
**Arisaig** Scotland 8 D1 56 51N 5 51W
**Arisaig, Sound of** Scotland 8 D1 56 50N 5 50W
**Arizona** *state* USA 90 D3 34 00N 112 00W
**Arkansas** *state* USA 91 H3 34 00N 93 00W
**Arkansas** *r.* USA 90 G4 36 00N 99 00W
**Arkansas City** USA 91 G4 37 03N 97 02W
**Arkhangel'sk** Russia 62 G9 64 32N 40 40E
**Arklow** RoI 21 E2 52 48N 6 09W
**Arles** France 50 G2 43 41N 4 38E
**Arlon** Belgium 48 D2 49 41N 5 49E
**Armadale** Highland Scotland 8 D2 57 05N 5 54W
**Armadale** West Lothian Scotland 11 F2 55 54N 3 41W
**Armagh** NI 10 C1 54 21N 6 39W
**Armagh** *district* NI 10 C1 54 20N 6 40W
**ARMENIA** 57 F3
**Armentières** France 50 F6 50 41N 2 53E
**Armthorpe** England 13 E2 53 32N 1 03W
**Arnhem** Neths 48 D4 52 00N 5 53E
**Arnold** England 15 E2 53 00N 1 09W
**Arran** *i.* Scotland 10 D2 55 35N 5 15W
**Arras** France 50 F6 50 17N 2 46E
**Artane** RoI 20 E3 53 23N 6 12W
**Arthur's Seat** *sum.* Scotland 11 F2 55 57N 3 11W
**Arthurstown** RoI 21 E2 52 15N 6 57W
**Aruba** *i.* Neths 96 C16
**Arun** *r.* England 16 C2 51 00N 0 30W
**Arundel** England 16 C1 50 51N 0 34W
**Arvika** Sweden 49 D2 59 41N 12 38E
**Ascoli Piceno** Italy 54 E5 42 52N 13 35E
**Ascot** England 16 C2 51 25N 0 41W
**Ash** England 17 E2 51 17N 1 16E
**Ashbourne** England 13 E2 53 01N 1 43W
**Ashbourne** RoI 20 E3 53 31N 6 24W
**Ashburton** England 18 D2 50 31N 3 45W
**Ashby-de-la-Zouch** England 15 E2 52 46N 1 28W
**Ashford** England 17 D2 51 09N 0 53E
**Ash Fork** USA 90 D4 35 13N 112 29W
**Ashgabat** Turkmenistan 62 H4 37 58N 58 24E
**Ashington** England 13 E4 55 11N 1 34W
**Ashkirk** Scotland 11 G2 55 29N 2 50W
**Ashton-in-Makerfield** England 12 D2 53 29N 2 39W
**Ashton-under-Lyne** England 12 D2 53 29N 2 06W
**Ashwell** England 16 C3 52 03N 0 09W
**Askeaton** RoI 21 C2 52 36N 8 58W
**Askern** England 13 E2 53 37N 1 09W
**Askrigg** England 12 D3 54 19N 2 04W
**Asmara** Eritrea 81 N5 15 20N 38 58E
**Aspatria** England 12 C3 54 46N 3 20W
**Assisi** Italy 54 E5 43 04N 12 37E
**Astana** Kazakhstan 62 K7 51 10N 71 28E
**Astee** RoI 21 B2 52 33N 9 34W
**Asti** Italy 54 C6 44 54N 8 13E
**Astorga** Spain 51 B5 42 27N 6 04W
**Astoria** USA 90 B6 46 12N 123 50W
**Astrakhan'** Russia 62 G6 46 22N 48 04E
**Asunción** Paraguay 96 F7 25 15S 57 40W
**Aswân** Egypt 81 M6 24 05N 32 56E
**Asyût** Egypt 81 M7 27 14N 31 07E
**Athabasca** Canada 88 N4 54 44N 113 15W
**Athabasca, Lake** Canada 88 P5 59 10N 109 30W
**Athboy** RoI 20 E3 53 37N 6 55W
**Athenry** RoI 20 C3 53 18N 8 45W
**Athens** Greece 55 K2 38 00N 23 44E
**Atherstone** England 15 E2 52 35N 1 31W
**Athleague** RoI 20 C3 53 34N 8 15W
**Athlone** RoI 20 D3 53 25N 7 56W
**Áthos** *mt.* Greece 55 L4 40 10N 24 19E
**Athy** RoI 21 E2 52 59N 6 59W
**Atlanta** USA 91 K3 33 45N 84 23W
**Atlantic City** USA 91 M4 39 23N 74 27W
**Atlantic Ocean** 104
**Attleborough** England 17 E3 52 31N 1 01E
**Attymass** RoI 20 B4 54 03N 9 04W
**Aubagne** France 50 G2 43 17N 5 35E
**Aubenas** France 50 G3 44 37N 4 24E
**Auch** France 50 E2 43 40N 0 36E
**Auchencairn** Scotland 11 F1 54 51N 3 53W
**Auchterarder** Scotland 11 F3 56 18N 3 43W
**Auchtermuchty** Scotland 11 F3 56 17N 3 15W
**Auckland** NZ 73 P3 36 51S 174 46E
**Audley** England 12 D2 53 03N 2 20W
**Augher** NI 10 B1 54 26N 7 08W
**Aughnacloy** NI 10 C1 54 25N 6 59W
**Aughrim** RoI 21 E2 52 52N 6 20W
**Aughton** England 13 E2 53 22N 1 19W
**Augsburg** Germany 52 E4 48 21N 10 54E
**Augusta** USA 91 K3 33 29N 82 00W
**Auldgirth** Scotland 11 F2 55 12N 3 43W
**Aurillac** France 50 F3 44 56N 2 26E
**Austin** USA 91 G3 30 18N 97 47W
**AUSTRALIA** 72/73
**AUSTRIA** 52 E3/G3
**Autun** France 50 G4 46 58N 4 18E
**Auxerre** France 50 F4 47 48N 3 35E
**Avallon** France 50 F4 47 30N 3 54E
**Avebury** England 16 B2 51 27N 1 51W
**Aveiro** Portugal 51 A4 40 38N 8 40W
**Aveley** England 17 D2 51 30N 0 15E
**Avesta** Sweden 49 E3 60 09N 16 10E
**Avezzano** Italy 54 E5 42 02N 13 26E
**Aviemore** Scotland 9 F2 57 12N 3 50W
**Avignon** France 50 G2 43 56N 4 48E
**Avila** Spain 51 C4 40 39N 4 42W
**Avoca** RoI 21 E2 52 52N 6 13W
**Avon** *r.* Warwickshire England 15 E2 52 10N 1 55W
**Avon** *r.* Hampshire England 16 B1 51 05N 1 55W
**Avon** *r.* Wiltshire England 16 A2 51 25N 2 05W
**Avon** *r.* Devon England 18 D2 50 20N 3 48W
**Avon** *r.* Scotland 9 F2 51 27N 1 40W
**Avonmouth** England 19 E3 51 31N 2 42W
**Avranches** France 50 D5 48 42N 1 21W
**Axbridge** England 19 E3 51 18N 2 49W

**Axe** *r.* Somerset England 19 E3 51 11N 2 43W
**Axe** *r.* Dorset England 19 E2 50 50N 2 50W
**Axminster** England 19 E2 50 47N 3 00W
**Ayamonte** Spain 51 B2 37 13N 7 24W
**Ayers Rock** *mt.* Australia 72 F5 25 18S 131 18E
**Aylesbury** England 16 C2 51 50N 0 50W
**Aylesford** England 17 D2 51 18N 0 30E
**Aylesham** England 17 E2 51 12N 1 11E
**Aylsham** England 17 E3 52 49N 1 15E
**Ayr** *r.* Scotland 11 E2 55 30N 4 10W
**Ayr** Scotland 11 E2 55 28N 4 38W
**Ayre, Point of** Isle of Man 12 B3 54 25N 4 22W
**Aysgarth** England 12 E3 54 17N 2 00W
**AZERBAIJAN** 57 G3
**Azov, Sea of** Asia 57 E4 46 00N 36 00E
**Azuaga** Spain 51 C3 38 16N 5 40W

## B

**Babbacombe Bay** England 18 D2 50 30N 3 30W
**Bacup** England 12 D2 53 42N 2 12W
**Badajoz** Spain 51 B3 38 53N 6 58W
**Badalona** Spain 51 G4 41 27N 2 15E
**Baden** Switz. 52 D3 47 28N 8 19E
**Bad Ischl** Austria 52 F3 47 43N 13 38E
**Baffin Bay** Canada/Greenland 89 X8/Z8 72 00N 65 00W
**Baffin Island** Canada 89 U8/W7 68 30N 70 00W
**Baggy Point** England 18 C3 51 09N 4 16W
**Baghdad** Iraq 64 D5 33 20N 44 26E
**BAHAMAS, THE** 95 J4
**Bahía Blanca** Argentina 97 E6 38 45S 62 15W
**Bahia de Campeche** Mexico 94 E4/F4 20 00N 95 00W
**BAHRAIN** 65 F4
**Bahrain, Gulf of** The Gulf 65 F4 25 55N 50 30E
**Bahr el Abiad** *r.* Sudan 81 M4 14 00N 32 20E
**Bahr el Azraq** *r.* Sudan 81 M4 13 30N 33 45E
**Baia Mare** Romania 53 L3 47 39N 23 36E
**Baie de la Seine** France 50 D5 49 40N 0 30W
**Baildon** England 13 E2 53 52N 1 46W
**Bailieborough** RoI 20 E3 53 54N 6 59W
**Baja California** *p.* Mexico 94 B5 27 30N 113 00W
**Bakersfield** USA 90 C4 35 25N 119 00W
**Bakewell** England 13 E2 53 13N 1 40W
**Baku** Azerbaijan 57 G3 40 22N 49 53E
**Bala** Wales 14 C2 52 54N 3 35W
**Balaton** *l.* Hungary 53 H3 47 00N 17 30E
**Balbriggan** RoI 20 E3 53 37N 6 11W
**Balcombe** England 16 C2 51 04N 0 08W
**Baldock** England 16 C2 51 59N 0 12W
**Balearic Islands** Med. Sea 51 F3/H3 40 00N 2 00E
**Baleshare** *i.* Scotland 8 B2 57 32N 7 22W
**Bali** *i.* Indonesia 71 E2/F2 8 30S 115 00E
**Ballachulish** Scotland 8 D1 56 40N 5 10W
**Ballagan Point** RoI 20 E4 54 00N 6 06W
**Ballaghaderreen** RoI 20 C3 53 55N 8 36W
**Ballantrae** Scotland 10 D2 55 06N 5 00W
**Ballater** Scotland 9 F2 57 03N 3 03W
**Ballaugh** Isle of Man 12 B3 54 18N 4 32W
**Ballina** RoI 20 B4 54 07N 9 09W
**Ballinafad** RoI 20 C4 54 02N 8 20W
**Ballinagleragh** RoI 20 C4 54 09N 8 01W
**Ballinalack** RoI 20 D3 53 38N 7 28W
**Ballinalee** RoI 20 D3 53 46N 7 39W
**Ballinamore** RoI 20 D4 54 03N 7 47W
**Ballinasloe** RoI 20 C3 53 20N 8 13W
**Ballincollig** RoI 21 C1 51 54N 8 35W
**Ballindine** RoI 20 C3 53 40N 8 57W
**Ballingry** Scotland 11 F3 56 11N 3 20W
**Ballinhassig** RoI 21 C1 51 49N 8 32W
**Ballinrobe** RoI 20 B3 53 37N 9 13W
**Ballinskelligs Bay** RoI 21 A1 51 46N 10 17W
**Ballintra** RoI 20 C4 54 35N 8 07W
**Ballon** RoI 21 E2 52 44N 6 46W
**Ballybay** RoI 20 E4 54 08N 6 54W
**Ballybofey** RoI 20 D4 54 48N 7 47W
**Ballybunnion** RoI 21 B2 52 31N 9 40W
**Ballycanew** RoI 21 E2 52 36N 6 18W
**Ballycastle** NI 10 C2 55 12N 6 15W
**Ballycastle** RoI 20 B4 54 17N 9 22W
**Ballyclare** NI 10 C1 54 45N 6 00W
**Ballycolla** RoI 21 D2 52 52N 7 26W
**Ballyconnell** RoI 20 D4 54 07N 7 35W
**Ballycotton** RoI 21 C1 51 50N 8 01W
**Ballycotton Bay** RoI 21 D1 51 51N 7 57W
**Ballycroy** RoI 20 B4 54 02N 9 49W
**Ballydavid Head** RoI 21 A2 52 14N 10 21W
**Ballydehob** RoI 21 B1 51 34N 9 28W
**Ballydesmond** RoI 21 B2 52 10N 9 13W
**Ballydonegan** RoI 21 A1 51 37N 10 12W
**Ballydonegan Bay** RoI 21 A1 51 29N 10 09W
**Ballygalley Head** NI 10 D1 54 54N 5 51W
**Ballygowan** NI 10 D1 54 30N 5 47W
**Ballyhaunis** RoI 20 C3 53 46N 8 46W
**Ballyheige** RoI 21 B2 52 24N 9 50W
**Ballyheige Bay** RoI 21 B2 52 23N 9 52W
**Ballyhoura Mountains** RoI 21 C2 52 17N 8 33W
**Ballyjamesduff** RoI 20 D3 53 52N 7 12W
**Ballykinler** NI 10 D1 54 17N 5 47W
**Ballylanders** RoI 21 C2 52 23N 8 21W
**Ballylongford** RoI 21 B2 52 33N 9 28W
**Ballymacarbry** RoI 21 D2 52 16N 7 12W
**Ballymahon** RoI 20 D3 53 34N 7 45W
**Ballymena** NI 10 C1 54 52N 6 17W
**Ballymena** *district* NI 10 C1 54 55N 6 20W
**Ballymoe** RoI 20 C3 53 42N 8 28W
**Ballymoney** NI 10 C2 55 04N 6 31W
**Ballymoney** *district* NI 10 C2 55 00N 6 30W
**Ballymore** Donegal RoI 20 D5 55 09N 7 56W
**Ballymore** Westmeath RoI 20 D3 53 29N 7 40W
**Ballymote** RoI 20 C4 54 06N 8 31W
**Ballynahinch** NI 10 D1 54 24N 5 54W
**Ballynamona** RoI 21 C2 52 05N 8 39W
**Ballynamult** RoI 21 D2 52 13N 7 40W
**Ballyquintin Point** NI 10 D1 54 19N 5 30W
**Ballyragget** RoI 21 D2 52 47N 7 20W
**Ballyshannon** RoI 20 C4 54 30N 8 11W
**Ballyteige Bay** RoI 21 E2 52 12N 6 45W
**Ballyvaghan** RoI 21 B3 53 07N 9 09W
**Ballyvourney** RoI 21 B1 51 56N 9 10W
**Ballywalter** NI 10 D1 54 33N 5 30W
**Balmaclellan** Scotland 11 E2 55 05N 4 07W
**Balmoral** Scotland 9 F2 57 02N 3 13W
**Balrath** RoI 20 E3 53 37N 6 29W
**Baltic Sea** Europe 49 E2/F2 55 15N 17 00E
**Baltimore** RoI 21 B1 51 29N 9 22W
**Baltimore** USA 91 L4 39 18N 76 38W
**Baltinglass** RoI 21 E2 52 55N 6 41W
**Baltray** RoI 20 E3 53 44N 6 16W
**Bamako** Mali 80 C4 12 40N 7 59W
**Bamberg** Germany 52 E4 49 54N 10 54E
**Bamburgh** England 11 H2 55 36N 1 42W
**Bampton** England 18 D2 51 00N 3 29W
**Banagher** RoI 21 D3 53 11N 7 59W
**Banbridge** NI 10 C1 54 21N 6 16W
**Banbridge** *district* NI 10 C1 54 15N 6 20W
**Banbury** England 16 B3 52 04N 1 20W
**Banchory** Scotland 9 G2 57 30N 2 30W
**Bandar Seri Begawan** Brunei 71 F4 4 53N 115 00E
**Bandon** RoI 21 C1 51 45N 8 45W
**Bandung** Indonesia 71 D2 6 57S 107 34E
**Banff** Canada 88 M4 51 10N 115 34W
**Banff** Scotland 9 G2 57 40N 2 33W
**Bangalore** India 66 D2 12 58N 77 35E
**Bangkok** Thailand 71 C6 13 44N 100 30E
**BANGLADESH** 67 F4/G4
**Bangor** NI 10 D1 54 40N 5 40W
**Bangor** RoI 20 B4 54 09N 9 44W
**Bangor** Wales 14 B3 53 13N 4 08W
**Bangui** CAR 81 J2 4 23N 18 37E
**Banja Luka** Bosnia-Herzegovina 55 G6 44 47N 17 11E
**Banjul** The Gambia 80 A4 13 28N 16 39W
**Bann** *r.* Coleraine NI 10 C2 55 05N 6 40W
**Bann** *r.* Banbridge NI 10 C1 54 20N 6 11W
**Bann** *r.* RoI 21 E2 52 39W 6 24W
**Bannockburn** Scotland 11 F3 56 05N 3 56W
**Bannow Bay** RoI 21 E2 52 13N 6 48W
**Bansha** RoI 21 C2 52 28N 8 04W
**Banstead** England 16 C2 51 19N 0 12W
**Banteer** RoI 21 C2 52 08N 8 54W
**Bantry** RoI 21 B1 51 41N 9 27W
**Bantry Bay** RoI 21 B1 51 35N 9 40W
**Baotou** China 69 J6 40 38N 109 59E
**Baracaldo** Spain 51 D5 43 17N 2 59W
**Baranavichy** Belarus 53 N6 53 09N 26 00E
**BARBADOS** 94 S11
**Barbastro** Spain 51 F5 42 02N 0 07E
**Barcelona** Spain 51 G4 41 25N 2 10E
**Barcs** Hungary 53 H2 45 58N 17 30E
**Bardsey Island** Wales 14 B2 52 46N 4 48W
**Barents Sea** Arctic Ocean 101 75 00N 40 00E
**Bargoed** Wales 14 C1 51 43N 3 15W
**Bari** Italy 55 G4 41 07N 16 52E
**Barisal** Bangladesh 67 G4 22 41N 90 20E
**Barking** England 17 D2 51 33N 0 06E
**Barle** *r.* England 18 D3 51 05N 3 36W
**Bar-le-Duc** France 50 G5 48 46N 5 10E
**Barletta** Italy 54 G4 41 20N 16 17E
**Barmouth** Wales 14 B2 52 43N 4 03W
**Barnard Castle** England 13 E3 54 33N 1 55W
**Barnatra** RoI 20 B4 54 13N 9 50W
**Barnet** England 16 C2 51 39N 0 12W
**Barnoldswick** England 12 D2 53 56N 2 16W
**Barnsley** England 13 E2 53 34N 1 28W
**Barnstaple** England 18 C3 51 05N 4 04W
**Barnstaple or Bideford Bay** England 18 C3 51 05N 4 25W
**Barquisimeto** Venezuela 96 D16 10 03N 69 18W
**Barra** *i.* Scotland 8 B1/B2 57 00N 7 25W
**Barra Head** Scotland 8 B1 56 47N 7 36W
**Barranquilla** Colombia 96 C16 11 10N 74 50W
**Barra, Sound of** Scotland 8 B2 57 06N 7 22W
**Barre** USA 91 M5 44 13N 72 31W
**Barreiras** Brazil 96 J11 12 09S 44 58W
**Barreiro** Portugal 51 A3 38 40N 9 05W
**Barrhead** Scotland 11 E2 55 48N 4 24W
**Barrow** *r.* RoI 21 E2 52 38N 6 58W
**Barrow-in-Furness** England 12 C3 54 07N 3 14W
**Barry** Wales 14 C1 51 24N 3 18W
**Barton-upon-Humber** England 13 F2 53 41N 0 27W
**Barvas** Scotland 8 C3 58 22N 6 32W
**Barysaw** Belarus 53 P7 54 09N 28 30E
**Basel** Switz. 52 C3 47 33N 7 36E
**Basildon** England 17 D2 51 34N 0 25W
**Basingstoke** England 16 B2 51 16N 1 05W
**Bassenthwaite Lake** England 12 C3 54 40N 3 13W
**Bastia** France 50 J2 42 14N 9 26E
**Bastogne** Belgium 48 D3 50 00N 5 43E
**Bath** England 19 E3 51 23N 2 22W
**Bath and North East Somerset** *u.a.* England 19 E3 51 25N 2 30W
**Bathgate** Scotland 11 F2 55 55N 3 39W
**Bathurst** Canada 89 X3 47 37N 65 40W
**Batley** England 13 E2 53 44N 1 37W
**Baton Rouge** USA 91 H3 30 30N 91 10W
**Battle** England 17 D1 50 55N 0 29E
**Bat Yam** Israel 64 N10 31 59N 34 45E
**Bawdsey** England 17 E3 52 01N 1 25E
**Bawtry** England 13 E2 53 26N 1 01W
**Bayeux** France 50 D5 49 16N 0 42W
**Bayonne** France 50 D2 43 30N 1 28W
**Bayreuth** Germany 52 E4 49 27N 11 35E
**Bayston Hill** *tn.* England 14 D2 52 40N 2 48W
**Beachy Head** England 17 D1 50 44N 0 16E
**Beaconsfield** England 16 C2 51 37N 0 39W
**Beadnell Bay** England 11 H2 55 32N 1 30W
**Beaminster** England 19 E2 50 49N 2 45W
**Bear Island** RoI 21 B1 51 40N 9 48W
**Bearsden** Scotland 11 E2 55 56N 4 20W
**Bearsted** England 17 D2 51 17N 0 35E
**Beatrice** USA 91 G5 40 17N 96 45W
**Beaulieu** England 16 B1 50 49N 1 27W
**Beauly** Scotland 9 E2 57 29N 4 29W
**Beauly** *r.* Scotland 9 E2 57 25N 4 33W
**Beauly Firth** Scotland 9 E2 57 30N 4 20W
**Beaumaris** Wales 14 B3 53 16N 4 05W
**Beaune** France 50 G4 47 02N 4 50E
**Beauvais** France 50 F5 49 26N 2 05E
**Bebington** England 12 D2 53 20N 2 59W
**Beccles** England 17 E3 52 28N 1 34E
**Béchar** Algeria 80 D8 31 35N 2 17W
**Bedale** England 13 E3 54 17N 1 35W
**Bedford** England 16 C3 52 08N 0 29W
**Bedfordshire** *co.* England 16 C2 52 08N 0 29W
**Bedlington** England 13 E4 55 08N 1 25W
**Bedwas** Wales 14 C1 51 35N 3 12W
**Bedworth** England 15 E2 52 29N 1 28W
**Beeston** England 15 E2 52 56N 1 12W
**Beighton** England 13 E2 53 21N 1 21W
**Beijing** China 69 L5 39 55N 116 26E
**Beinn Dearg** *mt.* Scotland 9 E2 57 47N 4 56W
**Beira** Mozambique 82 F4 19 49S 34 52E
**Beirut** Lebanon 64 C5 33 52N 35 30E
**Beith** Scotland 11 E2 55 45N 4 38W
**Beja** Portugal 51 B3 38 01N 7 52W
**Bejaïa** Algeria 80 F9 36 49N 5 03E
**Béjar** Spain 51 C4 40 24N 5 45W
**BELARUS** 56 B5/6
**Belderg** RoI 20 B4 54 18N 9 33W
**Belém** Brazil 96 H12 1 27S 48 29W
**Belfast** NI 10 D1 54 35N 5 55W
**Belfast** *district* NI 10 C1/D1 54 40N 6 05W
**Belfast Lough** *est.* NI 10 D1 54 40N 5 50W
**Belford** England 11 H2 55 36N 1 49W
**Belfort** France 50 H4 47 38N 6 52E
**BELGIUM** 48 B3/D3
**Belgrade** Serbia & Montenegro 55 J6 44 50N 20 30E
**BELIZE** 94 G3
**Bellac** France 50 E4 46 07N 1 04E
**Bellacorick** RoI 20 B4 54 07N 9 34W
**Bellanagh** RoI 20 D3 53 56N 7 24W
**Bellavary** RoI 20 B3 53 53N 9 12W
**Bellingham** England 12 D4 55 09N 2 16W
**Bellsbank** Scotland 11 E2 55 19N 4 24W
**Bellshill** Scotland 11 E2 55 49N 4 02W
**Belmont** Scotland 7 D3 60 41N 0 58W
**Belmopan** Belize 94 G3 17 13N 88 48W
**Belmullet** RoI 20 B4 54 14N 10 00W
**Belo Horizonte** Brazil 96 J10 19 54S 43 54W
**Belper** England 13 E2 53 01N 1 29W
**Belton** England 17 E3 52 34N 1 40E
**Beltra Lough** RoI 20 B3 53 55N 9 25W
**Belturbet** RoI 20 D4 54 06N 7 28W
**Belvoir, Vale of** England 15 F2 52 58N 0 55W
**Bembridge** England 16 B1 50 41N 1 05W
**Ben Alder** *mt.* Scotland 9 E1 56 49N 4 28W
**Benavente** Spain 51 C4 42 00N 5 40W
**Benbane Head** NI 10 C2 55 14N 6 29W
**Benbecula** *i.* Scotland 8 B2 57 25N 7 20W
**Ben Cleuch** *mt.* Scotland 11 F3 56 11N 3 47W
**Ben Cruachan** *mt.* Scotland 8 D1 56 26N 5 09W
**Benevento** Italy 54 F4 41 08N 14 46E
**Bengal, Bay of** Indian Ocean 67 F3 17 00N 88 00E
**Benghazi** Libya 81 K8 32 07N 20 04E
**Benguela** Angola 82 B5 12 34S 13 24E
**Ben Hope** *mt.* Scotland 9 E3 58 24N 4 37W
**Benicarló** Spain 51 F4 40 25N 0 25E
**Benidorm** Spain 51 E3 38 33N 0 09W
**BENIN** 80 E4
**Benin City** Nigeria 80 F3 6 19N 5 41E
**Ben Klibreck** *mt.* Scotland 9 E3 58 14N 4 22W
**Ben Lawers** *mt.* Scotland 9 E1 56 33N 4 15W
**Benllech** Wales 14 B3 53 19N 4 15W
**Ben Lomond** *mt.* Scotland 11 E3 56 12N 4 38W
**Ben Loyal** *mt.* Scotland 9 E3 58 24N 4 26W
**Ben Macdui** *mt.* Scotland 9 F2 57 04N 3 40W
**Ben More** *mt.* Highland Scotland 8 C1 56 25N 6 02W
**Ben More** *mt.* Stirling Scotland 9 E1 56 23N 4 31W
**Ben More Assynt** *mt.* Scotland 9 E3 58 07N 4 52W
**Bennettsbridge** RoI 21 D2 52 35N 7 10W
**Ben Nevis** *mt.* Scotland 9 E1 56 40N 5 00W
**Benson** USA 90 D3 31 58N 110 19W
**Bentley** England 13 E2 53 33N 1 09W
**Benwee Head** RoI 20 B4 54 21N 9 48W
**Ben Wyvis** *mt.* Scotland 9 E2 57 40N 4 35W
**Beppu** Japan 70 B1 33 18N 131 30E
**Bérgamo** Italy 54 C6 45 42N 9 40E
**Bergen** Norway 49 C3 60 23N 5 20E
**Bergerac** France 50 E3 44 50N 0 29E
**Bergisch Gladbach** Germany 52 C5 50 59N 7 10E
**Bering Sea** Pacific Ocean 100 H12/J13 60 00N 175 00W
**Bering Strait** Russia/USA 88 B7 69 00N 169 00W
**Berkeley** England 15 D1 51 42N 2 27W
**Berkhamsted** England 16 C2 51 46N 0 35W
**Berkshire Downs** England 16 B2 51 30N 1 25W
**Berlin** Germany 52 F6 52 32N 13 25E
**Bern** Switz. 52 C3 46 57N 7 26E
**Bernalda** Italy 55 G4 40 24N 16 44E
**Berner Alpen** *mts.* Switz. 52 C3/D3 46 25N 7 30E
**Berneray** *i.* Scotland 8 B2 57 45N 7 10W
**Bernina Pass** Switz. 52 E3 46 25N 10 02E
**Berriedale** Scotland 9 F3 58 11N 3 33W
**Berry Head** England 18 D2 50 24N 3 29W
**Berwick-upon-Tweed** England 11 G2 55 46N 2 00W
**Besançon** France 50 G4 47 14N 6 02E
**Bessacarr** England 13 E2 53 30N 1 05W
**Bessbrook** NI 10 C1 54 12N 6 24W
**Betanzos** Spain 51 A5 43 17N 8 13W
**Bethersden** England 17 D2 51 07N 0 45W
**Bethesda** Wales 14 B3 53 11N 4 03W
**Bethlehem** Middle East 64 C5 31 42N 35 12E
**Béthune** France 50 F6 50 32N 2 38E
**Bettyhill** Scotland 9 E3 58 32N 4 14W
**Betws-y-Coed** Wales 14 C3 53 05N 3 48W
**Beverley** England 13 F2 53 51N 0 26W
**Bewdley** England 15 D2 52 22N 2 19W
**Bewl Water** *l.* England 17 D2 51 02N 0 23E
**Bexhill** England 17 D1 50 50N 0 29E
**Bexley** England 17 D2 51 27N 0 09E
**Béziers** France 50 F2 43 21N 3 13E
**Bharatpur** India 66 D5 27 14N 77 29E
**Bhopal** India 66 D4 23 17N 77 28E
**Bhubaneshwar** India 67 F4 20 13N 85 50E
**BHUTAN** 67 G5
**Białystok** Poland 53 L6 53 09N 23 10E
**Biarritz** France 50 D2 43 29N 1 33W
**Bicester** England 16 B2 51 54N 1 09W
**Biddeford** USA 91 M5 43 29N 70 27W
**Biddulph** England 12 D2 53 08N 2 10W
**Bideford** England 18 C3 51 01N 4 13W
**Biel** Switz. 52 C3 46 27N 8 13E
**Bielefeld** Germany 52 D6 52 02N 8 32E
**Bielsko-Biała** Poland 53 J4 49 50N 19 00E
**Bigbury Bay** England 18 D2 50 17N 4 00W
**Biggar** Scotland 11 F2 55 38N 3 32W
**Biggleswade** England 16 C3 52 05N 0 17W
**Bihać** Bosnia-Herzegovina 54 F6 44 49N 15 53E
**Bikaner** India 66 C5 28 01N 73 22E
**Bila Tserkva** Ukraine 53 Q4 49 49N 30 10E
**Bilbao** Spain 51 D5 43 15N 2 56W
**Billericay** England 17 D2 51 38N 0 25E
**Billinge** England 12 D2 53 30N 2 42W
**Billingham** England 13 E3 54 36N 1 17W
**Billingshurst** England 16 C2 51 01N 0 28W
**Bill of Portland** *c.* England 19 E2 50 31N 2 27W
**Bingham** England 15 F2 52 57N 0 57W
**Bingley** England 13 E2 53 51N 1 50W
**Birchwood** *tn.* England 13 F2 53 10N 0 33W
**Birkenhead** England 12 C2 53 24N 3 02W
**Birmingham** England 15 E2 52 30N 1 50W
**Birmingham** USA 91 J3 33 30N 86 55W
**Birr** RoI 21 D3 53 05N 7 54W
**Birstall** England 15 E2 52 40N 1 08W
**Biscay, Bay of** Atlantic Ocean 50 C3 45 30N 2 50W
**Bishkek** Kyrgyzstan 62 K5 42 53N 74 46E

| | | | | |
|---|---|---|---|---|
| **Bishop Auckland** England | 13 | E3 | 54 40N | 1 40W |
| **Bishopbriggs** Scotland | 11 | E2 | 55 54N | 4 14W |
| **Bishop's Castle** tn. England | 14 | C2 | 52 29N | 3 00W |
| **Bishop's Cleeve** England | 15 | D1 | 51 57N | 2 04W |
| **Bishop's Lydeard** England | 19 | D3 | 51 04N | 3 12W |
| **Bishop's Stortford** England | 17 | D2 | 51 53N | 0 09E |
| **Bishopston** Wales | 14 | B1 | 51 35N | 4 03W |
| **Bishop's Waltham** England | 16 | B1 | 50 58N | 1 12W |
| **Bishopton** Scotland | 11 | E2 | 55 54N | 4 31W |
| **Bissau** Guinea-Bissau | 80 | A4 | 11 52N | 15 39W |
| **Blaby** England | 15 | E2 | 52 35N | 1 09W |
| **Black** r. RoI | 20 | D3 | 53 55N | 7 42W |
| **Blackburn** England | 12 | D2 | 53 45N | 2 29W |
| **Blackdown Hills** England | 19 | D2 | 50 55N | 3 10W |
| **Blackhall Colliery** tn. England | 13 | E3 | 54 43N | 1 20W |
| **Black Head** RoI | 21 | B3 | 53 09N | 9 16W |
| **Blackhope Scar** mt. Scotland | 11 | F2 | 55 58N | 3 07W |
| **Black Isle** Scotland | 9 | E2 | 57 35N | 4 15W |
| **Blackmoor Vale** England | 19 | E2 | 50 56N | 2 19W |
| **Black Mountains** Wales | 14 | C1 | 51 55N | 3 10W |
| **Black Mountain** Wales | 14 | C1 | 51 50N | 3 55W |
| **Blackpool** England | 12 | C2 | 53 50N | 3 03W |
| **Black Sea** Europe | 57 | D3/E3 | 43 00N | 35 00E |
| **Blacksod Bay** RoI | 20 | A3/A4 | 54 05N | 10 00W |
| **Blackstairs Mountains** RoI | 21 | E2 | 52 33N | 6 49W |
| **Blackwater** r. Hampshire England | 16 | C2 | 51 22N | 0 54W |
| **Blackwater** r. Essex England | 17 | D2 | 51 45N | 0 50E |
| **Blackwater** r. Waterford RoI | 21 | D2 | 52 02N | 7 51W |
| **Blackwater** r. Meath RoI | 20 | E3 | 53 41N | 6 44W |
| **Blackwaterfoot** Scotland | 10 | D2 | 55 30N | 5 19W |
| **Blackwater** r. NI | 10 | C1 | 54 20N | 6 50W |
| **Blackwood** Wales | 14 | C1 | 51 41N | 3 13W |
| **Blaenau Ffestiniog** Wales | 14 | C2 | 52 59N | 3 56W |
| **Blaenau Gwent** u.a. Wales | 14 | C1 | 51 45N | 3 10W |
| **Blaenavon** Wales | 14 | C1 | 51 48N | 3 05W |
| **Blagoveshchensk** Russia | 63 | Q7 | 50 19N | 127 30E |
| **Blaina** Wales | 14 | C1 | 51 48N | 3 11W |
| **Blair Atholl** Scotland | 9 | F1 | 56 46N | 3 51W |
| **Blairgowrie** Scotland | 9 | F1 | 56 36N | 3 21W |
| **Blakeney Point** England | 17 | E3 | 52 58N | 1 00E |
| **Blandford Forum** England | 19 | E2 | 50 52N | 2 11W |
| **Blantyre** Malawi | 82 | F4 | 15 46S | 35 00E |
| **Blantyre** Scotland | 11 | E2 | 55 47N | 4 06W |
| **Blarney** RoI | 21 | C1 | 51 56N | 8 34W |
| **Blennerville** RoI | 21 | B2 | 52 15N | 9 43W |
| **Blessington** RoI | 21 | E3 | 53 10N | 6 32W |
| **Bletchley** England | 16 | C2 | 52 00N | 0 46W |
| **Blidworth** England | 13 | E2 | 53 09N | 1 11W |
| **Blithe** r. England | 15 | E2 | 52 45N | 1 50W |
| **Blois** France | 50 | E4 | 47 36N | 1 20E |
| **Bloody Foreland** RoI | 20 | C5 | 55 10N | 8 15W |
| **Bludenz** Austria | 52 | D3 | 47 10N | 9 50E |
| **Blue Stack Mountains** RoI | 20 | C4 | 54 44N | 8 03W |
| **Blyth** Nottinghamshire England | 13 | E2 | 53 23N | 1 03W |
| **Blyth** Northumberland England | 13 | E4 | 55 07N | 1 30W |
| **Blyth** r. England | 13 | E4 | 55 08N | 1 45W |
| **Blyth** r. England | 17 | E3 | 52 28N | 1 30E |
| **Boa Vista** Brazil | 96 | E14 | 3 23S | 55 30W |
| **Bocholt** Germany | 52 | C5 | 51 49N | 6 37E |
| **Bochum** Germany | 52 | C5 | 51 28N | 7 11E |
| **Boddam** Scotland | 9 | H2 | 57 28N | 1 48W |
| **Boden** Sweden | 49 | F4 | 65 50N | 21 44E |
| **Bodensee** l. Switz. | 52 | D3 | 47 40N | 9 30E |
| **Bodmin** England | 18 | C2 | 50 29N | 4 43W |
| **Bodmin Moor** England | 18 | C2 | 50 35N | 4 40W |
| **Boggeragh Mountains** RoI | 21 | B2/C2 | 52 00N | 8 50W |
| **Bogie** r. Scotland | 9 | G2 | 57 25N | 2 50W |
| **Bognor Regis** England | 16 | C1 | 50 47N | 0 41W |
| **Bogotá** Colombia | 96 | C14 | 4 38N | 74 05W |
| **Böhmer Wald** mts. Germany | 52 | F4 | 49 00N | 13 00E |
| **Boise City** USA | 90 | F4 | 36 44N | 102 31W |
| **Boldon** England | 13 | E4 | 54 57N | 1 27W |
| **BOLIVIA** | 96D10/E10 | | | |
| **Bollington** England | 12 | D2 | 53 18N | 2 06W |
| **Bologna** Italy | 54 | E6 | 44 30N | 11 20E |
| **Bolsover** England | 13 | E2 | 53 15N | 1 26W |
| **Bolton** England | 12 | D2 | 53 35N | 2 26W |
| **Bolus Head** RoI | 21 | A1 | 51 45N | 10 10W |
| **Bolzano** Italy | 54 | D7 | 46 30N | 11 22E |
| **Bonar Bridge** tn. Scotland | 9 | E2 | 57 53N | 4 21W |
| **Bo'ness** Scotland | 11 | F3 | 56 01N | 3 37W |
| **Bonifacio** France | 50 | J1 | 41 23N | 9 10E |
| **Bonifacio, Strait of** France/Italy | 50 | J1 | 41 20N | 8 45E |
| **Bonn** Germany | 52 | C5 | 50 44N | 7 06E |
| **Bonnybridge** Scotland | 11 | F2 | 55 59N | 3 54W |
| **Bonnyrigg** Scotland | 11 | F2 | 55 52N | 3 08W |
| **Boolakennedy** RoI | 21 | C2 | 52 20N | 8 03W |
| **Bootle** Cumbria England | 12 | C2 | 54 17N | 3 23W |
| **Bootle** Merseyside England | 12 | C2 | 53 28N | 3 00W |
| **Bordeaux** France | 50 | D3 | 44 50N | 0 34W |
| **Bordon Camp** England | 16 | C2 | 51 07N | 0 53W |
| **Borneo** i. Indonesia/Malaysia | 71 | D3/F5 | 1 00N | 113 00E |
| **Bornholm** i. Denmark | 49 | D2 | 55 02N | 15 00E |
| **Boroughbridge** England | 13 | E3 | 54 05N | 1 24W |
| **Borris in Ossory** RoI | 21 | D2 | 52 57N | 7 37W |
| **Borrisokane** RoI | 21 | C2 | 53 00N | 8 08W |
| **Borrisoleigh** RoI | 21 | D2 | 52 45N | 8 00W |
| **Borrowash** England | 15 | E2 | 52 55N | 1 24W |
| **Borth** Wales | 14 | B2 | 52 29N | 4 03W |
| **Boscastle** England | 18 | C2 | 50 41N | 4 42W |
| **BOSNIA-HERZEGOVINA** | 55 | G6/H6 | | |
| **Boston** England | 13 | F1 | 52 29N | 0 01W |
| **Boston** USA | 91 | M5 | 42 20N | 71 05W |
| **Boston Spa** England | 13 | E2 | 53 54N | 1 21W |
| **Bothnia, Gulf of** Finland/Sweden | 49 | E3/F3 | 61 00N | 19 10E |
| **Botoşani** Romania | 53 | N3 | 47 44N | 26 41E |
| **BOTSWANA** | 82 | D3/E3 | | |
| **Bottrop** Germany | 52 | C5 | 51 31N | 6 55E |
| **Boulder** USA | 90 | E5 | 40 02N | 105 16W |
| **Boulogne-sur-Mer** France | 50 | E6 | 50 43N | 1 37E |
| **Bourg-en-Bresse** France | 50 | G4 | 46 12N | 5 13E |
| **Bourges** France | 50 | F4 | 47 05N | 2 23E |
| **Bourne** England | 15 | F2 | 52 46N | 0 23W |
| **Bourne End** England | 16 | C2 | 51 34N | 0 42W |
| **Bournemouth** England | 19 | F2 | 50 43N | 1 54W |
| **Bournemouth** u.a. England | 19 | F2 | 50 43N | 1 54W |
| **Bourton-on-the-Water** England | 15 | E1 | 51 53N | 1 46W |
| **Bovey** r. England | 18 | D2 | 50 40N | 3 50W |
| **Bovey Tracey** England | 18 | D2 | 50 36N | 3 40W |
| **Bowes** England | 12 | D3 | 54 30N | 2 01W |
| **Bowland, Forest of** England | 12 | D2/D3 | 54 00N | 2 40W |
| **Bowmore** Scotland | 10 | C2 | 55 45N | 6 17W |
| **Bowness-on-Windermere** England | 12 | D3 | 54 22N | 2 55W |
| **Boxtel** Neths | 48 | D4 | 51 36N | 5 20E |
| **Boyle** RoI | 20 | C3 | 53 58N | 8 18W |
| **Boyne** r. RoI | 20 | E3 | 53 45N | 6 25W |
| **Bozeman** USA | 90 | D6 | 45 40N | 111 00W |
| **Brabourne Lees** England | 17 | D2 | 51 08N | 1 00E |
| **Brackley** England | 15 | E2 | 52 02N | 1 09W |
| **Bracknell** England | 16 | C2 | 51 26N | 0 46W |
| **Bracknell Forest** u.a. England | 16 | C1 | 50 25N | 0 50W |
| **Bradford** England | 13 | E2 | 53 48N | 1 45W |
| **Bradford-on-Avon** England | 16 | A2 | 51 22N | 2 15W |
| **Bradwell-on-Sea** England | 17 | D2 | 51 44N | 0 54E |
| **Bradworthy** England | 18 | C2 | 50 54N | 4 29W |
| **Braemar** Scotland | 9 | F2 | 57 01N | 3 23W |
| **Braga** Portugal | 51 | A4 | 41 32N | 8 26W |
| **Bragança** Portugal | 51 | B4 | 41 47N | 6 46W |
| **Brahmaputra** r. India/Bangladesh | 67 | G5 | 26 40N | 93 00E |
| **Brăila** Romania | 53 | N2 | 45 17N | 27 58E |
| **Brailsford** England | 15 | E2 | 52 59N | 1 36W |
| **Braintree** England | 17 | D2 | 51 53N | 0 32E |
| **Brampton** England | 12 | D3 | 54 57N | 2 43W |
| **Brandenburg** Germany | 52 | F6 | 52 25N | 12 34E |
| **Brandon** Durham England | 13 | E3 | 54 46N | 1 38W |
| **Brandon** Suffolk England | 17 | D3 | 52 27N | 0 37E |
| **Brandon Bay** RoI | 21 | A2 | 52 15N | 10 05W |
| **Brandon Mountain** RoI | 21 | A2 | 52 10N | 10 10W |
| **Brandon Point** RoI | 21 | A2 | 52 15N | 10 05W |
| **Brant** r. England | 13 | F2 | 53 05N | 0 35W |
| **Brásília** Brazil | 96 | H10 | 15 45S | 47 57W |
| **Braşov** Romania | 53 | M2 | 45 39N | 25 35E |
| **Bratislava** Slovakia | 53 | H4 | 48 10N | 17 10E |
| **Braunau** Austria | 52 | F4 | 48 16N | 13 03E |
| **Braunschweig** Germany | 52 | E6 | 52 15N | 10 30E |
| **Braunton** England | 18 | C3 | 51 07N | 4 10W |
| **Bray** RoI | 20 | E3 | 53 12N | 6 06W |
| **Bray Head** RoI | 21 | A2 | 53 12N | 6 05W |
| **BRAZIL** | 96F11/H11 | | | |
| **Brazzaville** Congo | 82 | C7 | 4 14S | 15 14E |
| **Breaston** England | 15 | E2 | 52 56N | 1 12W |
| **Brechin** Scotland | 9 | G1 | 56 44N | 2 40W |
| **Brecon** Wales | 14 | C1 | 51 57N | 3 24W |
| **Breda** Neths | 48 | C4 | 51 35N | 4 46E |
| **Bregenz** Austria | 52 | D3 | 47 31N | 9 46E |
| **Bremen** Germany | 52 | D6 | 53 05N | 8 48E |
| **Bremerhaven** Germany | 52 | D6 | 53 33N | 8 35E |
| **Brent** England | 16 | C2 | 51 34N | 0 17W |
| **Brentwood** England | 17 | D2 | 51 38N | 0 18E |
| **Brescia** Italy | 54 | D6 | 45 33N | 10 13E |
| **Bressay** i. Scotland | 7 | C3 | 60 08N | 1 05W |
| **Brest** Belarus | 56 | B5 | 52 08N | 23 40E |
| **Brest** France | 50 | B5 | 48 23N | 4 30W |
| **Briançon** France | 50 | H3 | 44 53N | 6 39E |
| **Bridgend** Wales | 14 | C1 | 51 31N | 3 35W |
| **Bridgend** u.a. Wales | 14 | C1 | 51 32N | 3 35W |
| **Bridge of Dee** tn. Scotland | 9 | F1 | 56 59N | 3 20W |
| **Bridge of Don** tn. Scotland | 9 | G2 | 57 11N | 2 05W |
| **Bridge of Ericht** tn. Scotland | 9 | E1 | 56 44N | 4 23W |
| **Bridge of Orchy** tn. Scotland | 9 | E1 | 56 30N | 4 46W |
| **Bridge of Weir** tn. Scotland | 11 | E2 | 55 52N | 4 35W |
| **Bridgeport** USA | 91 | M5 | 41 12N | 73 12W |
| **Bridgetown** RoI | 21 | E2 | 52 14N | 6 33W |
| **Bridgnorth** England | 15 | D2 | 52 33N | 2 25W |
| **Bridgwater** England | 19 | D3 | 51 08N | 3 00W |
| **Bridgwater Bay** England | 19 | D3 | 51 15N | 3 20W |
| **Bridlington** England | 13 | F3 | 54 05N | 0 12W |
| **Bridlington Bay** England | 13 | F3 | 54 03N | 0 10W |
| **Bridport** England | 19 | E2 | 50 44N | 2 46W |
| **Brierfield** England | 12 | D2 | 53 50N | 2 14W |
| **Brig** Switz. | 52 | C3 | 46 19N | 8 00E |
| **Brigg** England | 13 | F2 | 53 34N | 0 30W |
| **Brighouse** England | 13 | E2 | 53 42N | 1 47W |
| **Brightlingsea** England | 17 | E2 | 51 49N | 1 02E |
| **Brighton** England | 16 | C1 | 50 50N | 0 10W |
| **Brighton and Hove** u.a. England | 16 | C1 | 50 51N | 0 09W |
| **Brignoles** France | 50 | H2 | 43 25N | 6 03E |
| **Bríndisi** Italy | 55 | G4 | 40 37N | 17 57E |
| **Brisbane** Australia | 73 | K5 | 27 30S | 153 00E |
| **Bristol** England | 19 | E3 | 51 27N | 2 35W |
| **Bristol** u.a. England | 19 | E3 | 51 30N | 2 38W |
| **Bristol Channel** UK | 18 | C1/D1 | 51 20N | 4 00W |
| **British Columbia** prov. Canada | 88 | K4/K5 | 56 50N | 125 30W |
| **British Isles** Europe | 6 | | | |
| **Briton Ferry** tn. Wales | 14 | C1 | 51 38N | 3 49W |
| **Brittas** RoI | 20 | E3 | 53 14N | 6 27W |
| **Brive-la-Gaillarde** France | 50 | E3 | 45 09N | 1 32E |
| **Brixham** England | 18 | D2 | 50 23N | 3 30W |
| **Brno** Czech Rep. | 53 | H4 | 49 13N | 16 40E |
| **Broad Bay** Scotland | 8 | C3 | 58 15N | 6 10W |
| **Broadford** RoI | 21 | C2 | 52 48N | 8 38W |
| **Broadford** Scotland | 8 | D2 | 57 14N | 5 54W |
| **Broad Haven** b. RoI | 20 | B4 | 54 18N | 9 55W |
| **Broadstairs** England | 17 | E2 | 51 22N | 1 27E |
| **Broadstone** England | 19 | E2 | 50 45N | 2 00W |
| **Broadway** England | 15 | E2 | 52 02N | 1 50W |
| **Brockenhurst** England | 16 | B1 | 50 49N | 1 34W |
| **Brodick** Scotland | 10 | D2 | 55 35N | 5 09W |
| **Bromley** England | 16 | D1 | 51 31N | 0 01E |
| **Bromsgrove** England | 15 | D2 | 52 20N | 2 03W |
| **Bromyard** England | 15 | D2 | 52 11N | 2 30W |
| **Brora** Scotland | 9 | F3 | 58 01N | 3 51W |
| **Brora** r. Scotland | 9 | E3 | 58 05N | 4 00W |
| **Broseley** England | 15 | D2 | 52 37N | 2 29W |
| **Brotton** England | 13 | F3 | 54 34N | 0 56W |
| **Brough** East Riding of Yorkshire England | 13 | F2 | 53 44N | 0 35W |
| **Brough** Cumbria England | 12 | D3 | 54 32N | 2 19W |
| **Brough Head** Scotland | 7 | A2 | 59 09N | 3 19W |
| **Broughshane** NI | 10 | C1 | 54 54N | 6 12W |
| **Broughton** Wales | 14 | D3 | 53 10N | 3 00W |
| **Broughton in Furness** England | 12 | C3 | 54 17N | 3 13W |
| **Broughty Ferry** tn. Scotland | 9 | G1 | 56 28N | 2 53W |
| **Brownhills** tn. England | 15 | E2 | 52 39N | 1 55W |
| **Broxburn** Scotland | 11 | F2 | 55 57N | 3 29W |
| **Bruay-en-Artois** France | 50 | F6 | 50 29N | 2 33E |
| **Brue** r. England | 19 | E3 | 51 10N | 2 50W |
| **Brugg** Switz. | 52 | D3 | 47 29N | 8 13E |
| **Brugge** Belgium | 48 | B4 | 51 13N | 3 14E |
| **BRUNEI** | 71 | E4/F4 | | |
| **Brunico** Italy | 54 | D7 | 46 47N | 11 57E |
| **Brunswick** USA | 91 | K3 | 31 09N | 81 30W |
| **Brussels** Belgium | 48 | C3 | 50 50N | 4 21E |
| **Bruton** England | 19 | E3 | 51 07N | 2 27W |
| **Bryher** i. England | 18 | A1 | 49 57N | 6 21W |
| **Brymbo** Wales | 14 | C3 | 53 05N | 3 03W |
| **Brynamman** Wales | 14 | C1 | 51 49N | 3 52W |
| **Brynmawr** Wales | 14 | C1 | 51 49N | 3 11W |
| **Bucharest** Romania | 53 | N2 | 44 25N | 26 07E |
| **Buchan Ness** Scotland | 9 | H2 | 57 28N | 1 47W |
| **Buckfastleigh** England | 18 | D2 | 50 29N | 3 46W |
| **Buckhaven** Scotland | 11 | F3 | 56 11N | 3 03W |
| **Buckie** Scotland | 9 | G2 | 57 40N | 2 58W |
| **Buckingham** England | 16 | C2 | 52 00N | 1 00W |
| **Buckinghamshire** co. England | 16 | C2 | 51 50N | 0 50W |
| **Buckley** Wales | 14 | C3 | 53 10N | 3 05W |
| **Budapest** Hungary | 53 | J3 | 47 30N | 19 03E |
| **Buddon Ness** Scotland | 9 | G1 | 56 28N | 2 45W |
| **Bude** England | 18 | C2 | 50 50N | 4 33W |
| **Bude Bay** England | 18 | C2 | 50 50N | 4 40W |
| **Budleigh Salterton** England | 19 | D2 | 50 38N | 3 20W |
| **Buenaventura** Colombia | 96 | B14 | 3 54N | 77 02W |
| **Buenos Aires** Argentina | 97 | F7 | 34 40S | 58 30W |
| **Buffalo** USA | 91 | L5 | 42 52N | 78 55W |
| **Builth Wells** Wales | 14 | C2 | 52 09N | 3 24W |
| **Bujumbura** Burundi | 83 | A3 | 3 22S | 29 19E |
| **Bukavu** CDR | 81 | L1 | 2 30S | 28 50E |
| **Bulawayo** Zimbabwe | 82 | E3 | 20 10S | 28 43E |
| **Bulford** England | 16 | B2 | 51 12N | 1 46W |
| **BULGARIA** | 55 | K5/M5 | | |
| **Bulkington** England | 15 | E2 | 52 28N | 1 25W |
| **Bull Point** England | 10 | C2 | 55 18N | 6 16W |
| **Bunclody** RoI | 21 | E2 | 52 38N | 6 40W |
| **Buncrana** RoI | 20 | D5 | 55 08N | 7 27W |
| **Bundoran** RoI | 20 | C4 | 54 28N | 8 17W |
| **Bungay** England | 17 | E3 | 52 28N | 1 26E |
| **Bunmahon** RoI | 21 | D2 | 52 08N | 7 23W |
| **Bunnahown** RoI | 20 | B3 | 53 24N | 9 47W |
| **Bunnyconnellan** RoI | 20 | B4 | 54 06N | 9 01W |
| **Buraydah** Saudi Arabia | 64 | D4 | 26 20N | 43 59E |
| **Bure** r. England | 17 | E3 | 52 47N | 1 20E |
| **Burgas** Bulgaria | 55 | M5 | 42 30N | 27 29E |
| **Burgess Hill** tn. England | 16 | C1 | 50 58N | 0 08W |
| **Burghead** tn. Scotland | 9 | F2 | 57 42N | 3 30W |
| **Burghead Bay** Scotland | 9 | F2 | 57 41N | 3 31W |
| **Burgh le Marsh** England | 13 | G2 | 53 10N | 0 15E |
| **Burgos** Spain | 51 | D5 | 42 21N | 3 41W |
| **BURKINA** | 80 | D4 | | |
| **Burley-in-Wharfedale** England | 13 | E2 | 53 55N | 1 45W |
| **Burnham Market** England | 17 | D3 | 52 57N | 0 44E |
| **Burnham-on-Crouch** England | 17 | D2 | 51 38N | 0 49E |
| **Burnham-on-Sea** England | 19 | D3 | 51 15N | 3 00W |
| **Burnley** England | 12 | D2 | 53 48N | 2 14W |
| **Burnmouth** Scotland | 11 | G2 | 55 50N | 2 04W |
| **Burnsall** England | 13 | E3 | 54 03N | 1 57W |
| **Burntisland** tn. Scotland | 11 | F3 | 56 03N | 3 15W |
| **Burntwood** England | 15 | E2 | 52 40N | 1 56W |
| **Burray** i. Scotland | 7 | B1 | 58 51N | 2 54W |
| **Burrow Head** Scotland | 11 | E1 | 54 41N | 4 24W |
| **Burry Port** Wales | 14 | B1 | 51 42N | 4 15W |
| **Burscough Bridge** tn. England | 12 | D2 | 53 37N | 2 51W |
| **Burton Agnes** England | 13 | F3 | 54 03N | 0 19W |
| **Burton Latimer** England | 15 | F2 | 52 22N | 0 40W |
| **Burton upon Trent** England | 15 | E2 | 52 48N | 1 36W |
| **BURUNDI** | 83 | A3/B3 | | |
| **Burwell** England | 17 | D2 | 52 16N | 0 19E |
| **Bury** England | 12 | D2 | 53 36N | 2 17W |
| **Bury St. Edmunds** England | 17 | D3 | 52 15N | 0 43E |
| **Bush** r. NI | 10 | C2 | 55 10N | 6 30W |
| **Bushey** England | 16 | C2 | 51 39N | 0 22W |
| **Bushmills** NI | 10 | C2 | 55 12N | 6 32W |
| **Bute** i. Scotland | 10 | D2 | 55 50N | 5 05W |
| **Bute, Sound of** Scotland | 10 | D2 | 55 45N | 5 10W |
| **Butler's Bridge** RoI | 20 | D4 | 54 03N | 7 22W |
| **Buttermere** l. England | 12 | C3 | 54 32N | 3 16W |
| **Buttevant** RoI | 21 | C2 | 52 14N | 8 40W |
| **Butt of Lewis** c. Scotland | 8 | C3 | 58 30N | 6 20W |
| **Buxton** England | 13 | E2 | 53 15N | 1 55W |
| **Bydgoszcz** Poland | 53 | H6 | 53 16N | 18 00E |
| **Byfield** England | 15 | E2 | 52 11N | 1 14W |
| **Bytom** Poland | 53 | J5 | 50 21N | 18 51E |

## C

| | | | | |
|---|---|---|---|---|
| **Cabo de Hornos** Chile | 97 | D2 | 56 00S | 67 15W |
| **Cabo Finisterre** Spain | 51 | A5 | 42 52N | 9 16W |
| **Cabrera** i. Spain | 51 | G3 | 39 00N | 2 59E |
| **Cáceres** Spain | 51 | B3 | 39 29N | 6 23W |
| **Cadair Idris** mt. Wales | 14 | C2 | 52 42N | 3 54W |
| **Cádiz** Spain | 51 | B2 | 36 32N | 6 18W |
| **Cádiz, Gulf of** Spain | 51 | B2 | 36 30N | 7 15W |
| **Caen** France | 50 | D5 | 49 11N | 0 22W |
| **Caerleon** Wales | 14 | D1 | 51 37N | 2 57W |
| **Caernarfon** Wales | 14 | B3 | 53 08N | 4 16W |
| **Caernarfon Bay** Wales | 14 | B3 | 53 05N | 4 30W |
| **Caerphilly** Wales | 14 | C1 | 51 35N | 3 14W |
| **Caerphilly** u.a. Wales | 14 | C1 | 51 31N | 3 5W |
| **Caersws** Wales | 14 | C2 | 52 31N | 3 26W |
| **Cágliari** Italy | 54 | C3 | 39 13N | 9 08E |
| **Caha Mountains** RoI | 21 | B1 | 51 40N | 9 40W |
| **Caher** RoI | 21 | D2 | 52 22N | 7 56W |
| **Caher Island** RoI | 20 | A3 | 53 43N | 10 01W |
| **Cahersiveen** RoI | 21 | A1 | 51 57N | 10 13W |
| **Cahore Point** RoI | 21 | E2 | 52 34N | 6 11W |
| **Cahors** France | 50 | E3 | 44 28N | 0 26E |
| **Cairn Gorm** mt. Scotland | 9 | F2 | 57 07N | 3 40W |
| **Cairngorms** mts. Scotland | 9 | F2 | 57 10N | 3 30W |
| **Cairnryan** Scotland | 10 | D2 | 54 58N | 5 02W |

**Cairns** Australia 73 J7 16 51S 145 43E
**Cairo** Egypt 81 M7 30 00N 31 15E
**Caister-on-Sea** England 17 E3 52 39N 1 44E
**Caistor** England 13 F2 53 30N 0 20W
**Calahorra** Spain 51 E5 46 19N 1 58W
**Calais** France 50 E6 50 57N 1 52E
**Calamocha** Spain 51 E4 40 54N 1 18W
**Calatayud** Spain 51 E4 41 21N 1 39W
**Caldew** *r.* England 12 C3 54 45N 3 00W
**Caldey Island** Wales 14 B1 51 38N 4 42W
**Caldicot** Wales 14 D1 51 36N 2 45W
**Caldwell** USA 90 C5 43 39N 116 40W
**Calgary** Canada 88 N4 51 05N 11 05W
**Cali** Colombia 96 B14 3 24N 76 30W
**California** *state* USA 90 C4 35 00N 119 00W
**Callan** RoI 21 D2 52 33N 7 23W
**Callander** Scotland 11 E3 56 15N 4 13W
**Callao** Peru 96 B11 12 05S 77 08W
**Calligarry** Scotland 8 D2 57 02N 5 58E
**Callington** England 18 C2 50 30N 4 18W
**Calne** England 16 A2 51 27N 2 00W
**Calverton** England 13 E2 53 03N 1 05W
**Calvi** France 50 J2 42 34N 8 44E
**Cam** *r.* England 17 D3 52 15N 0 11E
**Camberley** England 16 C2 51 21N 0 45W
**CAMBODIA** 71 C6/D6
**Camborne** England 18 B2 50 12N 5 19W
**Cambrai** France 50 F6 50 10N 3 14E
**Cambrian Mountains** Wales 14 C2 52 15N 3 45W
**Cambridge** England 17 D3 52 12N 0 07E
**Cambridge** USA 91 L4 38 34N 76 04W
**Cambridgeshire** *co.* England 16 C3 52 30N 0 00
**Camel** *r.* England 18 C2 50 31N 4 44W
**Camelford** England 18 C2C 50 37N 4 41W
**CAMEROON** 80 G3
**Campbeltown** Scotland 10 D2 55 26N 5 36W
**Campina Grande** Brazil 96 K12 7 15S 35 50W
**Campinas** Brazil 96 H9 22 54S 47 06W
**Campobasso** Italy 54 F4 41 33N 14 39E
**Campo Grande** Brazil 96 G9 20 24S 54 35W
**Campos** Brazil 96 J9 21 46S 41 21W
**Campsie Fells** Scotland 11 E3 56 00N 4 15W
**CANADA** 88/89
**Canadian** *r.* USA 90 F4 35 00N 104 00W
**Canary Islands** Spain 80 A7/B7 28 30N 15 10W
**Canaveral, Cape** USA 91 K2 28 28N 80 28W
**Canberra** Australia 73 J3 35 18S 149 08E
**Cancún** Mexico 94 G4 21 09N 86 45W
**Canisp** *mt.* Scotland 9 D3 58 07N 5 03W
**Canna** *i.* Scotland 8 C2 57 05N 6 35W
**Canna, Sound of** Scotland 8 C2 57 00N 6 30W
**Cannes** France 50 H2 43 33N 7 00E
**Cannich** Scotland 9 E2 57 12N 4 46W
**Cannington** England 19 D3 51 09N 3 04E
**Cannock** England 15 D2 52 42N 2 01W
**Cannock Chase** England 15 D2/E2 52 45N 2 04W
**Canonbie** Scotland 11 G2 55 05N 2 57W
**Canterbury** England 17 E2 51 17N 1 05E
**Canvey Island** England 17 D2 51 32N 0 33E
**Cap de la Hague** France 50 D5 49 44N 1 56W
**Cape Clear** RoI 21 B1 51 30N 9 30W
**Cape Cornwall** England 18 B2 50 07N 5 44W
**Capel** England 16 C2 51 10N 0 20W
**Cape Town** RSA 82 C1 33 56S 18 28E
**CAPE VERDE** 102
**Cape York Peninsula** Australia 73 H8 12 30S 142 30E
**Cap Ferret** *c.* France 50 D3 44 42N 1 16W
**Cappagh White** RoI 21 C2 52 40N 8 10W
**Cappamore** RoI 21 C2 52 37N 8 20W
**Cappoquin** RoI 21 D2 52 08N 7 50W
**Capri** *i.* Italy 54 F4 40 33N 14 15E
**Caprivi Strip** Namibia 82 D4 17 30S 27 50E
**Caracas** Venezuela 96 D16 10 35N 66 56W
**Carcassonne** France 50 F2 43 13N 2 21E
**Cardiff** Wales 14 C1 51 30N 3 13W
**Cardiff** *u.a.* Wales 14 C1 51 32N 3 14W
**Cardigan** Wales 14 B2 52 06N 4 40W
**Cardigan Bay** Wales 14 B2 52 30N 4 30W
**Cargenbridge** Scotland 11 F2 55 05N 3 41W
**Caribbean Sea** Central America 95 J3/L3 15 00N 75 00W
**Carlingford** RoI 20 E4 54 02N 6 11W
**Carlingford Lough** NI/RoI 10 C1 54 05N 6 10W
**Carlisle** England 12 D3 54 54N 2 55W
**Carlow** RoI 21 E2 52 50N 6 55W
**Carlow** *co.* RoI 21 E2 52 40N 6 55W
**Carloway** Scotland 8 C3 58 17N 6 48W
**Carlton in Lindrick** England 13 E2 53 22N 1 06W
**Carlton-on-Trent** England 13 F2 53 10N 0 48W
**Carluke** Scotland 11 F2 55 45N 3 51W
**Carmarthen** Wales 14 B1 51 51N 4 20W
**Carmarthen Bay** Wales 14 B1 51 40N 4 30W
**Carmarthenshire** *u.a.* Wales 14 B1 51 55N 4 00W
**Carmel Head** Wales 14 B3 53 24N 4 34W
**Carndonagh** RoI 20 D5 55 15N 7 15W
**Carn Eige** *mt.* Scotland 9 D2 57 22N 5 07W
**Carnforth** England 12 D3 54 08N 2 46W
**Carnlough** NI 10 D1 54 59N 5 59W
**Carnlough Bay** NI 10 D1 54 59N 5 57W
**Carnoustie** Scotland 9 G1 56 30N 2 44W
**Carnsore Point** RoI 21 E2 52 10N 6 22W
**Carpathians** *mts.* Europe 53 J4/M3 49 00N 22 00E
**Carpentras** France 50 G3 44 03N 5 03E
**Carradale** Scotland 10 D2 55 35N 5 28E
**Carrauntoohil** *mt.* RoI 21 B2 52 00N 9 45W
**Carrbridge** Scotland 9 F2 57 17N 3 49W
**Carrick** RoI 20 C4 54 39N 8 38W
**Carrickboy** RoI 20 D3 53 37N 7 41W
**Carrickfergus** NI 10 D1 54 43N 5 49W
**Carrickfergus** *district* NI 10 D1 54 45N 5 50W
**Carrickmacross** RoI 20 E3 53 58N 5 49W
**Carrick-on-Shannon** RoI 20 C3 53 57N 8 05W
**Carrick-on-Suir** RoI 21 D2 52 21N 7 25W
**Carrigaline** RoI 21 C1 51 48N 8 24W
**Carriganimmy** RoI 21 B1 51 57N 9 02W
**Carrigtohill** RoI 21 C2 51 55N 8 16W
**Carron** *r.* Scotland 9 E2 57 53N 4 30W
**Carronbridge** Scotland 11 F2 55 16N 3 48W
**Carrutherstown** Scotland 11 F2 55 04N 3 24W
**Carryduff** NI 10 D1 54 31N 5 53W
**Carsethorn** Scotland 11 F1 54 55N 3 35W
**Carsphairn** Scotland 11 E2 55 13N 4 16W
**Cartagena** Colombia 96 B16 10 24N 75 33W
**Cartagena** Spain 51 E2 37 36N 0 59W
**Carterton** England 16 B2 51 45N 1 35W
**Casablanca** Morocco 80 C8 33 39N 7 35W
**Cascade Range** N. America 90 B5/6 48 00N 121 00W
**Caserta** Italy 54 F4 41 04N 14 20E
**Cashel** RoI 21 D2 52 31N 7 53W
**Casper** USA 90 E5 42 50N 106 20W
**Caspian Sea** Asia 57 G3/H3 41 00N 50 00E
**Cassino** Italy 54 E4 41 29N 13 50E
**Castellane** France 50 H2 43 50N 6 30E
**Castellón de la Plana** Spain 51 E3 39 59N 0 03W
**Castelnaudary** France 50 E2 43 18N 1 57E
**Castelo Branco** Portugal 51 B3 39 50N 7 30W
**Castelvetrano** Italy 54 E2 37 41N 12 47E
**Castlebar** RoI 20 B3 53 52N 9 17W
**Castlebay** Scotland 8 B1 56 57N 7 28W
**Castleblaney** RoI 20 E4 54 07N 6 44W
**Castle Cary** England 19 E3 51 06N 2 31W
**Castlecomer** RoI 21 D2 52 48N 7 12W
**Castleconnell** RoI 21 C2 52 43N 8 30W
**Castledawson** NI 10 C1 54 47N 6 33W
**Castlederg** NI 10 B1 54 42N 7 36W
**Castledermot** RoI 21 E2 52 55N 6 50W
**Castle Donington** England 15 E2 52 51N 1 19W
**Castle Douglas** Scotland 11 F1 54 57N 3 56W
**Castlefin** RoI 20 D4 54 47N 7 35W
**Castleford** England 13 E2 43 44N 1 21W
**Castleisland** RoI 21 B2 52 14N 9 27W
**Castle Kennedy** Scotland 10 E1 54 54N 4 57W
**Castlemaine Harbour** RoI 21 B2 52 08N 9 33W
**Castlepollard** RoI 20 D3 53 41N 7 17W
**Castlerea** RoI 20 C3 53 46N 8 29W
**Castlereagh** *district* NI 10 D1 54 35N 5 55W
**Castletown** Isle of Man 12 B3 54 04N 4 38W
**Castletown** Scotland 9 F3 58 35N 3 23W
**Castletown Bearhaven** RoI 21 B2 51 39N 9 55W
**Castletownroche** RoI 21 C2 52 10N 8 28W
**Castlewellan** NI 10 D1 54 16N 5 57W
**Castres** France 50 F2 43 36N 2 14E
**Castrovillari** Italy 54 G3 39 48N 16 12E
**Catánia** Italy 54 F2 37 31N 15 06E
**Catanzaro** Italy 55 G3 38 54N 16 36E
**Caterham** England 16 C2 51 17N 0 04W
**Catlemartyr** RoI 21 C1 51 55N 8 03W
**Catrine** Scotland 11 E2 55 30N 4 20W
**Catshill** *tn.* England 15 D2 52 22N 2 03W
**Catterick Garrison** England 13 E3 54 23N 1 40W
**Caucasus** *mts.* Asia 57 F3/G3 43 00N 43 00E
**Cavan** RoI 20 D3 53 58N 7 21W
**Cavan** *co.* RoI 20 D3/4 54 02N 7 10W
**Cawood** England 13 E2 53 50N 1 07W
**Caxias** Brazil 96 J13 4 53N 43 20W
**Caxias do Sul** Brazil 97 G8 29 14S 51 10W
**Cayenne** French Guiana 96 G14 4 55N 52 18W
**Cedar Rapids** *tn.* USA 91 H5 41 59N 91 39W
**Cefalù** Italy 54 F3 38 03N 14 03E
**Cefn-mawr** Wales 14 C2 52 59N 3 02W
**Cellar Head** Scotland 8 C3 58 25N 6 11W
**Celle** Germany 52 E6 52 37N 10 05E
**Cemaes** Wales 14 B3 53 24N 4 27W
**Cemaes Head** Wales 14 B2 53 10N 4 50W
**CENTRAL AFRICAN REPUBLIC** 81 J3/K3
**Ceredigion** *u.a.* Wales 14 B2/C2 52 14N 4 00W
**Cergy-Pontoise** France 50 F5 49 02N 2 04E
**Cerignola** Italy 54 F4 41 16N 15 54E
**Cerne Abbas** England 19 E2 50 49N 2 29W
**Cesena** Italy 54 E6 44 09N 12 15E
**České Budějovice** Czech Rep. 52 G4 48 58N 14 29E
**Ceuta** *territory* Spain 51 C1 35 53N 5 19W
**Cévennes** *mts.* France 50 F3 44 20N 3 30E
**CHAD** 81 J4/K4
**Chadderton** England 12 D2 53 33N 2 08W
**Chad, Lake** West Africa 81 G4 13 50N 14 00E
**Chagford** England 18 D2 50 41N 3 50W
**Chalfont St. Giles** England 16 C2 51 39N 0 35W
**Chalkida** Greece 55 K3 38 28N 23 36E
**Chalkidikí** *p.* Greece 55 K4 40 30N 23 00E
**Chalmer** *r.* England 17 D2 51 54N 0 21E
**Châlons-sur-Marne** France 50 G5 48 58N 4 22E
**Chalon-sur-Saône** France 50 G4 46 47N 4 51E
**Chambéry** France 50 G3 45 34N 5 55E
**Chamonix** France 50 H3 45 55N 6 52E
**Champlain, Lake** USA 93 F2 45 00N 73 00W
**Chandigarh** India 66 D6 30 44N 76 54E
**Changchun** China 69 N6 43 53N 125 20E
**Chang Jiang** *r.* China 69 J4 31 00N 110 00E
**Changsha** China 69 K3 28 10N 113 00E
**Chaniá** Greece 55 L1 35 31N 24 01E
**Channel Islands** British Isles 19 E1 49 30N 2 30W
**Chapel-en-le-Frith** England 13 E2 53 20N 1 54W
**Chapel St. Leonards** England 13 G2 53 14N 0 20E
**Chapeltown** England 13 E2 53 28N 1 27W
**Chard** England 19 E2 50 53N 2 58W
**Charing** England 17 D2 51 13N 0 48E
**Charlbury** England 16 B2 51 53N 1 29W
**Charleroi** Belgium 48 C3 50 25N 4 27E
**Charleston** USA 91 K4 38 23N 81 40W
**Charlestown** RoI 20 C3 53 58N 8 47W
**Charlestown of Aberlour** Scotland 9 F2 57 28N 3 14W
**Charleville-Mézières** France 50 G5 49 46N 4 43E
**Charlotte** USA 91 K4 35 03N 80 50W
**Charlottesville** USA 91 L4 38 02N 78 29W
**Charlton Kings** England 15 D2 51 53N 2 03W
**Chartres** France 50 E5 48 27N 1 30E
**Châteaubriant** France 50 D4 47 43N 1 22W
**Châteaudun** France 50 E5 48 04N 1 20E
**Châteauroux** France 50 E4 46 49N 1 41E
**Château-Thierry** France 50 F5 49 03N 3 24E
**Châtellerault** France 50 E4 46 49N 0 33E
**Chatham** England 17 D2 51 23N 0 32E
**Chatham** Canada 89 U2 42 24N 82 11W
**Châtillon-sur-Seine** France 50 G4 47 52N 4 35E
**Chattanooga** USA 91 J4 35 02N 85 18W
**Chatteris** England 16 D3 52 27N 0 03E
**Chaumont** France 50 G3 48 07N 5 08E
**Cheadle** Greater Manchester England 12 D2 53 24N 2 13W
**Cheadle** Staffordshire England 15 E2 52 59N 1 59W
**Cheddar** England 19 E3 51 17N 2 46W
**Chelmsford** England 17 D2 51 44N 0 28E
**Cheltenham** England 15 D2 51 54N 2 04W
**Chelyabinsk** Russia 62 J8 55 12N 61 25E
**Chemnitz** Germany 52 F5 50 50N 12 55E
**Chengdu** China 69 H4 30 37N 104 06E
**Chennai** India 66 E2 13 05N 80 18E
**Chepstow** Wales 15 E1 51 39N 2 41W
**Cherbourg** France 50 D5 49 38N 1 37W
**Cherkasy** Ukraine 56 D4 49 27N 32 04E
**Chernivtsi** Ukraine 53 M4 48 19N 25 52E
**Chertsey** England 16 C2 51 23N 0 30W
**Cherwell** *r.* England 16 B2 51 55N 1 18W
**Chesapeake** USA 91 L4 36 45N 76 15W
**Chesham** England 16 C2 51 43N 0 38W
**Cheshire** *co.* England 12 D2 53 30N 2 30W
**Cheshire Plain** England 12 D2 53 12N 2 50W
**Cheshunt** England 16 C2 51 43N 0 02W
**Chesil Beach** England 19 E2 50 38N 2 50W
**Chester** England 12 D2 53 12N 2 54W
**Chesterfield** England 13 E2 53 15N 1 25W
**Chester-le-Street** England 13 E3 54 52N 1 34W
**Cheviot Hills** England/Scotland 11 G2 55 25N 2 30W
**Cheviot, The** *mt.* England 11 G2 55 29N 2 20W
**Chew** *r.* England 19 E3 51 21N 2 40W
**Chiba** Japan 70 D2 35 38N 140 07E
**Chicago** USA 91 J5 41 50N 87 45W
**Chichester** England 16 C1 50 50N 0 48W
**Chiclayo** Peru 96 B12 6 47S 79 47W
**Chieti** Italy 54 F5 42 21N 14 10E
**Chigwell** England 16 D2 51 37N 0 05E
**CHILE** 96/97 C5/C8
**Chillán** Chile 97 C6 36 37S 72 10W
**Chiltern Hills** England 16 B2/C2 51 40N 1 00W
**Chimborazo** *mt.* Ecuador 96 B13 1 29S 78 52W
**Chimbote** Peru 96 B12 9 04S 78 34W
**CHINA** 68/69
**Chinnor** England 16 C2 51 43N 0 56W
**Chioggia** Italy 54 E6 45 13N 12 17E
**Chíos** *i.* Greece 55 E3 38 00N 26 00E
**Chippenham** England 16 A2 51 28N 2 07W
**Chipping Campden** England 15 E2 52 03N 1 46W
**Chipping Norton** England 16 B2 51 56N 1 32W
**Chipping Ongar** England 17 D2 51 43N 0 15E
**Chipping Sodbury** England 15 D1 51 33N 2 24W
**Chirk** Wales 14 C2 52 66N 3 03W
**Chisinău** Moldova 53 P3 47 00N 28 50E
**Chita** Russia 63 P7 52 03N 113 35E
**Chittagong** Bangladesh 67 G4 22 20N 91 48E
**Chitterne** England 16 A2 51 11N 2 02W
**Chobham** England 16 C2 51 21N 0 37W
**Cholet** France 50 D4 47 04N 0 53W
**Chongqing** China 69 J3 29 30N 106 35E
**Chorley** England 12 D2 53 40N 2 38W
**Chorleywood** England 16 C2 51 40N 0 29W
**Chornobyl** Ukraine 53 Q5 51 17N 30 15E
**Christchurch** England 19 F2 50 44N 1 45W
**Christchurch** NZ 73 P2 43 32S 172 38E
**Chudleigh** England 18 D2 50 36N 3 45W
**Chudskoye Ozero** *l.* Russia/Estonia 56 C6 58 40N 27 30E
**Chulmleigh** England 18 D2 50 55N 3 52W
**Chum** *r.* England 15 D1 51 46N 1 58W
**Chur** Switz. 52 D3 46 52N 9 32E
**Churchill** Canada 89 S5 58 45N 94 00W
**Church Stoke** Wales 14 C2 52 32N 3 05W
**Church Stretton** England 14 D2 52 32N 2 49W
**Chute-aux-Outardes** Canada 93 G3 49 17N 67 57W
**Cincinnati** USA 91 K4 39 10N 83 30W
**Cinderford** England 15 D1 51 50N 2 29W
**Cirencester** England 15 E1 51 44N 1 59W
**Citlaltépetl** *mt.* Mexico 94 E3 19 00N 97 18W
**City of Edinburgh** *u.a.* Scotland 11 F2 55 55N 3 15W
**City of Kingston upon Hull** *u.a.* England 13 F2 53 45N 0 25W
**Ciudad Bolívar** Venezuela 96 E15 8 06N 63 36W
**Ciudadela** Spain 51 G3 40 00N 3 50E
**Ciudad Guayana** Venezuela 96 E15 8 22N 62 37W
**Ciudad Juárez** Mexico 94 C6 31 42N 106 29W
**Ciudad Obregón** Mexico 94 C5 27 28N 109 59W
**Ciudad Real** Spain 51 D3 38 59N 3 55W
**Ciudad Rodrigo** Spain 51 B4 40 36N 6 33W
**Civitavecchia** Italy 54 D5 42 05N 11 47E
**Clachan** Scotland 10 D2 55 45N 5 34W
**Clackmannan** Scotland 11 F3 56 05N 3 46W
**Clackmannanshire** *u.a.* Scotland 11 F3 56 10N 3 40W
**Clacton on Sea** England 17 E2 51 48N 1 09E
**Clane** RoI 20 E3 53 18N 6 41W
**Claonaig** Scotland 10 D2 55 46N 5 22W
**Clara** RoI 20 D3 53 20N 7 36W
**Clare** *r.* RoI 20 C3 53 20N 9 00W
**Clare** *co.* RoI 21 B2/C2 52 45N 9 00W
**Clarecastle** RoI 21 C2 52 49N 8 57W
**Claregalway** RoI 20 C3 53 21N 8 57W
**Clare Island** RoI 20 A3/B3 53 49N 10 00W
**Claremorris** RoI 20 C3 53 44N 9 00W
**Clarksville** USA 91 J4 36 31N 87 21W
**Clashmore** RoI 21 D2 52 00N 7 49W
**Clay Cross** England 13 E2 53 10N 1 24W
**Claydon** England 17 E3 52 06N 1 07E
**Clay Head** Isle of Man 12 B3 54 11N 4 25W
**Clayton-le-Moors** England 12 D2 53 47N 2 23W
**Clayton West** England 13 E2 53 36N 1 37W
**Clear Island** RoI 21 B1 51 26N 9 30W
**Clearwater** USA 91 K2 27 57N 82 48W
**Cleator Moor** England 12 C3 54 31N 3 30W
**Cleckheaton** England 13 E2 53 44N 1 43W
**Cleethorpes** England 13 F2 53 34N 0 02W
**Cleobury Mortimer** England 15 D2 52 23N 2 29W
**Clermont-Ferrand** France 50 F3 45 47N 3 05E
**Clevedon** England 19 E3 51 27N 2 51W
**Cleveland** USA 91 K5 43 15N 75 50W
**Cleveland Hills** England 13 E3 54 25N 1 15W
**Cleveleys** England 12 C2 53 53N 3 03W
**Clew Bay** RoI 20 B3 53 50N 9 50W
**Clifden** RoI 20 A3 53 29N 10 01W
**Clinton** USA 91 H5 41 51N 90 12W
**Clisham** *mt.* Scotland 8 C2 57 58N 6 50W
**Clitheroe** England 12 D2 53 53N 2 23W
**Cloghan** RoI 20 D3 53 13N 7 53W
**Clogheen** RoI 21 D2 52 17N 8 00W
**Clogher** NI 10 B1 54 25N 7 10W
**Clogherhead** RoI 20 E3 53 45N 6 15W
**Clogher Head** RoI 20 E3 53 48N 6 13W
**Cloghjordan** RoI 21 C2 52 57N 8 02W
**Clonakilty** RoI 21 C1 51 37N 8 54W
**Clonakilty Bay** RoI 21 C1 51 33N 8 50W
**Clonaslee** RoI 21 D3 53 09N 7 31W
**Clondalkin** RoI 20 E3 53 19N 6 24W
**Clonee** RoI 20 E3 53 25N 6 26W
**Clones** RoI 20 D4 54 11N 7 15W
**Clonmany** RoI 20 D5 55 16N 7 25W

**Clonmel** Rol 21 D2 52 21N 7 42W
**Clonroche** Rol 21 E2 52 27N 6 43W
**Clontarf** Rol 20 E3 53 22N 6 12W
**Cloondara** Rol 20 D3 53 43N 7 54W
**Cloppenburg** Germany 52 D6 52 52N 8 02E
**Cloughton** England 13 F3 54 20N 0 27W
**Clovelly** England 18 C2 51 00N 4 24W
**Clowne** England 13 E2 53 17N 1 15W
**Cluj-Napoca** Romania 53 L3 46 47N 23 37E
**Clwydian Range** Wales 14 C3 53 12N 3 20W
**Clydach** Wales 14 C1 51 42N 3 56W
**Clyde** *r.* Scotland 11 F2 55 40N 3 50W
**Clydebank** Scotland 11 E2 55 54N 4 24W
**Clyde River** *tn.* Canada 89 X8 70 30N 68 30W
**Coachford** Rol 21 C1 51 55N 8 47W
**Coalisland** NI 10 C1 54 32N 6 42W
**Coalville** England 15 E2 52 44N 1 20W
**Coast Ranges** USA 90 B4/6 41 00N 123 00W
**Coatbridge** Scotland 11 E2 55 52N 4 01W
**Cobh** Rol 21 C1 51 51N 8 17W
**Coburg** Germany 52 E5 50 15N 10 58E
**Cochin** India 66 D1 9 56N 76 15E
**Cockburnspath** Scotland 11 G2 55 56N 2 21W
**Cockenzie** Scotland 11 G2 55 58N 2 58W
**Cockermouth** England 12 C3 54 40N 3 21W
**Codó** Brazil 96 J13 4 28S 43 51W
**Codsall** England 15 D2 52 37N 2 12W
**Coggeshall** England 17 D2 51 52N 0 41E
**Cognac** France 50 D3 45 42N 0 19W
**Coimbra** Portugal 51 A4 40 12N 8 25W
**Colchester** England 17 D2 51 54N 0 54E
**Coldingham** Scotland 11 G2 55 53N 2 10W
**Coldstream** Scotland 11 G2 55 39N 2 15W
**Coleford** England 15 D1 51 48N 2 37W
**Coleraine** NI 10 C2 55 08N 6 40W
**Coleraine** *district* NI 10 C2 55 00N 6 40W
**Coleshill** England 15 E2 52 30N 1 42W
**Colgrave Sound** Scotland 7 D3 60 35N 0 55W
**Colintraive** Scotland 10 D2 55 56N 5 09W
**Coll** *i.* Scotland 8 C1 56 40N 6 35W
**Collin** Scotland 11 F2 55 04N 3 32W
**Collon** Rol 20 E3 53 47N 6 29W
**Collooney** Rol 20 C4 54 11N 8 29W
**Colmar** France 50 H5 48 05N 7 21E
**Colmonell** Scotland 10 E2 55 08N 4 55W
**Coln** *r.* England 15 E1 51 46N 1 50W
**Colne** England 12 D2 53 32N 2 09W
**Colne** *r.* England 17 D2 51 53N 0 45E
**Cologne** Germany 52 C5 50 56N 6 57E
**COLOMBIA** 96 C14
**Colombo** Sri Lanka 66 D1 6 55N 79 52E
**Colonsay** *i.* Scotland 10 C3 56 05N 6 15W
**Colorado** *r.* N. America 90 D3 33 00N 114 00W
**Colorado** *state* USA 90 E4/F4 39 00N 106 00W
**Colorado Springs** USA 90 F4 38 50N 104 50W
**Colsterworth** England 15 F2 52 48N 0 37W
**Coltishall** England 17 E3 52 44N 1 22E
**Columbia** *r.* N. America 90 B6 46 00N 120 00W
**Columbus** USA 91 K5 39 59N 83 03W
**Colwyn Bay** *tn.* Wales 14 C3 53 18N 3 43W
**Comacchio** Italy 54 E6 44 42N 12 11E
**Combe Martin** England 18 C3 51 13N 4 02W
**Comber** NI 10 D1 54 33N 5 45W
**Comeragh Mountains** Rol 21 D2 52 19N 7 40W
**Como** Italy 54 C6 45 48N 9 05E
**Comodoro Rivadavia** Argentina 97 D4 45 56S 67 30W
**COMOROS** 82 H5
**Compiègne** France 50 F5 49 25N 2 50E
**Comrie** Scotland 9 F1 56 23N 4 00W
**Conakry** Guinea 80 B3 9 30N 13 43W
**Concepción** Chile 97 C6 36 50S 73 03W
**Concordia** USA 91 G4 39 35N 97 39W
**Condom** France 50 E2 43 58N 0 23E
**Conflans-Ste. Honorine** France 50 F5 49 01N 2 09E
**Cong** Rol 20 B3 53 33N 9 16W
**Congleton** England 12 D2 53 10N 2 13W
**CONGO** 82 C7
**Congo** *r.* Congo/CDR 82 J2 2 00S 17 00E
**CONGO DEMOCRATIC REPUBLIC** 82 D7/E7
**Congresbury** England 19 E3 51 23N 2 48W
**Conisbrough** England 13 E2 53 29N 1 13W
**Coniston** England 12 C3 54 22N 3 05W
**Coniston Water** England 12 C3 54 20N 3 05W
**Connah's Quay** Wales 14 C3 53 13N 3 03W
**Connecticut** *state* USA 91 M5 41 00N 73 00W
**Conon Bridge** Scotland 9 E2 57 33N 4 26W
**Consett** England 13 E3 54 51N 1 49W
**Constanța** Romania 53 P2 44 12N 28 40E
**Constantine** Algeria 80 F9 36 22N 6 40W
**Conwy** Wales 14 C3 53 17N 3 50W
**Conwy** *r.* Wales 14 C3 53 10N 3 49W
**Conwy** *u.a.* Wales 14 C3 53 10N 3 40W
**Cookham** England 16 C2 51 34N 0 43W
**Cook, Mount** NZ 73 P2 43 36S 170 09E
**Cookstown** *district* NI 10 C1 54 35N 6 55W
**Coolgrange** Rol 21 D2 53 38N 7 08W
**Cootehill** Rol 20 D4 54 04N 7 05W
**Copeland Island** NI 10 D1 54 40N 5 31W
**Copenhagen** Denmark 49 D2 55 43N 12 34E
**Copiapo** Chile 97 C8 27 20S 70 23W
**Copper Harbor** USA 91 J6 47 28N 87 54W
**Coppull** England 12 D2 53 37N 2 40W
**Coquet** *r.* England 11 G2 55 20N 2 05W
**Coquimbo** Chile 97 C8 29 57S 71 25W
**Corbally** Rol 20 B4 54 10N 9 03W
**Corbeil-Essonnes** France 50 F5 48 36N 2 29E
**Corbridge** England 12 D3 54 58N 2 01W
**Corby** England 15 F2 52 49N 0 32W
**Cordillera Cantabrica** *mts.* Spain 51 B5/C5 43 00N 5 30W
**Córdoba** Argentina 97 E7 31 25S 64 11W
**Córdoba** Mexico 94 E3 18 55N 96 55W
**Córdoba** Spain 51 C2 37 53N 4 46W
**Corfe Castle** England 19 E2 50 38N 2 04W
**Corfu** Greece 55 H3 39 38N 19 55E
**Corinth** USA 91 J3 34 58N 88 30W
**Cork** Rol 21 C1 51 54N 8 28W
**Cork** *co.* Rol 21 C1 51 58N 8 40W
**Cork Harbour** Rol 21 C1 51 50N 8 10W
**Corraun Peninsula** Rol 20 B3 53 54N 9 51W
**Corrientes** Argentina 97 F8 27 30S 58 48W
**Corsewall Point** Scotland 10 D2 55 00N 5 10W
**Corsham** England 16 A2 51 26N 2 11W
**Corsica** *i.* France 50 J1/J2 42 00N 9 00E
**Corwen** Wales 14 C2 52 59N 3 22W
**Cosenza** Italy 54 G3 39 17N 16 16E
**COSTA RICA** 95 H1/2
**CÔTE D'IVOIRE** 80 C3/D3
**Cotgrave** England 15 E3 52 54N 1 03W
**Cotopaxi** *mt.* Ecuador 96 B13 0 40S 78 28W
**Cotswold Hills** England 15 D1/E1 51 40N 2 10W
**Cottbus** Germany 52 G5 51 43N 14 21E
**Cottenham** England 17 D3 52 18N 0 09E
**Coulommiers** France 50 F5 48 49N 3 05E
**Countesthorpe** England 15 E2 52 33N 1 09W
**Coupar Angus** Scotland 9 F1 56 33N 3 17W
**Courtown** Rol 21 E2 52 38N 6 13W
**Cove Bay** *tn.* Scotland 9 G2 57 05N 2 05W
**Coventry** England 15 E2 52 25N 1 30W
**Coverack** England 18 B2 50 01N 5 05W
**Cowbridge** Wales 14 C1 51 28N 3 27W
**Cowdenbeath** Scotland 11 F3 56 07N 3 21W
**Cowes** England 16 B1 50 45N 1 18W
**Cox's Bazar** Bangladesh 67 G4 21 25N 91 59E
**Craigavon** NI 10 C1 54 28N 6 25W
**Craigavon** *district* NI 10 C1 54 30N 6 30W
**Craigellachie** Scotland 9 F2 57 29N 3 12W
**Craighouse** Scotland 10 D2 55 51N 5 57W
**Craignure** Scotland 8 D1 56 28N 5 42W
**Crail** Scotland 11 G3 56 16N 2 38W
**Craiova** Romania 53 L2 44 18N 23 47E
**Cramlington** England 13 E4 55 05N 1 35W
**Cranleigh** England 16 C2 51 09N 0 30W
**Crathie** Scotland 9 F2 57 02N 3 12W
**Craughwell** Rol 20 C3 53 14N 8 44W
**Craven Arms** England 14 D2 52 26N 2 50W
**Crawford** Scotland 11 F2 55 28N 3 40W
**Crawley** England 16 C2 51 07N 0 12W
**Credenhill** England 14 D2 52 05N 2 48W
**Crediton** England 18 D2 50 47N 3 39W
**Cree** *r.* Scotland 11 E2 55 00N 4 30W
**Creeslough** Rol 20 D5 55 07N 7 55W
**Creetown** Scotland 11 E1 54 54N 4 23W
**Cregganbaun** Rol 20 B3 53 42N 9 51W
**Cremona** Italy 54 D6 45 08N 10 01E
**Creswell** England 13 E2 53 16N 1 12W
**Crêteil** France 50 F5 48 47N 2 28E
**Crewe** England 12 D2 53 05N 2 27W
**Crewkerne** England 19 E2 50 53N 2 48W
**Crianlarich** Scotland 9 E1 56 23N 4 37W
**Criccieth** Wales 14 B2 52 55N 4 14W
**Crickhowell** Wales 14 C1 51 53N 3 09W
**Cricklade** England 16 B2 51 39N 1 51W
**Crieff** Scotland 9 F1 56 23N 3 52W
**Crinan** Scotland 10 D3 56 06N 5 35W
**CROATIA** 55 G6
**Crofton** England 13 E2 53 39N 1 26W
**Cromarty** Scotland 9 E2 57 40N 4 02W
**Cromarty Firth** Scotland 9 E2 57 40N 4 20W
**Cromer** England 17 E3 52 56N 1 18E
**Crook** England 13 E3 54 43N 1 44W
**Crookhaven** Rol 21 B1 51 28N 9 43W
**Crookstown** Rol 21 C1 51 50N 8 50W
**Croom** Rol 21 C2 52 31N 8 43W
**Crosby** England 12 C2 53 30N 3 02W
**Cross Fell** *mt.* England 12 D3 54 43N 2 29W
**Crossgar** NI 10 D1 54 24N 5 45W
**Cross Hands** Wales 14 B1 51 48N 4 05W
**Crosshaven** Rol 21 C1 51 48N 8 17W
**Crossmaglen** NI 10 C1 54 05N 6 37W
**Crossmolina** Rol 20 B4 54 06N 9 20W
**Crouch** *r.* England 17 D2 51 38N 0 42E
**Crowborough** England 17 D2 51 03N 0 09E
**Crowland** England 16 C3 52 41N 0 11W
**Crowle** England 13 F2 53 37N 0 49W
**Crowthorne** England 16 C2 51 23N 0 49W
**Croyde** England 18 C3 51 08N 4 15W
**Croydon** England 16 C2 51 23N 0 06W
**Crumlin** NI 10 C1 54 37N 6 14W
**Crummock Water** England 12 C3 54 34N 3 18W
**Crusheen** Rol 21 C2 52 56N 8 53W
**Cruzeiro do Sul** Brazil 96 C12 7 40S 72 39W
**Crymych** Wales 14 B1 51 59N 4 40W
**CUBA** 95 J4
**Cuckfield** England 16 C2 51 00N 0 09W
**Cúcuta** Colombia 96 C15 7 55N 73 31W
**Cudworth** England 13 E2 53 35N 1 25W
**Cuenca** Ecuador 96 B13 2 54S 79 00W
**Cuenca** Spain 51 D4 40 04N 2 07W
**Cuiabá** Brazil 96 F10 15 32S 56 05W
**Cuillin Hills** Scotland 8 C2 57 15N 6 15W
**Culdaff** Rol 20 D5 55 17N 7 10W
**Cullen** Scotland 9 G2 57 41N 2 49W
**Cullompton** England 18 D2 50 52N 3 24W
**Cullybackey** NI 10 C1 54 53N 6 21W
**Culm** *r.* England 19 D2 50 55N 3 13W
**Culross** Scotland 11 F3 56 03N 3 35W
**Cumana** Venezuela 96 E16 10 29N 64 12W
**Cumbernauld** Scotland 11 E2 55 57N 4 00W
**Cumbria** *co.* England 12 C3/D3 54 35N 3 00W
**Cumbrian Mountains** England 12 C3/D3 54 30N 3 00W
**Cumnock** Scotland 11 E2 55 27N 4 16W
**Cuneo** Italy 54 B6 44 24N 7 33E
**Cupar** Scotland 9 F1 56 19N 3 01W
**Curitiba** Brazil 96 H8 25 25S 49 25W
**Cushendall** NI 10 C2 55 04N 6 04W
**Cushendun** NI 10 C2 55 07N 6 03W
**Cuxhaven** Germany 52 D6 53 52N 8 42E
**Cuzco** Peru 96 C11 13 32S 1 57W
**Cwmbran** Wales 14 C1 51 39N 3 00W
**CYPRUS** 57 D1/D2
**CZECH REPUBLIC** 52/53 G4/H4
**Częstochowa** Poland 53 J5 50 49N 19 07E

## D

**Dachau** Germany 52 E4 48 15N 11 26E
**Daingean** Rol 20 D3 53 17N 7 17W
**Dakar** Senegal 80 A4 14 38N 17 27W
**Dalbeattie** Scotland 11 F1 54 56N 3 49W
**Dalby** Isle of Man 12 B3 54 10N 4 44W
**Dalgety Bay** *tn.* Scotland 11 F3 56 03N 3 18W
**Dalian** China 69 M5 38 53N 121 37E
**Dalkeith** Scotland 11 F2 55 51N 3 04W
**Dallas** USA 91 G3 32 47N 96 48W
**Dalmally** Scotland 9 E1 56 24N 4 58W
**Dalmellington** Scotland 11 E2 55 19N 4 24W
**Dalry** Scotland 11 E2 55 43N 4 44W
**Dalton** USA 91 K3 34 46N 84 59W
**Dalton-in-Furness** England 12 C3 54 09N 3 11W
**Dalwhinnie** Scotland 9 E1 56 56N 4 14W
**Damascus** Syria 64 C5 33 30N 36 19E
**Danbury** England 17 E2 51 44N 0 33E
**Danville** USA 91 J5 40 09N 87 37W
**Dardanelles** *sd.* Turkey 57 C3 40 08N 26 10E
**Dar es Salaam** Tanzania 83 F1 6 51S 39 18E
**Darfield** England 13 E2 53 34N 1 22W
**Darjiling** India 67 F5 27 02N 88 20E
**Darling** *r.* Australia 73 H4 30 30S 144 00E
**Darlington** England 13 E3 54 31N 1 34W
**Darlington** *u.a.* England 13 E3 54 30N 1 30W
**Darmstadt** Germany 52 D4 49 52N 8 39E
**Dart** *r.* England 18 D2 50 30N 3 50W
**Dartford** England 17 D2 51 27N 0 13E
**Dartmouth** England 18 D2 50 21N 3 35W
**Darton** England 13 E2 53 36N 1 32W
**Darvel** Scotland 11 E2 55 37N 4 18W
**Darwen** England 12 D2 53 42N 2 28W
**Darwin** Australia 72 F8 12 23S 130 44E
**Daugava** *r.* Europe 49 F2 56 45N 24 30E
**Daugavpils** Latvia 49 G2 55 52N 26 31E
**Davao** Philippines 71 H5 7 05N 125 38E
**Davenport** USA 91 H5 41 32N 90 36W
**Daventry** England 15 E2 52 16N 1 09W
**Davis Strait** Canada/Greenland 89 Y7/Z6 69 00N 60 00W
**Davos** Switz. 52 D3 46 47N 9 50E
**Dawley** England 15 D2 52 40N 2 28W
**Dawlish** England 18 D2 50 35N 3 28W
**Dawros Head** Rol 20 C4 54 49N 8 34W
**Dead Sea** Israel/Jordan 64 C5 31 35N 35 30E
**Deal** England 17 E2 51 14N 1 24E
**Deel** *r.* Rol 21 C2 52 32N 8 57W
**Death Valley** USA 90 C4 36 00N 117 00W
**Deben** *r.* England 17 E3 52 02N 1 25E
**Debenham** England 17 E3 52 13N 1 11E
**Debrecen** Hungary 53 K3 47 30N 21 37E
**Decatur** USA 91 J4 39 51N 88 57W
**Deccan** *plat.* India 66 D3 18 00N 78 00E
**Deddington** England 16 B2 51 59N 1 19W
**Dee** *r.* England/Wales 14 C3 53 15N 3 10W
**Dee** *r.* Rol 20 E3 53 49N 6 38W
**Dee** *r.* Aberdeenshire Scotland 9 G2 57 05N 2 10W
**Dee** *r.* Dumfries and Galloway Scotland 11 E1 54 50N 4 00W
**Deel** *r.* Rol 20 D3 53 33N 7 07W
**Deer Sound** Scotland 7 B1 58 57N 2 47W
**Delaware** *state* USA 91 L4 39 00N 75 00W
**Delft** Neths 48 C5 52 00N 4 22E
**Delfzijl** Neths 48 E6 53 19N 6 56E
**Delhi** India 66 D5 28 40N 77 14E
**Delvin** Rol 20 D3 53 37N 7 05W
**Denbigh** Wales 14 C3 53 11N 3 25W
**Denbighshire** *u.a.* Wales 14 C3 53 05N 3 30W
**Denby Dale** *tn.* England 13 E2 53 35N 1 38W
**Den Helder** Neths 48 C5 52 58N 4 46E
**Denholme** England 13 E2 53 47N 1 53W
**Denison** USA 91 G3 33 47N 96 34W
**Denizli** Turkey 57 C2 37 46N 29 05E
**DENMARK** 49 C2/D2
**Denny** Scotland 11 F3 56 02N 3 55W
**Denpasar** Indonesia 71 F2 8 40S 115 14E
**Denton** England 12 D2 53 27N 2 07W
**Denver** USA 90 E4 39 45N 105 00W
**De Panne** Belgium 48 A4 51 06N 2 35E
**Derby** England 15 E2 52 55N 1 30W
**Derby City** *u.a.* England 15 E2 52 50N 1 30W
**Derbyshire** *co.* England 13 E2 52 55N 1 30W
**Derg** *r.* NI 10 B1 54 45N 7 30E
**Derriana Lough** Rol 21 A1/B1 51 54N 10 01W
**Derrybrien** Rol 21 C3 53 04N 8 37W
**Dersingham** England 17 D3 52 51N 0 30E
**Derwent** *r.* North Yorkshire England 13 F3 54 10N 0 35W
**Derwent** *r.* Cumbria England 12 C3 54 40N 3 30W
**Derwent** *r.* Derbyshire England 13 E2 53 15N 1 40W
**Derwent** *r.* Durham England 12 D3 54 50N 2 05W
**Derwent Reservoir** England 13 E2 53 25N 1 45W
**Derwent Water** England 12 C3 54 35N 3 09W
**Desborough** England 15 F2 52 27N 0 49W
**Desierto de Atacama** Chile 96 C9/D10 22 30S 70 00W
**Des Moines** USA 91 H5 41 35N 93 35W
**Dessau** Germany 52 F5 51 51N 12 15E
**Detroit** USA 91 K5 42 23N 83 05W
**Deveron** *r.* Scotland 9 G2 57 35N 2 35W
**Devil's Bridge** Wales 14 C2 52 23N 3 51W
**Devil's Lake** *tn.* USA 91 G6 48 03N 98 57W
**Devizes** England 16 B2 51 22N 1 59W
**Devon** *co.* England 18 D2 50 45N 3 50W
**Dewsbury** England 13 E2 53 42N 1 37W
**Dezfūl** Iran 65 E5 32 23N 48 28E
**Dhaka** Bangladesh 67 G4 23 42N 90 22E
**Didcot** England 16 B2 51 37N 1 15W
**Dieppe** France 50 E5 49 55N 1 05E
**Digby** Canada 89 X2 44 37N 65 47W
**Digne-les-Bains** France 50 H3 44 05N 6 14E
**Dijon** France 50 G4 47 20N 5 02E
**Dili** East Timor 71 H2 8 33S 125 34E
**Dinan** France 50 C5 48 27N 2 02W
**Dinara Planina** *mts.* Europe 55 G5 44 00N 17 00E
**Dinard** France 50 C5 48 38N 2 03W
**Dinas-Mawddwy** Wales 14 C2 52 43N 3 41W
**Dinas Powys** Wales 14 C1 51 25N 3 14W
**Dingle** Rol 21 A2 52 08N 10 15W
**Dingle Bay** Rol 21 A2 52 05N 10 15W
**Dingle Peninsula** Rol 21 A2 52 12N 10 15W
**Dingwall** Scotland 9 E2 57 35N 4 29W
**Dinnington** England 13 E2 53 22N 1 12W
**Dire Dawa** Ethiopia 81 P3 9 35N 41 50E
**Diss** England 17 E3 52 23N 1 06W
**DJIBOUTI** 81 P4
**Djibouti** Djibouti 81 P4 11 35N 43 11E
**Dnestr** *r.* Ukraine 56 C4 48 00N 27 30E
**Dnipro** *r.* Europe 57 D4 47 30N 33 00E
**Dnipropetrovs'k** Ukraine 56 D4 48 29N 35 00E
**Dobrich** Bulgaria 55 M5 43 34N 27 51E
**Dochart** *r.* Scotland 9 E1 56 25N 4 25W
**Docking** England 17 D3 52 55N 0 38E
**Dodekánisos** *is.* Greece 55 M2 37 00N 26 00E
**Dodman Point** England 18 C2 50 13N 4 48W
**Dodoma** Tanzania 83 D1 6 10S 35 40E
**Doha** Qatar 65 F4 25 15N 51 36E
**Dôle** France 50 G4 47 05N 5 30E
**Dolgellau** Wales 14 C2 52 44N 3 53W
**Dollar** Scotland 11 F3 56 09N 3 41W
**Dombas** Norway 49 C3 62 05N 9 07E
**DOMINICA** 94 Q8
**DOMINICAN REPUBLIC** 95 L3
**Don** *r.* England 13 E2 53 37N 1 02W
**Don** *r.* Russia 62 G7 50 00N 41 00E
**Don** *r.* Scotland 9 F2 57 15N 2 40W
**Donabate** Rol 20 E3 53 30N 6 09W
**Donaghadee** NI 10 D1 54 39N 5 33W
**Donau** *r.* Europe 52 D4 48 00N 16 00E
**Doncaster** England 13 E2 53 32N 1 07W
**Donegal** Rol 20 C4 54 39N 8 07W
**Donegal** *co.* Rol 20 C4/D4 54 55N 8 00W
**Donegal Bay** Rol 20 C4 54 30N 8 30W
**Donegal Point** Rol 21 B2 52 44N 9 38W
**Donets'k** Ukraine 56 E4 48 00N 37 50E
**Donnington** England 15 D2 52 43N 2 25W
**Donoughmore** Rol 21 C1 51 57N 8 45W
**Dooagh** Rol 20 A3 53 58N 10 07W
**Doon** *r.* Scotland 11 E2 55 20N 4 30W
**Doonbeg** Rol 21 B2 52 44N 9 32W
**Dorchester** England 19 E2 50 43N 2 26W
**Dorchester** England 16 B2 51 39N 1 10W

Dordogne *r.* France 50 E3 44 55N 0 30E
Dordrecht Neths 48 C4 51 48N 4 40E
Dorking England 16 C2 51 14N 0 20W
Dornie Scotland 8 D2 57 17N 5 31W
Dornoch Scotland 9 57 52N 4 02W
Dornoch Firth Scotland 9 E2/F2 57 55N 3 55W
Dorset *co.* England 19 E2 50 50N 2 20W
Dortmund Germany 52 C5 51 32N 7 27E
Douai France 50 F6 50 22N 3 05E
Douala Cameroon 80 F2 4 04N 9 43E
Douglas Isle of Man 12 B3 54 09N 4 29W
Douglas Scotland 11 F2 55 33N 3 51W
Douglas *r.* England 12 D2 53 38N 2 48W
Doulus Head Rol 21 A1 51 57N 10 19W
Doune Scotland 11 E3 56 11N 4 04W
Douro *r.* Portugal/Spain 51 B4 41 00N 8 30W
Dove *r.* England 13 E2 53 05N 1 55W
Dove Dale *v.* England 13 E2 53 05N 1 45W
Dover England 17 E2 51 08N 1 19E
Dover USA 91 L4 39 10N 75 32W
Dover, Strait of
  English Channel 17 E1/E2 51 00N 1 20W
Dovrefjell *mts.* Norway 49 C3 62 15N 9 10E
Down *district* NI 10 D1 54 25N 5 55W
Downham Market
  England 17 D3 52 36N 0 23E
Downpatrick NI 10 D1 54 20N 5 43W
Downpatrick Head Rol 20 B4 54 20N 9 15W
Downton England 16 B2 51 00N 1 44W
Dowra Rol 20 C4 54 11N 8 02W
Draguignan France 50 H2 43 32N 6 28E
Drakensberg *mts.* RSA 82 E1/2 30 00S 28 00E
Drammen Norway 49 D2 59 45N 10 15E
Draperstown NI 10 C1 54 48N 6 47W
Drava *r.* Europe 53 H3 46 00N 18 00E
Dreghorn Scotland 11 E2 55 37N 4 37W
Dresden Germany 52 F5 51 03N 13 45E
Drimoleague Rol 21 B1 51 40N 9 15W
Drobeta-Turnu-Severin
  Romania 53 L2 44 36N 22 39E
Drogheda Rol 20 E3 53 43N 6 21W
Droichead Nua Rol 20 E3 53 11N 6 45W
Droitwich England 15 D2 52 16N 2 09W
Dromcolliher Rol 21 C2 52 20N 8 55W
Dromore NI 10 B1 54 31N 7 28W
Dromore NI 10 C1 54 25N 6 09W
Dromore West Rol 20 C4 54 15N 8 53W
Dronfield England 13 E2 53 19N 1 27W
Drongan Scotland 11 E2 55 29N 4 30W
Drumkeeran Rol 20 C4 54 10N 8 08W
Drummore Scotland 10 E1 54 42N 4 54W
Drumnadrochit
  Scotland 9 E2 57 20N 4 30W
Drumshanbo Rol 20 C4 54 03N 8 02W
Drumsna Rol 20 D3 53 56N 8 00W
Dryden Canada 89 S3 49 48N 92 48W
Dubai UAE 65 G4 25 14N 55 17E
Dublin Rol 20 E3 53 20N 6 15W
Dublin *co.* Rol 20 E3 53 25N 6 10W
Dublin USA 91 K3 32 31N 82 54W
Dublin Bay Rol 20 E3 53 20N 6 05W
Dubrovnik Croatia 55 H5 42 40N 18 07E
Duddon *r.* England 12 C3 54 20N 3 20W
Dudley England 15 D2 52 30N 2 05W
Dudley England 13 E4 55 10N 1 20W
Duero *r.* Spain/Portugal 51 B4 41 25N 6 30W
Duffield England 15 E2 52 59N 1 29W
Dufftown Scotland 9 F2 57 26N 3 08W
Duisburg Germany 52 C5 51 26N 6 45E
Duleek Rol 20 E3 53 39N 6 25W
Dulnain *r.* Scotland 9 F2 57 20N 3 57W
Duluth USA 91 H6 46 45N 92 10W
Dulverton England 18 D3 51 03N 3 33W
Dumbarton Scotland 11 E2 55 57N 4 35W
Dumfries Scotland 11 F2 55 04N 3 37W
Dumfries and Galloway
  *u.a.* Scotland 11 E2/F2 55 10N 4 10W
Duna *r.* Hungary 53 J2/J3 46 00N 19 00E
Dunaff Head Rol 20 D5 55 17N 7 31W
Dunany Point Rol 20 E3 53 51N 6 14W
Dunav *r.* Europe 55 K5 45 00N 20 00E
Dunbar Scotland 11 G2 56 00N 2 31W
Dunbeath Scotland 9 F3 58 15N 3 25W
Dunblane Scotland 11 F3 56 12N 3 59W
Dunboyne Rol 20 E3 53 25N 6 29W
Duncannon Rol 21 E2 51 13N 6 55W
Duncansby Head
  Scotland 9 F3 58 39N 3 01W
Dunchurch England 15 E2 52 20N 1 16W
Dundalk Rol 20 E4 54 01N 6 25W
Dundalk Bay Rol 20 E3 53 55N 6 15W
Dundee Scotland 9 G1 56 28N 3 00W
Dundee City *u.a.*
  Scotland 9 F1/G1 56 30N 2 55W
Dundonald NI 10 D1 54 36N 5 48W
Dundrennan Scotland 11 F1 54 49N 3 57W
Dundrum Rol 20 E3 53 17N 6 15W
Dundrum Bay NI 10 D1 54 13N 5 45W
Dunedin NZ 73 P1 45 53S 170 30E
Dunfermline Scotland 11 F3 56 04N 3 29W
Dungannon NI 10 C1 54 31N 6 46W
Dungannon *district* NI 10 B1/C1 54 30N 7 05W
Dungarvan Rol 21 D2 52 05N 7 37W
Dungarvan Harbour
  Rol 21 D2 52 04N 7 35W
Dungeness *c.* England 17 D1 50 55N 0 58E
Dungiven NI 10 C1 54 55N 6 55W

Dunglow Rol 20 C4 54 57N 8 22W
Dungourney Rol 21 C1 51 58N 8 05W
Dunholme England 13 F2 53 18N 0 29W
Dunkeld Scotland 9 F1 56 34N 3 35W
Dunkerque France 50 F6 51 02N 2 23E
Dunkery Beacon *sum.*
  England 18 D3 51 11N 3 35W
Dún Laoghaire Rol 20 E3 53 17N 6 08W
Dunleer Rol 20 E3 53 50N 6 24W
Dunmanus Bay Rol 21 B1 51 30N 9 50W
Dunmanway Rol 21 B1 51 43N 9 06W
Dunmore Rol 20 C3 53 37N 8 44W
Dunmurry NI 10 C1 54 33N 6 00W
Dunnamaggan Rol 21 D2 52 29N 7 17W
Dunnet Scotland 9 F3 58 37N 3 21W
Dunnet Bay Scotland 9 F3 58 36N 3 21W
Dunnett Head Scotland 9 F3 58 40N 3 25W
Dunoon Scotland 10 E2 55 57N 4 56W
Duns Scotland 11 G2 55 47N 2 20W
Dunshaughlin Rol 20 E3 53 31N 6 33W
Dunstable England 16 C2 51 53N 0 32W
Dunster England 18 D3 51 11N 3 28W
Dunvegan Scotland 8 C2 57 26N 6 35W
Dunvegan Head
  Scotland 8 C2 57 30N 6 42W
Durban RSA 82 F2 29 53S 31 00E
Durham England 13 E3 54 47N 1 34W
Durham USA 91 L4 36 00N 78 54W
Durlston Head England 19 E2 50 35N 1 57W
Durness Scotland 9 E3 58 33N 4 45W
Durrington England 16 B2 51 12N 1 48W
Durrow Rol 21 D2 52 51N 7 23W
Dursey Head Rol 21 A1 51 20N 10 10W
Dursey Island Rol 21 A1 51 36N 10 12W
Dursley England 15 D1 51 42N 2 21W
Dushanbe Tajikistan 62 J4 38 38N 68 51E
Düsseldorf Germany 52 C5 51 13N 6 47E
Dyce Scotland 9 G2 57 12N 2 11W
Dyfi *r.* Wales 14 C2 52 38N 3 50W
Dyke *r.* Scotland 9 F3 58 25N 4 00W
Dymchurch England 17 D2 51 02N 1 00E
Dysart Scotland 11 F3 56 08N 3 08W

**E**

Eaglescliffe England 13 E3 54 31N 1 22W
Eaglesfield Scotland 11 F2 55 04N 3 12W
Ealing England 16 C2 51 31N 0 18W
Earby England 12 D2 53 56N 2 08W
Earls Colne England 17 D2 51 56N 0 42E
Earl Shilton England 15 E2 52 35N 1 19W
Earlston Scotland 11 G2 55 39N 2 40W
Earn *r.* Scotland 9 F1 56 20N 3 40W
Easington England 13 E3 54 47N 1 21W
Easington England 13 G2 53 40N 0 07E
Easingwold England 13 E3 54 07N 1 11W
Easky Rol 20 C4 54 17N 8 58W
East Ayrshire *u.a.*
  Scotland 11 E2 55 30N 4 30W
Eastbourne England 17 D1 50 46N 0 17E
East Burra *i.* Scotland 7 C3 60 04N 1 19W
East Cowes England 16 B1 50 45N 1 16W
East Dean England 17 D1 50 45N 0 13E
East Dereham England 17 D2 52 41N 0 56E
East Dunbartonshire
  *u.a.* Scotland 11 E2 55 55N 4 15W
Eastern Ghats *mts.*
  India 66 D2/E3 15 00N 80 00E
Eastfield England 13 F3 54 15N 0 25W
East Grand Forks USA 92 A3 47 56N 96 59W
East Ilsley England 16 B2 51 32N 1 17W
Eastington England 15 D1 51 45N 2 20W
East Kilbride Scotland 11 E2 55 46N 4 10W
East Leake England 15 E3 52 50N 1 11W
Eastleigh England 16 B1 50 58N 1 22W
East Linton Scotland 11 G2 55 59N 2 39W
East Loch Tarbert
  Scotland 8 C2 57 53N 6 48W
East London RSA 82 E1 33 00S 27 54E
East Lothian *u.a.*
  Scotland 11 G2 55 56N 2 42W
Eastmain Canada 89 V4 52 10N 78 30W
East Markham England 13 F2 53 15N 0 54W
Easton England 19 E2 50 32N 2 26W
East Renfrewshire *u.a.*
  Scotland 11 E2 55 45N 4 25W
East Riding of Yorkshire
  *u.a.* England 13 F2 53 54N 0 40W
Eastriggs Scotland 11 F1 54 59N 3 10W
East Suisnish Scotland 8 C2 57 21N 6 03W
East Sussex *co.*
  England 17 D1 50 55N 0 10E
East Timor 71 H2
East Wittering England 16 C1 50 41N 0 53W
Eastwood England 13 E2 53 01N 1 18W
Eaton Socon England 16 C3 52 13N 0 18W
Ebbw Vale Wales 14 C1 51 47N 3 12W
Ebro *r.* Spain 51 D5 43 00N 4 30W
Ecclefechan Scotland 11 F2 55 03N 3 17W
Eccles England 12 D2 53 29N 2 21W
Eccleshall England 15 D2 52 52N 2 15W
Eckington England 13 E2 53 19N 1 21W
ECUADOR 96 B13
Eday *i.* Scotland 7 B2 59 11N 2 47W
Eday Sound Scotland 7 B2 59 10N 2 43W
Ede Neths 48 D5 52 03N 5 40E
Eden *r.* Cumbria England 12 D3 54 50N 2 45W
Eden *r.* Kent England 17 D2 51 12N 0 06E

Eden *r.* Scotland 11 F3 56 20N 3 15W
Eden USA 91 L4 36 30N 79 46W
Edenbridge England 16 D2 51 12N 0 04E
Edenderry Rol 20 D3 53 21N 7 35W
Eden Mouth Scotland 9 G1 56 22N 2 36W
Edessa Greece 55 K4 40 48N 22 03E
Edgeworthstown Rol 20 D3 53 42N 7 36W
Edinburgh Scotland 11 F2 55 57N 3 13W
Edmonton Canada 88 N4 53 34N 113 25W
Eeklo Belgium 48 B4 51 11N 3 34E
Eems *r.* Neths/Germany 48 E6 53 25N 6 55E
Efyrnwy *r.* Wales 14 C2 52 40N 3 15W
Egham England 16 C2 51 26N 0 34W
Egilsay *i.* Scotland 7 B2 59 09N 2 56W
Eglinton NI 10 B2 55 01N 7 11W
Egremont England 12 C3 54 29N 3 33W
EGYPT 81 L7/M7
Eigg *i.* Scotland 8 C1 56 55N 6 10W
Eindhoven Neths 48 D4 51 26N 5 30E
Elat Israel 64 B4 29 33N 34 57E
Elâzığ Turkey 57 E2 38 41N 39 14E
Elbe *est.* Europe 52 E6 54 00N 9 00E
Elbeuf France 50 E5 49 17N 1 01E
Elbląg Poland 53 J7 54 10N 19 25E
El'brus *mt.* Russia 57 F3 43 21N 42 29E
Elburg Neths 48 D5 52 27N 5 50E
Elburz Mountains Iran 65 F6 36 15N 51 00E
Elche Spain 51 E3 38 16N 0 41W
El Dorado USA 91 H3 33 12N 92 40W
Elgin Scotland 9 F2 57 39N 3 20W
El Gîza Egypt 81 M7 30 01N 31 12E
Elgol Scotland 8 C2 57 09N 6 06W
Elie Scotland 11 G3 56 12N 2 50W
Elland England 13 E2 53 41N 1 50W
Ellen *r.* England 12 C3 54 45N 3 25W
Ellesmere England 14 D2 52 54N 2 54W
Ellesmere Island
  Canada 89 U9/X10 77 30N 82 30W
Ellesmere Port England 12 D2 53 17N 2 54W
Ellon Scotland 9 G2 57 22N 2 05W
Elmira USA 91 L5 42 06N 76 50W
El Minya Egypt 81 M7 28 06N 30 45E
Elmira USA 91 L5 42 06N 76 50W
El Obeid Sudan 81 M4 13 11N 30 10E
El Paso USA 90 E3 31 45N 106 30W
EL SALVADOR 94 G2
Ely England 17 D3 52 24N 0 16E
Elyria USA 91 K5 41 22N 82 06W
Embleton England 11 H2 55 30N 1 37W
Emden Germany 52 C6 53 23N 7 13E
Emmeloord Neths 48 D5 52 43N 5 46E
Emmen Neths 48 E5 52 47N 6 55E
Emsworth England 16 C1 50 51N 0 56W
Emvyale Rol 20 E4 54 21N 6 58W
Enard Bay Scotland 8 D3 58 10N 5 25W
Enfield England 16 C2 51 39N 0 05W
England *country* UK 53 00N 2 00W
English Channel
  UK/France 6 50 00N 2 00W
Enid USA 91 G4 36 24N 97 54W
Ennis Rol 21 C2 52 50N 8 59W
Enniscorthy Rol 21 E2 52 30N 6 34W
Enniskean Rol 21 C1 51 44N 8 56W
Enniskerry Rol 20 E3 53 12N 6 10W
Enniskillen NI 10 B1 54 21N 7 38W
Ennistimon Rol 21 B2 52 57N 9 15W
Enschede Neths 48 E5 52 13N 6 55E
Entebbe Uganda 83 C5 0 04N 32 27E
Épernay France 50 F5 49 02N 3 58E
Épinal France 50 H5 48 10N 6 28E
Epping England 17 D2 51 42N 0 08E
Epsom England 16 C2 51 20N 0 16W
Epworth England 13 F2 53 31N 0 50W
EQUATORIAL GUINEA 80 F2/G2
Erfurt Germany 52 E5 50 58N 11 02E
Erie USA 91 K5 42 07N 80 05W
Erie, Lake Canada/USA 93 D2 42 15N 81 00W
Eriskay *i.* Scotland 8 B2 57 05N 7 10W
ERITREA 81 P4 14 40N 40 15E
Erlangen Germany 52 E4 49 36N 11 02E
Errigal Mountain Rol 20 C5 55 02N 8 07W
Errill Rol 21 D2 52 52N 7 40W
Erris Head Rol 20 A4 54 20N 10 00W
Erzgebirge *mts.* Europe 52 F5 50 00N 13 00E
Erzurum Turkey 57 F2 39 57N 41 17E
Esbjerg Denmark 49 C2 55 20N 8 20E
Escondido USA 90 C3 33 07N 117 05W
Eşfahān Iran 65 F5 32 41N 51 41E
Esha Ness *c.* Scotland 7 C3 60 29N 1 37W
Esher England 16 C2 51 22N 0 22W
Esk *r.* North Yorkshire
  England 13 F3 54 25N 0 40W
Esk *r.* Cumbria England 12 C3 54 25N 3 10W
Esk *r.* Scotland/England 11 F2 55 10N 3 05W
Eskişehir Turkey 57 D2 39 46N 30 30E
Espoo Finland 49 F3 60 10N 24 40E
Esquel Argentina 97 C5 42 55S 71 20W
Essen Germany 52 C5 51 27N 6 57E
Essex *co.* England 17 D2 51 46N 0 30E
Eston England 13 E3 54 34N 1 07W
ESTONIA 49 G2
Estremoz Portugal 51 B3 38 50N 7 35W
Étampes France 50 F5 48 26N 2 10E
ETHIOPIA 81 N3/P3
Etna, Mount Italy 54 F2 37 45N 15 00E
Eton England 16 C2 51 31N 0 37W
Ettrick Scotland 11 F2 55 25N 3 05W
Ettrickbridge Scotland 11 G2 55 30N 2 58W

Eugene USA 90 B5 44 03N 123 04W
Euphrates *r.* Asia 64 D5 34 40N 42 00E
Europoort Neths 48 C4 51 55N 4 10E
Evanton Scotland 9 E2 57 40N 4 20W
Evercreech England 19 E3 51 09N 2 30W
Everest, Mount
  China/Nepal 67 F5 27 59N 86 56E
Evesham England 15 E2 52 06N 1 56W
Evora Portugal 51 B3 38 46N 7 41W
Evreux France 50 E5 49 03N 1 11E
Evvoia *i.* Greece 55 L3 38 00N 24 00E
Exe *r.* England 18 D3 51 08N 3 36W
Exeter England 18 D2 50 43N 3 31W
Exmouth England 18 D2 50 37N 3 25W
Eyam England 13 E2 53 24N 1 42W
Eye England 17 E3 52 19N 1 09E
Eyemouth Scotland 11 G2 55 25N 2 06W
Eye Peninsula Scotland 8 C3 58 10N 6 10W
Eyeries Rol 21 B1 51 41N 9 57W
Eynsham England 16 B2 51 46N 1 28W

**F**

Fairbourne Wales 14 B2 52 41N 4 03W
Fair Head NI 10 C2 55 15N 6 10W
Fair Isle Scotland 7 C2 59 32N 1 38W
Fairlight England 17 D1 50 52N 0 38E
Fairmont USA 91 K4 39 28N 80 08W
Faisalabad Pakistan 66 C6 31 25N 73 09E
Fakenham England 17 D3 52 50N 0 51E
Fal *r.* England 18 C2 50 16N 4 56W
Falcarragh Rol 20 C5 55 08N 8 06W
Falfurrias USA 91 G2 27 17N 98 10W
Falkirk Scotland 11 F2 55 59N 3 48W
Falkirk *u.a.* Scotland 11 F2 55 55N 3 45W
Falkland Scotland 11 F3 56 15N 3 13W
Falkland Islands
  S Atlantic Ocean 97 E2/F2 52 30S 60 00W
Falmouth England 18 B2 50 08N 5 04W
Fanad Head Rol 20 D5 55 17N 7 38W
Fane *r.* Rol 20 E3 53 58N 6 30W
Fano Italy 54 E5 43 51N 13 01E
Fareham England 16 B1 50 51N 1 10W
Fargo USA 91 G6 46 52N 96 49W
Faringdon England 16 B2 51 40N 1 35W
Farnborough England 16 C2 51 17N 0 46W
Farne Islands England 11 H2 55 38N 1 38W
Farnham England 16 C2 51 13N 0 49W
Farnworth
  England 16 C2 51 33N 0 38W
Farnworth England 12 D2 53 33N 2 24W
Faro Portugal 51 B2 37 01N 7 56W
Farranfore Rol 21 B2 52 10N 9 33W
Farrar *r.* Scotland 9 E2 57 15N 4 41W
Faughan *r.* NI 10 B1 54 54N 7 09W
Faversham England 17 D2 51 20N 0 53E
Fawley England 16 B1 50 49N 1 20W
Feale *r.* Rol 21 B2 52 26N 9 22W
Featherstone England 13 E2 53 41N 1 21W
Fécamp France 50 E5 49 45N 0 23E
FEDERATED STATES OF MICRONESIA
  100 F8
Feira de Santana Brazil 96 K11 12 17S 38 53W
Feldkirch Austria 52 D3 47 15N 9 38E
Felixstowe England 17 E2 51 58N 1 20E
Felling *tn.* England 13 E3 54 57N 1 33W
Felton England 11 H2 55 18N 1 42W
Feltwell England 17 D3 52 29N 0 33E
Fens, The England 16 C3/D3 52 45N 0 05E
Feolin Ferry *tn.*
  Scotland 10 C2 55 51N 6 06W
Ferbane Rol 20 D3 53 15N 7 49W
Fermanagh *district* NI 10 B1 54 25N 7 45W
Fermo Italy 54 E5 43 09N 13 44E
Fermoy Rol 21 C2 52 08N 8 16W
Ferndown England 19 F2 50 48N 1 55W
Ferns Rol 21 E2 52 35N 6 30W
Ferrara Italy 54 E6 44 50N 11 38E
Ferryhill *tn.* England 13 E3 54 41N 1 33W
Fès Morocco 80 C8 34 05N 5 00W
Fethaland, Point of
  Scotland 7 C3 60 35N 1 18W
Fethard Rol 21 D2 52 27N 7 41W
Fethard Rol 21 E2 52 11N 6 50W
Fetlar *i.* Scotland 7 D3 60 37N 0 52W
Fettercairn Scotland 9 G1 56 51N 2 34W
Ffestiniog Wales 14 C2 52 58N 3 55W
Fforest Fawr Wales 14 C1 51 50N 3 40W
Ffostrasol Wales 14 B2 52 06N 4 23W
Fianarantsoa
  Madagascar 82 J3 21 27S 47 05E
Fiddown Rol 21 D2 52 20N 7 19W
Fife *u.a.* Scotland 11 G3 56 10N 3 00W
Fife Ness *c.* Scotland 11 G3 56 17N 2 35W
Figeac France 50 F3 44 32N 2 01E
Figueira da Foz
  Portugal 51 A4 40 09N 8 51W
Figueres Spain 51 G5 42 16N 2 57E
FIJI 100 F8
Filey England 13 F3 54 12N 0 17W
Filton England 15 D1 51 31N 2 35W
Findhorn *r.* Scotland 9 F2 57 25N 3 55W
Findhorn Scotland 9 F2 57 39N 3 36W
Findhorn Bay Scotland 9 F2 57 39N 3 36W
Findochty Scotland 9 G2 57 42N 2 55W
Finglas Rol 20 E3 53 24N 6 18W
FINLAND 49 G3

**Finland, Gulf of** Finland 49 F2/G2 59 40N 23 30E
**Finnea** RoI 20 D3 53 47N 7 23W
**Fintagh Bay** RoI 20 C4 54 35N 8 33W
**Fintona** RoI 10 B1 54 30N 7 19W
**Fintown** RoI 20 C4 54 52N 8 07W
**Fionn Loch** Scotland 8 D2 57 45N 5 25W
**Fionnphort** Scotland 8 C1 56 19N 6 23W
**Firat** *r.* Asia 57 E2 37 30N 38 00E
**Firth of Clyde** Scotland 10 D2/E2 55 30N 5 00W
**Firth of Forth** Scotland 11 F3/G3 56 05N 3 00W
**Firth of Lorn** Scotland 10 D3 56 15N 6 00W
**Firth of Tay** Scotland 9 F1 56 25N 3 00W
**Fishbourne** England 16 B1 50 44N 1 12W
**Fishguard** Wales 14 B1 51 59N 4 59W
**Flagstaff** USA 90 D4 35 12N 111 38W
**Flamborough** England 13 F3 54 07N 0 07W
**Flamborough Head** England 13 F3 54 06N 0 04W
**Fleet** England 16 C2 51 16N 0 50W
**Fleetwood** England 12 C2 53 56N 3 01W
**Flekkefjord** Norway 49 C2 58 17N 6 40E
**Flimby** England 12 C3 54 41N 3 31W
**Flint** Wales 14 C3 53 15N 3 07W
**Flintshire** *u.a.* Wales 14 C3 53 15N 3 10W
**Flitwick** England 16 C3 52 00N 0 29W
**Florence** Italy 54 D5 43 47N 11 15E
**Florianópolis** Brazil 97 H8 27 35S 48 31W
**Florida** *state* USA 91 K2 28 00N 82 00W
**Florida Keys** *is.* USA 91 K1 25 00N 80 00W
**Flotta** *i.* Scotland 7 A1 58 49N 3 07W
**Fochabers** Scotland 9 F2 57 37N 3 05W
**Focşani** Romania 53 N2 45 41N 27 12E
**Fóggia** Italy 55 F4 41 28N 15 33E
**Folkestone** England 17 E2 51 05N 1 11E
**Folkingham** England 15 F2 52 54N 0 24W
**Fontainebleau** France 50 F5 48 24N 2 42E
**Fontstown** RoI 21 E3 53 03N 6 53W
**Fordham** England 17 D3 52 19N 0 24E
**Fordingbridge** England 16 B1 50 56N 1 47W
**Foreland** *p.* England 16 B1 50 41N 1 04W
**Foreland Point** England 18 D3 51 16N 3 47W
**Foreness Point** England 17 E2 51 25N 1 27E
**Forest** Channel Islands 19 E1 49 25N 2 35W
**Forest Row** England 16 D2 51 06N 0 02E
**Forfar** Scotland 9 G1 56 38N 2 54W
**Forlí** Italy 54 E6 44 13N 12 02E
**Formby** England 12 C2 53 34N 3 05W
**Formentera** *i.* Spain 51 F3 38 41N 1 30E
**FORMER YUGOSLAV REPUBLIC OF MACEDONIA (FYROM)** 55 J4/K4
**Formosa** Argentina 96 F8 26 07S 58 14W
**Forres** Scotland 9 F2 57 37N 3 38W
**Fortaleza** Brazil 96 K13 3 45S 38 35W
**Fort Augustus** Scotland 9 E2 57 09N 4 41W
**Fort George** Scotland 9 E2 57 35N 4 05W
**Forth** *r.* Scotland 11 E3 56 05N 4 05W
**Fort Lauderdale** USA 91 K2 26 08N 80 08W
**Fort Liard** Canada 88 L6 60 14N 123 28W
**Fortrose** Scotland 9 E2 57 34N 4 09W
**Fort Saskatchewan** Canada 88 N4 53 42N 113 12W
**Fortuneswell** England 19 E2 50 33N 2 27W
**Fort William** Scotland 9 D1 56 49N 5 07W
**Fort Worth** USA 91 G3 32 45N 97 20W
**Foula** *i.* Scotland 7 B3 60 08N 2 05W
**Foulness Island** England 17 D2 51 35N 0 57E
**Foulness Point** England 17 D2 51 37N 0 57E
**Fowey** England 18 C2 50 20N 4 38W
**Fowey** *r.* England 18 C2 50 27N 4 38W
**Foxford** RoI 20 B3 53 58N 9 08W
**Foyle** *r.* NI/RoI 20 D4 54 55N 7 25W
**Foynes** RoI 21 B2 52 37N 9 06W
**Foz** Spain 51 B5 43 34N 7 15W
**Framlingham** England 17 E3 52 13N 1 21E
**Frampton Cotterell** England 15 D1 51 32N 2 30W
**FRANCE** 50
**Francistown** Botswana 82 E3 21 11S 27 32E
**Frankfurt am Main** Germany 52 D5 50 06N 8 41E
**Frankfurt an der Oder** Germany 52 G6 52 20N 14 32E
**Fraser** *r.* Canada 88 L4 52 00N 122 00W
**Fraserburgh** Scotland 9 G2 57 42N 2 00W
**Freckleton** England 12 D2 53 46N 2 52W
**Frederick** USA 93 E1 39 25N 77 25W
**Fredericksburg** USA 91 L4 38 18N 77 30W
**Frederikshavn** Denmark 49 D2 57 28N 10 33E
**Fredrikstad** Norway 49 D2 59 20N 10 50E
**Freemount** RoI 21 C2 52 17N 8 53W
**Freeport** Canada 89 X2 44 17N 66 19W
**Freetown** Sierra Leone 80 B3 8 30N 13 17W
**Freiburg im Breisgau** Germany 52 C3 48 00N 7 52E
**Fréjus** France 50 H2 43 26N 6 44E
**French Guiana** *territory* France 96 G14 5 00N 53 00W
**Frenchpark** RoI 20 C3 53 53N 8 24W
**Freshwater** England 16 B1 50 40N 1 30E
**Fresno** USA 90 C4 36 41N 119 47W
**Fribourg** Switz. 52 C3 46 50N 7 10E
**Friedrichshafen** Germany 52 D3 47 39N 9 29E
**Frinton on Sea** England 17 E2 51 50N 1 14E
**Frodsham** England 12 D2 53 18N 2 44W
**Frogmore** England 16 C2 51 20N 0 50W
**Frome** England 19 E 51 14N 2 20W
**Frome** *r.* Herefordshire England 15 D2 52 05N 2 30W
**Frome** *r.* Somerset England 19 E3 51 10N 2 20W
**Frome** *r.* Dorset England 19 E2 50 42N 2 16W
**Fuday** *i.* Scotland 8 B2 57 05N 7 20W
**Fuenlabrada** Spain 51 D4 40 16N 3 49W
**Fuertaventura** *i.* Spain 80 B7 28 25N 14 00W
**Fuji** Japan 70 C2 35 10N 138 37E
**Fuji-san** *mt.* Japan 70 C2 35 23N 138 42E
**Fukuoka** Japan 70 B1 33 39N 130 21E
**Furnace** Scotland 10 D3 56 09N 5 12W
**Fürth** Germany 52 E4 49 28N 11 00E
**Fushun** China 69 M6 41 50N 123 54E
**Fuzhou** China 69 L3 26 09N 119 17E

# G

**GABON** 82 B7
**Gaborone** Botswana 82 E3 24 45S 25 55E
**Gadsden** USA 91 J3 34 00N 86 00W
**Gaeta** Italy 54 E4 41 13N 13 36E
**Gainsborough** England 13 F2 53 24N 0 48W
**Gairloch** *tn.* Scotland 8 D2 57 42N 5 40W
**Gairsay** *i.* Scotland 7 B2 59 05N 2 58W
**Galashiels** Scotland 11 G2 55 37N 2 49W
**Galaţi** Romania 53 P2 45 27N 28 02E
**Gallan Head** Scotland 8 B3 58 14N 7 01W
**Galley Head** RoI 21 C1 51 35N 8 55W
**Gallipoli** Italy 55 H4 40 03N 17 59E
**Gallup** USA 90 E4 32 32N 108 46W
**Galston** Scotland 11 E2 55 36N 4 24W
**Galveston** USA 91 H2 29 17N 94 48W
**Galway** *co.* RoI 20 B3/C3 53 25N 8 45W
**Galway** RoI 20 B3 53 16N 9 03W
**Galway Bay** RoI 20 B3 53 15N 9 15W
**Gambia** *r.* Senegal/The Gambia 80 B4 13 45N 13 15W
**GAMBIA, THE** 80 A4
**Gandía** Spain 51 E3 38 59N 0 11W
**Ganga** *r.* India 66 E5 25 00N 83 30E
**Gangdisê Shan** *mts.* China 68 D4 31 00N 82 30E
**Gap** France 50 H3 44 33N 6 05E
**Garda** Italy 54 D6 45 34N 10 43E
**Garden City** USA 90 F4 37 57N 100 54W
**Gardenstown** Scotland 9 G2 57 40N 2 20W
**Gare Loch** Scotland 11 E3 56 03N 4 39W
**Garelochhead** Scotland 11 E3 56 05N 4 50W
**Garforth** England 13 E2 53 48N 1 22W
**Garlieston** Scotland 11 E1 54 48N 4 22W
**Garonne** *r.* France 50 D3 44 45N 0 15E
**Garrison** NI 10 A1 54 25N 8 05W
**Garron Point** NI 10 D2 55 05N 6 00W
**Garstang** England 12 D2 53 55N 2 47W
**Garth** Wales 14 C2 52 08N 3 32W
**Garthorpe** England 13 F2 53 40N 0 42W
**Garve** Scotland 9 E2 57 37N 4 42W
**Gaspé** Canada 89 Y3 48 50N 64 30W
**Gastonia** USA 91 K4 35 14N 81 12W
**Gatehouse of Fleet** Scotland 11 E1 54 53N 4 11W
**Gateshead** England 13 E3 54 58N 1 35W
**Gatineau** Canada 93 E3 45 29N 75 40W
**Gatty Mountains** RoI 21 C2 52 21N 8 11W
**Gaydon** England 15 E2 52 45N 0 34E
**Gaza** Middle East 64 N10 31 30N 34 28E
**Gaziantep** Turkey 57 E2 37 04N 37 21E
**Gdańsk** Poland 53 J7 54 22N 18 41E
**Gdynia** Poland 53 J7 54 31N 18 30E
**Geleen** Neths 48 D3 50 58N 5 45E
**Gelligaer** Wales 14 C1 51 39N 3 15W
**Gelsenkirchen** Germany 52 C5 51 30N 7 05E
**Geneva** Switz. 52 C3 46 13N 6 09E
**Genk** Belgium 48 D3 50 58N 5 30E
**Genova** Italy 54 C6 44 24N 8 56E
**Gent** Belgium 48 B4 51 02N 3 42E
**Georgetown** Guyana 96 F15 6 46N 58 10W
**George Town** Malaysia 71 C5 5 25N 100 20E
**GEORGIA** 57 F3
**Georgia** *state* USA 91 K3 33 00N 83 00W
**Gera** Germany 52 F5 50 51N 12 11E
**GERMANY** 52
**Gerona** Spain 51 G4 41 59N 2 49E
**Gerrards Cross** England 16 C2 51 35N 0 34W
**Getafe** Spain 51 D4 40 18N 3 44W
**GHANA** 80 D3
**Giant's Causeway** NI 10 C2 55 10N 6 30W
**Gibraltar** *territory* U.K. 51 C2 36 09N 5 21W
**Gigha** *i.* Scotland 10 D2 55 40N 5 45W
**Gijón** Spain 51 C5 43 32N 5 40W
**Gila Bend** USA 90 D3 32 56N 112 42W
**Gilfach Goch** Wales 14 C1 51 38N 3 30W
**Gilford** NI 10 C1 54 23N 6 22W
**Gillingham** Medway England 17 D2 51 24N 0 33E
**Gillingham** Dorset England 19 E3 51 02N 2 17W
**Gilsland** England 12 D3 55 00N 2 35W
**Gironde** *r.* France 50 D3 45 30N 0 45W
**Girvan** Scotland 10 E2 55 15N 4 51W
**Givet** Belgium 48 C3 50 08N 4 49E
**Glanaman** Wales 14 C1 51 47N 3 55W
**Glanton** England 11 H2 55 26N 1 54W
**Glasgow** Scotland 11 E2 55 53N 4 15W
**Glasgow City** *u.a.* Scotland 11 E2 55 52N 4 16W
**Glass** *r.* Scotland 9 E2 57 15N 4 40W
**Glassan** RoI 20 D3 53 28N 7 52W
**Glasson** England 12 D2 54 00N 2 51W
**Glastonbury** England 19 E3 51 09N 2 43W
**Glenaffric** *r.* Scotland 8 D2 57 15N 5 15W
**Glenbeigh** RoI 21 B2 52 03N 9 56W
**Glencoe** Scotland 9 D1 56 40N 5 04W
**Glencolumbkille** RoI 20 C4 54 42N 8 44W
**Glenfinnan** Scotland 8 D1 56 53N 5 27W
**Glengad Head** RoI 20 D5 55 20N 7 11W
**Glengarriff** RoI 21 B1 51 45N 9 33W
**Glenluce** Scotland 11 E1 54 53N 4 49W
**Glenamaddy** RoI 20 C3 53 37N 8 35W
**Glenridding** England 12 D3 54 33N 2 58W
**Glenrothes** Scotland 11 F3 56 12N 3 10W
**Glen Shee** *r.* Scotland 9 F1 46 45N 3 25W
**Glenties** RoI 20 C4 54 47N 8 17W
**Glin** RoI 21 B2 52 34N 9 17W
**Glossop** England 13 E2 53 27N 1 57W
**Gloucester** England 15 D1 51 53N 2 14W
**Gloucestershire** *co.* England 15 D1/E1 51 50N 2 20W
**Glusburn** England 12 D2 53 55N 2 01W
**Glydar Fawr** *mt.* Wales 14 B3 53 05N 4 03W
**Glyncorrwg** Wales 14 C1 51 41N 3 38W
**Glyn-neath** Wales 14 C1 51 46N 3 38W
**Gniezno** Poland 53 H6 52 32N 17 32E
**Goat Fell** *mt.* Scotland 10 D2 55 39N 5 11W
**Gobi Desert** Mongolia 68/69 G6/J6 48 30N 100 00E
**Godalming** England 16 C2 51 11N 0 37W
**Godmanchester** England 16 C3 52 19N 0 11W
**Goiânia** Brazil 96 H10 16 43S 49 18W
**Goiás** Brazil 96 G10 15 57S 50 07W
**Golam Head** RoI 20 B3 53 14N 9 44W
**Golfo de California** Mexico 94 B5 27 00N 111 00W
**Golfo di Genova** Italy 54 C6 44 00N 9 00E
**Golfo di Táranto** Italy 55 G3/G4 40 00N 17 00E
**Golfo di Venézia** Italy 54 E6 45 00N 13 00E
**Golspie** Scotland 9 F2 57 58N 3 58W
**Gomera** *i.* Spain 80 A7 28 08N 17 14W
**Good Hope, Cape of** RSA 82 C1 34 30S 19 00E
**Goodland** USA 90 F4 39 20N 101 43W
**Goodwick** Wales 14 B1 52 00N 5 00W
**Goole** England 13 F2 53 42N 0 52W
**Gorebridge** Scotland 11 F2 55 51N 3 02W
**Gorey** RoI 21 E2 52 40N 6 18W
**Gorinchem** Neths 48 C4 51 50N 4 59E
**Goring** England 16 B2 51 32N 1 09W
**Gorizia** Italy 54 E6 45 57N 13 37E
**Gorseinon** Wales 14 B1 51 41N 4 02W
**Gort** RoI 21 C3 53 04N 8 50W
**Gorteen** RoI 20 C3 53 22N 8 35W
**Gortmore** RoI 20 B3 53 36N 9 36W
**Gorumna Island** RoI 20 B3 53 15N 9 55W
**Gosforth** England 13 E4 55 01N 1 37W
**Gosport** England 16 B1 50 48N 1 08W
**Göteborg** Sweden 49 D2 57 45N 12 00E
**Gotha** Germany 52 E5 50 57N 10 43E
**Gotland** *i.* Sweden 49 E2 57 30N 18 40E
**Göttingen** Germany 52 D5 51 32N 9 57E
**Gouda** Neths 48 C5 52 00N 4 42E
**Gourock** Scotland 11 E2 55 58N 4 49W
**Governador Valadares** Brazil 96 J10 18 49S 41 57W
**Gower** *p.* Wales 14 B1 51 35N 4 10W
**Gowran** RoI 21 D2 52 38N 7 04W
**Graemsay** *i.* Scotland 7 A1 58 56N 3 17W
**Grafham Water** England 16 C3 52 18N 0 20W
**Graiguenamanagh** RoI 21 E2 52 32N 6 56W
**Grain** England 17 D2 51 28N 0 43E
**Grampian Mountains** Scotland 9 E1/F1 56 45N 4 00W
**Granada** Spain 51 D2 37 10N 3 35W
**Granard** RoI 20 D3 53 47N 7 30W
**Gran Canaria** *i.* Spain 80 A7 28 00N 15 35W
**Grand Canal** RoI 20 D3 53 18N 7 15W
**Grand Canyon** USA 90 D4 36 04N 112 07W
**Grand Cayman** *i.* Caribbean Sea 95 H3 19 20N 81 15W
**Grand Rapids** *tn.* USA 91 J5 42 57N 86 40W
**Grange** RoI 20 C4 54 24N 8 31W
**Grangemouth** Scotland 11 F3 56 01N 3 44W
**Grange-over-Sands** *tn.* England 12 D3 54 33N 3 09W
**Grantham** England 15 F2 52 55N 0 39W
**Grantown-on-Spey** Scotland 9 F2 57 20N 3 58W
**Granville** France 50 D5 48 50N 1 35W
**Grasmere** England 12 C3 54 28N 3 02W
**Grasse** France 50 H2 43 40N 5 56E
**Grassington** England 12 E3 54 04N 1 59W
**Gravelines** France 48 A3 50 59N 2 08E
**Gravesend** England 17 D2 51 27N 0 24E
**Gray** France 50 G4 47 27N 5 35E
**Grays** England 17 D2 51 29N 0 20E
**Graz** Austria 52 G3 47 05N 15 22E
**Greasby** England 12 C2 53 23N 3 10W
**Great Bardfield** England 17 D2 51 57N 0 26E
**Great Bear Lake** Canada 88 L7/M7 66 00N 120 00W
**Great Broughton** England 13 E3 54 27N 1 110W
**Great Chesterford** England 17 D3 52 04N 0 11E
**Great Coates** England 13 F2 53 35N 0 08W
**Great Cumbrae** *i.* Scotland 10 E2 55 46N 4 55W
**Great Dividing Range** Australia 73 H8/J3 35 00N 148 00E
**Great Driffield** England 13 F3 54 01N 0 26W
**Great Dunmow** England 17 D2 51 53N 0 22E
**Great Eccleston** England 12 D2 53 51N 2 53W
**Greater Antilles** *is.* West Indies 95 H4/L3 19 00N 78 00W
**Greater London** *co.* England 16 C2 51 30N 0 10W
**Greater Manchester** *co.* England 12 D2 53 30N 2 15W
**Great Harwood** England 12 D2 53 48N 2 24W
**Great Malvern** England 15 D2 52 07N 2 19W
**Great Missenden** England 16 C2 51 43N 0 43W
**Great Ormes Head** Wales 14 C3 53 21N 3 53W
**Great Ouse** *r.* England 17 D3 52 40N 0 20E
**Great Salt Lake** USA 90 D5 41 10N 112 40W
**Great Shelford** England 17 D3 52 08N 0 09E
**Great Slave Lake** Canada 88 N6 62 00N 114 00W
**Great Torrington** England 18 C2 51 57N 4 09W
**Great Victoria Desert** Australia 72 E5/F5 28 00S 130 00E
**Great Whernside** *sum.* England 12 E3 54 10N 1 59W
**Great Wyrley** England 15 D2 52 27N 2 02W
**Great Yarmouth** England 17 E3 52 37N 1 44E
**GREECE** 55 J3/K3
**Greencastle** NI 10 C1 54 02N 6 06W
**Greenisland** *tn.* NI 10 D1 54 42N 5 52W
**GREENLAND** 89BB8/EE8
**Greenlaw** Scotland 11 G2 55 43N 2 28W
**Greenock** Scotland 11 E2 55 57N 4 45W
**Greenore** RoI 20 E4 54 02N 6 08W
**Greenore Point** RoI 21 E2 52 15N 6 18W
**Greenstone Point** Scotland 8 D2 57 55N 5 37W
**GRENADA** 94 R11
**Grenoble** France 50 G3 45 11N 5 43E
**Greta** *r.* England 12 D3 54 10N 2 30W
**Gretna** Scotland 11 F1 54 59N 3 04W
**Gretna Green** Scotland 11 F1 55 00N 3 04W
**Greystones** RoI 21 E3 53 09N 6 04W
**Grimethorpe** England 13 E2 53 34N 1 23W
**Grimsby** England 13 F2 53 35N 0 05W
**Groningen** Neths 48 E6 53 13N 6 35E
**Grosseto** Italy 54 D5 42 46N 11 07E
**Grove** England 16 B2 51 37N 1 26W
**Groznyy** Russia 62 G5 43 21N 45 42E
**Gruinard Bay** Scotland 8 D2 57 55N 5 30W
**Guadalajara** Mexico 94 D4 20 40N 103 20W
**Guadalajara** Spain 51 D4 40 37N 3 10W
**Guadalquivir** *r.* Spain 51 C2 37 45N 5 30W
**Guadeloupe** *i.* Lesser Antilles 94 Q9 16 30N 61 30W
**Guadiana** *r.* Spain/Portugal 51 B3 38 30N 7 30W
**GUAM** 100 E9
**Guangzhou** China 69 K2 23 08N 113 20E
**Guantánamo** Cuba 95 J4 20 09N 75 14W
**Guarda** Portugal 51 B4 40 32N 7 17W
**GUATEMALA** 94 F3
**Guatemala City** Guatemala 94 F2 14 38N 90 22W
**Guayaquil** Ecuador 96 A13 2 13S 79 54W
**Guéret** France 50 E4 46 10N 1 52E
**Guernsey** *i.* Channel Islands 19 E1 46 27N 2 35W
**Guide Post** England 13 E4 55 10N 1 36W
**Guildford** England 16 C2 51 14N 0 35W
**Guimarães** Portugal 51 A4 41 26N 8 19W
**GUINEA** 80 B4
**GUINEA-BISSAU** 80 A4/B4
**Guisborough** England 13 E3 54 32N 1 04W
**Guiseley** England 13 E2 53 53N 1 42W
**Guiyang** China 69 J3 26 35N 106 40E
**Gujranwala** Pakistan 66 C6 32 06N 74 11E
**Gulf, The** Middle East 65 F4 27 20N 51 00E
**Gullane** Scotland 11 G3 56 02N 2 50W
**Gunnislake** England 18 C2 50 31N 4 12W
**Gutcher** Scotland 7 C3 60 40N 1 00W
**Guthrie** USA 91 G4 35 53N 97 26W
**GUYANA** 96 F14
**Gweebarra Bay** RoI 20 C4 54 55N 8 30W
**Gweedore** RoI 20 C5 55 03N 8 14W
**Gwynedd** *u.a.* Wales 14 C2 52 45N 3 55W
**Györ** Hungary 53 H3 47 41N 17 40E

## H

Haarlem Neths 48 C5 52 23N 4 39E
Hacketstown RoI 21 E2 52 52N 6 33W
Haddington Scotland 11 G2 55 58N 2 47W
Hadleigh England 17 D3 52 02N 0 57E
Hadleigh England 17 D2 51 34N 0 36E
Hadley England 15 D2 52 43N 2 28W
Hagen Germany 52 C5 51 22N 7 27E
Hagley England 15 D2 52 26N 2 08W
Hagondange France 50 H5 49 16N 6 11E
Hags Head RoI 21 B2 52 56N 9 30W
Haguenau France 50 H5 48 49N 7 47E
Haifa Israel 64 B5 32 49N 34 59E
Hailsham England 17 D1 50 52N 0 16E
Hainan Dao i. China 69 J1/K1 18 50N 109 50E
Hai Phong Vietnam 69 J2 20 50N 106 41E
HAITI 95 K3
Hakodate Japan 70 D3 41 46N 140 44E
Hale England 12 D2 53 22N 2 20W
Halesowen England 15 D2 52 26N 2 05W
Halesworth England 17 E3 52 21N 1 30E
Halifax England 13 E2 53 44N 1 52W
Halifax Canada 89 Y2 44 38N 63 35W
Halkirk Scotland 9 F3 58 30N 3 30W
Halladale r. Scotland 9 F3 58 30N 3 55W
Halle Germany 52 E5 51 28N 11 58E
Hallein Austria 52 F3 47 41N 13 06E
Halmstad Sweden 49 D2 56 41N 12 55E
Halstead England 17 D2 51 57N 0 38E
Haltwhistle England 12 D3 54 58N 2 27W
Hamäh Syria 64 C6 35 10N 36 45E
Hamamatsu Japan 70 C1 34 42N 137 42E
Hamble-le-Rice England 16 B1 50 52N 1 19W
Hambleton Hills England 13 E3 54 15N 1 15W
Hamburg Germany 52 E6 53 33N 10 00E
Hämeenlinna Finland 49 F3 61 00N 24 25E
Hamilton Canada 89 U2 43 15N 79 50W
Hamilton NZ 73 Q3 37 47S 175 17E
Hamilton Scotland 11 E2 55 47N 4 03W
Hamm Germany 52 C5 51 40N 7 49E
Hammerfest Norway 49 F5 70 40N 23 44E
Hampshire co. England 16 B2 51 10N 1 15W
Hampshire Downs England 16 B2 51 15N 1 10W
Hamstreet England 17 D2 51 05N 0 52E
Hangzhou China 69 M4 30 18N 120 07E
Hanko Finland 49 F2 50 50N 23 00E
Hanley England 12 D2 53 01N 2 10W
Hannover Germany 52 D6 52 23N 9 44E
Hanoi Vietnam 69 J2 21 01N 105 52E
Harare Zimbabwe 82 F4 17 50S 31 03E
Harbin China 69 N7 45 45N 126 41E
Hargeysa Somalia 81 P3 9 31N 44 02E
Harlech Wales 14 B2 52 52N 4 07W
Harleston England 17 E3 52 24N 1 18E
Harlingen Neths 48 D6 53 10N 5 25E
Harlow England 17 D2 51 47N 0 08E
Haroldswick Scotland 7 D3 60 47N 0 50W
Harpenden England 16 C2 51 49N 0 22W
Harris i. Scotland 8 C2 57 50N 6 55W
Harris, Sound of Scotland 8 B2 57 40N 9 12W
Harrogate England 13 E2 54 00N 1 33W
Harrow England 16 C2 51 34N 0 20W
Hartford USA 91 M5 41 00N 72 00W
Hartington England 13 E2 53 09N 1 48W
Hartland England 18 C2 50 59N 4 29W
Hartland Point England 18 C3 51 02N 4 31W
Hartlepool England 13 E3 54 41N 1 13W
Hartlepool u.a. England 13 E3 54 41N 1 13W
Hartshill England 15 E2 52 33N 1 30W
Harwell England 16 B2 51 37N 1 18W
Harwich England 17 E2 51 57N 1 17E
Harworth England 13 E2 53 26N 1 04W
Harz mts. Europe 52 E5 51 40N 10 00E
Hascosay i. Scotland 7 D3 60 36N 1 00W
Haslemere England 16 C2 51 06N 0 43W
Haslingden England 12 D2 53 43N 2 18W
Hasselt Belgium 48 D3 50 56N 5 20E
Hastings England 17 D1 50 51N 0 36E
Hatfield South Yorkshire England 13 F2 53 36N 0 59W
Hatfield Hertfordshire England 16 C2 51 46N 0 13W
Hatherleigh England 18 C2 50 49N 4 04W
Hathersage England 13 E2 53 19N 1 38W
Hatton Scotland 9 H2 57 25N 1 55W
Haugesund Norway 49 C2 59 25N 5 16E
Haut Atlas mts. Morocco 80 C8 30 45N 6 50W
Havana Cuba 95 H4 23 07N 82 25W
Havant England 16 C1 50 51N 0 59W
Haverfordwest Wales 14 B1 51 49N 4 58W
Haverhill England 17 D3 52 05N 0 26E
Havering England 17 D2 51 34N 0 14E
Hawaiian Islands Pacific Ocean 100 J10/L9 25 00N 166 00W
Hawarden Wales 14 C3 53 11N 3 02W
Hawes England 12 D3 54 18N 2 12W
Hawick Scotland 11 G2 55 25N 2 47W
Hawkesbury Canada 89 W3 45 36N 74 38W
Hawkhurst England 17 D2 51 03N 0 30E
Haworth England 13 E2 53 50N 1 57W
Haxby England 13 E3 54 01N 1 05W

Haydon Bridge tn. England 12 D3 54 58N 21 4W
Hayle England 18 B2 50 10N 5 25W
Hay-on-Wye Wales 14 C2 52 04N 3 07W
Haywards Heath England 16 C2 51 00N 0 06W
Hazel Grove England 12 D2 53 23N 2 08W
Hazelmere England 16 C2 51 38N 0 44W
Heacham England 17 D3 52 55N 0 30E
Headford RoI 20 B3 53 28N 9 06W
Heanor England 13 E2 53 01N 1 22W
Heathfield England 17 D1 50 58N 0 15E
Hebden Bridge England 12 D2 53 45N 2 00W
Hebi China 69 K5 35 57N 114 08E
Heckington England 15 F2 52 59N 0 18W
Heckmondwike England 13 E2 53 43N 1 45W
Hedge End England 16 B1 50 55N 1 18W
Hednesford England 15 D2 52 43N 2 00W
Hedon England 13 F2 53 44N 0 12W
Heerenveen Neths 48 D5 52 57N 5 55E
Heerlen Neths 48 D3 50 53N 5 59E
Hefei China 69 L4 31 55N 117 18E
Heidelberg Germany 52 D4 49 25N 8 42E
Heighington England 13 F2 54 36N 1 37W
Heilbronn Germany 52 D4 49 08N 9 14E
Heisker is. Scotland 8 B2 57 30N 7 40W
Hekla mt. Iceland 49 L6 64 00N 19 41W
Helchteren Belgium 48 D4 51 03N 5 23E
Helena USA 90 D6 46 35N 112 00W
Helen's Bay tn. NI 10 D1 54 41N 5 50W
Helensburgh Scotland 11 E3 56 01N 4 44W
Hellifield England 12 D3 54 01N 2 12W
Helmsdale r. Scotland 9 F3 58 10N 3 50W
Helmsdale Scotland 9 F3 58 07N 3 40W
Helmsley England 13 E3 54 14N 1 04W
Helsingborg Sweden 49 D2 56 03N 12 43E
Helsinki Finland 49 F3 60 08N 25 00E
Helston England 18 B2 50 05N 5 16W
Helvellyn mt. England 12 C3 54 32N 3 02W
Hemel Hempstead England 16 C2 51 46N 0 28W
Hemsworth England 13 E2 53 38N 1 21W
Henfield England 16 C1 50 56N 0 17W
Hengelo Neths 48 E5 52 16N 6 46E
Henley-in-Arden England 15 E2 52 17N 1 46W
Henley-on-Thames England 16 C2 51 32N 0 56W
Herbertstown RoI 21 C2 52 32N 8 28W
Hereford England 14 C2 52 04N 2 43W
Herefordshire co. England 14/15 C2 52 10N 2 45W
Herentals Belgium 48 C4 51 11N 4 50E
Herm i. Channel Islands 19 E1 49 28N 2 27W
Herne Germany 48 F4 51 32N 7 12E
Herne Bay tn. England 17 E2 51 23N 1 08E
Herning Denmark 49 C2 56 08N 8 59E
Hertfordshire co. England 16 C2 51 50N 0 05W
Hessle England 13 F2 53 44N 0 26W
Hetton-le-Hole England 13 E3 54 50N 1 27W
Hexham England 12 D3 54 58N 2 06W
Heysham England 12 D3 54 02N 2 54W
Heywood England 12 D2 53 36N 2 13W
Higham Ferrers England 15 F2 52 18N 0 36W
High Bentham England 12 D3 54 08N 2 30W
Higher Walton England 12 D2 53 45N 2 37W
Highland u.a. Scotland 8/9 D2/E2 57 00N 5 00W
Highworth England 16 B2 51 38N 1 43W
High Wycombe England 16 C2 51 38N 0 46W
Hildesheim Germany 52 D6 52 09N 9 58E
Hillingdon England 16 C2 51 32N 0 27W
Hillsborough NI 10 C1 54 28N 6 05W
Hilversum Neths 48 D5 52 14N 5 10E
Himalaya mts. Asia 66/67 D6/G5 28 00N 85 00E
Hinckley England 15 E2 52 33N 1 21W
Hindhead England 16 C2 51 07N 0 44W
Hindley England 12 D2 53 32N 2 35W
Hindu Kush mts. Afghanistan 66 B7/C7 35 00N 70 00E
Hiroshima Japan 70 B1 34 23N 132 27E
Hirwaun Wales 14 C1 51 45N 3 30W
Histon England 17 D3 52 15N 0 06E
Hitachi Japan 70 D2 36 35N 140 40E
Hitchin England 16 C2 51 57N 0 17W
Hobart Australia 73 J2 42 54S 147 18E
Hồ Chi Minh Vietnam 71 D6 10 46N 106 43E
Hockley England 17 D2 51 35N 0 39E
Hodder r. England 12 D2 53 55N 2 30W
Hoddesdon England 16 C2 51 46N 0 01W
Hodnet England 15 D2 52 51N 2 35W
Hoek van Holland Neths 48 C5 51 59N 4 08E
Hohhot China 69 K6 40 49N 117 37E
Hokkaidō i. Japan 70 D3 43 30N 143 00E
Holbeach England 15 F2 52 48N 0 01E
Holderness p. England 13 F2 53 45N 0 05W
Hollyford RoI 21 C2 52 38N 8 06W
Hollywood RoI 21 E2 53 06N 6 35W
Holmes Chapel tn. England 12 D2 53 12N 2 22W
Holme-on-Spalding-Moor England 13 F2 53 49N 0 46W
Holmfirth England 13 E2 53 35N 1 46W
Holmhead Scotland 11 E2 55 29N 4 17W

Holsworthy England 18 C2 50 49N 4 21W
Holt England 17 E3 52 55N 1 05E
Holycross Limerick RoI 21 C2 52 38N 7 52W
Holycross Tipperary RoI 21 D2 52 38N 7 52W
Holyhead Wales 14 B3 53 19N 4 38W
Holy Island England 11 H2 55 41N 1 48W
Holy Island Wales 14 B3 53 16N 4 39W
Holywell Wales 14 C3 53 17N 3 13W
Holywood NI 10 D1 54 38N 5 50W
Homs Syria 64 C5 34 42N 36 40E
Homyel' Belarus 53 Q6 52 25N 31 00E
Honddu r. Wales 14 C1 51 50N 2 50W
HONDURAS 94 G2
Hong Kong China 69 K2 23 00N 114 00E
Honiton England 19 D2 50 48N 3 13W
Honshū i. Japan 70 C2 37 15N 139 00E
Hoo England 17 D2 51 26N 0 34E
Hoogeveen Neths 48 E5 52 43N 6 29E
Hook England 16 C2 51 17N 0 58W
Hook Head RoI 21 E2 52 10N 6 55W
Hope Wales 14 C3 53 07N 3 03W
Hopton on Sea England 17 E3 52 33N 1 43E
Horbury England 13 E2 53 41N 1 33W
Horley England 16 C2 51 11N 0 11W
Horncastle England 13 F2 53 13N 0 07W
Horn Head RoI 20 D5 55 13N 7 59W
Hornsea England 13 F2 53 55N 0 10W
Horsforth England 13 E2 53 51N 1 39W
Horsham England 16 C2 51 04N 0 21W
Horwich England 12 D2 53 37N 2 33W
Hospital RoI 21 C2 52 29N 8 25W
Hospitalet Spain 51 G4 41 21N 2 06E
Houghton-le-Spring England 13 E3 54 51N 1 28W
Hounslow England 16 C2 51 28N 0 21W
Houston USA 91 G2 29 45N 95 25W
Hove England 16 C1 50 49N 0 11W
Hovingham England 13 F3 54 10N 0 59W
Howden England 13 F2 53 45N 0 52W
Howth RoI 20 E3 53 23N 6 04W
Hoy i. Scotland 7 A1 58 48N 3 20W
Hoylake England 12 C2 53 23N 3 11W
Hoyland England 13 E2 53 30N 1 27W
Hoy Sound Scotland 7 A1 58 56N 3 20W
Hrodna Belarus 53 L6 53 40N 23 50E
Huambo Angola 82 C5 12 44S 15 47E
Huancayo Peru 96 B11 12 05S 75 12W
Huang He r. China 69 K5 38 00N 111 00E
Hucknall England 13 E2 53 02N 1 11W
Huddersfield England 13 E2 53 39N 1 47W
Hudson Bay Canada 89 T5/T6 58 00N 89 00W
Huelva Spain 51 C2 37 15N 6 56W
Huesca Spain 51 E4 42 08N 0 25W
Hugh Town England 18 A1 49 55N 6 19W
Hull r. England 13 F2 53 55N 0 20W
Hullbridge England 17 D2 51 37N 0 37E
Humaitá Brazil 96 E12 7 33S 63 01W
Humberston England 13 F2 53 32N 0 02W
HUNGARY 53 H3/K3
Hungerford England 16 B2 51 26N 1 30W
Hunmanby England 13 F3 54 11N 0 19W
Hunstanton England 17 D3 52 57N 0 30E
Huntford Scotland 11 G2 55 22N 2 28W
Huntingdon England 16 C3 52 20N 0 12W
Huntington USA 91 K4 38 24N 82 26W
Huntly Scotland 9 G2 57 27N 2 47W
Huntsville USA 91 J3 34 44N 86 35W
Huron, Lake Canada/USA 93 D2/3 45 00N 83 00W
Hurstpierpoint England 16 C1 50 56N 0 11W
Husbands Bosworth England 15 E2 52 27N 1 03W
Husum Germany 52 D7 54 29N 9 04E
Hutton Cranswick England 13 F2 53 57N 0 27W
Hyde England 12 D2 53 27N 2 04W
Hyderabad India 66 D3 17 22N 78 26E
Hyderabad Pakistan 66 B5 25 23N 68 24E
Hythe Hampshire England 16 B1 50 51N 1 24W
Hythe Kent England 17 E2 51 05N 1 05E

## I

Iaşi Romania 53 N3 47 09N 27 38E
Ibadan Nigeria 80 E3 7 23N 3 56E
Ibagué Colombia 96 B14 4 25N 75 20W
Ibiza i. Spain 51 F3 39 00N 1 20E
Ibstock England 15 E2 52 42N 1 23W
Ica Peru 96 B11 14 02S 75 48W
ICELAND 49 L7
Idaho state USA 90 D5 44 00N 115 00W
Idrigill Point Scotland 8 C2 57 21N 6 35W
Ieper Belgium 48 A3 50 51N 2 53E
Iglesias Italy 54 C3 39 19N 8 32E
IJsselmeer l. Neths 48 D5 52 45N 5 15E
Ilam England 13 E2 53 04N 1 49W
Ilchester England 19 E2 51 00N 2 41W
Ilebo CDR 82 D7 4 20S 20 35E
Îles Loyauté is. Pacific Ocean 73 N6/N7 21 00S 167 00E
Ilfracombe England 18 C3 51 13N 4 08W
Ilhéus Brazil 96 K11 14 50S 39 06W
Ilkeston England 13 E2 52 59N 1 18W
Ilkley England 13 E2 53 55N 1 50W
Illinois state USA 91 J5 40 00N 89 00W
Ilminster England 19 E2 50 56N 2 55W

Immingham England 13 F2 53 36N 0 11W
Imola Italy 54 E6 44 22N 11 43E
Imperatriz Brazil 96 H12 5 32S 47 28W
Imperia Italy 54 C5 43 53N 8 03E
Imphal India 67 G4 24 47N 93 55E
Inch Island i. RoI 20 D5 55 04N 7 30W
Inchon S. Korea 69 N5 37 30N 126 38E
Independence USA 91 G4 37 13N 95 43W
INDIA 66/67
Indiana state USA 91 J5 40 00N 86 00W
Indianapolis USA 91 J4 39 45N 86 10W
Indian Ocean 105
INDONESIA 71 C3/H3
Indore India 66 D4 22 42N 75 54E
Indus r. Pakistan 66 B5 28 00N 69 00E
Ingatestone England 17 D2 51 41N 0 22E
Ingleborough sum. England 12 D3 54 10N 2 23W
Ingleton England 12 D3 54 10N 2 27W
Ingolstadt Germany 52 E4 48 46N 11 27E
Inishannon RoI 21 C1 51 45N 8 38W
Inishbofin i. RoI 20 A3 53 38N 10 12W
Inishcrone RoI 20 B4 54 13N 9 05W
Inisheer i. RoI 21 B3 53 03N 9 30W
Inishkea North i. RoI 20 A4 54 08N 10 11W
Inishkea South i. RoI 20 A4 54 08N 10 11W
Inishmaan i. RoI 21 B3 53 05N 9 35W
Inishmore i. RoI 21 B3 53 07N 9 45W
Inishmurray i. RoI 20 C4 54 25N 8 35W
Inishowen Peninsula RoI 20 D5 55 15N 7 15W
Inishshark i. RoI 20 A3 53 37N 10 16W
Inishturk i. RoI 20 A3 53 43N 10 05W
Inn r. Europe 52 E3 48 00N 12 00E
Inner Hebrides is. Scotland 8 C1/C2 56 45N 6 45W
Innerleithen Scotland 11 F2 55 38N 3 05W
Inner Sound Scotland 8 D2 57 25N 5 55W
Innfield RoI 20 E3 53 25N 6 50W
Innsbruck Austria 52 E3 47 17N 11 25E
Inny r. RoI 21 A1 51 54N 10 06W
Interlaken Switz. 52 C3 46 42N 7 52E
Inuvik Canada 88 J7 68 16N 133 40W
Inverary Scotland 10 D3 56 13N 5 05W
Inver Bay RoI 20 C4 54 36N 8 20W
Inverbervie Scotland 9 G1 56 51N 2 17W
Invercargill NZ 73 N1 46 25S 168 22E
Inverclyde u.a. Scotland 11 E2 55 55N 4 45W
Invergarry Scotland 9 E2 57 02N 4 47W
Invergordon Scotland 9 E2 57 42N 4 10W
Inverkeithing Scotland 11 F3 56 02N 3 25W
Invermoriston Scotland 9 E2 57 13N 4 38W
Inverness Scotland 9 E2 57 27N 4 15W
Inverquharity Scotland 9 G1 56 44N 2 59W
Inverurie Scotland 9 G2 57 17N 2 23W
Ioánnina Greece 55 J3 39 40N 20 51E
Iona i. Scotland 8 C1 56 19N 6 25W
Iónia Nisiá Greece 55 H3 39 00N 20 00E
íos i. Greece 55 L2 36 00N 25 00E
Iowa state USA 91 H5 42 00N 94 00W
Iowa City USA 91 H5 41 39N 91 31W
Ipswich England 17 E3 52 04N 1 10E
Iquique Chile 96 C9 20 14S 70 07W
Iquitos Peru 96 C13 3 51S 73 13W
IRAN 65 F5/G5
IRAQ 64 D5
Irbid Jordan 64 C5 32 33N 35 51E
Irfon r. Wales 14 C2 52 10N 3 25W
Irish Sea British Isles 6 53 30N 5 30W
Irkutsk Russia 63 N7 52 18N 104 15E
Ironbridge England 15 D2 52 38N 2 30W
Irrawaddy r. Myanmar 71 A8 20 00N 95 00E
Irthing r. England 12 D3 55 00N 2 40W
Irthlingborough England 15 F2 52 20N 0 37W
Irtysh r. Asia 62 K8 57 30N 72 30E
Irún Spain 51 E5 43 20N 1 48W
Irvine Scotland 11 E2 55 37N 4 40W
Irvinestown NI 10 B1 54 29N 7 38W
Irving USA 91 G3 32 49N 96 57W
Ise r. England 15 F2 52 25N 0 50W
Isernia Italy 54 F4 41 35N 14 14E
Iskenderun Turkey 57 D2 36 37N 36 08E
Isla r. Scotland 9 F1 56 35N 3 10W
Isla d'Elba Italy 54 D5 42 00N 10 00E
Isla Grande de Tierra del Fuego Chile/Argentina 97 C1 54 00S 67 30W
Islamabad Pakistan 66 C6 33 40N 73 08E
Island Magee NI 10 D1 54 45N 5 40W
Islay i. Scotland 10 C2 55 48N 6 12W
Isleham England 17 D3 52 21N 0 25E
Isle of Man British Isles 12 B3 54 15N 4 15W
Isle of Portland England 19 E2 50 33N 2 27W
Isle of Purbeck England 19 E2 50 38N 2 05W
Isle of Whithorn tn. Scotland 11 E1 54 42N 4 22W
Isle of Wight u.a. England 16 B1 50 40N 1 15W
ISRAEL 64 B5/C5
Istanbul Turkey 57 C3 41 02N 28 57E
Itabuna Brazil 96 K11 14 48S 39 18W
ITALY 54 D5/E5
Itchen r. England 16 B2 51 05N 1 15W
Ithon r. Wales 14 C2 52 10N 3 25W
Itzehoe Germany 52 D6 53 56N 9 32E
Ivybridge England 18 D2 50 23N 3 56W
Ixworth England 17 D3 52 18N 0 50E

**Izhevsk** Russia 62 H8 56 49N 53 11E
**Izmir** Turkey 57 C2 38 25N 27 10E

## J

**Jackson** USA 91 H3 32 20N 90 11W
**Jacksonville** USA 91 K3 30 20N 81 40W
**Jaén** Spain 51 D2 37 46N 3 48W
**Jaipur** India 66 D5 26 53N 75 50E
**Jakarta** Indonesia 71 D2 6 08S 106 45E
**Jakobstad** Finland 49 F3 63 40N 22 42E
**JAMAICA** 95 J3
**James Bay** Canada 89 U4 53 45N 81 00W
**Jamestown** USA 91 G6 46 54N 98 42W
**Jammu** India 66 C6 32 43N 74 54E
**JAPAN** 70
**Jarrow** England 13 E3 54 59N 1 29W
**Jasper** Canada 88 M4 52 55N 118 05W
**Jastrowie** Poland 53 H6 53 25N 16 50E
**Jawa** *i.* Indonesia 71 D2/E2 7 00S 110 00E
**Jedburgh** Scotland 11 G2 55 29N 2 34W
**Jedda** Saudi Arabia 64 C3 21 30N 39 10E
**Jefferson City** USA 91 H4 38 33N 92 10W
**Jena** Germany 52 E5 50 56N 11 35E
**Jerez de la Frontera**
Spain 51 C2 36 41N 6 08W
**Jericho** Middle East 64 N10 31 51N 35 27E
**Jersey** *i.*
Channel Islands 19 E1 49 13N 2 07W
**Jerusalem**
Israel/Jordan 64 C5 31 47N 35 13E
**Jilin** China 69 N6 43 53N 126 35E
**Jinan** China 69 L5 36 41N 117 00E
**Jinja** Uganda 83 C5 0 27N 33 14E
**João Pessoa** Brazil 96 K12 7 06S 34 53W
**Jodhpur** India 66 C5 26 18N 73 08E
**Joensuu** Finland 49 G3 62 35N 29 46E
**Johannesburg** RSA 82 E2 26 10S 28 02E
**John o'Groats** Scotland 9 F3 58 38N 3 05W
**Johnstone** Scotland 11 E2 55 50N 4 31W
**Joinville** Brazil 96 H8 26 20S 48 55W
**Jönköping** Sweden 49 D2 57 45N 14 10E
**Jonquière** Canada 89 W3 48 25N 71 16W
**JORDAN** 64 C5
**Jordan** *r.* Middle East 64 N11 32 15N 32 10E
**Joyce Country** RoI 20 B3 53 32N 9 33W
**Juàzeiro** Brazil 96 J12 9 25S 40 30W
**Juba** Sudan 81 M2 4 50N 31 35E
**Juliaca** Peru 96 C10 15 29S 70 09W
**Juneau** USA 88 J5 58 20N 134 20W
**Jungfrau** *mt.* Switz. 52 C3 46 33N 7 58E
**Junsele** Sweden 49 E3 63 40N 16 55E
**Jura** *mts.* France/Switz. 50 G4/H4 46 30N 6 00E
**Jura** *i.* Scotland 10 D2/D3 55 55N 6 00W
**Jura, Sound of**
Scotland 10 D2 55 45N 5 55W
**Jylland** *p.* Denmark 49 C2 55 00N 9 00E
**Jyväskylä** Finland 49 G3 62 16N 25 50E

## K

**K2** *mt.* China/India 68 C5 35 47N 76 30E
**Kābul** Afghanistan 66 B6 34 30N 69 10E
**Kagoshima** Japan 70 B1 31 37N 130 32E
**Kaiserslautern**
Germany 52 C4 49 27N 7 47E
**Kajaani** Finland 49 G3 64 14N 27 37E
**Kalahari Desert**
Southern Africa 82 D3 23 30S 23 00E
**Kalamáta** Greece 55 K2 37 02N 22 07E
**Kalémié** CDR 82 E6 5 57S 29 10E
**Kaliavesi** *l.* Finland 49 G3 63 00N 27 20E
**Kaliningrad** Russia 49 F1 54 40N 20 30E
**Kalmar** Sweden 49 E2 56 39N 16 20E
**Kalmthout** Belgium 48 C4 51 23N 4 29E
**Kamchatka** *p.* Russia 63 T8 57 30N 160 00E
**Kames** Scotland 10 D2 55 54N 5 15W
**Kamina** CDR 82 E6 8 46S 25 00E
**Kamloops** Canada 88 L4 50 39N 120 24W
**Kampala** Uganda 83 C5 0 19N 32 35E
**Kâmpóng Cham**
Cambodia 71 D6 11 59N 105 26E
**Kananga** CDR 82 D6 5 53S 22 26E
**Kankakee** USA 91 J5 41 08N 87 52W
**Kannapolis** USA 91 K4 35 30N 80 36W
**Kano** Nigeria 80 F4 12 00N 8 31E
**Kanpur** India 66 E5 26 27N 80 14E
**Kansas** *state* USA 90/91 G4 38 00N 98 00W
**Kansas City** USA 91 H4 39 02N 94 33W
**Kanturk** RoI 21 C2 52 10N 8 55W
**Kaohsiung** Taiwan 69 M2 22 36N 120 17E
**Kapukasing** Canada 89 U3 49 25N 82 26W
**Karachi** Pakistan 66 B4 24 51N 67 02E
**Kara Sea** Russia 101 75 00N 70 00E
**Karasjok** Norway 49 G4 69 27N 25 30E
**Karcag** Hungary 53 K3 47 19N 20 53E
**Karlovac** Croatia 54 F6 45 30N 15 34E
**Karlskoga** Sweden 49 D2 59 19N 14 33E
**Karlskrona** Sweden 49 E2 56 10N 15 35E
**Karlsruhe** Germany 52 D4 49 00N 8 24E
**Karlstad** Sweden 49 D2 59 24N 13 32E
**Kaskinen** Finland 49 F3 62 23N 21 10E
**Kassala** Sudan 81 N5 15 24N 36 30E
**Kassel** Germany 52 D5 51 18N 9 30E
**Kathmandu** Nepal 67 F5 27 42N 85 19E
**Katowice** Poland 53 J5 50 15N 18 59E
**Katrineholm** Sweden 49 E2 58 59N 16 15E
**Kattegat** *sd.*
Denmark/Sweden 49 D2 57 00N 11 00E
**Kaunas** Lithuania 49 F1 54 52N 23 55E
**Kawasaki** Japan 70 C2 35 30N 139 45E
**Kayseri** Turkey 57 E2 38 42N 35 28E
**KAZAKHSTAN** 62 H6/J6
**Kazan'** Russia 62 G8 55 45N 49 10E
**Keady** NI 10 C1 54 15N 6 42W
**Kealkill** RoI 21 B1 51 45N 9 20W
**Kecskemét** Hungary 53 J3 46 56N 19 43E
**Kefallonia** *i.* Greece 55 J3 38 00N 20 00E
**Keighley** England 13 E2 53 52N 1 54W
**Keith** Scotland 9 G2 57 32N 2 57W
**Kells** RoI 20 E3 53 44N 6 53W
**Kelowna** Canada 88 M3 49 50N 119 29W
**Kelso** Scotland 11 G2 55 36N 2 25W
**Kelvedon** England 17 D2 51 51N 0 42E
**Kemnay** Scotland 9 G2 57 14N 2 27W
**Kempston** England 16 C3 52 07N 0 30W
**Kempten** Germany 52 E3 47 44N 10 19E
**Kendal** England 12 D3 54 20N 2 45W
**Kenilworth** England 15 E2 52 21N 1 34W
**Kenmare** RoI 21 B1 51 53N 9 35W
**Kennacraig** Scotland 10 D2 55 50N 5 27W
**Kennet** *r.* England 16 B2 51 30N 1 45W
**Kennington** England 17 D2 51 09N 0 55E
**Kennoway** Scotland 11 F3 56 12N 3 03W
**Kenora** Canada 89 S3 49 47N 94 26W
**Kent** *co.* England 17 D2 51 10N 0 45E
**Kentford** England 17 D3 52 16N 0 30E
**Kentucky** *state* USA 91 J4 37 00N 85 00W
**KENYA** 83 E5/F5
**Kerkrade** Neths 48 E3 50 52N 6 04E
**Kérkyra** *i.* Greece 55 H3 39 00N 19 00E
**Kerpen** Germany 48 E3 50 52N 6 42E
**Kerrera** *i.* Scotland 8 D1 56 25N 5 34W
**Kerry** *co.* RoI 21 B2 52 10N 9 30W
**Kerry Head** RoI 21 B2 52 25N 9 55W
**Kesh** NI 10 B1 54 31N 7 43W
**Kessingland** England 17 E3 52 25N 1 42E
**Keswick** England 12 C3 54 37N 3 08W
**Kettering** England 15 F2 52 24N 0 44W
**Keynsham** England 19 E3 51 26N 2 30W
**Key West** USA 91 K1 24 34N 81 48W
**Keyworth** England 15 E2 52 52N 1 05W
**Khabarovsk** Russia 63 R6 48 32N 135 08E
**Kharkiv** Ukraine 56 E4 50 00N 36 15E
**Khartoum** Sudan 81 M5 15 33N 32 35E
**Kherson** Ukraine 57 D4 46 39N 32 38E
**Khmel'nyts'kyy**
Ukraine 53 N4 49 25N 26 59E
**Khulna** Bangladesh 67 F4 22 49N 89 34E
**Khyber Pass**
Afghanistan/Pakistan 66 C6 34 06N 71 05E
**Kidderminster** England 15 D2 52 23N 2 14W
**Kidlington** England 16 B2 51 50N 1 17W
**Kidsgrove** England 12 D2 53 05N 2 14W
**Kidwelly** Wales 14 B1 51 45N 4 18W
**Kiel** Germany 52 E7 54 20N 10 08E
**Kielce** Poland 53 K5 50 51N 20 39E
**Kielder Water** England 12 D4 55 10N 2 30W
**Kiev** Ukraine 53 Q5 50 25N 30 30E
**Kigali** Rwanda 83 B4 1 56S 30 04E
**Kigoma** Tanzania 83 A2 4 52S 29 36E
**Kilbaha** RoI 21 B2 52 35N 9 52W
**Kilbeggan** RoI 20 D3 53 22N 7 30W
**Kilberry** RoI 20 E3 53 42N 6 41W
**Kilbirnie** Scotland 11 E2 55 46N 4 41W
**Kilbrannan Sound**
Scotland 10 D2 55 30N 5 30W
**Kilcavan** RoI 20 D3 53 11N 7 21W
**Kilchoan** Scotland 8 C1 56 43N 6 06W
**Kilcock** RoI 20 E3 53 24N 6 40W
**Kilcormac** RoI 21 D3 53 10N 7 43W
**Kilcreggan** Scotland 11 E2 56 01N 4 48W
**Kilcullen** RoI 21 E3 53 08N 6 45W
**Kildare** RoI 21 E3 53 10N 6 55W
**Kildare** *co.* RoI 21 E3 53 10N 6 55W
**Kildorrery** RoI 21 C2 52 15N 8 20W
**Kilfinnane** RoI 21 C2 52 22N 8 28W
**Kilgetty** Wales 14 B1 51 45N 4 44W
**Kinglass Lough** RoI 20 C3 53 50N 8 01W
**Kilimanjaro** *mt.*
Tanzania 83 E3 3 04S 37 22E
**Kilkeary** RoI 21 C2 52 50N 8 07W
**Kilkee** RoI 21 B2 52 41N 9 38W
**Kilkeel** NI 10 C1 54 04N 6 00W
**Kilkelly** RoI 20 C3 53 53N 8 51W
**Kilkenny** *co.* RoI 21 D2 52 30N 7 10W
**Kilkenny** RoI 21 D2 52 39N 7 15W
**Kilkhampton** England 18 C2 50 53N 4 29W
**Kilkieran Bay** RoI 20 B3 53 18N 9 41W
**Kilkishen** RoI 21 C2 52 48N 8 45W
**Kill** RoI 20 E3 53 15N 6 35W
**Killala** RoI 20 B4 54 13N 9 13W
**Killaloe** RoI 21 C2 52 48N 8 27W
**Killamarsh** England 13 E2 53 20N 1 20W
**Killarney** RoI 21 B2 52 03N 9 30W
**Killary Harbour** RoI 20 B3 53 38N 9 55W
**Killeigh** RoI 20 D3 53 13N 7 27W
**Killenaule** RoI 21 D2 52 34N 7 40W
**Killimer** RoI 21 B2 52 37N 9 24W
**Killimor** RoI 21 C3 53 10N 8 17W
**Killin** Scotland 9 E1 56 28N 4 19W
**Killorglin** RoI 21 B2 52 06N 9 47W
**Killybegs** RoI 20 C4 54 38N 8 27W
**Killyleagh** NI 10 D1 54 24N 5 39W
**Kilmacolm** Scotland 11 E2 55 54N 4 38W
**Kilmacrenan** RoI 20 D5 55 02N 7 47W
**Kilmacthomas** RoI 21 D2 52 12N 7 25W
**Kilmallock** RoI 21 C2 52 23N 8 34W
**Kilmaluag** Scotland 8 C2 57 41N 6 17W
**Kilmarnock** Scotland 11 E2 55 36N 4 30W
**Kilmona** RoI 21 C1 51 58N 8 33W
**Kilmory** Scotland 10 D2 55 55N 5 41W
**Kilrea** NI 10 C1 54 57N 6 33W
**Kilrenny** Scotland 11 G3 56 15N 2 41W
**Kilrush** RoI 21 B2 52 39N 9 30W
**Kilsheelan** RoI 21 D2 52 22N 7 35W
**Kilsyth** Scotland 11 E2 55 59N 4 04W
**Kiltamagh** RoI 20 B3 53 51N 9 00W
**Kiltealy** RoI 21 E2 52 34N 6 45W
**Kiltullagh** RoI 20 C3 53 35N 8 35W
**Kilwinning** Scotland 11 E2 55 40N 4 42W
**Kimberley** Canada 88 M3 49 40N 115 58W
**Kimberley** England 15 E2 53 00N 1 17W
**Kimberley** England 15 E2 53 00N 1 17W
**Kimbolton** RSA 82 D2 28 45S 24 46E
**Kinbrace** Scotland 9 F3 58 15N 3 56W
**Kineton** England 15 E2 52 10N 1 30W
**Kingsbridge** England 18 D2 50 17N 3 46W
**Kingsclere** England 16 B2 51 20N 1 14W
**Kingscourt** RoI 20 E3 53 53N 6 48W
**Kingskerswell** England 18 D2 50 30N 3 37W
**Kings Langley** England 16 C2 51 43N 0 28W
**King's Lynn** England 17 D3 52 45N 0 24E
**Kings Muir** Scotland 11 F2 55 40N 3 12W
**Kingsport** USA 91 K4 36 33N 82 34W
**Kingsteignton** England 18 D2 50 33N 3 35W
**Kingston** Jamaica 95 U13 17 58N 76 48W
**Kingston upon Hull**
England 13 F2 53 45N 0 20W
**Kingston-upon-Thames**
England 16 C2 51 25N 0 18W
**Kingswood** England 15 D1 51 28N 2 30W
**Kington** England 14 C2 52 12N 3 01W
**Kingussie** Scotland 9 E2 57 05N 4 03W
**Kinloch** Scotland 8 C2 57 01N 6 17W
**Kinlocheil** Scotland 8 D1 56 51N 5 20W
**Kinlochewe** Scotland 8 D2 57 36N 5 20W
**Kinlochleven** Scotland 9 E1 56 42N 4 58W
**Kinloch Rannoch**
Scotland 9 E1 56 42N 4 11W
**Kinloss** Scotland 9 F2 57 38N 3 33W
**Kinlough** RoI 20 C4 54 27N 8 17W
**Kinnaird Head** Scotland 9 G2 57 40N 2 00W
**Kinnegad** RoI 20 D3 53 27N 7 05W
**Kinnitty** RoI 21 D3 53 06N 7 43W
**Kinross** Scotland 11 F3 56 13N 3 27W
**Kinsale** RoI 21 C1 51 42N 8 32W
**Kinsale Harbour** RoI 21 C1 51 40N 8 30W
**Kinshasa** CDR 82 C7 4 18S 15 18E
**Kintore** Scotland 9 G2 57 13N 2 21W
**Kintyre** *p.* Scotland 10 D2 55 30N 5 35W
**Kinvarra** RoI 21 C3 53 08N 8 56W
**Kippax** England 13 E2 53 46N 1 22W
**Kippford** Scotland 11 F1 54 52N 3 49W
**Kirby Muxloe** England 15 E2 52 38N 1 14W
**Kircubbin** NI 10 D1 55 24N 5 28W
**KIRIBATI** 100 K7
**Kirikkale** Turkey 57 D2 39 51N 33 32E
**Kirkbean** Scotland 11 F1 54 55N 3 36W
**Kirkburton** England 13 E2 53 37N 1 42W
**Kirkby** England 12 D2 53 29N 2 54W
**Kirkby in Ashfield**
England 13 E2 53 13N 1 15W
**Kirkby Londsale**
England 12 D3 54 13N 2 36W
**Kirkbymoorside**
England 13 F3 54 16N 0 55W
**Kirkby Stephen**
England 12 D3 54 28N 2 20W
**Kirkcaldy** Scotland 11 F3 56 07N 3 10W
**Kirkcolm** Scotland 10 D1 54 58N 5 04W
**Kirkconnel** Scotland 11 F2 55 23N 4 00W
**Kirkcowan** Scotland 11 E1 54 55N 4 36W
**Kirkcudbright** Scotland 11 E1 54 50N 4 03W
**Kirkgunzeon** Scotland 11 F1 54 58N 3 46W
**Kirkham** England 12 D2 53 47N 2 53W
**Kirkintilloch** Scotland 11 E2 55 57N 4 10W
**Kirkmuirhill** Scotland 11 F2 55 39N 3 55W
**Kirkpatrick Durham**
Scotland 11 F2 55 01N 3 54W
**Kirk Sandall** England 13 E2 53 34N 1 04W
**Kirkwall** Scotland 7 B1 58 59N 2 58W
**Kirov** Russia 62 G8 58 00N 49 38E
**Kirovohrad** Ukraine 56 D4 48 31N 32 15E
**Kirriemuir** Scotland 9 F1 56 41N 3 01W
**Kirton** England 16 C3 52 56N 0 04W
**Kisangani** CDR 81 L2 0 33N 25 14E
**Kismaayo** Somalia 81 P1 0 25S 42 31E
**Kisumu** Kenya 83 D4 0 08S 34 47E
**Kita-Kyūshu** Japan 70 B1 33 52N 130 49E
**Kitami** Japan 70 D3 43 51N 143 54E
**Kitchener** Canada 89 U2 43 27N 80 30W
**Kitimat** Canada 88 K4 54 05N 128 38W
**Kitwe** Zambia 82 E5 0 08S 30 30E
**Kitzbühel** Austria 52 F3 47 27N 12 23E
**Klaipėda** Lithuania 49 F2 55 43N 21 07E
**Klamath Falls** *tn.* USA 90 B5 42 14N 121 47W
**Klöfta** Norway 49 D3 60 04N 11 06E
**Knapdale** Scotland 10 D2 55 52N 5 31W
**Knaresborough**
England 13 E3 54 00N 1 27W
**Knighton** England 14 C2 52 21N 3 03W
**Knock** RoI 20 C3 53 47N 8 55W
**Knockadoon Head** RoI 21 D1 51 50N 7 50W
**Knockanevin** RoI 21 C2 52 17N 8 21W
**Knockdrislagh** RoI 21 C2 52 03N 8 41W
**Knocktopher** RoI 21 D2 52 29N 7 13W
**Knokke-Heist** Belgium 48 B4 51 21N 3 19E
**Knottingley** England 13 E2 53 43N 1 14W
**Knowle** England 15 E2 52 23N 1 43W
**Knoxville** USA 91 K4 36 00N 83 57W
**Knutsford** England 12 D2 53 18N 2 23W
**Kōbe** Japan 70 C1 34 40N 135 12E
**Koblenz** Germany 52 C5 50 21N 7 36E
**Kōfu** Japan 70 C2 35 42N 138 34E
**Kokkola** Finland 49 F3 62 45N 30 06E
**Kolhapur** India 66 C3 16 40N 74 20E
**Kolkata** India 67 F4 22 30N 88 20E
**Komatsu** Japan 70 C2 36 25N 136 27E
**Komotiní** Greece 55 L4 41 06N 25 25E
**Köniz** Switz. 52 C3 46 56N 7 25E
**Konstanz** Germany 52 D3 47 40N 9 10E
**Konya** Turkey 57 D2 37 51N 32 30E
**Konz** Germany 48 E2 49 42N 6 35E
**Koper** Slovenia 54 E6 45 31N 13 44E
**Korbach** Germany 52 D5 51 16N 8 53E
**Korinthiakós Kólpos** *g.*
Greece 55 K3 38 00N 22 00E
**Kórinthos** Greece 55 K2 37 56N 22 55E
**Kortrijk** Belgium 48 B3 50 50N 3 17E
**Kós** *i.* Greece 55 M2 36 45N 27 10E
**Košice** Slovakia 53 K4 48 44N 21 15E
**Koszalin** Poland 53 H7 54 10N 16 10E
**Kotka** Finland 49 G3 60 26N 26 55E
**Kovrov** Russia 56 F6 56 23N 41 21E
**Kozáni** Greece 55 J4 40 18N 21 48E
**Kragujevac**
Serbia & Montenegro 55 J5 44 01N 20 55E
**Kraków** Poland 53 J5 50 03N 19 55E
**Kranj** Slovenia 54 F7 46 15N 14 20E
**Krefeld** Germany 52 C5 51 20N 6 32E
**Krems** Austria 52 G4 48 25N 15 36E
**Kristiansand** Norway 49 C2 58 08N 8 01E
**Kristianstad** Sweden 49 D2 56 02N 14 10E
**Kríti** *i.* Greece 55 L1 35 00N 25 00E
**Kryvyy Roh** Ukraine 56 D4 47 55N 33 24E
**Kuala Lumpur**
Malaysia 71 C4 3 09N 101 42E
**Kuching** Malaysia 71 E4 1 35N 110 21E
**Kufstein** Austria 52 F3 47 36N 12 11E
**Kumamoto** Japan 70 B1 32 50N 130 42E
**Kunlun Shan** *mts.*
China 68 D5/E5 36 30N 85 00E
**Kunming** China 69 H3 25 04N 102 41E
**Kurashiki** Japan 70 B1 34 36N 133 43E
**Kurgan** Russia 62 J8 55 30N 65 20E
**Kuril Islands** Russia 63 T6 50 00N 155 00E
**Kursk** Russia 62 F7 51 45N 36 14E
**Kushiro** Japan 70 D3 42 58N 144 24E
**KUWAIT** 65 E4
**Kuwait** Kuwait 65 E4 29 20N 48 00E
**Kuytun** China 68 D6 44 30N 85 00E
**Kwangju** S. Korea 69 N5 35 07N 126 52E
**Kyleakin** Scotland 8 D2 57 16N 5 44W
**Kyle of Lochalsh**
Scotland 8 D2 57 17N 5 43W
**Kyles of Bute** Scotland 10 D2 55 52N 5 13W
**Kylestrome** Scotland 9 D3 58 16N 5 02W
**Kyōto** Japan 70 C2 35 02N 135 45E
**KYRGYZSTAN** 62 K5
**Kythira** *i.* Greece 55 K2 36 00N 23 00E
**Kýthnos** *i.* Greece 55 L2 37 25N 24 25E
**Kyūshū** *i.* Japan 70 B1 32 20N 131 00E

## L

**Laâyoune**
Western Sahara 80 B7 27 10N 13 11W
**Labrador** *geog. reg.*
Canada 89 Y4 54 00N 63 00W
**la Chaux-de-Fonds**
Switz. 52 C3 47 07N 6 51E
**la Ciotat** France 50 G2 43 10N 5 36E
**Lac Léman** Switzerland 52 C3 46 20N 6 20E
**La Crosse** USA 91 H5 43 48N 91 04W
**Lacul Razim** *l.* Romania 53 P2 45 00N 29 00E
**Ladozhskoye Ozero** *l.*
Russia 62 F9 61 00N 30 00E
**Ladybank** Scotland 11 F3 56 17N 3 08W
**Lagan** *r.* NI 10 C1 54 30N 6 05W
**Laggan Bay** *b.* Scotland 10 C2 55 38N 6 17W
**Lago di Como** *l.* Italy 54 C6 46 00N 9 00E
**Lago di Garda** *l.* Italy 54 D6 45 40N 10 00E
**Lago Maggiore** *l.* Italy 54 C6 46 00N 8 00E
**Lagos** Nigeria 80 E3 6 27N 3 28E
**Lagos** Portugal 51 A2 37 05N 8 40W
**Lago Titicaca** *l.*
Peru/Bolivia 96 C10/D10 16 00S 69 30W
**La Grande** USA 90 C6 45 21N 118 05W
**La Grande Rivière**
Canada 89 W4 54 00N 74 00W
**Lahore** Pakistan 66 C6 31 34N 74 22E
**Lahti** Finland 49 G3 61 00N 25 40E
**Lairg** Scotland 9 E3 58 01N 4 25W
**Lakenheath** England 17 D3 52 25N 0 31E
**La Línea de la Concepción**
Spain 51 C2 36 10N 5 21W
**La Maddalena** Italy 54 C4 41 13N 9 25E

**Lamar** USA 90 F4 38 04N 102 37W
**Lambay Island** RoI 20 E3 53 29N 6 01W
**Lambourn** England 16 B2 51 31N 1 31W
**Lambourn** *r.* England 16 B2 51 27N 1 28W
**Lamego** Portugal 51 B4 41 05N 7 49W
**Lamía** Greece 55 K3 38 55N 22 26E
**Lamlash** Scotland 10 D2 55 32N 5 08W
**Lammermuir Hills** Scotland 11 G2 55 50N 2 45W
**Lampeter** Wales 14 B2 52 07N 4 05W
**Lanark** Scotland 11 F2 55 41N 3 48W
**Lancashire** *co.* England 12 D2 53 50N 2 30W
**Lancaster** England 12 D3 54 03N 2 48W
**Lanchester** England 13 E3 54 50N 1 44W
**Lancing** England 16 C1 50 50N 0 19W
**Landerneau** France 50 B5 48 27N 4 16W
**Land's End** *c.* England 18 B2 50 03N 5 44W
**Landshut** Germany 52 F4 48 31N 12 10E
**Landskrona** Sweden 49 D2 55 53N 12 50E
**Langdon** USA 92 A3 48 46N 98 21W
**Langholm** Scotland 11 F2 55 09N 3 00W
**Langport** England 19 E3 51 02N 2 50W
**Langres** France 50 G4 47 53N 5 20E
**Lannion** France 50 C5 48 44N 3 27W
**L'Anse** USA 92 C3 46 45N 88 27W
**Lansing** USA 91 K5 42 44N 85 34W
**Lanzarote** *i.* Spain 80 B7 29 00N 13 38W
**Lanzhou** China 69 H5 36 01N 103 45E
**Laois** *co.* RoI 21 D2 52 59N 7 25W
**Laon** France 50 F5 49 34N 3 37E
**LAOS** 71 C7/D7
**La Paz** Bolivia 96 D10 16 30S 68 10W
**Lapford** England 18 D2 50 52N 3 47W
**La Plata** Argentina 97 F6 34 52S 57 55W
**Lappland** *geog. reg.* Finland/Sweden 49 F4/G4 67 30N 20 05E
**Lapua** Finland 49 F3 62 57N 23 00E
**L'Aquila** Italy 54 E5 42 22N 13 24E
**Laragh** RoI 21 E3 53 00N 6 18W
**Laredo** USA 90 G2 27 32N 99 22W
**Largs** Scotland 10 E2 55 48N 4 52W
**Lárisa** Greece 55 K3 39 38N 22 25E
**Larkhall** Scotland 11 F2 55 45N 3 59W
**Larne** NI 10 D1 54 51N 5 49W
**Larne** *district* NI 10 D1 54 50N 6 00W
**Larne Lough** *b.* NI 10 D1 54 49N 5 49W
**la Rochelle** France 50 D4 46 10N 1 10W
**la Roche-sur-Yon** France 50 D4 46 40N 1 25W
**La Roda** Spain 51 D3 39 13N 2 15W
**La See d'Urgel** Spain 51 F5 42 22N 1 27E
**la Seyne-sur-Mer** France 50 G2 43 06N 5 53E
**Las Palmas** Spain 80 A7 28 08N 15 27W
**La Spézia** Italy 54 C6 44 07N 9 48E
**Las Vegas** USA 90 C4 36 10N 115 10W
**Latakia** Syria 64 C6 35 31N 35 47E
**Latheron** Scotland 9 F3 58 17N 3 22W
**Latina** Italy 54 E4 41 28N 12 53E
**LATVIA** 49 G2
**Lauder** Scotland 11 G2 55 43N 2 45W
**Laugharne** Wales 14 B1 51 47N 4 28W
**Launceston** Australia 73 J2 41 25S 147 07E
**Launceston** England 18 C2 50 38N 4 21W
**Lauragh** RoI 21 B1 51 45N 9 46W
**Laurel** USA 91 J3 31 41N 89 09W
**Laurencekirk** Scotland 9 G1 56 50N 2 29W
**Laurencetown** RoI 20 C3 53 14N 8 10W
**Lauria** Italy 54 F4 40 03N 15 50E
**Laurieston** Scotland 11 E1 54 57N 4 03W
**Laurieston** Scotland 11 F2 55 59N 3 45W
**Lausanne** Switz. 52 C3 46 32N 6 39E
**Laval** France 50 D5 48 04N 0 45W
**Lavenham** England 17 D3 52 06N 0 47E
**Lawrence** USA 91 G4 38 58N 95 15W
**Lawton** USA 91 G3 34 36N 98 25W
**Laxey** Isle of Man 12 B3 54 14N 4 24W
**Laxford Bridge** *tn.* Scotland 9 D3 58 22N 5 01W
**Lay** *r.* France 50 D4 46 32N 1 15W
**Lazonby** England 12 D3 54 46N 2 41W
**Lea** *r.* England 16 C2 51 47N 0 17W
**Leach** *r.* England 15 E1 51 46N 1 44W
**Leadhills** Scotland 11 F2 55 25N 3 47W
**Leadon** *r.* England 15 D1 51 57N 2 21W
**Leatherhead** England 16 C2 51 18N 0 20W
**LEBANON** 64 C5
**Lebanon** USA 91 H4 37 40N 92 40W
**le Blanc** France 50 E4 46 38N 1 04E
**Lebrija** Spain 51 B2 36 55N 6 10W
**Lecarrow** RoI 20 C3 53 33N 8 02W
**Lecce** Italy 55 H4 40 21N 18 11E
**Lechlade** England 15 E1 51 43N 1 41W
**le Creusot** France 50 G4 46 48N 4 27E
**Ledbury** England 15 D2 52 02N 2 25W
**Ledmore** Scotland 9 E3 58 03N 4 58W
**Leeds** England 13 E2 53 50N 1 35W
**Leek** England 12 D2 53 06N 2 01W
**Leenaun** RoI 20 B3 53 36N 9 41W
**Leeuwarden** Neths 48 D6 53 12N 5 48E
**Leeward Islands** Lesser Antilles 95 M3 17 30N 64 00W
**Lefkáda** *i.* Greece 55 J3 38 45N 20 40E
**Leganés** Spain 51 D4 40 20N 3 46WE
**Legnica** Poland 53 H5 51 12N 16 10E
**le Havre** France 50 E5 49 30N 0 06E
**Leicester** England 15 E2 52 38N 1 05W
**Leicester City** *u.a.* England 15 E2 52 38N 1 05W
**Leicestershire** *co.* England 15 E2 52 40N 1 20W
**Leiden** Neths 48 C5 52 10N 4 30E
**Leigh** England 12 D2 53 30N 2 33W
**Leighton Buzzard** England 16 C2 51 55N 0 41W
**Leipzig** Germany 52 F5 51 20N 12 25E
**Leiria** Portugal 51 A3 39 45N 8 49W
**Leiston** England 17 E3 52 13N 1 34E
**Leith Hill** *tn.* England 16 C2 51 11N 0 23W
**Leitrim** *co.* RoI 20 C4 54 05N 8 00W
**Leivadiá** Greece 55 K3 38 26N 22 53E
**Leixlip** RoI 20 E3 53 22N 6 30W
**le Mans** France 50 E4 48 00N 0 12E
**Lemybrien** RoI 21 D2 52 10N 7 31W
**Lena** *r.* Russia 63 Q10 70 00N 125 00E
**Lens** France 50 F6 50 26N 2 50E
**Leoben** Austria 52 G3 47 23N 15 06E
**Leominster** England 14 D2 52 14N 2 45W
**León** Mexico 94 D4 21 10N 101 42W
**León** Spain 51 C5 42 35N 5 34W
**Leonding** Austria 52 G4 48 19N 14 17E
**le Puy** France 50 F3 45 03N 3 53E
**Lerwick** Scotland 7 C3 60 09N 1 09W
**Lesmahagow** Scotland 11 F2 55 38N 3 54W
**LESOTHO** 82 E2
**les Sables-d'Olonne** France 50 D4 46 30N 1 47W
**Lesser Antilles** *is.* W. Indies 95 M2/M3 18 00N 65 00W
**Lesser Slave Lake** Canada 88 M5 55 25N 115 30W
**Lésvos** *i.* Greece 55 M3 39 00N 26 00E
**Leswalt** Scotland 10 D1 54 46N 5 05W
**Letchworth** England 16 C2 51 58N 0 14W
**le Touquet-Paris-Plage** France 50 E6 50 13N 1 36E
**le Tréport** France 50 E6 50 04N 1 22E
**Letterfrack** RoI 20 B3 53 33N 9 57W
**Letterkenny** RoI 20 D4 54 57N 7 44W
**Leuchars** Scotland 9 G1 56 23N 2 53W
**Levanto** Italy 54 C6 44 10N 9 37E
**Leven** England 13 F2 53 54N 0 19W
**Leven** Scotland 11 G3 56 11N 3 00W
**Leverburgh** Scotland 8 B2 57 45N 7 00W
**Leverkusen** Germany 52 C5 51 02N 6 59E
**Lewes** England 16 D1 50 52N 0 01E
**Lewis** *i.* Scotland 8 C3 58 15N 6 30W
**Lexington** USA 91 K4 38 03N 84 30W
**Lexington** USA 93 E1 37 47N 79 27W
**Leyburn** England 13 E3 54 19N 1 49W
**Leyland** England 12 D2 53 42N 2 42W
**Leysdown on Sea** England 17 D2 51 24N 0 55E
**Lhasa** China 68 F3 29 41N 91 10E
**LIBERIA** 80 B3/C3
**Libourne** France 50 D3 44 55N 0 14W
**Libreville** Gabon 80 F2 0 30N 9 25E
**LIBYA** 81 G7/K7
**Libyan Desert** North Africa 81 K7/L6 25 00N 25 00E
**Lichfield** England 15 E2 52 42N 1 48W
**LIECHTENSTEIN** 52 D3
**Liège** Belgium 48 D3 50 38N 5 35E
**Lienz** Austria 52 F3 46 51N 12 50E
**Liepāja** Latvia 49 F2 56 30N 21 00E
**Liffey** *r.* RoI 20 E3 53 16N 6 37W
**Lifford** RoI 20 D4 54 50N 7 29W
**Likasi** CDR 82 E5 10 58S 26 47E
**Lille** France 50 F6 50 39N 3 05E
**Lillehammer** Norway 49 D3 61 06N 10 27E
**Lilongwe** Malawi 82 F5 13 58S 33 49E
**Lima** Peru 96 B11 12 06S 8 40W
**Limavady** NI 10 C2 55 03N 6 57W
**Limavady** *district* NI 10 B1/C1 54 50N 7 00W
**Limburg** Germany 52 D5 50 23N 8 04E
**Limerick** RoI 21 C2 52 04N 8 38W
**Limerick** *co.* RoI 21 C2 52 30N 8 40W
**Limoges** France 50 E3 45 50N 1 15E
**Limoux** France 50 F2 43 03N 2 13E
**Limpopo** *r.* Southern Africa 82 F3 22 30S 32 00E
**Linares** Spain 51 D3 38 05N 3 38W
**Lincoln** England 13 F2 53 14N 0 33W
**Lincoln** USA 91 G5 40 49N 96 41W
**Lincolnshire** *co.* England 13 F2 53 10N 0 20W
**Lincoln Wolds** England 13 F2 53 25N 0 05W
**Lingen** Germany 52 C6 52 32N 7 19E
**Lingfield** England 16 C2 51 11N 0 01W
**Linköping** Sweden 49 E2 58 25N 15 35E
**Linlithgow** Scotland 11 F2 55 59N 3 37W
**Linslade** England 16 C2 51 55N 0 42W
**Linton** England 17 D3 52 06N 0 17E
**Linz** Austria 52 G4 48 19N 14 18E
**Liphook** England 16 C2 51 05N 0 49W
**Lippstadt** Germany 52 D5 51 41N 8 20E
**Lisbellaw** NI 10 B1 54 19N 7 32W
**Lisbon** Portugal 51 A3 38 44N 9 08W
**Lisburn** NI 10 C1 54 31N 6 03W
**Lisburn** *co.* NI 10 C1 54 35N 6 10W
**Liscannor Bay** RoI 21 B2 52 55N 9 30W
**Lisdoonvarna** RoI 21 B3 53 02N 9 17W
**Lisieux** France 50 E5 49 09N 0 14E
**Liskeard** England 18 C2 50 28N 4 28W
**Lismore** RoI 21 D2 52 08N 7 55W
**Lismore** *i.* Scotland 8 D1 56 30N 5 30W
**Lisnaskea** NI 10 B1 54 15N 7 27W
**Liss** England 16 C2 51 03N 0 55W
**Lisse** Neths 48 C5 52 16N 4 34E
**Lissycasey** RoI 21 B2 52 44N 9 12W
**Listowel** RoI 21 B2 52 27N 9 29W
**Litherland** England 12 D2 53 28N 2 59W
**LITHUANIA** 49 F2
**Littleborough** England 12 D2 53 39N 2 05W
**Littlehampton** England 16 C1 50 48N 0 33W
**Little Minch** *sd.* Scotland 8 C2 57 45N 6 30W
**Little Ouse** *r.* England 17 D3 52 30N 0 30E
**Littleport** England 17 D3 52 28N 0 19E
**Liverpool** England 12 D2 53 25N 2 55W
**Liverpool Bay** England 12 C2 53 28N 3 15W
**Liversedge** England 13 E2 53 43N 1 41W
**Livingston** Scotland 11 F2 55 53N 3 32W
**Livorno** Italy 54 D5 43 33N 10 18E
**Lizard** England 18 B1 49 57N 5 13W
**Lizard Point** England 18 B1 49 56N 5 13W
**Ljubljana** Slovenia 54 F7 46 04N 14 30E
**Ljusdal** Sweden 49 E3 61 57N 16 05E
**Llanberis** Wales 14 B3 53 07N 4 06W
**Llanbister** Wales 14 C2 52 21N 3 19W
**Llandeilo** Wales 14 C1 51 53N 3 59W
**Llandovery** Wales 14 C1 51 59N 3 48W
**Llandrindod Wells** Wales 14 C2 52 15N 3 23W
**Llandudno** Wales 14 C3 53 19N 3 49W
**Llandysul** Wales 14 B2 52 02N 4 19W
**Llanelli** Wales 14 B1 51 42N 4 10W
**Llanerchymedd** Wales 14 B3 53 20N 4 22W
**Llanes** Spain 51 C3 43 25N 4 45W
**Llanfair Caereinion** Wales 14 C2 52 39N 3 20W
**Llanfairfechan** Wales 14 C3 53 15N 3 58W
**Llanfyllin** Wales 14 C2 52 46N 3 17W
**Llangadfan** Wales 14 C2 52 41N 3 28W
**Llangadog** Wales 14 C1 51 57N 3 53W
**Llangefni** Wales 14 B3 53 16N 4 18W
**Llangollen** Wales 14 C2 52 58N 3 10W
**Llangurig** Wales 14 C2 52 25N 3 36W
**Llangynog** Wales 14 C2 52 50N 3 25W
**Llanharan** Wales 14 C1 51 32N 3 28W
**Llanidloes** Wales 14 C2 52 27N 3 32W
**Llanrhystud** Wales 14 B2 52 18N 4 09W
**Llanrwst** Wales 14 C3 53 08N 3 48W
**Llantrisant** Wales 14 C1 51 33N 3 23W
**Llantwit Major** Wales 14 C1 51 25N 3 30W
**Llanuwchllyn** Wales 14 C2 52 52N 3 41W
**Llanwrtyd Wells** Wales 14 C2 52 07N 3 38W
**Llanybydder** Wales 14 B2 52 04N 4 09W
**Lleida** Spain 51 F4 41 37N 0 38E
**Lleyn Peninsula** Wales 14 B2 52 53N 4 30W
**Llyn Brianne Resevoir** Wales 14 C2 52 09N 3 43W
**Llyn Celyn** Wales 14 C2 53 04N 4 10W
**Llyn Efyrnwy** Wales 14 C2 52 50N 3 30W
**Llyn Tegid** Wales 14 C2 52 53N 3 38W
**Loanhead** Scotland 11 F2 55 53N 3 09W
**Loch Affric** Scotland 9 D2 57 15N 5 02W
**Lochailort** *tn.* Scotland 8 D1 56 53N 5 40W
**Lochaline** *tn.* Scotland 8 D1 56 32N 5 47W
**Loch Alsh** Scotland 8 D2 57 15N 5 35W
**Locharbriggs** Scotland 11 F2 55 06N 3 35W
**Loch Assynt** Scotland 9 D3/E3 58 05N 5 00W
**Loch Awe** Scotland 10 D3 56 15N 5 17W
**Lochboisdale** *tn.* Scotland 8 B2 57 09N 7 19W
**Loch Bracadale** Scotland 8 C2 57 20N 6 30W
**Loch Broom** Scotland 8 D2 57 52N 5 08W
**Loch Caolisport** Scotland 10 D2 55 53N 5 38W
**Lochcarron** *tn.* Scotland 8 D2 57 24N 5 30W
**Loch Carron** Scotland 8 D2 57 40N 5 45W
**Loch Chroisg** Scotland 8 D2 57 30N 5 10W
**Loch Craignish** Scotland 10 D3 56 05N 5 30W
**Loch Creran** Scotland 8 D1 56 30N 5 20W
**Lochdon** *tn.* Scotland 8 D1 56 25N 5 35W
**Loch Duich** Scotland 8 D2 57 15N 5 30W
**Loch Earn** Scotland 9 F1 56 25N 4 10W
**Lochearnhead** Scotland 9 E1 56 23N 4 18W
**Loch Eil** Scotland 8 D1 56 50N 5 25W
**Loch Eishort** Scotland 8 C2/D2 57 08N 6 00W
**Loch Eriboll** Scotland 9 E3 58 31N 4 41W
**Loch Ericht** Scotland 9 E1 56 48N 4 25W
**Loch Erisort** Scotland 8 C3 58 05N 6 30W
**Loches** France 50 E4 47 08N 1 00E
**Loch Etive** Scotland 8 D1 56 30N 5 10W
**Loch Ewe** Scotland 8 D2 57 50N 5 40W
**Loch Eynort** Scotland 8 B2 57 10N 7 15W
**Loch Fannich** Scotland 9 E2 57 40N 5 00W
**Loch Fyne** Scotland 10 D3 56 00N 5 20W
**Loch Gairloch** Scotland 8 D2 57 45N 5 45W
**Loch Garry** Scotland 9 D2 57 05N 5 00W
**Lochgelly** Scotland 11 F3 56 08N 3 19W
**Lochgilphead** Scotland 10 D3 56 03N 5 26W
**Loch Glass** Scotland 9 E2 57 40N 4 30W
**Loch Goil** Scotland 10 E3 56 05N 4 55W
**Loch Gorm** Scotland 10 C2 55 49N 6 24W
**Loch Gruinart** Scotland 10 C2 55 51N 6 21W
**Loch Hope** Scotland 9 E3 58 25N 4 37W
**Loch Indaal** Scotland 10 C2 55 45N 6 21W
**Lochinver** Scotland 8 D3 58 09N 5 15W
**Loch Katrine** Scotland 11 E3 56 16N 4 30W
**Loch Ken** Scotland 11 E2 55 02N 4 07W
**Loch Laggan** Scotland 9 E1 56 55N 4 30W
**Loch Langavat** Scotland 8 C3 58 05N 6 50W
**Loch Leven** Scotland 9 D1/E1 56 45N 5 05W
**Loch Leven** Scotland 9 D1 56 15N 3 23W
**Loch Leven** Scotland 11 F3 56 15N 3 23W
**Loch Linnhe** Scotland 8 D1 56 35N 5 25W
**Loch Lochy** Scotland 9 E1 56 58N 4 55W
**Loch Lomond** Scotland 11 E3 56 10N 4 35W
**Loch Long** Scotland 10 E3 56 05N 4 55W
**Loch Lyon** Scotland 9 E1 56 30N 4 35W
**Lochmaben** Scotland 11 F2 55 08N 3 27W
**Lochmaddy** Scotland 8 B2 57 36N 7 08W
**Loch Maree** Scotland 8 D2 57 40N 5 30W
**Loch Meig** Scotland 9 E2 57 05N 4 45W
**Loch Melfort** Scotland 10 D3 56 10N 5 30W
**Loch Monar** Scotland 9 D2 57 25N 5 02W
**Loch Morar** Scotland 8 D1 57 55N 5 45W
**Loch More** Scotland 9 E3 58 15N 4 55W
**Loch Mullardoch** Scotland 9 D2 57 27N 4 58W
**Lochnagar** *mt.* Scotland 9 F1 56 57N 3 16W
**Loch Nan Clar** Scotland 9 E3 58 15N 4 05W
**Loch Ness** Scotland 9 E2 57 02N 4 30W
**Loch Nevis** Scotland 8 D1/D2 57 00N 5 40W
**Loch of Harray** Scotland 7 A2 59 02N 3 14W
**Loch Rannoch** Scotland 9 E1 56 41N 4 18W
**Lochranza** *tn.* Scotland 10 D2 55 42N 5 18W
**Loch Roag** Scotland 8 C3 58 10N 6 50W
**Loch Ryan** Scotland 10 D2 55 00N 5 05W
**Loch Scridain** Scotland 8 C1 56 20N 6 05W
**Loch Seaforth** Scotland 8 C2 58 00N 6 40W
**Loch Shell** Scotland 8 C2 58 00N 6 25W
**Loch Shiel** Scotland 8 D1 56 45N 5 35W
**Loch Shin** Scotland 9 E3 58 05N 4 30W
**Loch Snizort** Scotland 8 C2 57 30N 6 30W
**Loch Striven** Scotland 10 D2 55 58N 5 07W
**Loch Sunart** Scotland 8 D1 56 40N 5 45W
**Loch Sween** Scotland 10 D2 55 58N 5 38W
**Loch Tarbert** Scotland 10 D2 55 57N 5 55W
**Loch Tay** Scotland 9 E1 56 31N 4 10W
**Loch Torridon** Scotland 8 D2 57 35N 5 46W
**Lockerbie** Scotland 11 F2 55 07N 3 22W
**Lockhart** USA 91 G2 29 54N 97 14W
**Locks Heath** England 16 B1 50 52N 1 17W
**Loddon** England 17 E3 52 32N 1 29E
**Łodz** Poland 53 J5 51 49N 19 28E
**Lofoten Islands** Norway 49 D4 68 30N 15 00E
**Lofthouse Gate** England 13 E2 53 44N 1 30W
**Loftus** England 13 F3 54 33N 0 53W
**Logan, Mount** Canada 88 G6 60 34N 140 25W
**Logroño** Spain 51 D5 42 28N 2 26W
**Loire** *r.* France 50 D4 47 20N 1 20W
**Loja** Spain 51 C2 37 10N 4 09W
**Lolland** *i.* Denmark 49 D1 54 45N 12 20E
**Lomé** Togo 80 E3 6 10N 1 21E
**London** Canada 89 U2 42 58N 81 15W
**London** England 16 C2 51 30N 0 10W
**Londonderry** NI 10 B1 54 59N 7 19W
**Londonderry** *district* NI 10 B1 54 55N 7 20W
**Londrina** Brazil 96 G9 23 18S 51 13W
**Long Beach** *tn.* USA 90 C3 33 47N 118 15W
**Longbenton** England 13 E4 55 02N 1 35W
**Long Crendon** England 16 C2 51 47N 0 59W
**Long Eaton** England 15 E2 52 54N 1 15W
**Longford** RoI 20 D3 53 44N 7 47W
**Longford** *co.* RoI 20 D3 53 40N 7 50W
**Longhorsley** England 13 E4 55 15N 1 46W
**Longhoughton** England 11 H2 55 26N 1 36W
**Long Melford** England 17 D3 52 05N 0 43E
**Longridge** England 12 D3 53 51N 2 36W
**Long Sutton** England 17 D3 52 47N 0 08E
**Longuyon** France 50 G5 49 27N 5 36E
**Longwy** France 50 G5 49 32N 5 46E
**Lons-le-Saunier** France 50 G4 46 41N 5 33E
**Looe** England 18 C2 51 21N 4 27W
**Loop Head** RoI 21 B2 52 30N 9 55W
**Lora del Rio** Spain 51 C2 37 39N 5 32W
**Lorca** Spain 51 E2 37 40N 1 41W
**Loreto** Italy 54 E5 43 36N 13 36E
**Lorient** France 50 C4 47 45N 3 21W
**Los Angeles** USA 90 C3 34 00N 118 15W
**Lossiemouth** Scotland 9 F2 57 43N 3 18W
**Lostwithiel** England 18 C2 50 25N 4 40W
**Loudéac** France 50 C5 48 11N 2 45W
**Lough Allen** RoI 20 C4/D4 54 15N 8 00W
**Lough Arrow** RoI 20 C4 54 05N 8 15W
**Lough Beg** NI 10 C1 54 48N 6 30W
**Loughborough** England 15 E2 52 47N 1 11W
**Lough Carra** RoI 20 B3 53 40N 9 15W
**Lough Conn** RoI 20 B4 54 05N 9 10W
**Lough Corrib** RoI 20 B3 53 10N 9 10W
**Lough Cullin** RoI 20 B3 53 58N 9 10W
**Lough Dan** RoI 21 E3 53 05N 6 17W
**Lough Derg** Donegal RoI 20 D4 54 35N 7 55W
**Lough Derg** Tipperary RoI 21 C2 52 55N 8 15W
**Lough Derravaragh** RoI 20 D3 53 40N 7 22W
**Lough Ennell** RoI 20 D3 53 28N 7 24W
**Lough Eske** RoI 20 C4 54 41N 8 05W
**Lough Feeagh** RoI 20 B3 53 57N 9 35W
**Lough Finn** RoI 20 C4 54 51N 8 08W
**Lough Graney** RoI 21 C2 52 59N 8 40W
**Lough Kinale** RoI 20 D3 53 46N 7 25W

**Lough Leane** RoI 21 B2 52 03N 9 33W
**Lough Macnean Lower** NI 10 B1 54 17N 7 50W
**Lough Macnean Upper** NI/RoI 10 B1 54 18N 7 55W
**Lough Mask** RoI 20 B3 53 40N 9 30W
**Lough Melvin** RoI/NI 20 C4 54 25N 8 05W
**Lough Neagh** NI 10 C1 54 35N 6 30W
**Lough Oughter** RoI 20 D4 54 00N 7 30W
**Loughrea** *tn.* RoI 20 C3 53 12N 8 34W
**Lough Rea** RoI 20 C3 53 11N 8 35W
**Lough Ree** RoI 20 D3 53 35N 8 00W
**Lough Sheelin** RoI 20 D3 53 48N 7 20W
**Lough Swilly** RoI 20 D5 55 20N 7 35W
**Loughton** England 16 D2 51 39N 0 03E
**Louisburgh** RoI 20 B3 53 46N 9 48W
**Louisiana** *state* USA 91 H3 32 00N 92 00W
**Louisville** USA 91 J4 38 13N 85 48W
**Lourdes** France 50 D2 43 06N 0 02W
**Louth** England 13 F2 53 22N 0 01W
**Louth** RoI 20 E3 53 57N 6 33W
**Louth** *co.* RoI 20 E3 53 55N 6 25W
**Louviers** France 50 E5 49 13N 1 11E
**Lower Lough Erne** NI 10 B1 54 30N 7 45W
**Lowestoft** England 17 E3 52 29N 1 45E
**Lowick** England 11 H2 55 38N 1 58W
**Lowther Hills** Scotland 11 F2 55 25N 3 45W
**Luanda** Angola 82 B6 8 50S 13 15E
**Luarca** Spain 51 B5 43 33N 6 31W
**Lubango** Angola 82 B5 14 55S 13 30E
**Lubbock** USA 90 F3 33 35N 101 53W
**Lübeck** Germany 52 E6 53 52N 10 40E
**Lublin** Poland 53 L5 51 18N 22 31E
**Lubumbashi** CDR 82 E5 11 41S 27 29E
**Lucan** RoI 20 E3 53 22N 6 27W
**Lucca** Italy 54 D5 43 50N 10 30E
**Luce Bay** Scotland 11 E1 54 47N 4 50W
**Lucena** Spain 51 C2 37 25N 4 29W
**Luckenwalde** Germany 52 F6 52 05N 13 11E
**Lucknow** India 66 E5 26 50N 80 54E
**Lüderitz** Namibia 82 C2 26 38S 15 10E
**Ludgershall** England 16 B2 51 16N 1 37W
**Ludhiana** India 66 D6 30 56N 75 52E
**Ludlow** England 14 D2 52 22N 2 43W
**Ludwigshafen** Germany 52 D4 49 29N 8 27E
**Ludwigslust** Germany 52 E6 53 20N 11 30E
**Lugano** Switz. 52 D2 46 01N 8 57E
**Lugnaquillia** *mt.* RoI 21 E2 52 58N 6 27W
**Lugo** Spain 51 B5 43 00N 7 33W
**Luhans'k** Ukraine 56 E4 48 35N 39 20E
**Luleå** Sweden 49 F4 65 35N 22 10E
**Lundy** *i.* England 18 C3 51 11N 4 40W
**Lüneburg** Germany 52 E6 53 15N 10 24E
**Lunéville** France 50 H5 48 35N 6 30E
**Lurgan** NI 10 C1 54 28N 6 20W
**Lusaka** Zambia 82 E4 15 26S 28 20E
**Lusk** RoI 20 E3 53 32N 6 10W
**Lustenau** Austria 52 D3 47 26N 9 42E
**Luton** England 16 C2 51 53N 0 25W
**Luton** *u.a.* England 16 C2 51 53N 0 25W
**Lutterworth** England 15 E2 52 28N 1 10W
**LUXEMBOURG** 48 D2/E2
**Luxembourg** Luxembourg 48 E2 49 37N 6 08E
**Luxor** Egypt 81 M7 25 41N 32 24E
**Luzern** Switz. 52 D3 47 03N 8 17E
**Luzhou** China 69 J3 28 55N 105 25E
**Luzon** *i.* Philippines 71 G7 15 00N 122 00E
**L'viv** Ukraine 53 M4 49 50N 24 00E
**Lybster** Scotland 9 F3 58 18N 3 13W
**Lydd** England 17 D1 50 57N 0 55E
**Lydham** England 14 D2 52 31N 2 59W
**Lydney** England 15 D1 51 43N 2 32W
**Lyme Bay** England 19 E2 50 40N 2 55W
**Lyme Regis** England 19 E2 50 44N 2 57W
**Lyminge** England 17 E2 51 08N 1 05E
**Lymington** England 16 B1 50 46N 1 33W
**Lymm** England 12 D2 53 23N 2 28W
**Lyndhurst** England 16 B1 50 52N 1 34W
**Lynher** *r.* England 18 C2 50 33N 4 25W
**Lynmouth** England 18 D3 51 15N 3 50W
**Lynn Lake** *tn.* Canada 89 Q5 56 51N 101 01W
**Lynton** England 18 D3 51 15N 3 50W
**Lyons** France 50 G3 45 46N 4 50E
**Lytham St. Anne's** England 12 C2 53 45N 3 01W

## M

**Maam Cross** RoI 20 B3 53 27N 9 33W
**Maas** *r.* Neths 48 E4 51 50N 5 32E
**Maastricht** Neths 48 D3 50 51N 5 42E
**Mablethorpe** England 13 G2 53 21N 0 15E
**Macao** China 69 K2 22 10N 113 40E
**Macapá** Brazil 96 G14 0 04N 51 04W
**Macarata** Italy 54 E5 43 18N 13 27E
**Macclesfield** England 12 D2 53 16N 2 07W
**Macdonnell Ranges** Australia 72 F6 24 00S 132 30E
**Macduff** Scotland 9 G2 57 40N 2 29W
**Maceió** Brazil 96 K12 9 40S 35 44W
**Macgillycuddy's Reeks** *mts.* RoI 21 B2 52 00N 9 27W
**The Machars** Scotland 11 E1 54 49N 4 31W
**Machrihanish** Scotland 10 D2 55 26N 5 44W
**Machynlleth** Wales 14 C2 52 35N 3 51W
**Mackenzie** *r.* Canada 88 J7 67 00N 132 00W
**Mackenzie Mountains** Canada 88 J7/L6 66 00N 130 00W
**McKinley, Mount** USA 88 E6 62 02N 151 01W
**McKinney** USA 91 G3 33 14N 96 37W
**Mâcon** France 50 G4 46 18N 4 50E
**Macon** USA 91 K3 32 49N 83 37W
**Macroom** RoI 21 C1 51 54N 8 57W
**MADAGASCAR** 82 H2/J5
**Madeira** *r.* Brazil 96 E12 6 00S 61 30W
**Madeira Islands** Atlantic Ocean 80 A8 32 45N 17 00W
**Madison** USA 91 J5 43 04N 89 22W
**Madrid** Spain 51 D4 40 25N 3 43W
**Maesteg** Wales 14 C1 51 37N 3 40W
**Magadan** Russia 63 T8 59 38N 150 50E
**Magdeburg** Germany 52 E6 52 08N 11 37E
**Maghera** NI 10 C1 54 51N 6 40W
**Magherafelt** NI 10 C1 54 45N 6 36W
**Magherafelt** *district* NI 10 C1 54 45N 6 50W
**Maghull** England 12 D2 55 32N 2 57W
**Mahajanga** Madagascar 82 J4 15 40S 46 20E
**Mahilyow** Belarus 53 Q6 53 54N 30 20E
**Mahón** Spain 51 H3 39 54N 4 15E
**Maidenhead** England 16 C2 51 32N 0 44W
**Maiden Newton** England 19 E2 50 46N 2 35W
**Maidstone** England 17 D2 51 17N 0 32E
**Maiduguri** Nigeria 80 G4 11 53N 13 16E
**Main** *r.* NI 10 C1 54 50N 6 25W
**Maine** *state* USA 91 N6 45 00N 70 00W
**Mainz** Germany 52 D4 50 00N 8 16E
**Malabo** Equatorial Guinea 80 F2 3 45N 8 48E
**Málaga** Spain 51 C2 36 43N 4 25W
**Malahide** RoI 20 E3 53 27N 6 09W
**Malange** Angola 82 C6 9 32S 16 20E
**Malatya** Turkey 57 E2 38 22N 38 18E
**MALAWI** 82 F5
**MALAYSIA** 71 C5/E5
**Mal Bay** RoI 21 B2 52 40N 9 30W
**MALDIVES** 103
**Maldon** England 17 D2 51 45N 0 40E
**Malé** Maldives 59 4 10N 73 29E
**MALI** 80 C4/D5
**Malin Head** RoI 20 D5 55 30N 7 20W
**Malin More** RoI 20 C4 54 41N 8 46W
**Mallaig** Scotland 8 D2 57 00N 5 50W
**Mallorca** *i.* Spain 51 G3 39 50N 2 30E
**Mallow** RoI 21 C2 52 08N 8 39W
**Malmesbury** England 16 A2 51 36N 2 06W
**Malmö** Sweden 49 D2 55 35N 13 00E
**Malpas** England 12 D2 53 01N 2 46W
**MALTA** 54 F1
**Maltby** England 13 E2 53 26N 1 11W
**Malton** England 13 F3 54 08N 0 48W
**Malvern Hills** England 15 D2 52 05N 2 22W
**Manacor** Spain 51 G3 39 35N 3 12E
**Managua** Nicaragua 94 G2 12 06N 86 18W
**Manama** Bahrain 65 F4 26 12N 50 38E
**Manaus** Brazil 96 F13 3 06S 60 00W
**Manchester** England 12 D2 53 30N 2 15W
**Manfredonia** Italy 54 F4 41 37N 15 55E
**Mangotsfield** England 15 D1 51 29N 2 31W
**Manhattan** USA 93 F2 40 48N 73 58W
**Manila** Philippines 71 G6 14 37N 120 58E
**Manitoba** *prov.* Canada 89 R5 55 15N 100 00W
**Manitoba, Lake** Canada 89 R4 50 30N 98 15W
**Manizales** Colombia 96 B15 5 03N 75 32W
**Mannheim** Germany 52 D4 49 30N 8 28E
**Manningtree** England 17 E2 51 57N 1 04E
**Manorhamilton** RoI 20 C4 54 18N 8 10W
**Manresa** Spain 51 F4 41 43N 1 50E
**Mansfield** England 13 E2 53 09N 1 11W
**Mansfield Woodhouse** England 13 E2 53 10N 1 11W
**Mantes-la-Jolie** France 50 E5 48 59N 1 43E
**Manuta** Italy 54 D6 45 10N 10 47E
**Manzanares** Spain 51 D3 39 00N 3 23W
**Maputo** Mozambique 82 F2 25 58S 32 35E
**Marabá** Brazil 96 H12 5 23S 49 10W
**Maracaibo** Venezuela 96 C16 10 44N 71 37W
**Marbella** Spain 51 C2 36 31N 4 53W
**Marburg** Germany 52 D5 50 49N 8 36E
**March** England 17 D3 52 33N 0 06E
**Mar del Plata** Argentina 97 F6 38 00S 57 32W
**Marden** England 17 D2 51 11N 0 30E
**Maresfield** England 17 D2 51 00N 0 05E
**Margam** Wales 14 C1 51 34N 3 44W
**Margate** England 17 E2 51 24N 1 24E
**Maribor** Slovenia 54 F7 46 34N 15 38E
**Marignane** France 50 G2 43 25N 5 12E
**Mariscal Estigarribia** Paraguay 96 E9 22 04S 60 34W
**Mariupol'** Ukraine 57 E4 47 05N 37 34E
**Market Deeping** England 15 F2 52 41N 0 19W
**Market Drayton** England 15 D2 52 54N 2 29W
**Market Harborough** England 15 F2 52 29N 0 55W
**Markethill** NI 10 C1 54 18N 6 31W
**Market Rasen** England 13 F2 53 24N 0 21W
**Market Weighton** England 13 F2 53 52N 0 40W
**Markinch** Scotland 11 F3 56 12N 3 09W
**Marlborough** England 16 B2 51 26N 1 43W
**Marlborough Downs** England 16 B2 51 28N 1 50W
**Marlow** England 16 C2 51 35N 0 48W
**Marmande** France 50 E3 44 30N 0 10E
**Marple** England 12 D2 53 24N 2 03W
**Marrakesh** Morocco 80 C8 31 49N 8 00W
**Marsala** Italy 54 E2 37 48N 12 27E
**Marsden** West Yorkshire England 13 E2 53 36N 1 55W
**Marsden** Tyne and Wear England 13 E3 54 58N 1 21W
**Marseilles** France 50 G2 43 18N 5 22E
**MARSHALL ISLANDS** 100 G9/H8
**Marshfield** England 15 D1 51 28N 2 19W
**Marske-by-the-Sea** England 13 E3 54 36N 1 01W
**Martigny** Switz. 52 C3 46 07N 7 05E
**Martigues** France 50 G2 43 24N 5 03E
**Martinique** *i.* Lesser Antilles 94 R12 14 30N 61 00W
**Martos** Spain 51 C2 37 44N 3 58W
**Maryland** *state* USA 91 L4 39 00N 77 00W
**Maryport** England 12 C3 54 43N 3 30W
**Maryport** Scotland 10 E1 54 43N 3 30W
**Maseru** Lesotho 82 E2 29 19S 27 29E
**Masham** England 13 E3 54 13N 1 40W
**Mashhad** Iran 65 G6 36 16N 59 34E
**Massa** Italy 54 D5 44 02N 10 09E
**Massachusetts** *state* USA 91 M5 42 00N 72 00W
**Massif Central** France 50 F3 45 00N 3 30E
**Massif de la Vanoise** France 50 H3 45 20N 6 20E
**Matadi** CDR 82 B6 5 50S 13 32E
**Mataró** Spain 51 G4 41 32N 2 27E
**Matera** Italy 55 G4 40 40N 16 37E
**Matlock** England 13 E2 53 08N 1 32W
**Matterhorn** *mt.* Switz. 52 C2 45 59N 7 39E
**Maubeuge** France 50 F6 50 17N 3 58E
**Mauchline** Scotland 11 E2 55 31N 4 24W
**MAURITANIA** 80 B5/C6C
**MAURITIUS** 103
**Maybole** Scotland 11 E2 55 21N 4 41W
**Mayenne** France 50 D5 48 18N 0 37W
**Maykop** Russia 62 G5 44 37N 40 48E
**Maynooth** RoI 20 E3 53 23N 6 35W
**Mayo** RoI 20 B3 54 35N 9 07W
**Mayo** *co.* RoI 20 B3 53 55N 9 20W
**Mazamet** France 50 F2 43 29N 2 22E
**Mazara del Vallo** Italy 54 E2 37 40N 12 34E
**Mazarrón** Spain 51 E2 37 36N 1 19W
**Mazatlán** Mexico 94 C4 23 11N 106 25W
**Mazirbe** Latvia 49 F2 57 40N 22 21E
**Mbabane** Swaziland 82 F2 26 20S 31 08E
**Mbandaka** CDR 81 J2 0 03N 18 28E
**Mbeya** Tanzania 82 F6 8 54S 33 29E
**Meath** *co.* RoI 20 E3 53 35N 6 30W
**Meaux** France 50 F5 48 58N 2 54E
**Mecca** Saudi Arabia 64 C3 21 26N 39 49E
**Medan** Indonesia 71 B4 3 35N 98 39E
**Medellín** Colombia 96 B15 6 15N 75 36W
**Mediterranean Sea** 54 B2/E2 35 00N 15 00E
**Medway** *r.* England 17 D2 51 24N 0 40E
**Medway** *u.a.* England 17 D2 51 25N 0 35E
**Meerut** India 66 D5 29 00N 77 42E
**Meig** *r.* Scotland 9 E2 57 35N 4 50W
**Mekong** *r.* Asia 71 C8 16 00N 105 00E
**Melbourn** England 16 D3 52 05N 0 01E
**Melbourne** Australia 73 H3 37 45S 144 58E
**Melbourne** England 15 E2 52 49N 1 25W
**Melilla** *territory* Spain 51 D1 35 17N 2 57W
**Melitopol'** Ukraine 57 E4 46 51N 35 22E
**Melksham** England 16 A2 51 23N 2 09W
**Melrose** Scotland 11 G2 55 36N 2 44W
**Meltham** England 13 E2 53 36N 1 52W
**Melton Mowbray** England 15 F2 52 46N 0 53W
**Melun** France 50 F5 48 32N 2 40E
**Melvich** Scotland 9 F3 58 33N 3 55W
**Memmingen** Germany 52 E3 47 59N 10 11E
**Memphis** USA 91 J4 35 10N 90 00W
**Menai Bridge** *tn.* Wales 14 B3 53 14N 4 10W
**Menai Strait** Wales 14 B3 53 14N 4 10W
**Mende** France 50 F3 44 32N 3 30E
**Mendip Hills** England 19 E3 51 18N 2 45W
**Mendoza** Argentina 97 D7 32 48S 68 52W
**Menorca** *i.* Spain 51 G4/H3 39 45N 4 15E
**Menton** France 50 H2 43 47N 7 30E
**Meppel** Neths 48 E5 52 42N 6 12E
**Merano** Italy 54 D7 46 41N 11 10E
**Mere** England 16 A2 51 06N 2 16W
**Mérida** Mexico 94 G4 20 59N 89 39W
**Mérida** Spain 51 B3 38 55N 6 20W
**Mérignac** France 50 D3 44 50N 0 36W
**Merrick** *mt.* Scotland 11 E2 55 08N 4 29W
**Merriott** England 19 E2 50 55N 2 49W
**Mersea Island** England 17 D2 51 46N 0 57
**Mersey** *r.* England 12 D2 53 20N 2 53W
**Merseyside** *co.* England 12 D2 53 30N 3 20W
**Merthyr Tydfil** Wales 14 C1 51 46N 3 23W
**Merthyr Tydfil** *u.a.* Wales 14 C1 51 48N 3 20W
**Merton** England 16 C2 51 25N 0 12W
**Messina** Italy 54 F3 38 13N 15 33E
**Metheringham** England 13 F2 53 08N 0 25W
**Methil** Scotland 11 F3 56 12N 3 01W
**Metz** France 50 H5 49 07N 6 11E
**Meuse** *r.* Belgium/France 48 D3 50 03N 4 40E
**Mevagissey** England 18 C2 50 16N 4 48W
**Mexborough** England 13 E2 53 30N 1 17W
**MEXICO** 94 C5/E3
**Mexico City** Mexico 94 E3 19 25N 99 10W
**Mexico, Gulf of** Mexico 94 F4/G4 25 00N 90 00W
**Miami** USA 91 K2 25 45N 80 15W
**Michigan** *state* USA 91 J5/K5 45 00N 85 00W
**Michigan, Lake** Canada/USA 92 C2/C3 45 00N 87 00W
**Middlesbrough** England 13 E3 54 35N 1 14W
**Middlesbrough** *u.a.* England 13 E3 54 33N 1 04W
**Middleton** England 12 D2 53 33N 2 12W
**Middleton-in-Teesdale** England 12 D3 54 38N 2 04W
**Middleton-on-the-Wolds** England 13 F2 53 56N 0 33W
**Middlewich** England 12 D2 53 11N 2 27W
**Midhurst** England 16 C1 50 59N 0 45W
**Midleton** RoI 21 C1 51 55N 8 10W
**Midlothian** *u.a.* Scotland 11 F2 55 50N 3 05W
**Midsomer Norton** England 19 E3 51 17N 2 30W
**Mieres** Spain 51 C5 43 15N 5 46W
**Mikkeli** Finland 49 G3 61 44N 27 15E
**Milan** Italy 54 C6 45 28N 9 12E
**Milborne Port** England 19 E2 50 58N 2 29W
**Mildenhall** England 17 D3 52 21N 0 30E
**Milestone** RoI 21 C2 52 40N 8 05W
**Milford** RoI 20 D5 55 07N 7 43W
**Milford Haven** Wales 14 A1 51 44N 5 02W
**Millau** France 50 F3 44 06N 3 05E
**Millisle** NI 10 D1 54 37N 5 30W
**Millom** England 12 C3 54 13N 3 18W
**Millport** Scotland 10 E2 55 46N 4 55W
**Millstreet** RoI 21 B2 52 03N 9 04W
**Milltown Malbay** RoI 21 B2 52 52N 9 23W
**Milngavie** Scotland 11 E2 55 57N 4 19W
**Milnthorpe** England 12 D3 54 14N 2 46W
**Milton Keynes** England 16 C3 52 02N 0 42W
**Milton Keynes** *u.a.* England 16 C3 52 02N 0 42W
**Milverton** England 19 D3 51 02N 3 16W
**Milwaukee** USA 91 J5 43 03N 87 56W
**Mimizan** France 50 D3 44 12N 1 14W
**Minch, The** *sd.* Scotland 8 C3/D3 58 00N 6 00W
**Mindanao** *i.* Philippines 71 G5/H5 8 00N 125 00E
**Minden** Germany 52 D6 52 18N 8 54E
**Mindoro** *i.* Philippines 71 G6 13 00N 121 00E
**Minehead** England 18 D3 51 13N 3 29W
**Mine Head** RoI 21 D1 52 00N 7 35W
**Mingulay** *i.* Scotland 8 B1 56 49N 7 38W
**Minneapolis** USA 91 H5 45 00N 93 15W
**Minnedosa** Canada 89 R4 50 14N 99 50W
**Minnesota** *state* USA 91 H6 47 00N 95 00W
**Minsk** Belarus 56 C5 53 51N 27 30E
**Minster** England 17 D2 51 26N 0 49E
**Minster** England 17 E2 51 20N 1 19E
**Minsterley** England 14 D2 52 39N 2 55W
**Mintlaw** Scotland 9 G2 57 31N 2 00W
**Mirfield** England 13 E2 53 41N 1 42W
**Miskolc** Hungary 53 K4 48 07N 20 47E
**Misratah** Libya 81 G8 32 23N 15 00E
**Mississippi** *r.* USA 91 H3 35 00N 90 00W
**Mississippi** *state* USA 91 H3/J3 32 00N 90 00W
**Missouri** *r.* USA 91 H4 39 00N 93 00W
**Missouri** *state* USA 91 H4 38 00N 93 00W
**Mitchelstown** RoI 21 C2 52 16N 8 16W
**Mito** Japan 70 D2 36 22N 140 29E
**Mizen Head** RoI 21 B1 51 30N 9 50W
**Moate** RoI 20 D3 53 24N 7 58W
**Mobile** USA 91 J3 30 40N 88 05W
**Moçambique** Mozambique 82 H4 15 03S 40 45E
**Modbury** England 18 D2 50 21N 3 53W
**Módena** Italy 54 D6 44 39N 10 55E
**Modesto** USA 90 B4 37 37N 121 00W
**Mödling** Austria 53 H4 48 06N 16 18E
**Moelfre** Wales 14 B3 53 21N 4 14W
**Moffat** Scotland 11 F2 55 20N 3 27W
**Mogadishu** Somalia 81 Q2 2 02N 45 21E
**Mohill** RoI 20 D3 53 54N 7 52W
**Mo-i-Rana** Norway 49 D4 66 18N 14 00E
**Mold** Wales 14 C3 53 10N 3 08W
**MOLDOVA** 53 P3
**Mole** *r.* England 16 C2 51 15N 0 20W
**Molfetta** Italy 55 G4 41 12N 16 36E
**Mölndal** Sweden 49 D2 57 40N 12 00E
**Mombasa** Kenya 83 F2 4 04S 39 40E
**MONACO** 50 H2
**Monadhliath Mountains** Scotland 9 E2 57 10N 4 00W
**Monaghan** RoI 20 E4 54 15N 6 58W
**Monaghan** *co.* RoI 20 D4/E4 54 10N 7 00W
**Monasterevin** RoI 21 D3 53 07N 7 02W
**Mönchengladbach** Germany 52 C5 51 12N 6 25E
**Moncton** Canada 89 Y3 46 04N 64 50W
**Mondovi** Italy 54 B6 44 23N 7 49E
**Moneen** RoI 20 C3 53 26N 8 54W
**Moneymore** NI 10 C1 54 42N 6 40W
**Monfalcone** Italy 54 E6 45 49N 13 32E

MONGOLIA 68/69 G7/K7
Moniaive Scotland 11 F2 55 12N 3 55W
Monifieth Scotland 9 G1 56 29N 2 49W
Monmouth Wales 14 D1 51 50N 2 43W
Monmouthshire u.a.
Wales 14 D1 51 47N 2 55W
Monopoli Italy 55 G4 40 57N 17 18E
Monrovia Liberia 80 B3 6 20N 10 46W
Montana Bulgaria 55 K5 43 25N 23 11E
Montana state USA 90 E6 47 00N 111 00W
Montargis France 50 F4 48 00N 2 44E
Montauban France 50 E3 44 01N 1 20E
Montbéliard France 50 H4 47 31N 6 48E
Mont Blanc France/Italy 50 H3 45 50N 6 52E
Montceau-les-Mines
France 50 G4 46 40N 4 23E
Mont-de-Marsan
France 50 D2 43 54N 0 30W
Montdidier France 50 F5 49 39N 2 35E
Monte Carlo Monaco 50 H2 43 44N 7 25E
Montélimar France 50 G3 44 33N 4 45E
Montepulciano Italy 54 D5 43 0 5N 11 46E
Monterrey Mexico 94 D5 25 40N 100 20W
Montes Claros Brazil 96 J10 16 45S 43 52W
Montevideo Uruguay 97 F6 34 55S 56 10W
Montgomery USA 91 J3 32 22N 86 20W
Montgomery Wales 14 C2 52 34N 3 10W
Montluçon France 50 F4 46 20N 2 36E
Montmirail France 50 F5 48 52N 3 34E
Montmorilion France 50 E4 46 26N 0 52E
Montoro Spain 51 C3 38 02N 4 23W
Montpelier Rol 21 C2 52 45N 8 30W
Montpellier France 50 F2 43 36N 3 53E
Montréal Canada 89 W3 45 32N 73 36W
Montreux Switz. 52 C3 46 27N 6 55E
Montrose Scotland 9 G1 56 43N 2 29W
Montserrat i.
Lesser Antilles 94 P9 16 45N 62 14W
Monza Italy 54 C6 45 35N 9 16E
Mooncoin Rol 21 D2 52 17N 7 15W
Moone Rol 21 E2 52 58N 6 49W
Moorfoot Hills Scotland 11 F2/G2 55 50N 3 00W
Moosonee Canada 89 U4 51 18N 80 39W
Mora Sweden 49 D3 61 00N 14 30E
Moray u.a. Scotland 9 F2 57 20N 3 10W
Moray Firth Scotland 9 F2 57 45N 3 45W
Morcenx France 50 D3 44 02N 0 55W
Morecambe England 12 D3 54 04N 2 53W
Morecambe Bay
England 12 D3 54 10N 2 55W
Morelia Mexico 94 D3 19 40N 101 11W
Moreton England 12 C2 53 24N 3 07W
Moretonhampstead
England 18 D2 50 40N 3 45W
Moreton-in-Marsh
England 15 E1 51 59N 1 42W
Morlaix France 50 C5 48 35N 3 50W
Morley England 13 E2 53 46N 1 36W
MOROCCO 80 C8/D8
Morón de la Frontera
Spain 51 C2 37 07N 5 27W
Moroni Comoros 82 H5 11 40S 43 16E
Morpeth England 13 E4 55 10N 1 41W
Morte Point England 18 C3 51 10N 4 15W
Mosborough England 13 E2 53 19N 1 22W
Moscow Russia 62 F8 55 45N 37 42E
Mosel r.
Germany/France 52 C4 50 00N 7 00E
Mossley NI 10 D1 54 42N 5 57W
Mossoró Brazil 96 K12 5 10S 37 18W
Mostar
Bosnia-Herzegovina 55 G5 43 20N 17 50E
Móstoles Spain 51 D4 40 19N 3 53W
Mosul Iraq 64 D6 36 21N 43 08E
Motherwell Scotland 11 F2 55 48N 3 59W
Motril Spain 51 D2 36 45N 3 31W
Moulins France 50 F4 47 00N 3 48E
Mountain Ash Wales 14 C1 51 42N 3 24W
Mount Bellew Rol 20 C3 53 28N 8 30W
Mount Leinster Rol 21 E2 52 36N 6 42W
Mountmellick tn. Rol 21 D3 53 07N 7 20W
Mountrath Rol 21 D3 53 00N 7 27W
Mounts Bay England 18 B2 50 05N 5 30W
Mountshannon Rol 21 C2 52 56N 8 26W
Mountsorrel England 15 E2 52 44N 1 07W
Moura Portugal 51 B3 38 08N 7 27W
Mourne Mountains NI 10 C1 54 05N 6 05W
Mousehole England 18 B2 50 04N 5 34W
Moville Rol 20 D5 55 11N 7 03W
Moy NI 10 C1 54 27N 6 42W
Moycullen Rol 20 B3 53 20N 9 10W
Moygashel NI 10 C1 54 31N 6 46W
Moyle district NI 10 C2 55 10N 6 14W
Moyvore Rol 20 D3 53 22N 7 37W
MOZAMBIQUE 82 F3/G5
Much Wenlock
England 15 D2 52 36N 2 34W
Muck i. Scotland 8 C1 56 50N 6 15W
Muckross Rol 21 B2 52 01N 9 29W
Muff Rol 20 D5 55 04N 7 16W
Muine Bheag Rol 21 E2 52 41N 6 57W
Muirkirk Scotland 11 F2 55 31N 4 04W
Muir of Ord Scotland 9 F2 57 31N 4 27W
Mulhacén mt. Spain 51 D2 37 04N 3 19W
Mülheim an der Ruhr
Germany 48 E4 51 25N 6 50E
Mulhouse France 50 H4 47 45N 7 21E

Mulhuddart Rol 20 E3 53 24N 6 24W
Mulkear r. Rol 21 C2 52 38N 8 24W
Mull i. Scotland 8 D1 56 25N 6 00W
Mullet Peninsula Rol 20 A4 54 07N 10 05W
Mull Head Scotland 7 B2 59 20N 2 58W
Mullinahone Rol 21 D2 52 30N 7 30W
Mullinavat Rol 21 D2 52 22N 7 10W
Mullingar Rol 20 D3 53 32N 7 20W
Mullion England 18 B2 50 01N 5 14W
Mull of Galloway
Scotland 10 E1 54 38N 4 55W
Mull of Kintyre
Scotland 10 D2 55 17N 5 55W
Mull of Oa Scotland 10 C2 55 35N 6 20W
Mull, Sound of Scotland 8 D1 56 30N 6 00W
Multan Pakistan 66 C6 30 10N 71 36E
Mumbai India 66 C3 18 56N 72 51E
Mundesley England 17 E3 52 53N 1 26E
Munich Germany 52 E4 48 08N 11 35E
Münster Germany 52 C5 51 58N 7 37E
Murcia Spain 51 E2 37 59N 1 08W
Murmansk Russia 62 F10 68 59N 33 08E
Murray r. Australia 73 H3 35 30S 144 00E
Murray Bridge Australia 73 G3 35 10S 139 17E
Murton England 13 E3 54 49N 1 23W
Muş Turkey 57 F2 38 45N 41 30E
Muscat Oman 65 G3 23 37N 58 38E
Mushin Nigeria 80 E3 6 30N 3 15E
Musselburgh Scotland 11 F2 55 57N 3 04W
Mutare Zimbabwe 82 F4 18 58N 32 40E
Mutton Island i. Rol 21 B2 52 49N 9 31W
Mwanza Tanzania 83 C3 2 31S 32 56E
MYANMAR 71 B7
Mybster Scotland 9 F3 58 27N 3 25W
Mykolayiv Ukraine 57 D4 46 57N 32 00E
Mynydd Eppynt Wales 14 C2 52 02N 3 35W
Mynydd Preseli Wales 14 B1 51 58N 4 45W
Mytilíni Greece 55 M3 39 06N 26 34E
Mytishchi Russia 56 E6 55 54N 37 47E

**N**

Naas Rol 20 E3 53 13N 6 39W
Naestved Denmark 49 D2 55 14N 11 47E
Náfplio Greece 55 K2 37 34N 22 48E
Nagano Japan 70 C2 36 39N 138 10E
Nagasaki Japan 70 A1 32 45N 129 52E
Nagles Mountains Rol 21 C2 52 10N 8 25W
Nagoya Japan 70 C2 35 08N 136 53E
Nagpur India 66 D4 21 10N 79 12E
Nailsea England 19 E3 51 26N 2 46W
Nailsworth England 15 D1 51 42N 2 14W
Nairn Scotland 9 F2 57 35N 3 53W
Nairobi Kenya 83 E4 1 17S 36 50E
Namib Desert Namibia 82 B3/C2 22 00S 14 00E
NAMIBIA 82 C3
Nampula Mozambique 82 G4 15 09S 39 14E
Namur Belgium 48 C3 50 28N 4 52E
Nanchang China 69 L3 28 33N 115 58E
Nancy France 50 H5 48 42N 6 12E
Nanjing China 69 L4 32 03N 118 47E
Nantes France 50 D4 47 14N 1 35W
Nantwich England 12 D2 53 04N 2 32W
Naples Italy 54 F4 40 50N 14 15E
Naples USA 91 K2 26 09N 81 48W
Narberth Wales 14 B1 51 48N 4 45W
Narbonne France 50 F2 43 11N 3 00E
Narborough England 15 E2 52 35N 1 11W
Narvik Norway 49 E4 68 26N 17 25E
Naseby England 15 F2 52 23N 0 59W
Nashville USA 91 J4 36 10N 86 50W
Nassau Bahamas 95 J5 25 05N 77 20W
Nasser, Lake Egypt 81 M6 22 35N 31 40E
Nässjö Sweden 49 D2 57 40N 14 40E
Natal Brazil 96 K12 5 46S 35 15W
Natchez USA 91 H3 31 32N 91 24W
Naul Rol 20 E3 53 35N 6 17W
NAURU 100 G7
Navan Rol 20 E3 53 39N 6 41W
Navenby England 13 F2 53 06N 0 32W
Naver r. Scotland 9 E3 58 30N 4 10W
Ndjamena Chad 81 J4 12 10N 14 59E
Néapoli Greece 55 K2 36 31N 23 03E
Neath Wales 14 C1 51 40N 3 48W
Neath Port Talbot u.a.
Wales 14 C1 51 37N 3 43W
Nebraska state USA 90 F5 42 00N 102 00W
Needham Market
England 17 E3 52 09N 1 03E
Needles, The England 16 B1 50 39N 1 35W
Nefyn Wales 14 B2 52 56N 4 32W
Negro r. Brazil 96 D13 0 05S 60 00W
Neiva Colombia 96 B14 2 58N 75 15W
Nelson England 12 D2 53 51N 2 13W
Neman r.
Lithuania/Russia 56 C5 55 00N 22 00E
Nenagh Rol 21 C2 52 52N 8 12W
Nene r. England 15 F2 52 25N 0 30W
Nene r. England 17 D3 52 45N 0 10E
NEPAL 66/67 E5/F5
Nephin mt. Rol 20 B4 54 01N 9 22W
Nephin Beg Range Rol 20 B3/B4 54 00N 9 34W
Nerva Spain 51 B2 37 41N 6 33W
Neston England 12 C2 53 18N 3 03W
NETHERLANDS 48 D5/E5
Nether Stowey
England 19 D3 51 09N 3 10W

Netzahualcóyotl
Mexico 94 E3 19 24N 99 02W
Neubrandenburg
Germany 52 F6 53 33N 13 16E
Neuchâtel Switz. 52 C3 46 55N 6 56E
Neufchâteau France 50 G5 48 21N 5 42E
Neufchâtel-en-Bray
France 50 E5 49 44N 1 26E
Neumünster Germany 52 D7 54 05N 9 59E
Neuruppin Germany 52 F6 52 56N 12 49E
Neuss Germany 52 C5 51 12N 6 42E
Neustrelitz Germany 52 F6 53 22N 13 05E
Nevada state USA 90 C4 39 00N 118 00W
Nevers France 50 F4 47 00N 3 09E
New Abbey Scotland 11 F1 54 59N 3 38W
New Alresford England 16 B2 51 06N 1 10W
Newark USA 91 M5 40 44N 74 24W
Newark-on-Trent
England 13 F2 53 05N 0 49W
New Ash Green
England 17 D2 51 21N 0 18E
Newbiggin-by-the-Sea
England 13 E4 55 11N 1 30W
Newbridge Wales 14 C1 51 41N 3 09W
Newbridge-on-Wye
Wales 14 C2 52 13N 3 27W
New Brunswick prov.
Canada 89 X3 47 30N 66 00W
New Buildings NI 10 B1 54 56N 7 21W
Newburgh Scotland 9 F1 56 21N 3 15W
Newburn England 13 E3 54 59N 1 43W
Newbury England 16 B2 51 25N 1 20W
New Caledonia i.
Pacific Ocean 73 M6/N6 22 00S 165 00E
Newcastle NI 10 D1 54 12N 5 54W
Newcastle Emlyn
Wales 14 B2 52 02N 4 28W
Newcastle-under-Lyme
England 12 D2 53 00N 2 14W
Newcastle upon Tyne
England 13 E3 54 59N 1 35W
Newcastle West Rol 21 B2 52 27N 9 03W
New Cumnock
Scotland 11 E2 55 24N 4 12W
New Deer Scotland 9 G2 57 30N 2 12W
New Delhi India 66 D5 28 37N 77 14E
Newfoundland i.
Canada 89 Z3 48 15N 57 00W
Newfoundland and Labrador prov.
Canada 89 Y5/AA3 52 30N 62 30W
New Galloway
Scotland 11 E2 55 05N 4 10W
New Guinea i.
Pacific Ocean 72/73 F10/J9 5 00S 141 00E
New Hampshire state
USA 91 M5 43 00N 72 00W
New Haven USA 91 M5 41 18N 72 55W
New Holland England 13 F2 53 42N 0 22W
Newick England 16 D1 50 58N 0 01E
New Jersey state USA 91 M4 40 00N 75 00W
New Liskeard Canada 89 V3 47 31N 79 41W
Newlyn England 18 B2 50 06N 5 34W
Newmachar Scotland 9 G2 57 16N 2 10W
Newmains Scotland 11 F2 55 47N 3 53W
Newmarket England 17 D3 52 15N 0 25E
Newmarket Rol 21 C2 52 13N 9 00W
Newmarket-on-Fergus
Rol 21 C2 52 45N 8 53W
New Mexico state USA 90 E3 35 00N 107 00W
New Milton England 16 B1 50 46N 1 40W
New Orleans USA 91 H2 30 00N 90 03W
Newport Telford and Wrekin
England 15 D2 52 47N 2 22W
Newport Isle of Wight
England 16 B1 50 42N 1 18W
Newport Essex
England 17 D2 51 58N 0 13E
Newport Mayo Rol 20 B3 53 53N 9 32W
Newport Tipperary Rol 21 C2 52 43N 8 25W
Newport Pembrokeshire
Wales 14 B2 52 01N 4 50W
Newport Newport
Wales 14 C1 51 35N 3 00W
Newport u.a. Wales 14 C1/D1 51 33N 3 00W
Newport Bay Rol 20 B3 53 51N 9 41W
Newport News USA 91 L4 36 59N 76 26W
Newport-on-Tay
Scotland 9 G1 56 27N 2 56W
Newport Pagnell
England 16 C3 52 05N 0 44W
Newquay England 18 B2 50 25N 5 05W
New Quay Wales 14 B2 52 13N 4 22W
New Radnor Wales 14 C2 52 15N 3 10W
New Romney England 17 D1 50 59N 0 57E
New Ross Rol 21 D2 52 24N 6 56W
New Rossington
England 13 E2 53 29N 1 04W
Newry NI 10 C1 54 11N 6 20W
Newry and Mourne
district NI 10 C1 54 10N 6 35W
New Sauchie Scotland 11 F3 56 08N 3 41W
New Scone Scotland 9 F1 56 25N 3 25W
Newton Abbot
England 18 D2 50 32N 3 36W
Newton Aycliffe
England 13 E3 54 37N 1 34W

Newton Mearns
Scotland 11 E2 55 45N 4 18W
Newtonmore Scotland 9 E2 57 04N 4 08W
Newton Stewart
Scotland 11 E1 54 57N 4 29W
Newton Rol 21 D2 52 52N 7 07W
Newtown Wales 14 C2 52 32N 3 19W
Newtownabbey NI 10 D1 54 40N 5 54W
Newtownabbey
district NI 10 C1/D1 54 40N 6 05W
Newtownards NI 10 D1 54 36N 5 41W
Newtown Cunningham
Rol 20 D4 54 59N 7 31W
Newtownhamilton NI 10 C1 54 12N 6 35W
Newtownmountkennedy
Rol 21 E3 53 06N 6 07W
Newtown St. Boswells
Scotland 11 G2 55 34N 2 40W
Newtownstewart NI 10 B1 54 43N 7 24W
New Tredegar Wales 14 C1 51 44N 3 15W
New York USA 91 M5 40 40N 73 50W
New York state USA 91 L5 43 00N 76 00W
NEW ZEALAND 73
Neyland Wales 14 B1 51 43N 4 57W
Niagara Falls tn.
Canada 93 E2 43 05N 79 06W
Niamey Niger 80 E4 13 32N 2 05E
NICARAGUA 95 G2
Nice France 50 H2 43 42N 7 16E
Nicobar Islands India 67 G1 8 30N 94 00E
Nicosia Cyprus 57 D2 35 11N 33 23E
Nienburg Germany 52 D6 52 38N 9 13E
NIGER 80 E5/G5
Niger r. West Africa 80 F3 5 30N 6 15E
NIGERIA 80 F4/G4
Nigg Scotland 9 G2 57 08N 2 03W
Nigg Bay Highland
Scotland 9 F2 57 42N 4 01W
Nigg Bay Aberdeenshire
Scotland 9 G2 57 08N 2 03W
Nijmegen Neths 48 D4 51 50N 5 52E
Nikopol' Ukraine 56 A5 34 34N 34 25E
Nile r. Sudan/Egypt 81 M7 27 30N 31 40E
Nîmes France 50 G4 43 50N 4 21E
Niort France 50 D4 46 19N 0 27W
Niterói Brazil 96 J9 22 54S 43 06W
Nith r. Scotland 11 F2 55 20N 3 50W
Nizhniy Novgorod
Russia 62 G8 56 20N 44 00E
Nordhausen Germany 52 E5 51 31N 10 48E
North Cape Norway 49 F5 71 11N 25 40E
Norfolk co. England 17 D3 52 45N 1 00E
Norfolk USA 91 L4 36 54N 76 18W
Norham England 11 G3 55 43N 2 10W
Normanton England 13 E2 53 42N 1 25W
Norrköping Sweden 49 E3 58 35N 16 10E
Northallerton England 13 E3 54 20N 1 26W
Northam England 18 C3 51 02N 4 14W
Northampton England 15 F2 52 14N 0 54W
Northamptonshire co.
England 15 E2/F2 52 20N 1 00W
North Ayrshire u.a.
Scotland 11 E2 55 30N 5 05W
North Baddesley
England 16 B1 50 58N 1 27W
North Ballachulish
Scotland 8 D1 56 42N 5 11W
North Berwick Scotland 11 G3 56 04N 2 44W
North Carolina state
USA 91 L4 36 00N 80 00W
North Channel
British Isles 10 D1/D2 55 20N 5 50W
North Dakota state
USA 90/91 F6/G6 47 00N 102 00W
North Dorset Downs
England 19 E2 50 40N 2 30W
North Down co. NI 10 D1 54 40N 5 40W
North Downs England 16 C2 51 13N 0 30W
North East Lincolnshire
u.a. England 13 F2 53 40N 0 05W
Northern Ireland UK 10 B1/C1 54 40N 7 00W
NORTHERN MARIANAS
100 F9
North Foreland England 17 E2 51 23N 1 27E
North Hykeham
England 13 F2 53 12N 0 34W
Northiam England 17 D1 50 59N 0 36E
North Island NZ 73 P3/Q3 39 00N §76 00E
North Kessock Scotland 9 F2 57 30N 4 15W
NORTH KOREA 69 N5/N6
North Lanarkshire u.a.
Scotland 11 E2/F2 55 50N 3 55W
Northleach England 15 E1 51 51N 1 50W
North Lincolnshire u.a.
England 13 F2 53 40N 0 30W
North Petherton
England 19 D3 51 06N 3 01W
North Pole
Arctic Ocean 101 90 00N
North Ronaldsay i.
Scotland 7 B2 59 23N 2 26W
North Shields England 13 E4 55 01N 1 26W
North Somercotes
England 13 G2 53 28N 0 08E
North Somerset u.a.
England 19 E3 51 15N 2 50W
North Sound Rol 21 B3 53 11N 9 43W

**North Sound, The** Scotland 7 B2 59 17N 2 45W
**North Tawton** England 18 D2 50 48N 3 53W
**North Tidworth** England 16 B2 51 16N 1 40W
**North Uist** *i.* Scotland 8 B2 57 04N 7 15W
**Northumberland** *co.* England 12 D4 55 10N 2 05W
**North Walsham** England 17 E3 52 50N 1 24E
**Northwest Highlands** Scotland 8/9 D2/E3 58 00N 5 00W
**Northwest Territories** *territory* Canada 88 L6/N6 65 15N 115 00W
**North Wheatley** England 13 F2 53 22N 0 52W
**Northwich** England 12 D2 53 16N 2 32W
**North Wingfield** England 13 E2 53 09N 1 24W
**North York Moors** England 13 F3 55 22N 0 45W
**North Yorkshire** *co.* England 13 E3/F3 54 10N 2 10W
**Norton** England 13 F3 54 08N 0 48W
**NORWAY** 49 C2/F4
**Norwich** England 17 E3 52 38N 1 18E
**Norwich** USA 93 F2 41 32N 72 05W
**Noss Head** Scotland 9 F3 58 28N 3 04W
**Nottingham** England 15 E2 52 58N 1 10W
**Nottinghamshire** *co.* England 13 E2/F2 53 20N 1 00W
**Nouadhibou** Mauritania 80 A6 20 54N 17 01W
**Nouakchott** Mauritania 80 A5 18 09N 15 58W
**Novara** Italy 54 C6 45 27N 8 37E
**Nova Scotia** *prov.* Canada 89 Y2/Y3 44 30N 65 00W
**Novgorod** Russia 62 F8 58 30N 31 20E
**Novi Sad** Serbia & Montenegro 55 H6 45 15N 19 51E
**Novosibirsk** Russia 63 L8 55 04N 83 05E
**Nubian Desert** Sudan 81 M6 21 00N 33 00E
**Nullarbor Plain** Australia 72 E4 32 00S 128 00E
**Nuneaton** England 15 E2 52 32N 1 28W
**Nuoro** Italy 54 C4 40 20N 9 21E
**Nuremberg** Germany 52 E4 49 27N 11 05E
**Nuuk** Greenland 89 AA6 64 10N 51 40W
**Nyasa, Lake** Southern Africa 82 F5 12 00S 35 00E
**Nykøbing** Denmark 49 D1 54 47N 11 53E
**Nyköping** Sweden 49 E2 58 45N 17 03E
**Nyons** France 50 G3 44 22N 5 08E

## O

**Oadby** England 15 E2 52 36N 1 04W
**Oakengates** England 15 D2 52 42N 2 28W
**Oakham** England 16 F2 52 40N 00 43W
**Oakland** USA 90 B4 37 50N 122 15W
**Oakley** England 16 B2 51 15N 1 11W
**Oakley** Scotland 11 F3 56 05N 3 35W
**Ob'** *r.* Russia 62 J10 65 30N 66 00E
**Oban** Scotland 8 D1 56 25N 5 29W
**Oberhausen** Germany 52 C5 51 27N 6 50E
**Ob', Gulf of** Russia 62 K10 68 00N 74 00E
**O'Briensbridge** RoI 21 C2 52 45N 8 30W
**Ochil Hills** Scotland 11 E3 56 15N 3 30W
**Odda** Norway 49 C3 60 03N 6 34E
**Oddsta** Scotland 7 D3 60 37N 0 59W
**Odense** Denmark 49 D2 55 24N 10 25E
**Oder** *r.* Europe 52 G6 52 00N 15 30E
**Odessa** Ukraine 57 D4 46 30N 30 46E
**Offaly** *co.* RoI 20 D3 53 15N 7 35W
**Offenbach am Main** Germany 52 D5 50 06N 8 46E
**Offenburg** Germany 52 C4 48 29N 7 57E
**Ohio** *state* USA 91 K5 40 00N 83 00W
**Ohio** *r.* USA 91 J4 38 00N 86 00W
**Oilgate** RoI 21 E2 52 25N 6 32W
**Ōita** Japan 70 B1 33 15N 131 36E
**Okehampton** England 18 C2 50 44N 4 00W
**Oklahoma** *state* USA 91 G4 36 00N 98 00W
**Oklahoma City** USA 91 G4 35 28N 97 33W
**Öland** *i.* Sweden 49 E2 56 45N 51 50E
**Olbia** Italy 54 C4 40 56N 9 30E
**Oldbury** England 15 D2 52 30N 2 00W
**Oldcastle** RoI 20 D3 53 46N 7 10W
**Oldenburg** Germany 52 D6 53 08N 8 13E
**Old Fletton** England 16 C3 52 34N 0 12W
**Oldham** England 12 D2 53 33N 2 07W
**Old Head of Kinsale** *c.* RoI 21 C1 51 40N 8 30W
**Oldmeldrum** Scotland 9 G2 57 20N 2 20W
**Olhâo** Portugal 51 B2 37 01N 7 50W
**Ollerton** England 13 E2 53 12N 1 00W
**Olney** England 16 C3 52 09N 0 43W
**Olsztyn** Poland 53 K6 53 48N 20 29E
**Olten** Switz. 52 C3 47 22N 7 55E
**Ólympos** *mt.* Greece 55 J4 40 05N 22 21E
**Omagh** NI 10 B1 54 36N 7 18W
**Omagh** *district* NI 10 B1 54 30N 7 30W
**Omaha** USA 91 G5 41 15N 96 00W
**OMAN** 65 F2/G3
**Oman, Gulf of** Iran/Oman 65 G3 24 30N 58 30E
**Omdurman** Sudan 81 M5 15 37N 32 29E
**Omeath** RoI 20 E4 54 06N 6 17W
**Omsk** Russia 62 K8 55 00N 73 22E
**Ōmuta** Japan 70 B1 33 02N 130 26E
**Onezhskoye Ozero** *l.* Russia 62 F9 62 00N 40 00E
**Ontario** *prov.* Canada 89 S4/U4 51 00N 91 00W
**Ontario, Lake** Canada/USA 93 E2 43 45N 78 00W
**Oola** RoI 21 C2 52 33N 8 16W
**Oostende** Belgium 48 A4 51 13N 2 55E
**Opava** Czech Rep. 53 H4 49 58N 17 55E
**Opole** Poland 53 H5 50 40N 17 56E
**Oporto** Portugal 51 A4 41 09N 8 37W
**Oran** Algeria 80 D9 35 45N 0 38W
**Orange** France 50 G3 44 08N 4 48E
**Oranmore** RoI 20 C3 53 16N 8 54W
**Orbetello** Italy 54 D5 42 27N 11 07E
**Örebro** Sweden 49 E2 59 17N 15 13E
**Oregon** *state* USA 90 B5/C5 44 00N 120 00W
**Oregon City** USA 90 B6 45 21N 122 36W
**Orel** Russia 62 F7 52 58N 36 04E
**Orenburg** Russia 62 H7 51 50N 55 00E
**Orense** Spain 51 B5 42 20N 7 52W
**Orford** England 17 E3 52 06N 1 31E
**Orihuela** Spain 51 E3 38 05N 0 56W
**Orinoco** *r.* Venezuela 96 E15 8 00N 64 00W
**Oristano** Italy 54 C3 39 54N 8 36E
**Orkney Islands** *u.a.* Scotland 7 A1/B2 59 00N 3 00W
**Orlando** USA 91 K2 28 33N 81 21W
**Orléans** France 50 E4 47 54N 1 54E
**Ormskirk** England 12 D2 53 35N 2 54W
**Oronsay** *i.* Scotland 10 C3 56 00N 6 15W
**Orrell** England 12 D2 53 33N 2 45W
**Orsha** Belarus 56 D5 54 30N 30 23E
**Orthez** France 50 D2 43 29N 0 46W
**Ortigueira** Spain 51 B5 43 43N 8 13W
**Ortona** Italy 54 F5 42 21N 14 24E
**Oruro** Bolivia 96 D10 17 59S 67 08W
**Orvieto** Italy 54 E5 42 43N 12 06E
**Orwell** *r.* England 17 E3 52 00N 1 15E
**Ōsaka** Japan 70 C1 34 40N 135 30E
**Oshawa** Canada 89 V2 43 53N 78 51W
**Oslo** Norway 49 D2 59 56N 10 45E
**Osmaniye** Turkey 57 E2 37 04N 36 15E
**Osnabrück** Germany 52 D6 52 17N 8 03E
**Osorno** Chile 97 C5 40 35S 73 14W
**Ossett** England 13 E2 53 41N 1 35W
**Östersund** Sweden 49 D3 63 10N 14 40E
**Östervall** Sweden 49 E3 62 20N 15 20E
**Ostrava** Czech Rep. 53 J4 49 50N 18 15E
**Oswaldtwistle** England 12 D2 53 44N 2 24W
**Oswestry** England 14 C2 52 52N 3 03W
**Otford** England 17 D2 51 19N 0 12E
**Otley** England 13 E2 53 54N 1 41W
**Otranto** Italy 55 H4 40 08N 18 30E
**Ottawa** Canada 89 V3 45 24N 75 38W
**Otterburn** England 12 D4 55 14N 2 10W
**Ottery St. Mary** England 19 D2 50 45N 3 17W
**Ouagadougou** Burkina 80 D4 12 20N 1 40W
**Oughterard** RoI 20 B3 53 26N 9 19W
**Oulart** RoI 21 E2 53 30N 6 23W
**Oulu** Finland 49 G4 65 02N 25 27E
**Oundle** England 15 F2 52 29N 0 29W
**Ouse** *r.* North Yorkshire England 13 E3 54 05N 1 15W
**Ouse** *r.* East Sussex England 16 D1 50 55N 0 03E
**Outer Hebrides** *is.* Scotland 8 B2/B3 58 00N 7 00W
**Outwell** England 17 D3 52 37N 0 14E
**Overton** England 16 B2 51 15N 1 15W
**Overton** Wales 14 D2 52 58N 2 56W
**Oviedo** Spain 51 C5 43 21N 7 18W
**Owen Sound** Canada 89 U2 44 3N 80 56W
**Oxford** England 16 B2 51 46N 1 15W
**Oxfordshire** *co.* England 16 B2 51 50N 1 25W
**Oxted** England 16 C2 51 15N 0 01W
**Oysterhaven** RoI 21 C1 51 41N 8 26W
**Ozero Balkhash** *l.* Kazakhstan 62 K6 46 00N 75 00E
**Ozero Baykal** *l.* Russia 63 N7 54 00N 109 00E
**Ozieri** Italy 54 C4 40 35N 9 01E

## P

**Pacific Ocean** 100
**Padang** Indonesia 71 C3 1 00S 100 21E
**Paddock Wood** *tn.* England 17 D2 51 11N 0 23E
**Paderborn** Germany 52 D5 51 43N 8 44E
**Padiham** England 12 D2 53 49N 2 19W
**Padstow** England 18 C2 50 33N 4 56W
**Padua** Italy 54 D6 45 24N 11 53E
**Paisley** Scotland 11 E2 55 50N 4 26W
**PAKISTAN** 66 B5/C5
**PALAU** 100 D8 7 30N 134 30E
**Palawan** *i.* Philippines 71 F5/F6 10 00N 119 00E
**Palembang** Indonesia 71 C3 2 59S 104 45E
**Palencia** Spain 51 C3 41 01N 4 32W
**Palermo** Italy 54 E3 38 08N 13 23E
**Palma de Mallorca** Spain 51 G3 39 35N 2 39E
**Palmer** USA 88 F6 61 35N 149 10W
**Palmyra** Syria 64 C5 34 40N 38 10E
**Pamiers** France 50 E2 43 07N 1 36E
**Pampas** *geog. reg.* Argentina 97 E6 36 00S 63 00W
**Pamplona** Spain 51 E5 42 49N 1 39W
**PANAMA** 95 H1/J1
**Panama Canal** Panama 95 J1 9 00N 80 00W
**Panama City** Panama 95 J1 8 57N 19 30W
**Pangbourne** England 16 B2 51 29N 1 05W
**Papa Sound** Scotland 7 B2 59 20N 2 54W
**Papa, Sound of** Scotland 7 C3 60 37N 1 40W
**Papa Stour** *i.* Scotland 7 C3 60 20N 1 42W
**Papa Westray** *i.* Scotland 7 B2 59 22N 2 54W
**PAPUA NEW GUINEA** 73 H9/K9
**PARAGUAY** 96 F9
**Paraguay** *r.* Paraguay/Argentina 96 F8 26 30S 58 00W
**Paramaribo** Suriname 96 F15 5 52N 55 14W
**Parana** Argentina 97 E7 31 45S 60 30W
**Paraná** *r.* Paraguay/Argentina 96 F7 27 00S 56 00W
**Paris** France 50 F5 48 52N 2 20E
**Parkano** Finland 49 F3 62 03N 23 00E
**Parma** Italy 54 D6 44 48N 10 19E
**Parnaíba** Brazil 96 J13 2 58S 41 46W
**Páros** Greece 55 L2 37 04N 25 06E
**Parrett** *r.* England 19 E3 51 05N 2 54W
**Parthenay** France 50 D4 46 39N 0 14W
**Partry** RoI 20 B3 53 42N 9 16W
**Partry Mountains** RoI 20 B3 53 39N 9 27W
**Paşcani** Romania 53 N3 47 14N 26 46E
**Passage East** RoI 21 E2 52 13N 6 59W
**Passage West** *tn.* RoI 21 C1 51 52N 8 20W
**Passau** Germany 52 F4 48 35N 13 28E
**Pasto** Colombia 96 B14 1 12N 77 17W
**Patagonia** *geog. reg.* Argentina 97 C2/D4 48 00S 70 00W
**Pateley Bridge** *tn.* England 13 E3 54 05N 1 45W
**Patna** Scotland 11 E2 55 20N 4 30W
**Patna** India 67 F5 25 37N 85 12E
**Patras** Greece 55 J3 38 14N 21 44E
**Patrickswell** RoI 21 C2 52 36N 8 42W
**Patrington** England 13 F2 53 41N 0 02W
**Pau** France 50 D2 43 18N 0 22W
**Pavia** Italy 54 C6 45 12N 9 09E
**Pavullo nel Frignano** Italy 54 D6 44 20N 10 49E
**Peacehaven** England 16 C1 50 47N 0 01E
**Peč** Serbia & Montenegro 55 J5 42 40N 20 19E
**Pécs** Hungary 53 J3 46 04N 18 15E
**Peebles** Scotland 11 F2 55 39N 3 12W
**Peel** Isle of Man 12 B3 54 14N 4 42W
**Pelopónnisos** *geog. reg.* Greece 55 J2/K2 37 00N 22 00E
**Pelotas** Brazil 97 G7 31 45S 52 20W
**Pembridge** England 14 D2 52 14N 2 53W
**Pembroke** Wales 14 B1 51 41N 4 55W
**Pembroke Dock** Wales 14 B1 51 42N 4 56W
**Pembrokeshire** *u.a.* Wales 14 A1/B1 51 50N 5 00W
**Pembury** England 17 D2 51 08N 0 19E
**Penarth** Wales 14 C1 51 27N 3 11W
**Pencoed** Wales 14 C1 51 31N 3 30W
**Pendine** Wales 14 B1 51 34N 4 35W
**Penicuik** Scotland 11 F2 55 50N 3 14W
**Penistone** England 13 E2 53 32N 1 37W
**Penkridge** England 15 D2 52 44N 2 07W
**Penmaenmawr** Wales 14 C3 53 16N 3 54W
**Pennines** England 12/13 D3/E2 54 30N 2 10W
**Pennsylvania** *state* USA 91 L5 41 00N 78 00W
**Penrhyndeudraeth** Wales 14 B2 52 56N 4 04W
**Penrith** England 12 D3 54 40N 2 44W
**Penryn** England 18 B2 50 09N 5 06W
**Pensilva** England 18 C2 50 30N 4 25W
**Pentland Firth** Scotland 7 A1 58 45N 3 10W
**Pentland Hills** Scotland 11 F2 55 45N 3 30W
**Pen-y-ghent** *sum.* England 12 D3 54 10N 2 14W
**Penygroes** Wales 14 B3 53 04N 4 17W
**Penzance** England 18 B2 50 07N 5 33W
**Périgueux** France 50 E3 45 12N 0 44E
**Perm'** Russia 62 H8 58 01N 56 10E
**Pernik** Bulgaria 55 K5 42 36N 23 03E
**Péronne** France 50 F5 49 56N 2 57E
**Perpignan** France 50 F2 42 42N 2 54E
**Perranporth** England 18 B2 50 20N 5 09W
**Perth** Australia 72 C4 31 58S 115 49E
**Perth** Scotland 9 F1 56 42N 3 28W
**Perth and Kinross** *u.a.* Scotland 9 F1 56 45N 3 50W
**PERU** 96 B11/C11
**Perugia** Italy 54 E5 43 07N 12 23E
**Pesaro** Italy 54 E5 43 54N 12 54E
**Pescara** Italy 54 F5 42 27N 14 13E
**Peshawar** Pakistan 66 C6 34 01N 71 40E
**Pessac** France 50 D3 44 49N 0 37W
**Peterborough** England 16 C3 52 35N 0 15W
**Peterculter** Scotland 9 G2 57 05N 2 16W
**Peterhead** Scotland 9 H2 57 30N 1 46W
**Peterlee** England 13 E3 54 46N 1 19W
**Petersfield** England 16 C2 51 00N 0 56W
**Petrolina** Brazil 96 J12 9 22S 40 30W
**Petworth** England 16 C1 50 59N 0 38W
**Pevensey** England 17 D1 50 49N 0 20E
**Pewsey** England 16 B2 51 21N 1 46W
**Pforzheim** Germany 52 D4 48 53N 8 41E
**Phenix City** USA 91 J3 32 28N 85 01W
**Philadelphia** USA 91 L4 40 00N 75 10W
**PHILIPPINES, THE** 71 G7/H5
**Phnom Penh** Cambodia 71 C6 11 35N 104 55E
**Phoenix** USA 90 D3 33 30N 112 03W
**Piacenza** Italy 54 C6 45 03N 9 41E
**Pickering** England 13 F3 54 14N 0 46W
**Píndhos** *mts.* Greece 55 J3 40 00N 21 00E
**Pinsk** Belarus 57 C5 52 08N 26 01E
**Pinwherry** Scotland 10 E2 55 09N 4 50W
**Piombino** Italy 54 D5 42 56N 10 32E
**Pirineos** *mts.* Spain/France 51 E5/F5 42 50N 0 30E
**Pirmasens** Germany 52 C4 49 12N 7 37E
**Pisa** Italy 54 D5 43 43N 10 24E
**Pistoia** Italy 54 D5 43 56N 10 55E
**Piteå** Sweden 49 F4 65 19N 21 30E
**Piteşti** Romania 53 M2 44 51N 24 51E
**Pitlochry** Scotland 9 F1 56 43N 3 45W
**Pitmedden** Scotland 9 G2 57 19N 2 11W
**Pittenweem** Scotland 11 G3 56 13N 2 44W
**Pittsburgh** USA 91 L5 40 26N 80 00W
**Piura** Peru 96 A12 5 15S 80 38W
**Planalto de Mato Grosso** *geog. reg.* Brazil 96 F11/G11 13 00S 56 00W
**Plasencia** Spain 51 B4 40 02N 6 05W
**Platte** *r.* USA 90 F5 41 00N 100 00W
**Plauen** Germany 52 F5 50 29N 12 08E
**Pleven** Bulgaria 55 L5 43 25N 24 04E
**Płock** Poland 53 J6 52 32N 19 40E
**Plockton** Scotland 8 D2 57 20N 5 40W
**Ploieşti** Romania 53 M2 44 57N 26 01E
**Plovdiv** Bulgaria 55 L5 42 08N 24 45E
**Plymouth** England 18 C2 50 23N 4 10W
**Plympton** England 18 C2 50 23N 4 03W
**Plymstock** England 18 C2 50 21N 4 07W
**Plynlimon** *mt.* Wales 14 C2 52 28N 3 47W
**Plzeň** Czech Rep. 52 F4 49 45N 13 25E
**Po** *r.* Italy 54 D6 45 00N 10 00E
**Pocklington** England 13 F2 53 56N 0 46W
**Podgorica** Serbia & Montenegro 55 H5 42 28N 19 17E
**Point-Noire** *tn.* Congo 82 B7 4 46S 11 53E
**Poitiers** France 50 E4 46 35N 0 20E
**POLAND** 53 J6/K6
**Polegate** England 17 D1 50 50N 0 15E
**Polesworth** England 15 E2 52 44N 1 36W
**Policoro** Italy 55 G4 40 12N 16 40E
**Pollensa** Spain 51 G3 39 52N 3 01E
**Polperro** England 18 C2 50 19N 4 31W
**Polýgyros** Greece 55 K4 40 23N 23 25E
**Pondicherry** India 66 D2 11 59N 79 50E
**Ponferrada** Spain 51 B5 42 33N 6 35W
**Pont** *r.* England 12 E4 55 02N 1 55W
**Ponta Grossa** Brazil 96 G8 25 07S 50 09W
**Pont-à-Mousson** France 50 H5 48 55N 6 03E
**Pontardawe** Wales 14 C1 51 44N 3 52W
**Pontardulais** Wales 14 B1 51 43N 4 02W
**Pontefract** England 13 E2 53 42N 1 18W
**Ponteland** England 13 E4 55 03N 1 44W
**Ponte Leccia** France 50 J2 42 28N 9 12E
**Ponterwyd** Wales 14 C2 52 25N 3 50W
**Pontevedra** Spain 51 A5 42 25N 8 39W
**Pontivy** France 50 C5 48 04N 2 58W
**Pontrilas** England 14 D2 51 57N 2 53W
**Pontyclun** Wales 14 C1 51 33N 3 23W
**Pontycymer** Wales 14 C1 51 36N 3 35W
**Pontypool** Wales 14 C1 51 43N 3 02W
**Pontypridd** Wales 14 C1 51 37N 3 22W
**Poole** England 19 F2 50 43N 1 59W
**Poole** *u.a.* England 19 F2 50 43N 1 59W
**Poole Bay** *b.* England 19 F3 50 41N 1 52W
**Poolewe** Scotland 8 D2 57 45N 5 37W
**Pooley Bridge** *tn.* England 12 D3 54 38N 2 49W
**Popacatepetl** *mt.* Mexico 94 E3 19 02N 98 38W
**Pordenone** Italy 54 E6 45 58N 12 39E
**Pori** Finland 49 F3 61 28N 21 45E
**Porlock** England 18 D3 51 14N 3 36W
**Portadown** NI 10 C1 54 26N 6 27W
**Portaferry** NI 10 D1 54 23N 5 33W
**Portalegre** Portugal 51 B3 39 17N 7 25W
**Portarlington** RoI 21 D3 53 10N 7 11W
**Port Askaig** Scotland 10 C2 55 51N 6 07W
**Port-au-Prince** Haiti 95 K3 18 33N 72 20W
**Portavogie** NI 10 D1 54 28N 5 25W
**Port Elizabeth** RSA 82 E1 33 58S 25 36E
**Port Ellen** Scotland 10 C2 55 39N 6 12W
**Port Erin** Isle of Man 12 B3 54 05N 4 45W
**Port-Eynon** Wales 14 B1 51 28N 4 12W
**Port Gentil** Gabon 82 A7 0 40S 8 50E
**Port Glasgow** Scotland 11 E2 55 56N 4 41W
**Portglenone** NI 10 C1 54 52N 6 29W
**Port Harcourt** Nigeria 80 F2 4 43N 7 05E
**Porthcawl** Wales 14 C1 51 29N 3 43W
**Porthleven** England 18 BB2 50 30N 5 21W
**Porthmadog** Wales 14 B2 52 55N 4 08W
**Portimaõ** Portugal 51 A2 37 08N 8 32W
**Port Isaac** England 18 C2 50 35N 4 49W
**Portishead** England 19 E3 51 30N 2 46W
**Portknockie** Scotland 9 G2 57 42N 2 52W
**Portland** USA 90 B6 45 32N 122 40W
**Portlaoise** RoI 21 D3 53 02N 7 17W
**Portlaw** RoI 21 D2 52 17N 7 19W

**Portlethen** Scotland 9 G2 57 03N 2 07W
**Portmahomack** Scotland 9 F2 57 49N 3 50W
**Port Moresby** PNG 73 J9 9 30S 147 07E
**Portnaguran** Scotland 8 C3 58 17N 6 13W
**Portnahaven** Scotland 10 C2 55 41N 6 31W
**Porto Alegre** Brazil 97 G7 30 03S 51 10W
**Port of Ness** Scotland 8 C3 58 29N 6 13W
**Port of Spain**
Trinidad and Tobago 95 V15 10 38N 61 31W
**Porto Novo** Benin 80 E3 6 30N 2 47E
**Porto Tórres** Italy 54 C4 40 51N 8 24E
**Porto-Vecchio** France 50 J1 41 35N 9 16E
**Pôrto Velho** Brazil 96 E12 8 45S 63 54W
**Portpatrick** Scotland 10 D2 54 51N 5 07W
**Portreath** England 18 B2 50 51N 5 17W
**Portree** Scotland 8 C2 57 24N 6 12W
**Portrush** NI 10 C2 55 12N 6 40W
**Port Said** Egypt 81 M8 31 17N 6 40W
**Portskerra** Scotland 9 F3 58 33N 3 55W
**Portsmouth** England 16 B1 50 48N 1 05W
**Portsmouth** u.a.
England 16 B1 50 48N 1 05W
**Portsmouth** USA 91 L4 36 50N 76 20W
**Portsoy** Scotland 9 G2 57 41N 2 41W
**Portstewart** NI 10 C2 55 11N 6 43W
**Port Sudan** Egypt 81 N5 19 38N 37 07E
**Port Talbot** Wales 14 C1 51 36N 3 47W
**PORTUGAL** 51 A3/B3
**Portumna** RoI 21 C3 53 06N 8 13W
**Port Wemyss** Scotland 10 C2 55 42N 6 29W
**Port William** Scotland 11 E1 54 46N 4 35W
**Potenza** Italy 54 F4 40 38N 15 48E
**Potomac** r. USA 91 L4 39 10N 77 30W
**Potsdam** Germany 52 F6 52 24N 13 04E
**Potter's Bar** England 16 C2 51 42N 0 11W
**Potton** England 16 C3 52 08N 0 14W
**Poulton-le-Fylde**
England 12 D2 53 51N 3 00W
**Powell River** tn. Canada 88 L3 49 54N 124 34W
**Power Head** RoI 21 C1 51 47N 8 10W
**Powfoot** Scotland 11 F1 54 59N 3 20W
**Powys** u.a. Wales 14 C2 52 10N 3 30W
**Poznań** Poland 53 H6 52 25N 16 53E
**Prague** Czech Rep. 52 G5 50 06N 14 26E
**Prato** Italy 54 D5 43 53N 11 06E
**Prawle Point** England 18 D2 50 12N 3 44W
**Preesall** England 12 D2 53 55N 3 00W
**Prešov** Slovakia 53 K4 49 00N 21 10E
**Prestatyn** Wales 14 C3 53 20N 3 24W
**Presteigne** Wales 14 C2 52 17N 3 00W
**Preston** England 12 D2 53 46N 2 42W
**Prestonpans** Scotland 11 G2 55 57N 3 00W
**Prestwich** England 12 D2 53 32N 2 17W
**Prestwick** Scotland 11 E2 55 30N 4 37W
**Pretoria** RSA 82 E2 25 45S 28 12E
**Příbram** Czech Rep. 52 G4 49 42N 14 01E
**Prince Albert** Canada 89 P4 53 13N 105 45W
**Prince Edward Island**
prov. Canada 89 Y3 46 30N 63 00W
**Prince Rupert** Canada 88 J4 54 09N 130 20W
**Princes Risborough**
England 16 C2 51 44N 0 51W
**Princetown** England 18 D2 50 33N 3 59W
**Pripyats'** r. Belarus 56 B5/C5 52 30N 27 00E
**Priština**
Serbia & Montenegro 55 J5 42 39N 21 20E
**Privas** France 50 G3 44 44N 4 36E
**Prokop'yevsk** Russia 63 L7 53 55N 86 45E
**Providence** USA 91 M5 41 50N 71 28W
**Provins** France 50 F5 48 34N 3 18E
**Prudhoe** England 13 E3 54 58N 1 51W
**Prudhoe Bay** tn. USA 88 F8 70 05N 148 20W
**Pruszków** Poland 53 K6 52 10N 20 47E
**Prut** r. Romania 53 P3 47 00N 28 00E
**Pskov** Russia 62 E8 57 48N 28 26E
**Pucallpa** Peru 96 C12 8 21S 74 33W
**Puddletown** England 19 E2 50 45N 2 21W
**Pudsey** England 13 E2 53 48N 1 40W
**Puebla** Mexico 94 E3 19 03N 98 10W
**Pueblo** USA 90 F4 38 17N 104 38W
**Puertollano** Spain 51 C3 38 41N 4 07W
**Puerto Montt** Chile 97 C5 41 28S 73 00W
**PUERTO RICO** 95 L3
**Pulborough** England 16 C1 50 58N 0 30W
**Pumsaint** Wales 14 C2 52 03N 3 58W
**Pune** India 66 C3 18 34N 73 58E
**Punta Arenas** Chile 97 C3 53 10S 70 56W
**Purfleet** England 17 D2 51 29N 0 14E
**Puri** India 67 F3 19 49N 85 54E
**Pusan** S. Korea 69 N5 35 05N 129 02E
**Puulavesi** l. Finland 49 G3 61 50N 26 40E
**Pwllheli** Wales 14 B2 52 53N 4 25W
**Pyle** Wales 14 C1 51 32N 3 42W
**Pyongyang** N. Korea 69 N5 39 00N 125 47E
**Pyrénées** mts.
France/Spain 50 E2 42 50N 0 30E
**Pýrgos** Greece 55 J2 37 40N 21 27E
**Pýrgos** Greece 55 L1 35 00N 25 10E

## Q

**QATAR** 65 F4
**Qingdao** China 69 M5 36 04N 120 22E
**Qiqihar** China 69 M7 47 23N 124 00E
**Quantock Hills** England 19 D3 51 10N 3 15W
**Québec** prov. Canada 89 V4/X4 50 00N 75 00W
**Québec** Canada 89 W3 46 50N 71 15W

**Queenborough** England 17 D2 51 26N 0 45E
**Queen Charlotte Islands**
Canada 88 J4 53 00N 132 30W
**Queen Elizabeth Islands**
Canada 89 N9/S9 77 30N 105 00W
**Queensbury** England 13 E2 53 46N 1 50W
**Queenstown** Australia 73 J2 42 07S 145 33E
**Querétaro** Mexico 94 D4 20 38N 100 23W
**Quetta** Pakistan 66 B6 30 15N 67 00E
**Quezon City** Philippines 71 G6 14 39N 121 02E
**Quiberon** France 50 C4 47 29N 3 07W
**Quigley's Point** RoI 20 D5 55 10N 7 08W
**Quimper** France 50 B4 48 00N 4 06W
**Quito** Ecuador 96 B13 0 14S 78 30W
**Quorn** England 15 E2 52 45N 1 10W

## R

**Raahe** Finland 49 F3 64 42N 24 30E
**Raasay** i. Scotland 8 C2 57 25N 6 05W
**Rabat-Salé** Morocco 80 C8 34 02N 6 51W
**Radcliffe** England 12 D2 53 34N 2 20W
**Radcliffe on Trent**
England 15 E3 52 57N 1 03W
**Radlett** England 16 C2 51 41N 0 19W
**Radom** Poland 53 K5 51 26N 21 10E
**Radstadt** Austria 52 F3 47 23N 13 28E
**Radstock** England 19 E3 51 18N 2 28W
**Raglan** Wales 14 D1 51 47N 2 51W
**Ragusa** Italy 54 F2 36 56N 14 44E
**Rainford** England 12 D2 53 30N 2 48W
**Rainier, Mount** USA 90 B6 46 25N 121 45W
**Rainworth** England 13 E2 53 08N 1 08W
**Raipur** India 66 E4 21 16N 81 42E
**Rame Head** England 18 C2 50 19N 4 13W
**Ramore Head** NI 10 C2 55 13N 6 39W
**Ramsbottom** England 12 D2 53 39N 2 19W
**Ramsey** England 16 C3 52 27N 0 07W
**Ramsey** Isle of Man 12 B3 54 19N 4 23W
**Ramsey Island** Wales 14 A1 51 53N 5 20W
**Ramsgate** England 17 E2 51 20N 1 25E
**Randalstown** NI 10 C1 54 45N 6 19W
**Rangpur** Bangladesh 67 F5 25 45N 89 21E
**Rannoch Moor** Scotland 9 E1 56 35N 4 50W
**Rannoch Station**
Scotland 9 E1 5642N 4 34W
**Rapallo** Italy 54 C6 44 21N 9 13E
**Raphoe** RoI 20 D4 54 52N 7 36W
**Rastatt** Germany 52 D4 48 51N 8 13E
**Rathangan** RoI 20 E3 53 12N 6 59W
**Rathcoole** RoI 20 E3 53 17N 6 28W
**Rathcormack** RoI 21 C2 52 05N 8 17W
**Rathdowney** RoI 21 D2 52 51N 7 35W
**Rathdrum** RoI 21 E2 52 56N 6 13W
**Rathfriland** NI 10 C1 54 14N 6 10W
**Rathkeale** RoI 21 C2 52 32N 8 56W
**Rathlackan** RoI 20 B4 54 18N 9 15W
**Rathlin Sound** NI 10 C2 55 20N 6 25W
**Rath Luirc** RoI 21 C2 52 21N 8 41W
**Rathmelton** RoI 20 D5 55 02N 7 38W
**Rathmore** RoI 21 B2 52 05N 9 13W
**Rathnew** RoI 21 E2 52 58N 6 05W
**Rathvilly** RoI 21 E2 52 53N 6 41W
**Ratoath** RoI 20 E3 53 31N 6 24W
**Raton** USA 90 F4 36 45N 104 27W
**Rattray** Scotland 9 F1 56 36N 3 20W
**Rattray Head** Scotland 9 H2 57 38N 1 45W
**Rauma** Finland 49 F3 61 09N 21 30E
**Raunds** England 15 F2 52 21N 0 33W
**Ravenglass** England 12 C3 54 21N 3 24W
**Ravenna** Italy 54 E6 44 25N 12 12E
**Ravensburg** Germany 52 D3 47 47N 9 37E
**Ravenshead** England 13 E2 53 40N 1 10W
**Rawalpindi** Pakistan 66 C6 33 40N 73 08E
**Rawmarsh** England 13 E2 53 27N 1 21W
**Rawtenstall** England 12 D2 53 42N 2 18W
**Rayleigh** England 17 D2 51 36N 0 36E
**Reading** England 16 C2 51 28N 0 59W
**Reading** u.a. England 16 B1/C1 50 25N 1 00W
**Reay Forest** Scotland 9 E3 58 30N 4 50W
**Recife** Brazil 96 L12 8 06S 34 53W
**Recklinghausen**
Germany 52 C5 51 37N 7 11E
**Red Bay** NI 10 C2 55 04N 6 01W
**Redbourn** England 16 C2 51 48N 0 24W
**Redbridge** England 16 C2 51 35N 0 05E
**Redcar** England 13 E3 54 37N 1 04W
**Redcar and Cleveland** u.a.
England 13 E3/F3 54 30N 1 00W
**Redditch** England 15 E2 52 19N 1 56W
**Redhill** tn. England 16 C2 51 14N 0 11W
**Redon** France 50 C4 47 39N 2 05W
**Redruth** England 18 B2 50 13N 5 14W
**Red Sea** Middle East 64 C4/C2 27 00N 35 00E
**Reepham** England 17 E3 52 46N 1 07E
**Regensburg** Germany 52 F4 49 01N 12 07E
**Reggio di Calabria** Italy 54 F3 38 06N 15 39E
**Regina** Canada 89 Q4 50 30N 104 38W
**Reigate** England 16 C2 51 14N 0 13W
**Reims** France 50 G5 49 15N 4 02E
**Reinosa** Spain 51 C3 43 01N 4 09W
**Rendsburg** Germany 52 D7 54 19N 9 39E
**Renfrew** Canada 89 V3 45 28N 76 44W
**Renfrew** Scotland 11 E2 55 53N 4 24W
**Renfrewshire** u.a.
Scotland 11 E2 55 50N 4 40W
**Rennes** France 50 D5 48 06N 1 40W

**Reno** USA 90 C4 39 32N 119 49W
**REPUBLIC OF IRELAND**
20/21
**REPUBLIC OF SOUTH AFRICA**
82 D1/D2
**Requena** Spain 51 E3 39 29N 1 08W
**Resistencia** Argentina 97 F8 27 28S 59 00W
**Resolute** Canada 89 S8 74 40N 95 00W
**Retford** England 13 F2 53 19N 0 56W
**Rethel** France 50 G5 49 31N 4 22E
**Reus** Spain 51 F4 41 10N 1 06E
**Reutlingen** Germany 52 D4 48 30N 9 13E
**Reykjavík** Iceland 49 K6 64 09N 21 58W
**Rhayader** Wales 14 C2 52 18N 3 30W
**Rheden** Neths 48 E5 52 01N 6 02E
**Rheidol** r. Wales 14 C2 52 23N 3 55W
**Rhein** r. Germany 52 C4 50 30N 8 00E
**Rheine** Germany 52 C6 52 17N 7 26E
**Rhode Island** state USA 91 M5 41 00N 71 00W
**Rhondda** Wales 14 C1 51 40N 3 30W
**Rhondda Cynon Taff** u.a.
Wales 14 C1 51 40N 3 30W
**Rhône** r. Switz./France 50 G3 45 00N 4 50E
**Rhoose** Wales 14 C1 51 26N 3 22W
**Rhosllanerchrugog**
Wales 14 C3 53 01N 3 04W
**Rhosneigr** Wales 14 B3 53 14N 4 31W
**Rhossili** Wales 14 B1 51 34N 4 17W
**Rhubodach** Scotland 10 D2 55 56N 5 09W
**Rhum** i. Scotland 8 C2 57 00N 6 20W
**Rhyl** Wales 14 C3 53 19N 3 29W
**Rhymney** Wales 14 C1 51 46N 3 18W
**Rhynie** Scotland 9 G2 57 19N 2 50W
**Ribble** r. England 12 D2 53 45N 2 30W
**Ribeirão Prêto** Brazil 96 H9 21 09S 47 48W
**Riccall** England 13 E3 53 51N 1 04W
**Richhill** NI 10 C1 54 22N 6 33W
**Richmond** England 13 E3 54 24N 1 44W
**Richmond upon Thames**
England 16 C2 51 28N 0 19W
**Rickmansworth**
England 16 C2 51 38N 0 29W
**Ried** Austria 52 F4 48 13N 13 29E
**Riesa** Germany 52 F5 51 18N 13 18E
**Rieti** Italy 54 E5 42 24N 12 51E
**Riga** Latvia 49 F2 56 53N 24 08E
**Rimini** Italy 54 E6 44 03N 12 34E
**Ringabella Bay** RoI 21 C1 51 46N 8 13W
**Ringaskiddy** RoI 21 C1 51 49N 8 18W
**Ringwood** England 16 B1 50 51N 1 47W
**Rio Branco** tn. Brazil 96 D12 9 59S 67 49W
**Rio de Janeiro** tn.
Brazil 96 I8 22 53S 43 17W
**Rio de la Plata** est.
Uruguay/Argentina 97 F5 35 00S 57 00W
**Rio Grande** tn. Brazil 97 G7 32 03S 52 08W
**Río Grande** r.
Mexico/USA 90 E3/F3 30 00N 105 00W
**Ripley** England 13 E2 53 03N 1 24W
**Ripon** England 13 E3 54 08N 1 31W
**Risca** Wales 14 C1 51 37N 3 07W
**Rishton** England 12 D2 53 47N 2 24W
**Riva del Garda** Italy 54 D6 45 53N 10 50E
**Riverton** Canada 89 R4 51 00N 97 00W
**Rivne** Ukraine 53 M5 50 39N 26 10E
**Riyadh** Saudi Arabia 65 E3 24 39N 46 46E
**Roanne** France 50 G4 46 02N 4 05E
**Robin Hood's Bay** tn.
England 13 F3 54 25N 0 33W
**Rochdale** England 12 D2 53 38N 2 09W
**Rochefort** France 50 D3 45 57N 0 58W
**Rochester** Northamptonshire
England 12 D4 55 16N 2 16W
**Rochester** Medway
England 17 D2 51 24N 0 30E
**Rochester** USA 91 L5 43 12N 77 37W
**Rochford** England 17 D2 51 36N 0 43E
**Rochfortbridge** RoI 20 D3 53 25N 7 18W
**Rockhampton** Australia 73 K6 23 22S 150 32E
**Rocky Mountains**
Canada/USA 88 K5/P2 55 00N 125 00W
**Rodel** Scotland 8 C2 57 41N 7 05W
**Rodez** France 50 F3 44 21N 2 34E
**Roding** r. England 17 D2 51 45N 0 15E
**Roermond** Neths 48 D4 51 12N 6 00E
**ROMANIA** 53 L2/M2
**Romans-sur-Isère**
France 50 G3 45 03N 5 03E
**Rome** Italy 54 E4 41 53N 12 30E
**Romney Marsh** England 17 D2 51 03N 0 58E
**Romorantin-Lanthenay**
France 50 E4 47 22N 1 44E
**Romsey** England 16 B1 50 59N 1 30W
**Rona** i. Scotland 8 D2 57 32N 5 59W
**Ronda** Spain 51 C2 36 45N 5 10W
**Ronse** Belgium 48 B3 50 45N 3 36E
**Roosendaal** Neths 48 C4 51 32N 4 28E
**Roosky** RoI 20 D3 53 50N 7 55W
**Roquetas de Mar** Spain 51 D2 36 46N 2 35W
**Rora Head** Scotland 7 A1 58 52N 3 26W
**Rosario** Argentina 97 E7 33 00S 60 40W
**Rosarno** Italy 54 F3 38 29N 15 59E
**Roscoff** France 50 C5 48 43N 3 59W
**Roscommon** RoI 20 C3 53 38N 8 11W
**Roscommon** co. RoI 20 C3 53 45N 8 10W
**Roscrea** RoI 21 D2 52 57N 7 47W
**Rosehearty** Scotland 9 G2 57 42N 2 07W

**Rosenheim** Germany 52 F3 47 51N 12 09E
**Roskilde** Denmark 49 D2 55 39N 12 05E
**Rossano** Italy 54 G3 39 35N 16 38E
**Rossan Point** RoI 20 C4 54 40N 8 50W
**Ross Carbery** RoI 21 B1 51 35N 9 02W
**Rosscarbery Bay** RoI 21 B1 51 32N 9 00W
**Rosslare** RoI 21 E2 52 15N 6 22W
**Rosslare Harbour** tn.
RoI 21 E2 52 15N 6 22W
**Rosslare Point** RoI 21 E2 52 19N 6 21W
**Ross-on-Wye** England 15 D1 51 55N 2 35W
**Rostock** Germany 52 F7 54 06N 12 09E
**Rostov-na-Donu**
Russia 62 F6 47 15N 39 45E
**Rostrevor** NI 10 C1 54 06N 6 12W
**Rothbury** England 11 H2 55 19N 1 55W
**Rother** r. England 16 C1 51 00N 0 50W
**Rother** r. England 17 D1 51 00N 0 17E
**Rotherham** England 13 E2 53 26N 1 20W
**Rothes** Scotland 9 F2 57 31N 3 13W
**Rothesay** Scotland 10 D2 55 51N 5 03W
**Rothwell** West Yorkshire
England 13 E2 53 45N 1 29W
**Rothwell** Northamptonshire
England 15 F2 52 25N 0 48W
**Rotterdam** Neths 48 C4 51 54N 4 28E
**Rottingdean** England 16 C1 50 48N 0 04W
**Roubaix** France 50 F6 50 42N 3 10E
**Rouen** France 50 E5 49 26N 1 05E
**Roundwood** RoI 21 E3 53 04N 6 13W
**Rousay** i. Scotland 7 A2 59 10N 3 00W
**Rovigo** Italy 54 D6 45 04N 11 47E
**Rowlands Gill** England 13 E3 54 54N 1 45W
**Royal Leamington Spa**
England 15 E2 52 18N 1 31W
**Royal Tunbridge Wells**
England 17 D2 51 08N 0 16E
**Royan** France 50 D3 45 38N 1 02W
**Royston** South Yorkshire
England 13 E2 53 37N 1 27W
**Royston** Hertfordshire
England 16 C3 52 03N 0 01W
**Royton** England 12 D2 53 34N 2 08W
**Ruabon** Wales 14 C2 52 59N 3 02W
**Ruddington** England 15 E2 52 46N 1 09W
**Rufford** England 12 D2 53 39N 2 49W
**Rugby** England 15 E2 52 23N 1 15W
**Rugeley** England 15 E2 52 46N 1 55W
**Ruhr** r. Germany 52 D5 51 00N 7 00E
**Rumney** Wales 14 C1 51 30N 3 10W
**Runcorn** England 12 D2 53 20N 2 44W
**Ruse** Bulgaria 55 M5 43 50N 25 59E
**Rush** RoI 20 E3 53 32N 6 06W
**Rushden** England 15 F2 52 17N 0 36W
**RUSSIAN FEDERATION**
62/63
**Ruthin** Wales 14 C3 53 07N 3 18W
**Rutland** u.a. England 15 F2 5241N 00 39W
**Rutland Water** England 15 F2 52 40N 0 37W
**RWANDA** 83 A3/B3
**Ryazan'** Russia 62 F7 54 37N 39 43E
**Rybinsk** Russia 62 F8 58 03N 38 50E
**Rybnik** Poland 53 J5 50 07N 18 30E
**Ryde** England 16 B1 50 44N 1 10W
**Rye** r. England 13 E3 54 15N 1 10W
**Rye** England 17 D1 50 57N 0 44E
**Ryton** England 13 E3 54 59N 1 57W
**Ryukyu Islands** Japan 69 N3 27 30N 127 30E

## S

**Saalfeld** Germany 52 E5 50 39N 11 22E
**Saarbrücken** Germany 52 C4 49 15N 6 58E
**Sabadell** Spain 51 G4 41 33N 2 07E
**Sacramento** USA 90 B4 38 32N 121 30W
**Sacriston** England 13 E3 54 50N 1 38W
**Saffron Walden**
England 17 D3 52 01N 0 15E
**Saga** Japan 70 B1 33 16N 130 18E
**Sagunto** Spain 51 E3 39 40N 0 17W
**Sahagún** Spain 51 C3 42 23N 5 02W
**St. Abbs** Scotland 11 G2 55 54N 2 08W
**St. Abb's Head** Scotland 11 G2 55 55N 2 09W
**St. Agnes** England 18 B2 50 18N 5 13W
**St. Albans** England 16 C2 51 46N 0 21W
**St Alban's or St Aldheim's Head**
England 19 E2 50 34N 2 04W
**St. Andrews** Scotland 9 G1 56 20N 2 48W
**St. Anne** Channel Islands 19 E1 49 43N 2 12W
**St. Asaph** Wales 14 C3 53 15N 3 26W
**St. Austell** England 18 C2 50 20N 4 48W
**St. Bees** England 12 C3 54 29N 3 35W
**St. Blazey** England 18 C2 50 22N 4 43W
**St. Bride's Bay** Wales 14 A1 51 50N 5 15W
**St-Brieuc** France 50 C5 48 31N 2 45W
**St. Catharines** Canada 89 V2 43 10N 79 15W
**St-Chamond** France 50 G3 45 29N 4 32E
**St. Clears** Wales 14 B1 51 50N 4 30W
**St. Columb Major**
England 18 C2 50 26N 4 56W
**St. David's** Wales 14 A1 51 54N 5 16W
**St. David's Head** Wales 14 A1 51 55N 5 19W
**St.-Dié** France 50 H5 48 17N 6 57E
**St.-Dizier** France 50 G5 48 38N 4 58E
**Saintes** France 50 D3 45 44N 0 38W
**St.-Étienne** France 50 G3 45 26N 4 23E
**Saintfield** NI 10 D1 54 28N 5 50W

St. Finnan's Bay RoI 21 A1 51 50N 10 23W
St. Gallen Switz. 52 D3 47 25N 9 23E
St.-Gaudens France 50 E2 43 07N 0 44E
St. George's Grenada 94 R11 12 04N 61 44W
St. George's Channel
 British Isles 21 E2 52 00N 6 00W
St.-German en-Laye
 France 50 F5 48 54N 2 04E
St-Girons France 50 E2 42 59N 1 08E
St. Helens England 12 D2 53 28N 2 44W
St. Helier
 Channel Islands 19 E1 49 12N 2 07W
St. Ives England 16 C3 52 20N 0 05W
St. Ives England 18 B2 50 12N 5 29W
St. Jean-de-Luz France 50 D2 43 23N 1 39W
Saint John Canada 89 X3 45 16N 66 03W
St. John's Canada 89 AA3 47 34N 52 41W
St. John's Point NI 10 D1 54 15N 5 35W
St Johnstown RoI 20 D4 54 56N 7 28W
St. John's Town of Dalry
 Scotland 11 E2 55 07N 4 10W
St. Just England 18 B2 50 07N 5 41W
St. Keverne England 18 B2 50 03N 5 06W
St. Kilda i. Scotland 8 A2 57 49N 8 34W
ST. KITTS AND NEVIS 94 P9
St. Lawrence r.
 Canada/USA 89 X3 46 55N 55 24W
St. Leonards England 19 F2 50 50N 1 50W
St-Lô France 50 D5 49 07N 1 05W
St. Louis USA 91 H4 38 40N 90 15W
ST. LUCIA 94R11/R12
St. Malo France 50 C5 48 39N 2 00W
St. Margaret's at Cliffe
 England 17 E2 51 10N 1 23E
St. Mawes England 18 B2 50 09N 5 01W
St. Monance Scotland 11 G3 56 13N 2 46W
St. Moritz Switz. 52 D3 46 30N 9 51E
St.-Nazaire France 50 C4 47 17N 2 12W
St. Neots England 16 C3 52 14N 0 17W
St.-Omer France 50 F6 50 45N 2 15E
St. Paul USA 91 H5 45 00N 93 10W
St. Peter Port
 Channel Islands 19 E1 49 27N 3 32W
St. Petersburg Russia 62 F8 59 55N 30 25E
Saint-Pierre & Miquelon
 is. Atlantic Ocean 89 Z3 47 00N 56 20W
St. Pölten Austria 53 G4 48 13N 15 37E
St.-Quentin France 50 F5 49 51N 3 17E
St.-Tropez France 50 H2 43 16N 6 39E
ST. VINCENT AND THE GRENADINES
 94 R11
Sakhalin i. Russia 63 S7 50 00N 143 00E
Salamanca Spain 51 C4 40 58N 5 40W
Salcombe England 18 D2 50 13N 3 47W
Sale England 12 D2 53 26N 2 19W
Salen Scotland 8 D1 56 43N 5 47W
Salen Scotland 8 D1 56 31N 5 57W
Salerno Italy 54 F4 40 40N 14 46E
Salford England 12 D2 53 30N 2 16W
Salisbury England 16 B2 51 05N 1 48W
Salisbury Plain
 England 16 B2 51 10N 1 55W
Salmon USA 90 D6 45 11N 113 55W
Salo Finland 49 F3 60 23N 23 10E
Salò Italy 54 D6 45 37N 10 31E
Salon-de-Provence
 France 50 G2 43 38N 5 06E
Salt Sudan 64 N11 32 03N 35 44E
Salta Argentina 96 D9 24 46S 65 28W
Saltash England 18 C2 50 24N 4 12W
Saltburn-by-the-Sea
 England 13 F3 54 35N 0 58W
Saltcoats Scotland 11 E2 55 38N 4 47W
Saltdal Norway 49 E4 67 06N 15 25E
Saltdean England 16 C1 50 49N 0 02W
Saltfleet England 13 G2 53 26N 0 10E
Saltillo Mexico 94 D5 25 30N 101 00W
Salt Lake City USA 90 D5 40 45N 111 55W
Salvador Brazil 96 K11 12 58S 38 29W
Salween r.
 China/Myanmar 71 B8 20 00N 103 00E
Salzburg Austria 52 F3 47 48N 13 03E
Salzgitter Germany 52 E6 52 13N 10 20E
Samara Russia 62 H7 53 10N 50 10E
SAMOA 100 J6
Samsun Turkey 57 E3 41 17N 36 22E
Sana Yemen 64 D2 15 23N 44 14E
San Antonio USA 91 G2 29 25N 98 30W
San Cristóbal
 Venezuela 96 C15 7 46N 72 15W
Sanda Island Scotland 10 D2 55 18N 5 35W
Sanday i. Scotland 7 B2 59 15N 2 30W
Sandbach England 12 D2 53 09N 2 22W
Sandhead Scotland 10 E1 54 48N 4 58W
Sandhurst England 16 C2 51 21N 0 48W
San Diego USA 90 C3 32 45N 117 10W
Sandown England 16 B1 50 39N 1 09W
Sandray i. Scotland 8 B1 56 53N 7 30W
Sandringham England 17 D3 52 50N 0 31E
Sandwich England 17 E2 51 17N 1 20E
Sandy England 16 C3 52 08N 0 18W
San Fernando Spain 51 B2 36 28N 6 12W
San Francisco USA 90 B4 37 45N 122 27W
San José Costa Rica 95 H1 9 59N 84 04W
San José USA 90 B4 37 20N 121 55W
San Juan Argentina 97 D7 31 33S 68 31W
San Juan Puerto Rico 95 L3 18 29N 66 08W

San Juan Bautista
 Spain 51 F3 39 05N 1 31E
SAN MARINO 54 E5 44 00N 12 00E
Sanmenxia China 69 K4 34 46N 111 17E
San Miguel de Tucumán
 Argentina 96 D8 26 48S 65 16W
Sanming China 69 L3 26 16N 117 35E
Sannox Scotland 10 D2 56 41N 5 08W
Sanquhar Scotland 11 F2 55 22N 3 56W
San Remo Italy 54 B5 43 48N 7 46E
San Salvador
 El Salvador 94 G2 13 40N 89 10W
San Sebastián Spain 51 E5 43 19N 1 59W
San Severo Italy 54 F4 41 41N 15 23E
Santa Barbara USA 90 C3 33 29N 119 01W
Santa Cruz Bolivia 96 E10 17 50S 63 10W
Santa Cruz Spain 80 A7 28 28N 16 15W
Sante Fé Argentina 97 E7 31 35S 60 50W
Santa Fe USA 90 E4 35 41N 105 57W
Santa Maria Brazil 97 G8 29 45S 53 40W
Santa Marta Colombia 96 C16 11 18N 74 10W
Santander Spain 51 D5 43 28N 3 48W
Sant'Antioco Italy 54 C3 39 04N 8 27E
Santarém Brazil 96 G13 2 26S 54 41W
Santarém Portugal 51 A3 39 14N 8 40W
Santa Teresa Gallura
 Italy 54 C4 41 14N 9 12E
Santiago Chile 97 C6 33 30S 70 40W
Santiago de Compostela
 Spain 51 A5 42 52N 8 33W
Santiago del Estero
 Argentina 97 E8 27 45S 64 19W
Santo Andre Brazil 96 H9 23 39S 46 29W
Santo Domingo
 Dom. Rep. 95 L3 18 30N 69 57W
Santoña Spain 51 D5 43 27N 3 26W
Santos Brazil 96 H9 23 56S 46 22W
São Francisco r. Brazil 96 J11 15 00S 44 00W
São Luís Brazil 96 J13 2 34S 44 16W
Saône r. France 50 G4 46 28N 4 55E
São Paulo Brazil 96 H9 23 33S 46 39W
SÃO TOMÉ AND PRINCIPE
 80 F2
Sapporo Japan 70 D3 43 05N 141 21E
Sarajevo
 Bosnia-Herzegovina 55 H5 43 52N 18 26E
Saratov Russia 62 G7 51 30N 45 55E
Sardinia i. Italy 54 C3/C4 40 00N 9 00E
Sark i. Channel Islands 19 E1 49 26N 2 22W
Sarnia Canada 89 U2 42 58N 82 23W
Sarrebourg France 50 H5 48 43N 7 03E
Sarreguemines France 50 H5 49 06N 6 55E
Sartène France 50 J1 41 37N 8 58E
Saskatchewan prov.
 Canada 88/89 P5/Q5 53 50N 109 00W
Saskatoon Canada 88 P4 52 10N 106 40W
Sassari Italy 54 C4 40 43N 8 34E
Sassnitz Germany 52 F7 54 32N 13 40E
Sassuolo Italy 54 D6 44 32N 10 47E
Satu Mare Romania 53 L3 47 48N 22 52E
SAUDI ARABIA 64/65 D3/F3
Sault Ste. Marie
 Canada 89 U3 46 31N 84 20W
Saumur France 50 D4 47 16N 0 05W
Saundersfoot Wales 14 B1 51 43N 4 43W
Savannah USA 91 K3 32 04N 81 07W
Savona Italy 54 C6 44 18N 8 28E
Sawbridgeworth
 England 17 D2 51 50N 0 09E
Sawel mt. NI 10 B1 54 49N 7 02W
Sawston England 17 D2 52 07N 0 10E
Saxilby England 13 F2 53 17N 0 40W
Saxmundham England 17 E3 52 13N 1 29E
Saxthorpe England 17 E3 52 50N 1 09E
Scafell Pike mt.
 England 12 C3 54 27N 3 14W
Scalasaig Scotland 10 C3 56 04N 6 12W
Scalby England 13 F3 54 18N 0 27W
Scalloway Scotland 7 C3 60 08N 1 17W
Scalpay i. Scotland 8 C2 57 52N 6 40W
Scalpay i. Scotland 8 D2 57 15N 6 00W
Scapa Scotland 7 B1 58 58N 2 59W
Scapa Flow sd. Scotland 7 A1 58 55N 3 00W
Scarborough England 13 F3 54 17N 0 24W
Scarinish Scotland 8 C1 56 29N 6 48W
Scarp i. Scotland 8 B3 58 02N 7 08W
Scarriff RoI 21 C2 52 55N 8 37W
Scartaglin RoI 21 B2 52 10N 9 26W
Schaffhausen Switz. 52 D3 47 42N 8 38E
Schefferville Canada 89 X4 54 50N 67 00W
Schleswig Germany 52 D7 54 32N 9 34E
Schwäbisch Gmünd
 Germany 52 D4 48 49N 9 48E
Schwarzwald mts.
 Germany
Schwaz Austria 52 E3 47 21N 11 44E
Schwechat Austria 53 H4 48 09N 16 27E
Schweinfurt Germany 52 E5 50 03N 10 16E
Schwerin Germany 52 E6 53 38N 11 25E
Schwyz Switz. 52 D3 47 02N 8 34E
Sciacca Italy 54 E2 37 31N 13 05E
Scilly, Isles of England 18 A1 49 56N 6 20W
Sconser Scotland 8 C2 57 18N 6 07W
Scotland UK 6 56 00N 4 00W
Scottish Borders u.a.
 Scotland 11 F2/G2 55 30N 2 55W
Scourie Scotland 8 D3 58 20N 5 08W

Scrabster Scotland 9 F3 58 37N 3 34W
Scunthorpe England 13 F2 53 35N 0 39W
Seaford England 17 D1 50 46N 0 06E
Seaham England 13 E3 54 50N 1 20W
Seahouses England 11 H2 55 35N 1 38W
Seascale England 12 C3 54 24N 3 29W
Seaton Devon England 19 D2 50 43N 3 05W
Seaton Cumbria
 England 12 C3 54 40N 3 30W
Seaton Delaval England 13 E4 55 04N 1 31W
Seattle USA 90 B6 47 35N 122 20W
Sedan France 50 G5 49 42N 4 57E
Sedbergh England 12 D3 54 20N 2 31W
Segovia Spain 51 C4 40 57N 4 07W
Seine r. France 50 E5 49 15N 1 15E
Sekondi Takoradi
 Ghana 80 D2 4 59N 1 43W
Selborne England 16 C2 51 06N 0 56W
Selby England 13 E2 53 48N 1 04W
Sélestat France 50 H5 48 16N 7 28E
Selkirk Scotland 11 G2 55 33N 2 50W
Selsey England 16 C1 50 44N 0 48W
Selsey Bill p. England 16 C1 50 43N 0 48W
Semarang Indonesia 71 E2 6 58S 110 29E
Sendai Japan 70 D2 38 16N 140 52E
SENEGAL 80 A4/B4
Senlis France 50 F5 49 12N 2 35E
Sennen England 18 B2 50 03N 5 42W
Sennybridge Wales 14 C1 51 57N 3 34W
Sens France 50 F5 48 12N 3 18E
Seoul S. Korea 69 N5 37 32N 127 00E
Sept-Îles tn. Canada 89 X4 50 10N 66 00W
SERBIA AND MONTENEGRO
 55 H6/K6
Serov Russia 62 J8 59 42N 60 32E
Serra Brazil 96 J10 20 06S 40 16W
Sérres Greece 55 K4 41 03N 23 33E
Sète France 50 F2 43 25N 3 43E
Settle England 12 D3 54 04N 2 16W
Setúbal Portugal 51 A3 38 31N 8 54W
Sevastopol' Ukraine 57 D4 44 36N 33 31E
Seven Head c. RoI 21 C1 51 34N 8 42W
Severn r.
 England/Wales 14 C2 52 30N 3 15W
Severodvinsk Russia 62 F9 64 35N 39 50E
Seville Spain 51 C2 37 24N 5 59W
Seward USA 88 F6 60 05N 149 34W
SEYCHELLES 103
Sézannes France 50 F5 48 44N 3 44E
Sfântu Gheorghe
 Romania 53 M2 45 51N 25 48E
Sfax Tunisia 80 G8 34 45N 10 43E
Sgurr Mór mt. Scotland 9 D2 57 42N 5 03W
Shaftesbury England 19 E3 51 01N 2 12W
Shaldon England 18 D2 51 35N 3 30W
Shanghai China 69 M4 31 06N 121 22E
Shanklin England 16 B1 50 38N 1 10W
Shannon RoI 21 C2 52 41N 8 55W
Shannon r. RoI 21 C2/B2 52 45N 8 57W
Shap England 12 D3 54 32N 2 41W
Shapinsay i. Scotland 7 B2 59 03N 2 51W
Sheerness England 17 D2 51 27N 0 45E
Sheffield England 13 E2 53 23N 1 30W
Shefford England 16 C3 52 02N 0 20W
Shelburne Canada 89 X2 43 37N 65 20W
Shenandoah USA 91 G5 40 48N 95 22W
Shenyang China 69 M6 41 50N 123 26E
Shepetovka Ukraine 53 N5 50 12N 27 01E
Shepperton England 16 C2 51 23N 0 28W
Sheppey, Isle of
 England 17 D2 51 25N 0 50E
Shepshed England 15 E2 52 47N 1 18W
Shepton Mallet
 England 19 E3 51 12N 2 33W
Sherborne England 19 E2 50 57N 2 31W
Sherbrooke Canada 89 W3 45 24N 71 54W
Shercock RoI 21 E3 53 59N 6 54W
Sheringham England 17 E3 52 57N 1 12E
Sherkin Island RoI 21 B1 51 28N 9 25W
's-Hertogenbosch
 Neths 48 D4 51 41N 5 19E
Shetland Islands u.a.
 Scotland 7 C2/D3 60 00N 1 15W
Shieldaig Scotland 8 D2 57 31N 5 39W
Shifnal England 15 D2 52 41N 2 22W
Shijiazhuang China 69 K5 38 04N 114 28E
Shilbottle England 11 H2 55 23N 1 42W
Shildon England 13 E3 54 38N 1 39W
Shillelagh RoI 21 E2 52 45N 6 32W
Shipley England 13 E2 53 50N 1 47W
Shipston-on-Stour
 England 15 E2 52 04N 1 37W
Shipton England 13 E3 54 01N 1 09W
Shirebrook England 13 E2 53 13N 1 13W
Shīrāz Iran 65 F4 29 38N 52 34E
Shoeburyness England 17 D2 51 32N 0 48E
Shoreham-by-Sea
 England 16 C1 50 49N 0 17W
Shotton Wales 14 C3 53 12N 3 02W
Shotts Scotland 11 F2 55 49N 3 48W
Shrewsbury England 15 D2 52 43N 2 45W
Shrewton England 16 B2 51 12N 1 55W
Shropshire co.
 England 14/15 D2 52 35N 2 25W
Shumen Bulgaria 55 M5 43 17N 26 55E
Sible Hedingham
 England 17 D2 51 58N 0 35E

Sicily i. Italy 54 E2/F2 37 00N 14 00E
Siderno Italy 54 G3 38 16N 16 17E
Sidi Bel Abbès Algeria 80 D9 35 15N 0 39W
Sidlaw Hills Scotland 9 F1/G1 56 30N 3 10W
Sidmouth England 19 D2 50 41N 3 15W
Siegen Germany 52 D5 50 52N 8 02E
Siena Italy 54 D5 43 19N 11 19E
SIERRA LEONE 80 B3
Sierra Madre Occidental
 mts. Mexico 94 C5/D4 26 00N 107 00W
Sierra Nevada mts.
 Spain 51 D2 37 00N 3 20W
Sierra Nevada mts.
 USA 90 C4 37 00N 119 00W
Sighişoara Romania 53 M3 46 12N 24 48E
Sigüenza Spain 51 D4 41 04N 2 38W
Sileby England 15 E2 52 44N 1 06W
Silistra Bulgaria 55 M6 44 06N 27 17E
Silloth England 12 C3 54 52N 3 23W
Silverton England 18 D2 50 48N 3 30W
Silves Portugal 51 A2 37 11N 8 26W
Simferopol' Ukraine 57 D3 44 57N 34 05E
Simonsbath England 18 D3 51 09N 3 45W
Sinclair's Bay Scotland 9 F3 58 30N 3 07
Sines Portugal 51 A2 37 58N 8 52W
SINGAPORE 71 C4
Sintra Portugal 51 A3 38 48N 9 22W
Sion Switz. 52 C3 46 14N 7 22E
Sion Mills NI 10 B1 54 47N 7 28W
Siracusa Italy 54 F2 37 04N 15 19E
Sisak Croatia 54 G6 45 30N 16 22E
Sisteron France 50 G3 44 16N 5 56E
Sittingbourne England 17 D2 51 21N 0 44E
Sixmilebridge RoI 21 C2 52 45N 8 46W
Sjaelland i. Denmark 49 D2 55 15N 11 30E
Skagerrak sd.
 Denmark/Norway 49 C2 57 30N 8 00E
Skegness England 13 G2 53 10N 0 21E
Skellefteå Sweden 49 F3 64 45N 21 00E
Skellingthorpe England 13 F2 53 14N 0 37W
Skelmanthorpe
 England 13 E2 53 36N 1 40W
Skelmersdale England 12 D2 53 55N 2 48W
Skelmorlie Scotland 10 E2 55 51N 4 53W
Skelton England 13 E3 54 33N 0 59W
Skerries RoI 20 E3 53 35N 6 07W
Skibbereen RoI 21 B1 51 33N 9 15W
Skiddaw mt. England 12 C3 54 40N 3 08W
Skien Norway 49 C2 59 14N 9 37E
Skipton England 12 D2 53 58N 2 01W
Skokholm Island Wales 14 A1 51 42N 5 16W
Skomer Island Wales 14 A1 51 45N 5 18W
Skopje FYROM 55 J4 42 00N 21 28E
Skövde Sweden 49 D2 58 24N 13 52E
Skull RoI 21 B1 51 32N 9 40W
Skye i. Scotland 8 C2 57 20N 6 15W
Slaidburn England 12 D2 54 00N 2 26W
Slane RoI 20 E3 54 43N 6 33W
Slaney r. RoI 21 E2 52 41N 6 33W
Slatina Romania 53 M2 44 26N 24 22E
Slave r. Canada 88 N5 59 20N 111 10W
Slavonski Brod Croatia 55 G6 45 09N 18 02E
Sleaford England 13 F2 53 00N 0 24W
Sleat, Point of Scotland 8 C2 57 01N 6 01W
Sleat, Sound of Scotland 8 D2 57 05N 5 55W
Sledmere England 13 F3 54 04N 0 35W
Sleights England 13 F3 54 26N 0 40W
Slieve Donard mt. NI 10 D1 54 11N 5 55W
Slieve Gamph hills RoI 20 C4 54 11N 8 53W
Slieve Mish Mountains
 RoI 21 B2 52 11N 10 00W
Sligo RoI 20 C4 54 17N 8 28W
Sligo co. RoI 20 C4 54 10N 8 40W
Sligo Bay b. RoI 20 C4 54 17N 8 40W
Sliven Bulgaria 55 M5 42 40N 26 19E
Slough England 16 C2 51 31N 0 36W
Slough u.a. England 16 C2 51 30N 0 40W
SLOVAKIA 53 J4/K4
SLOVENIA 54 F6
Smethwick England 15 E2 52 30N 1 58W
Smolensk Russia 62 F7 54 49N 32 04E
Smolyan Bulgaria 55 L4 41 34N 24 42E
Snaefell mt. Isle of Man 12 B3 54 16N 4 28W
Snake r. USA 90 C5 44 00N 118 00W
Sneek Neths 48 D6 53 02N 5 40E
Sneem RoI 21 B1 51 50N 9 45W
Snodland England 17 D2 51 20N 0 27E
Snowdon mt. Wales 14 B3 53 04N 4 05W
Snowy Mountains
 Australia 73 J3 36 50S 147 00E
Soar r. England 15 E2 52 50N 1 15W
Soay i. Scotland 8 C2 57 09N 6 13W
Socotra i. Yemen 65 F1 12 05N 54 10E
Söderhamn Sweden 49 E3 61 19N 17 10E
Sodertälje Sweden 49 E2 59 11N 17 39E
Sofia Bulgaria 55 K5 42 40N 23 18E
Soham England 17 D3 52 20N 0 20E
Soissons France 50 F5 49 23N 3 20E
Solapur India 66 D3 17 43N 75 56E
Solent, The sd. England 16 B1 50 45N 1 25W
Soligorsk Belarus 53 N6 52 50N 27 32E
Solihull England 15 E2 52 25N 1 45W
Sollefteå Sweden 49 E3 63 09N 17 15E
Söller Spain 51 F3 39 46N 2 42E
SOLOMON ISLANDS 73 L9/M9
Solothurn Switz. 52 C3 47 13N 7 32E
Soltau Germany 52 D6 52 59N 9 50E

**Solway Firth** Scotland 11 F1 54 45N 3 40W
**SOMALIA** 81 Q2/Q3
**Somerset** *co.* England 19 D3/E3 51 10N 3 00W
**Somerton** England 19 E3 51 03N 2 44W
**Sompting** England 16 C1 50 50N 0 21W
**Sønderborg** Denmark 49 C1 54 55N 9 48E
**Sondrio** Italy 54 C7 46 11N 9 52E
**Sonning Common** *tn.* England 16 C2 51 31N 0 59W
**Sopron** Hungary 53 H3 47 40N 16 35E
**Sora** Italy 54 E4 41 43N 13 37E
**Soria** Spain 51 D4 41 46N 2 28W
**Soroca** Moldova 53 P4 48 08N 28 12E
**Sosnowiec** Poland 53 J5 50 16N 19 07E
**Souillac** France 50 E3 44 53N 1 29E
**Southam** England 15 E2 52 15N 1 23W
**Southampton** England 16 B1 50 55N 1 25W
**Southampton** *u.a.* England 16 B1 50 55N 1 25W
**Southampton Water** *sd.* England 16 B1 50 50N 1 20W
**South Ayrshire** *u.a.* Scotland 11 E2 55 20N 4 40W
**South Ballachulish** Scotland 8 D1 56 42N 5 09W
**South Bank** *tn.* England 13 E3 54 35N 1 10W
**South Benfleet** England 17 D2 51 33N 0 34E
**Southborough** England 17 D2 51 10N 0 15E
**South Brent** England 18 D2 50 25N 3 50W
**South Carolina** *state* USA 91 K3 34 00N 81 00W
**South Cave** *tn.* England 13 F2 53 46N 0 35W
**South Cerney** England 15 E1 51 41N 1 57W
**South Dakota** *state* USA 90 F5 45 00N 102 00W
**South Dorset Downs** England 19 E2 50 40N 2 30W
**South Downs** England 16 C1 50 50N 0 30W
**Southend** *u.a.* England 17 D2 51 32N 0 50E
**Southend** Scotland 10 D2 55 20N 5 38W
**Southend-on-Sea** England 17 D2 51 33N 0 43E
**Southerness Point** Scotland 11 F1 54 53N 3 37W
**Southern Ocean** 101
**South Esk** *r.* Scotland 9 F1 56 50N 3 10W
**South Foreland** *c.* England 17 E2 51 09N 1 23E
**South Gloucestershire** *u.a.* England 15 D1 51 30N 2 30W
**South Hayling** England 16 C1 50 47N 0 59W
**South Island** NZ 73 N1/P2 42 30S 172 00E
**South Kirkby** England 13 E2 53 34N 1 20W
**SOUTH KOREA** 69 N5
**South Lanarkshire** *u.a.* Scotland 11 E2/F2 55 35N 3 55W
**Southminster** England 17 D2 51 40N 0 50E
**South Molton** England 18 D3 51 01N 3 50W
**South Normanton** England 13 E2 53 06N 1 20W
**South Ockendon** England 17 D2 51 32N 0 18E
**South Petherton** England 19 E2 50 58N 2 49W
**South Pole** Antarctica 101 90 00S
**Southport** England 12 C2 53 39N 3 01W
**South Queensferry** Scotland 11 F2 55 59N 3 25W
**South Shields** England 13 E3 55 00N 1 25W
**South Tyne** *r.* England 12 D3 54 52N 2 30W
**South Uist** *i.* Scotland 8 B2 57 20N 7 15W
**South Walls** *i.* Scotland 7 A1 58 47N 3 11W
**Southwick** England 16 C1 51 01N 0 14W
**Southwold** England 17 E3 52 20N 1 40E
**South Woodham Ferrers** England 17 D2 51 38N 0 38E
**South Yorkshire** *co.* England 13 E2 53 30N 1 30W
**Sowerby Bridge** England 13 E2 53 43N 1 54W
**SPAIN** 51 C3/E3
**Spalding** England 16 C3 52 47N 0 10W
**Spárti** Greece 55 K2 37 05N 22 25E
**Spean Bridge** *tn.* Scotland 9 E1 56 53N 4 54W
**Spennymoor** England 13 E3 54 42N 1 35W
**Sperrin Mountains** NI 10 B1 54 49N 7 06W
**Spey** *r.* Scotland 9 F2 57 15N 3 45W
**Spiddle** RoI 20 B3 53 15N 9 18W
**Spilsby** England 13 G2 53 11N 0 05E
**Spink** RoI 21 D2 52 53N 7 12W
**Spithead** *sd.* England 16 B1 50 45N 1 10W
**Spitsbergen** *i.* Arctic Ocean 62 D12 79 00N 15 00E
**Spittal an der Drau** Austria 52 F3 46 48N 13 30E
**Split** Croatia 55 G5 43 31N 16 28E
**Spokane** USA 90 C6 47 40N 117 25W
**Spoleto** Italy 54 E5 42 44N 12 44E
**Springfield** Illinois USA 91 J4 39 49N 89 39W
**Springfield** Massachusetts USA 91 M5 42 07N 72 35W
**Spurn Head** England 13 G2 53 36N 0 07E
**SRI LANKA** 66 E1
**Srinagar** Kashmir 66 C6 34 08N 74 50E
**Stacks Mountains** RoI 21 B2 52 19N 9 33W
**Staffa** *i.* Scotland 8 C1 56 30N 6 15W
**Stafford** England 15 D2 52 48N 2 07W
**Staffordshire** *co.* England 15 D2/E2 52 50N 2 00W
**Staines** England 16 C2 51 26N 0 30W
**Stainforth** England 13 E2 53 36N 1 01W
**Stalbridge** England 19 E2 50 58N 2 23W
**Stalybridge** England 15 D3 53 29N 2 04W
**Stamford Bridge** *tn.* England 13 F2 53 59N 0 55W
**Standish** USA 93 D2 43 59N 83 58W
**Stane** Scotland 11 F2 55 50N 3 47W
**Stanhope** England 12 D3 54 45N 2 01W
**Stanley** England 13 E2 53 44N 1 28W
**Stanley** England 13 E3 54 53N 1 42W
**Stanley** Falkland Islands 97 F3 51 45S 57 56W
**Stansted Mountfitchet** England 17 D2 51 54N 0 12E
**Stapleford** England 15 E2 52 56N 1 15W
**Staplehurst** England 17 D2 51 10N 0 33E
**Starachowice** Poland 53 K5 51 03N 21 00E
**Stara Zagora** Bulgaria 55 L5 42 25N 25 37E
**Start Bay** England 18 D2 50 16N 3 35W
**Start Point** England 18 D2 50 13N 3 38W
**Staunton** USA 91 L4 38 10N 79 05W
**Stavanger** Norway 49 C2 58 58N 5 45E
**Staveley** England 13 E2 53 16N 1 20W
**Stavropol'** Russia 62 G6 45 03N 41 59E
**Staxton** England 13 F3 54 11N 0 26W
**Steeping** *r.* England 13 G2 53 10N 0 10E
**Stendal** Germany 52 E6 52 36N 11 52E
**Stenhousemuir** Scotland 11 F3 56 02N 3 49W
**Stevenage** England 16 C2 51 55N 0 14W
**Stevenston** Scotland 11 E2 55 39N 4 45W
**Stewart** Canada 88 J5 55 56N 130 01W
**Stewarton** Scotland 11 E2 55 41N 4 31W
**Stewartstown** NI 10 C1 54 35N 6 41W
**Steyning** England 16 C1 50 53N 0 20W
**Steyr** Austria 52 G3 48 04N 14 25E
**Stilton** England 16 C3 52 29N 0 17W
**Stinchar** *r.* Scotland 11 E2 55 10N 4 50W
**Stirling** Canada 88 N3 49 34N 112 30W
**Stirling** *u.a.* Scotland 11 E3 56 10N 4 15W
**Stirling** Scotland 11 F3 56 07N 3 57W
**Stockbridge** England 16 B2 51 07N 1 29W
**Stockerau** Austria 53 H4 48 24N 16 13E
**Stockholm** Sweden 49 E2 59 20N 18 05E
**Stockport** England 12 D2 53 25N 2 10W
**Stocksbridge** England 13 E2 53 27N 1 34W
**Stockton-on-Tees** England 13 E3 54 34N 1 19W
**Stockton-on-Tees** *u.a.* England 13 E3 54 33N 1 24W
**Stoer, Point of** Scotland 8 D3 58 25N 5 25W
**Stoke-on-Trent** *u.a.* England 12 D2 53 00N 2 10W
**Stokenchurch** England 16 C2 51 40N 0 55W
**Stoke-on-Trent** England 12 D2 53 00N 2 10W
**Stoke Poges** England 16 C2 51 33N 0 35W
**Stokesley** England 13 E3 54 28N 1 11W
**Stone** England 15 D2 52 54N 2 10W
**Stonehaven** Scotland 9 G1 56 38N 2 13W
**Stonehouse** Scotland 11 F2 55 43N 3 59W
**Stony Stratford** England 16 C3 52 04N 0 52W
**Størdal** Norway 49 D3 63 18N 11 48E
**Støren** Norway 49 D3 63 03N 10 16E
**Stornoway** Scotland 8 C3 58 12N 6 23W
**Storrington** England 16 C1 50 55N 0 28W
**Storr, The** *mt.* Scotland 8 C2 57 30N 6 11W
**Storuman** Sweden 49 E4 65 05N 17 10E
**Stotfold** England 16 C3 52 01N 0 15W
**Stour** *r.* Warwickshire England 15 E2 52 05N 1 45E
**Stour** *r.* Suffolk England 17 D3 52 10N 0 30E
**Stour** *r.* Kent England 17 E2 51 10N 1 10E
**Stour** *r.* Dorset England 19 E2 50 50N 2 10W
**Stourbridge** *tn.* England 15 D2 52 27N 2 09W
**Stourport-on-Severn** England 15 D2 52 21N 2 16W
**Stow** Scotland 11 G2 55 42N 2 51W
**Stowmarket** England 17 E3 52 11N 1 00E
**Stow-on-the-Wold** England 15 E1 51 56N 1 44W
**Strabane** NI 10 B1 54 49N 7 27W
**Strabane** *district* NI 10 B1 54 45N 7 30W
**Strachur** Scotland 10 D3 56 10N 5 04W
**Stradbally** RoI 21 D3 53 00N 7 08W
**Stradone** RoI 20 D3 53 59N 7 14W
**Stralsund** Germany 52 F7 54 18N 13 06E
**Strangford Lough** NI 10 D1 54 30N 5 35W
**Stranorlar** RoI 20 D4 54 48N 7 46W
**Stranraer** Scotland 10 D2 54 55N 5 02W
**Strasbourg** France 50 H5 48 35N 7 45E
**Stratford-upon-Avon** England 15 E2 52 12N 1 41W
**Strathaven** Scotland 11 E2 55 41N 4 05W
**Strathcarron** Scotland 8 D2 57 26N 5 27W
**Strathpeffer** Scotland 9 E2 57 35N 4 33W
**Strathy** Scotland 9 F3 58 33N 3 59W
**Strathy** *r.* Scotland 9 E3 58 30N 4 00W
**Strathy Point** Scotland 9 E3 58 35N 4 02W
**Stratton** England 18 C2 50 50N 4 31W
**Straubing** Germany 52 F4 48 53N 12 35E
**Street** England 19 3 51 03N 2 45W
**Stretford** England 12 D2 53 27N 2 19W
**Strímonas** *r.* Greece 55 K4 41 00N 23 00E
**Strokestown** RoI 20 C3 53 47N 8 06W
**Stroma** *i.* Scotland 7 A1 58 40N 3 08W
**Stromboli** *mt.* Italy 54 F3 38 48N 15 15E
**Stromeferry** Scotland 8 D2 57 21N 5 34W
**Stromness** Scotland 7 A1 58 57N 3 18W
**Stronsay** *i.* Scotland 7 B2 59 07N 2 37W
**Stronsay Firth** Scotland 7 B2 59 00N 2 55W
**Strontian** Scotland 8 D1 56 41N 5 34W
**Strood** England 17 D2 51 24N 0 28E
**Stroud** England 15 D1 51 45N 2 12W
**Strumble Head** Wales 14 A2 52 02N 5 04W
**Stryy** Ukraine 53 L4 49 16N 23 51E
**Stubbington** England 16 B1 50 49N 12 13W
**Studley** England 15 E2 52 16N 1 52W
**Sturminster Marshall** England 19 E2 50 48N 2 04W
**Sturminster Newton** England 19 E2 50 55N 2 18W
**Stuttgart** Germany 52 D4 48 47N 9 12E
**Subotica** Serbia & Montenegro 55 H7 46 04N 19 41E
**Sucre** Bolivia 96 D10 19 05S 65 15W
**SUDAN** 81 L4/M4
**Sudbury** Canada 89 U3 46 30N 81 01W
**Sudbury** England 17 E3 52 02N 0 44E
**Suffolk** *co.* England 17 D3/E3 52 10N 1 00E
**Suhl** Germany 52 E5 50 37N 10 43E
**Sulawesi** *i.* Indonesia 71 F3/G3 2 00S 120 00E
**Sumatera** *i.* Indonesia 71 B4/C3 0 00 100 00E
**Sumburgh Head** Scotland 7 C2 59 51N 1 16W
**Summerhill** RoI 20 E3 53 29N 6 44W
**Sumy** Ukraine 56 D5 50 55N 34 49E
**Sunderland** England 13 E3 54 55N 1 23W
**Sundsvall** Sweden 49 E3 62 22N 17 20E
**Sunningdale** England 16 C2 51 24N 0 37W
**Superior, Lake** Canada/USA 92 C3 48 00N 88 00W
**Surabaya** Indonesia 71 E2 7 14S 112 45E
**Surat** India 66 C4 21 10N 72 54E
**SURINAME** 96 F14
**Surrey** *co.* England 16 C2 51 15N 0 30W
**Sutton** England 16 C2 51 22N 0 12W
**Sutton Coldfield** England 15 E2 52 34N 1 48W
**Sutton in Ashfield** England 13 E2 53 08N 1 15W
**Sutton on Sea** England 13 G2 53 19N 0 17E
**Suwalki** Poland 53 L7 54 06N 22 56E
**Suzhou** China 69 M4 31 21N 120 40E
**Svilengrad** Bulgaria 55 M4 41 45N 26 14E
**Swadlincote** England 15 E2 52 46N 1 33W
**Swaffham** England 17 D3 52 39N 0 41E
**Swale** *r.* England 12 D3 54 20N 2 00W
**Swanage** England 19 F2 50 37N 1 58W
**Swanley** England 17 D2 51 24N 0 12E
**Swanlinbar** RoI 20 D4 54 12N 7 42W
**Swanscombe** England 17 D2 51 26N 0 18E
**Swansea** Wales 14 C1 51 38N 3 57W
**Swansea Bay** Wales 14 C1 51 35N 3 55W
**Swansea** *u.a.* Wales 14 B1/C1 51 35N 4 10W
**SWAZILAND** 82 F2
**SWEDEN** 49 D2/E4
**Swiebodzin** Poland 53 G6 52 15N 15 31E
**Swilly** *r.* RoI 20 D4 54 55N 7 45W
**Swindon** England 16 B2 51 34N 1 47W
**Swindon** *u.a.* England 16 B2 51 34N 1 47W
**Swinford** RoI 20 C3 53 57N 8 57W
**Swinoujście** Poland 52 G6 53 55N 14 18E
**Swinton** England 13 E2 53 30N 1 19W
**SWITZERLAND** 52 C3/D3
**Swords** RoI 20 E3 53 28N 6 13W
**Sydney** Australia 73 K4 33 55S 151 10E
**Sylt** Germany 52 D7 54 00N 8 00E
**Symbister** Scotland 7 C3 60 21N 1 02W
**Symington** Scotland 11 F2 55 37N 3 35W
**Syracuse** USA 91 L5 43 03N 76 10W
**SYRIA** 64 C6
**Syzran'** Russia 62 G7 53 10N 48 29E
**Szczecin** Poland 52 G6 53 25N 14 32E
**Szeged** Hungary 53 K3 46 15N 20 09E
**Szombathely** Hungary 53 H3 47 14N 16 38E

## T

**Tábor** Czech Rep. 52 G4 49 25N 14 39E
**Tabrīz** Iran 65 E6 38 05N 46 18E
**Tacoma** USA 90 B6 47 16N 122 30W
**Tadcaster** England 13 E2 53 53N 1 16W
**Tadley** England 16 B2 51 21N 1 08W
**Taegu** S. Korea 69 N5 35 52N 128 36E
**Taejon** S. Korea 69 N5 36 20N 127 26E
**Tafalla** Spain 51 E5 42 32N 1 41W
**Taff** *r.* Wales 14 C1 51 30N 3 10W
**Taghmon** RoI 21 E2 52 19N 6 39W
**Tain** Scotland 9 E2 57 48N 4 04W
**Taipei** Taiwan 69 M3 25 05N 121 32E
**TAIWAN** 69 M2
**Taiyuan** China 69 K5 37 50N 112 30E
**TAJIKISTAN** 62 J4/K4
**Tajo** *r.* Spain/Portugal 51 B3 39 00N 7 00W
**Takamatsu** Japan 70 B1 34 20N 134 01E
**Talca** Chile 97 C6 35 28S 71 40W
**Talcahuano** Chile 97 C6 36 40S 73 10W
**Talgarth** Wales 14 C1 52 00N 3 15W
**Tallaght** RoI 20 E3 53 17N 6 21W
**Tallahassee** USA 91 K3 30 26N 84 16W
**Tallinn** Estonia 49 F2 59 22N 24 48E
**Tallow** RoI 21 C2 52 05N 8 01W
**Tal-y-bont** Wales 14 C2 52 29N 3 59W
**Tamanrasset** Algeria 80 F6 22 50N 5 28E
**Tamar** *r.* England 18 C2 50 32N 4 13W
**Tambov** Russia 62 G7 52 44N 41 28E
**Tame** *r.* England 15 E2 52 40N 1 45W
**Tampa** USA 91 K2 27 58N 82 38W
**Tampere** Finland 49 F3 61 32N 23 45E
**Tamworth** England 15 E2 52 39N 1 40W
**Tandragee** NI 10 C1 54 22N 6 25W
**Tanganyika, Lake** East Africa 83 B1 7 00S 30 00E
**Tangier** Morocco 80 C9 35 48N 5 45W
**Tangshan** China 69 L5 39 37N 118 05E
**TANZANIA** 83 B1/D1
**Tapachula** Mexico 94 F2 14 54N 92 15W
**Taransay** *i.* Scotland 8 B2 57 55N 7 00W
**Táranto** Italy 55 G4 40 28N 17 15E
**Tarazona** Spain 51 E4 41 54N 1 44W
**Tarbat Ness** *c.* Scotland 9 F2 57 50N 3 45W
**Tarbert** RoI 21 B2 52 32N 9 23W
**Tarbert** Western Isles Scotland 8 C2 57 54N 6 49W
**Tarbert** Argyle and Bute Scotland 10 D2 55 52N 5 26W
**Tarbes** France 50 E2 43 14N 0 05E
**Tarbet** Scotland 11 E3 56 12N 4 43W
**Tarbolton** Scotland 11 E2 55 31N 4 29W
**Târgovişte** Romania 53 M2 44 56N 25 27E
**Târgu Mureş** Romania 53 M3 46 33N 24 34E
**Tarija** Bolivia 96 D9 21 33S 65 02W
**Tarleton** England 12 D2 53 41N 2 50W
**Tarporley** England 12 D2 53 10N 2 40W
**Tarragona** Spain 51 F4 41 07N 1 15E
**Tarrasa** Spain 51 G4 41 34N 2 00E
**Tarsus** Turkey 57 D2 26 25N 34 52E
**Tartu** Estonia 49 G2 58 20N 26 44E
**Tashkent** Uzbekistan 62 J5 41 16N 69 13E
**Taunton** England 19 D3 51 01N 3 06W
**Taunus** *mts.* Germany 52 D5 50 00N 8 00E
**Tauranga** NZ 73 Q3 37 41S 176 10E
**Taverham** England 17 E3 52 41N 1 10E
**Tavira** Portugal 51 B2 37 07N 7 39W
**Tavistock** England 18 C2 50 33N 4 08W
**Tavy** *r.* England 18 C2 50 36N 4 04W
**Taw** *r.* England 18 D2 50 46N 3 55W
**Tay** *r.* Scotland 9 F1 56 30N 3 25W
**Taynuilt** Scotland 8 D1 56 25N 5 14W
**Tayport** Scotland 9 G1 56 27N 2 53W
**T'bilisi** Georgia 57 F3 41 43N 44 48E
**Tczew** Poland 53 J7 54 05N 18 46E
**Tebay** England 12 D3 54 26N 2 35W
**Tees** *r.* England 12 D3 54 40N 1 55W
**Tegucigalpa** Honduras 94 G2 14 05N 87 14W
**Tehran** Iran 65 F6 35 40N 51 26E
**Teifi** *r.* Wales 14 B2 52 03N 4 30W
**Teign** *r.* England 18 D2 50 42N 3 48W
**Teignmouth** England 18 D2 50 33N 3 30W
**Teith** *r.* Scotland 11 E3 56 10N 4 08W
**Tejo** *r.* Portugal 51 B3 39 30N 8 15W
**Tel Aviv-Yafo** Israel 64 B5 32 05N 34 46E
**Telford** England 15 D2 52 42N 2 28W
**Teme** *r.* England 15 D2 52 15N 2 25W
**Tempio Pausania** Italy 54 C4 40 54N 9 07E
**Templemore** RoI 21 D2 52 48N 7 50W
**Templenoe** RoI 21 B1 51 51N 9 41W
**Temuco** Chile 97 C6 38 45S 72 40W
**Tenbury Wells** England 15 D2 52 19N 2 35W
**Tenby** Wales 14 B1 51 41N 4 43W
**Tenerife** *i.* Spain 80 A7 28 15N 16 35W
**Tennessee** *r.* USA 91 J4 35 00N 88 00W
**Tennessee** *state* USA 91 J4 35 00N 87 00W
**Tenterden** England 17 D2 51 05N 0 41E
**Teramo** Italy 54 E5 42 40N 13 43E
**Teresina** Brazil 96 J12 5 09S 42 46W
**Termim Imerese** Italy 54 E2 27 59N 13 42E
**Termoli** Italy 54 F4 42 00N 15 00E
**Terni** Italy 54 E5 42 34N 12 39E
**Terracina** Italy 54 E4 41 17N 13 15E
**Terrasini** Italy 54 E3 38 09N 13 05E
**Terre Haute** USA 91 J4 39 27N 87 24W
**Teruel** Spain 51 E4 40 21N 1 06W
**Teslin** Canada 88 J6 60 10N 132 42W
**Test** *r.* England 16 B2 51 05N 1 30W
**Tetbury** England 15 D1 51 39N 2 10W
**Tetney** England 13 F2 53 30N 0 01W
**Tétouan** Morocco 80 C9 35 34N 5 22W
**Teviotdale** Scotland 11 G2 55 20N 3 00W
**Teviothead** Scotland 11 G2 55 19N 2 56W
**Tewkesbury** England 15 D1 51 59N 2 09W
**Texas** *state* USA 90 G3 31 00N 100 00W
**THAILAND** 71 B7/C7
**Thame** *r.* England 16 B2 51 45N 1 05W
**Thame** England 16 C2 51 45N 0 59W
**Thames** *r.* England 16 C2 51 32N 0 50W
**Thane** India 66 C3 19 14N 73 02E
**Thar Desert** India 66 C5 27 30N 72 00E
**Thatcham** England 16 B2 51 25N 1 15W
**Thaxted** England 17 D2 51 57N 0 20E
**The Downs** *tn.* RoI 20 D3 53 30N 7 14W
**The Everglades** USA 91 K2 26 00N 81 00W
**The Mumbles** *tn.* Wales 14 C1 51 35N 3 59W
**The Peak** *mt.* England 13 E2 53 24N 1 51W
**Thermaïkós Kólpos** *g.* Greece 55 K3/K4 40 00N 22 50E

The Rower *tn.* RoI 21 E2 52 27N 6 57W
Thessaloníki Greece 55 K4 40 38N 22 58E
Thetford England 17 D3 52 25N 0 45E
Thiers France 50 F3 45 51N 3 33E
Thimphu Bhutan 67 F5 27 32N 89 43E
Thionville France 50 H5 49 22N 6 11E
Thíra *i.* Greece 55 L2 36 00N 25 00E
Thirsk England 13 E3 54 14N 1 20W
Thíva Greece 55 K3 38 19N 23 19E
Thomastown RoI 21 D2 52 31N 7 08W
Thompson Canada 89 R5 55 45N 97 54W
Thornaby-on-Tees
England 13 E3 54 34N 1 18W
Thornbury England 15 D1 51 37N 2 32W
Thorne England 13 F2 53 37N 0 58W
Thorney England 16 C3 52 37N 0 07W
Thornhill Scotland 11 F2 55 15N 3 46W
Thornton England 12 D2 53 53N 3 00W
Thorpe-le-Soken
England 17 E2 51 52N 1 10E
Thouars France 50 D4 46 59N 0 13W
Thrapston England 15 F2 52 24N 0 32W
Thuin Belgium 48 C3 50 21N 4 18E
Thun Switz. 52 C3 46 46N 7 38E
Thurcroft England 13 E2 53 24N 1 16W
Thurles RoI 21 D2 52 41N 7 49W
Thurrock *u.a.* England 17 D2 51 30N 0 25E
Thursby England 12 C3 54 51N 3 03W
Thurso Scotland 9 F3 58 35N 3 32W
Thurso *r.* Scotland 9 F3 58 20N 3 40W
Thurso Bay Scotland 9 F3 58 37N 3 34W
Tianjin China 69 L5 39 08N 117 12E
Tiberias, Lake Israel 64 N11 32 45N 35 30E
Tibesti *mts.* Chad 81 J6 21 00N 17 00E
Tickhill England 13 E2 53 26N 1 06W
Tighnabruaich Scotland 10 D2 55 56N 5 14W
Tigris *r.* Turkey/Iraq 65 E5 32 00N 46 00E
Tijuana Mexico 94 A6 32 29N 117 10W
Tilburg Neths 48 D4 51 34N 5 05E
Tilbury England 17 D2 51 28N 0 23E
Till *r.* England 11 G2 55 38N 2 08W
Till *r.* England 13 F2 53 20N 0 40W
Tillicoultry Scotland 11 F3 56 09N 2 45W
Tilt *r.* Scotland 9 F1 56 50N 3 45W
Timişoara Romania 53 K2 45 45N 21 15E
Timor *i.* Indonesia 71 G1/H2 9 00S 125 00E
Tinahely RoI 21 E2 52 48N 6 28W
Tingwall Scotland 7 A2 59 04N 3 03W
Tintagel England 18 C2 50 40N 4 45W
Tintagel Head England 18 C2 50 41N 4 46W
Tipperary RoI 21 C2 52 29N 8 10W
Tipperary *co.* RoI 21 D2 52 30N 8 00W
Tiptree England 17 D2 51 49N 0 45E
Tiranë Albania 55 H4 41 20N 19 49E
Tiree *i.* Scotland 8 C1 56 30N 6 55W
Titu Romania 53 M2 44 40N 25 32E
Tiverton England 18 D2 50 55N 3 29W
Tivoli Italy 54 E4 41 58N 12 48E
Tlemcen Algeria 80 D8 34 53N 1 21W
Toamasina Madagascar 82 J4 18 10S 49 23E
Tobercurry RoI 20 C4 54 03N 8 43W
Tobermore NI 10 C1 54 48N 6 42W
Tobermory Scotland 8 C1 56 37N 6 05W
Todmorden England 12 D2 53 43N 2 05W
Toe Head RoI 21 B1 51 30N 9 12W
Toe Head Scotland 8 B2 57 50N 7 07W
TOGO 80 E3
Tokelau Islands
Pacific Ocean 100 J7 9 00S 168 00W
Tokyo Japan 70 C2 35 40N 139 45E
Toledo Spain 51 C3 39 52N 4 02W
Toledo USA 91 K5 41 40N 83 35W
Toliara Madagascar 82 H3 23 20N 43 41E
Tolmezzo Italy 54 E7 46 24N 13 01E
Tolosa Spain 51 D5 43 09N 2 04W
Tolsta Head Scotland 8 C3 58 20N 6 10W
Toluca Mexico 94 E3 19 20N 99 40W
Tomar Portugal 51 A3 39 36N 8 25W
Tomatin Scotland 9 F2 57 20N 3 59W
Tombouctou Mali 80 D5 16 49N 2 59W
Tomelloso Spain 51 D3 39 09N 3 01W
Tomintoul Scotland 9 F2 57 14N 3 22W
Tomsk Russia 63 L8 56 30N 85 05E
TONGA 100 J5
Tongue Scotland 9 E3 58 28N 4 25W
Tønsberg Norway 49 D2 59 16N 10 25E
Tonyrefail Wales 14 C1 51 36N 3 25W
Toomyvara RoI 21 C2 52 51N 8 02W
Topsham England 18 D2 50 42N 3 27W
Tor Bay England 18 D2 50 27N 3 30W
Torbay England 18 D2 50 27N 3 30W
Tordesillas Spain 51 C4 41 30N 5 00W
Tore Scotland 9 E2 57 27N 4 21W
Torfaen *u.a.* Wales 14 C1 51 38N 3 02W
Tornio Finland 49 F4 65 50N 24 10E
Toronto Canada 89 V2 43 42N 79 46W
Torpoint England 18 C2 50 22N 4 11W
Torre del Greco Italy 54 F4 40 46N 14 22E
Torrelavega Spain 51 C5 43 21N 4 03W
Torremolinos Spain 51 C2 36 38N 4 30W
Torrente Spain 51 E3 39 27N 0 28W
Torres Vedras Portugal 51 A3 39 05N 9 15W
Torrevieja Spain 51 E2 37 59N 0 40W
Torridge *r.* England 18 C2 50 56N 4 22W
Torridon Scotland 8 D2 57 33N 5 31W
Tortosa Spain 51 F4 40 49N 0 31E

Toruń Poland 53 J6 53 01N 18 35E
Tory Island RoI 20 C5 55 16N 8 14W
Tory Sound RoI 20 C5 55 15N 8 05W
Totland England 16 B1 50 40N 1 32W
Totnes England 18 D2 50 25N 3 41W
Tottington England 12 D2 53 37N 2 20W
Totton England 16 B1 50 56N 1 29W
Touggourt Algeria 80 F8 33 08N 6 04E
Toul France 50 G5 48 41N 5 54E
Toulon France 50 G2 43 07N 5 55E
Toulouse France 50 E2 43 33N 1 24E
Tourcoing France 50 F6 50 44N 310E
Tours France 50 E4 47 23N 0 42E
Toward Scotland 10 E2 55 54N 4 58W
Towcester England 15 F2 52 08N 1 00W
Townsville Australia 73 J7 19 13S 146 48E
Toyama Japan 70 C2 36 42N 137 14E
Toyota Japan 70 C2 35 05N 137 09E
Tralee RoI 21 B2 52 16N 9 42W
Tralee Bay RoI 21 B2 52 20N 9 55W
Tramore RoI 21 D2 52 10N 7 10W
Tramore Bay RoI 21 D2 52 08N 7 08W
Tranent Scotland 11 G2 55 57N 2 57W
Trani Italy 55 G4 41 17N 16 25E
Trápani Italy 54 E3 38 02N 12 32E
Traun Austria 52 G4 48 14N 14 15E
Trawsfynydd Wales 14 C2 52 54N 3 55W
Tredegar Wales 14 C1 51 47N 3 16W
Tregaron Wales 14 C2 52 13N 3 55W
Tregony England 18 C2 50 16N 4 55W
Tremadog Bay Wales 14 B2 52 54N 4 15W
Trent *r.* England 13 F2 53 30N 0 40W
Trento Italy 54 D7 46 04N 11 08E
Tresco *i.* England 18 A1 49 57N 6 20W
Treviso Italy 54 E6 45 40N 12 15E
Trevose Head England 18 B2 50 33N 5 01W
Trier Germany 52 C4 49 45N 6 39E
Trieste Italy 54 E6 45 39N 13 47E
Tríkala Greece 55 J3 39 33N 21 46E
Trim RoI 20 E3 53 34N 6 47W
Tring England 16 C2 51 48N 0 40W
TRINIDAD AND TOBAGO 95 M2
Trípoli Greece 55 K2 37 31N 22 22E
Tripoli Lebanon 64 C5 34 27N 35 50E
Tripoli Libya 81 G8 32 54N 13 11E
Trois-Rivières *tn.*
Canada 89 W3 46 21N 72 34W
Trollhättan Sweden 49 D2 58 17N 12 20E
Tromsø Norway 49 E4 69 42N 19 00E
Trondheim Norway 49 D3 63 36N 10 23E
Troon Scotland 11 E2 55 32N 4 40W
Trotternish *p.* Scotland 8 C2 57 00N 6 00W
Troup Head Scotland 9 G2 57 41N 2 19W
Trouville France 50 E5 49 22N 0 05E
Trowbridge England 16 A2 51 20N 2 13W
Troyes France 50 G5 48 18N 4 05E
Trujillo Peru 96 B12 8 06S 79 00W
Trujillo Spain 51 C3 39 28N 5 53W
Truro England 18 B2 50 16N 5 03W
Tuam RoI 20 C3 53 31N 8 50W
Tübingen Germany 52 D4 48 32N 9 04E
Tubruq Libya 81 K8 32 05N 23 59E
Tucson USA 90 D3 32 15N 110 57W
Tudela Spain 51 E4 42 04N 1 37W
Tudweiliog Wales 14 B2 52 55N 4 39W
Tula Russia 62 F7 54 11N 37 38E
Tulcea Romania 53 P2 45 10N 28 50E
Tulla RoI 21 C2 52 52N 8 45W
Tullamore RoI 20 D3 53 16N 7 30W
Tulle France 50 E3 45 16N 1 46E
Tullow RoI 21 E2 52 48N 6 44W
Tulsa USA 91 G4 36 07N 95 58W
Tulsk RoI 20 C3 53 47N 8 15W
Tummel *r.* Scotland 9 F1 56 40N 4 00W
Tummel Bridge
Scotland 9 F1 56 43N 3 58W
Tunis Tunisia 80 G9 36 50N 10 13E
TUNISIA 80 F8/G8
Tŭrgovishte Bulgaria 55 M5 43 14N 26 37E
Turgutlu Turkey 55 M3 38 30N 27 43E
Turin Italy 54 B6 45 04N 7 40E
Turkana, Lake
Ethiopia/Kenya 83 D6/E6 4 00N 36 00E
TURKEY 57 C2/E2
TURKMENISTAN 57 H4/J4
Turks and Caicos Islands
W. Indies 95 K4 21 30N 72 00W
Turku Finland 49 F3 60 27N 22 15E
Turnberry Scotland 10 E2 55 20N 4 50W
Turriff Scotland 9 G2 57 32N 2 28W
Tutbury England 15 E2 52 52N 1 40W
Tuttlingen Germany 52 D4 47 59N 8 49E
TUVALU 100 H7
Túy Spain 51 A5 42 03N 8 39W
Tuz Gölü *l.* Turkey 57 D2 38 40N 33 35E
Tuzla
Bosnia-Herzegovina 55 H6 44 33N 18 41E
Tver' Russia 62 F8 56 49N 35 57E
Tweed *r.*
Scotland/England 11 G2 55 45N 2 10W
Tweedmouth England 11 G3 55 47N 2 00W
Twin Falls *tn.* USA 90 D5 42 34N 114 30W
Twyford England 16 C2 51 29N 0 53W
Twyford England 16 B2 51 01N 1 19W
Tyldesley England 12 D2 53 32N 2 29W
Tylers Green England 16 C2 51 37N 0 42W

Tyndrum Scotland 9 E1 57 27N 4 44W
Tyne *r.* Scotland 11 G2 55 58N 2 43W
Tyne and Wear *co.*
England 13 E3 55 00N 1 10W
Tynemouth *tn.* England 13 E4 55 01N 1 24W
Tynset Norway 49 D3 62 17N 10 47E
Tywi *r.* Wales 14 B1 51 50N 4 25W
Tywyn Wales 14 B2 52 35N 4 05W

**U**

Uaupés Brazil 96 D13 0 07S 67 05W
Uberaba Brazil 96 H10 19 47S 47 57W
Uberlândia Brazil 96 H10 18 57S 48 17W
Uckfield England 17 D1 50 58N 0 05E
Uddevalla Sweden 49 D2 58 20N 11 56E
Udine Italy 54 E7 46 04N 13 14E
Uelzen Germany 52 E6 52 58N 10 34E
Ufa Russia 62 H7 54 45N 55 58E
Uffculme England 19 D2 50 54N 3 21W
UGANDA 83 B6/C6
Ugie *r.* 9 H2 57 31N 1 55W
Uig Scotland 8 C2 57 35N 6 22W
Uithuizen Neths 48 E6 53 24N 6 41E
Ujung Pandang
Indonesia 71 F2 5 09S 119 28E
UKRAINE 56 C4/D4
Ulan Bator Mongolia 69 J7 47 54N 106 52E
Ulan-Ude Russia 63 N7 51 55N 107 40E
Ullapool Scotland 8 D2 57 54N 5 10W
Ullswater *l.* England 12 D3 54 35N 2 55W
Ulm Germany 52 D4 48 24N 10 00E
Ulva *i.* Scotland 8 C1 56 30N 6 15W
Ulverston England 12 C3 54 12N 3 06W
Umeå Sweden 49 F3 63 50N 20 15E
Unapool Scotland 9 D3 58 14N 5 01W
Ungava Bay Canada 89 X5 59 00N 67 30W
Unimak Island USA 88 C4 55 00N 164 00W
UNITED ARAB EMIRATES 65 F3
UNITED KINGDOM 6
UNITED STATES OF AMERICA 90/91
Unst *i.* Scotland 7 D3 60 45N 0 55W
Upavon England 16 B2 51 18N 1 49W
Upper Lough Erne NI 10 B1 54 15N 7 30W
Upper Tean England 15 E2 52 57N 2 00W
Uppingham England 15 F2 52 35N 0 43W
Uppsala Sweden 49 E2 59 55N 17 38E
Upton West Yorkshire
England 13 E2 53 37N 1 17W
Upton Dorset England 19 E2 50 42N 2 02W
Ural *r.* Asia 62 H6 48 00N 52 00E
Ural Mountains Russia 62 H7/H9 57 00N 60 00E
Ure *r.* England 13 E3 54 20N 1 55W
Urlingford RoI 21 D2 52 43N 7 35W
Urmston England 12 D2 53 27N 2 21W
Urr *r.* Scotland 11 F2 55 05N 3 57W
URUGUAY 97 F6
Uruguay *r.*
Uruguay/Argentina 97 F6 32 00S 57 40W
Ürümqi China 68 E6 43 43N 87 38E
Ushaw Moor *tn.*
England 13 E3 54 47N 1 39W
Usk *r.* Wales 14 C1 51 55N 3 40W
Ussel France 50 F3 45 32N 2 18E
Ust'-Ilimsk Russia 63 N8 58 03N 102 39E
Utah *state* USA 90 D4 39 00N 112 00W
Utah Lake USA 90 D5 40 10N 111 50W
Utrecht Neths 48 D5 52 05N 5 07E
Utrera Spain 51 C2 37 10N 5 47W
Uttoxeter England 15 E2 52 54N 1 52W
UZBEKISTAN 62 J5

**V**

Vác Hungary 53 J3 47 46N 19 08E
Vadodara India 66 C4 22 19N 73 14E
Valdepeñas Spain 51 D3 38 46N 3 24W
Valdivia Chile 97 C6 39 46S 73 15W
Val-d'Or Canada 89 V3 48 07N 77 47W
Valence France 50 G3 44 56N 4 54E
Valencia Spain 51 E3 39 29N 0 24W
Valencia Venezuela 96 D16 10 14N 67 59W
Valencia, Gulf of Spain 51 F3 39 30N 0 20E
Valencia Island *i.* RoI 21 A1 51 52N 10 20W
Valenciennes France 50 F6 50 22N 3 32E
Vale of Glamorgan, The
*u.a.* Wales 14 C1 51 30N 3 30W
Valladolid Mexico 94 G4 20 40N 88 11W
Valladolid Spain 51 C4 41 39N 4 45W
Valletta Malta 54 F1 35 54N 14 32E
Valley *tn.* Wales 14 B3 53 17N 4 34W
Valparaíso Chile 97 C7 33 05S 71 40W
Valverde del Camino
Spain 51 B2 37 35N 6 45W
Vanadzor Armenia 57 F3 40 49N 44 30E
Vancouver Canada 88 L3 49 13N 123 06W
Vänern *l.* Sweden 49 D2 59 00N 13 30E
Van Gölü *l.* Turkey 57 F2 38 33N 42 46E
Vanna *i.* Norway 49 E5 70 05N 19 50E
Vännäs Sweden 49 E3 63 56N 19 50E
Vannes France 50 C4 47 40N 2 44W
Vantaa Finland 49 F3 60 20N 24 50E
VANUATU 73 N7
Varanasi India 66 E5 25 20N 83 00E

Vardø Norway 49 K6 70 22N 31 06E
Varese Italy 54 C6 45 49N 8 49E
Varkhaus Finland 49 G3 62 20N 27 57E
Varna Bulgaria 55 M5 43 12N 27 57E
Värnamo Sweden 49 D2 57 11N 14 03E
Vaslui Romania 53 N3 46 37N 27 46E
Västerås Sweden 49 E2 59 36N 16 32E
Västervik Sweden 49 E2 57 45N 16 40E
Vasto Italy 54 F5 42 07N 14 43E
Vatersay *i.* Scotland 8 B1 56 55N 7 32W
Vatra Dornei Romania 53 M3 47 20N 25 21E
Vättern *l.* Sweden 49 D2 58 20N 14 20E
Växjö Sweden 49 D2 56 52N 14 50E
Vejle Denmark 49 C2 55 43N 9 33E
Velebit *mts.* Croatia 54 F6 44 00N 15 00E
Vélez-Málaga Spain 51 C2 36 47N 4 06W
Vélez Rubio Spain 51 D2 37 39N 2 04W
Vementry *i.* Scotland 7 C3 60 20N 1 28W
Vendôme France 50 E4 47 48N 1 04E
VENEZUELA 96 D15
Venice Italy 54 E6 45 26N 12 20E
Venissieux France 50 G3 45 42N 4 46E
Venlo Netherland 48 E4 51 22N 6 10E
Ventnor England 16 B1 50 36N 1 11W
Ventura USA 90 C3 34 16N 119 18W
Vera Spain 51 E2 37 15N 1 51W
Veracruz Mexico 94 E3 19 11N 96 10W
Vercelli Italy 54 CC6 45 19N 8 26E
Verdun-sur-Meuse
France 50 G5 49 10N 5 24E
Verin Spain 51 B4 41 55N 7 26W
Vermont *state* USA 91 M5 44 00N 73 00W
Verona Italy 54 D6 45 26N 11 00E
Versailles France 50 F5 48 48N 2 07E
Verviers Belgium 48 D3 50 36N 5 52E
Verwood England 19 E2 50 53N 1 52W
Vesoul France 50 H4 47 38N 6 09E
Vesuvio *vol.* Italy 54 F4 40 49N 14 26E
Veszprém Hungary 53 H3 47 06N 17 54E
Vetlanda Sweden 49 E2 57 26N 15 05E
Vevey Switz. 52 C3 46 28N 6 51E
Viano do Castelo
Portugal 51 A4 41 41N 8 50W
Viareggio Italy 54 D5 43 52N 10 15E
Viborg Denmark 49 C2 56 28N 9 25E
Vibo Valentia Italy 54 G3 38 40N 16 06E
Vicenza Italy 54 D6 45 33N 11 32E
Vichy France 50 F4 46 07N 3 25E
Victoria Canada 88 L3 48 26N 123 20W
Victoria, Lake
East Africa 83 C4 2 00S 33 00E
Vienna Austria 53 H4 48 13N 16 22E
Vienne France 50 G3 45 32N 4 54E
Vientiane Laos 71 C7 17 59N 102 38E
Vierzon France 50 F4 47 14N 2 03E
VIETNAM 71 D7
Vigo Spain 51 A5 42 15N 8 44W
Vila Nova de Gaia
Portugal 51 A4 41 08N 8 37W
Vila Real Portugal 51 B4 41 17N 7 45W
Vilhelmina Sweden 49 E4 64 38N 16 40E
Villach Austria 52 F3 46 37N 13 51E
Villalba Spain 51 B5 43 17N 7 41W
Villarrobledo Spain 51 D3 39 16N 2 36W
Villefranche-sur-Saône
France 50 G3 46 00N 4 43E
Villena Spain 51 E3 38 39N 0 52W
Villeneuve-sur-Lot
France 50 E3 44 25N 0 43E
Villeurbanne France 50 G3 45 46N 4 54E
Vilnius Lithuania 49 G1 54 40N 25 19E
Viña del Mar Chile 97 C7 33 02S 71 35W
Vinaroz Spain 51 F4 40 29N 0 28E
Vipiteno Italy 54 D7 46 54N 11 27E
Virginia RoI 20 D3 53 49N 7 04W
Virginia USA 91 H6 47 30N 92 28W
Virginia *state* USA 91 L4 38 00N 77 00W
Virginia Water *tn.*
England 16 C2 51 24N 0 34W
Virgin Islands W. Indies 95 M3 18 00N 64 30W
Visby Sweden 49 E2 57 32N 18 15E
Viseu Portugal 51 B4 40 40N 7 55W
Vishakhapatnam India 66 E3 17 42N 83 24E
Viterbo Italy 54 E5 42 26N 12 07E
Vitória Brazil 96 J9 20 20S 40 18W
Vitoria Gasteiz Spain 51 D5 42 51N 2 40W
Vitry-le-François
France 50 G5 48 44N 4 36E
Vitsyebsk Belarus 56 D6 55 10N 30 14E
Vittoria Italy 54 F2 36 58N 14 32E
Vladimir Russia 62 G8 56 08N 40 25E
Vladivostok Russia 63 R5 43 09N 131 53E
Vlissingen Neths 48 B4 51 27N 3 35E
Voitsberg Austria 52 G3 47 04N 15 09E
Volga *r.* Russia 62 G6 50 00N 45 00E
Volgograd Russia 62 G6 48 45N 44 30E
Volta, Lake Ghana 80 D3 7 30N 0 30W
Vóreioi Sporádes *is.*
Greece 55 K3/L3 39 00N 24 00E
Vosges *mts.* France 50 H5 48 10N 6 50E
Voss Norway 49 C3 60 38N 6 25E
Voxnan Sweden 49 E3 61 22N 15 39E
Vukovar Croatia 55 H6 45 19N 19 01E

**W**

Waal *r.* Neths 48 D4 51 50N 5 07E

**Waalwijk** Neths 48 D4 51 42N 5 04E
**Waddesdon** England 16 C2 51 51N 0 56W
**Waddington** England 13 F2 53 10N 0 32W
**Wadebridge** England 18 C2 50 32N 4 50W
**Wadhurst** England 17 D2 51 04N 0 21E
**Wadi Halfa** Sudan 81 M6 21 55N 31 20E
**Wainfleet All Saints** England 13 G2 53 06N 0 15E
**Wakefield** England 13 E2 53 42N 1 29W
**Wales** UK 14 C2 52 40N 3 30W
**Wallasey** England 12 C2 53 26N 3 03W
**Wallingford** England 16 B2 51 37N 1 08W
**Wallis and Futuna** *is.* Pacific Ocean 100 J6 13 16S 176 15W
**Walls** Scotland 7 C3 60 14N 1 34W
**Wallsend** England 13 E3 55 00N 1 31W
**Walney, Isle of** England 12 C3 54 05N 3 10W
**Walsall** England 15 E2 52 35N 1 58W
**Waltham Abbey** England 16 D2 51 41N 0 00
**Waltham Forest** England 16 C2 51 36N 0 00
**Waltham on the Wolds** England 15 F2 52 49N 0 49W
**Walton-le-Dale** England 12 D2 53 45N 2 41W
**Walton-on-the-Naze** England 17 E2 51 51N 1 16E
**Walvis Bay** *tn.* Namibia 82 B3 22 59S 14 31E
**Wanganui** NZ 73 P3 39 56S 175 03E
**Wanlockhead** Scotland 11 F2 55 24N 3 47W
**Wantage** England 16 B2 51 36N 1 25W
**Warboys** England 16 C3 52 24N 0 06W
**Ward Hill** Scotland 7 A1 58 54N 3 20W
**Ware** England 16 C2 51 49N 0 02W
**Wareham** England 19 E2 50 41N 2 07W
**Warley** England 15 E2 52 30N 1 59W
**Warlingham** England 16 C2 51 19N 0 04W
**Warminster** England 16 A2 51 13N 2 12W
**Warrenpoint** *tn.* NI 10 C1 54 06N 6 15W
**Warrington** England 12 D2 53 24N 2 37W
**Warsaw** Poland 53 K6 52 15N 21 00E
**Warsop** England 13 E2 53 13N 1 10W
**Warton** England 12 D2 53 46N 2 54W
**Warwick** England 15 E2 52 17N 1 34W
**Warwickshire** *co.* England 15 E2 52 15N 1 40W
**Washingborough** England 13 F2 53 14N 0 28W
**Washington** England 13 E3 54 54N 1 31W
**Washington** *state* USA 90 B6/C6 47 00N 120 00W
**Washington D.C.** USA 91 L4 38 55N 77 00W
**Wash, The** *b.* England 17 D3 52 55N 0 10E
**Watchet** England 19 D3 51 12N 3 20W
**Waterbeach** England 17 D3 52 16N 0 11E
**Waterford** RoI 21 D2 52 15N 7 06W
**Waterford** *co.* RoI 21 D2 52 10N 7 30W
**Waterford Harbour** RoI 21 E2 52 10N 7 00W
**Watergrasshill** RoI 21 C2 52 00N 8 20W
**Waterlooville** England 16 B1 50 53N 1 02W
**Waterville** RoI 21 A1 51 50N 10 10W
**Watford** England 16 C2 51 39N 0 24W
**Wath upon Dearne** England 13 E2 53 31N 1 21W
**Watlington** England 16 C2 51 39N 1 00W
**Watton** England 17 D3 52 34N 0 50E
**Waveney** *r.* England 17 E3 52 30N 1 30E
**Wavre** Belgium 48 C3 50 43N 4 37E
**Wear** *r.* England 12 D3 54 45N 2 05W
**Weaverham** England 12 D2 53 16N 2 35W
**Wedmore** England 19 E3 51 14N 2 49W
**Weedon Bec** England 15 E2 52 14N 1 05W
**Weiden** Germany 52 F4 49 40N 12 10E
**Weimar** Germany 52 E5 50 59N 11 20E
**Weissenfels** Germany 52 E5 51 12N 11 58E
**Welland** *r.* England 16 C3 52 50N 0 10
**Welland** *r.* England 15 F2 52 30N 0 50W
**Wellingborough** England 15 F2 52 19N 0 42W
**Wellington** Telford and Wrekin England 15 D2 52 43N 2 31W
**Wellington** Somerset England 19 D2 50 59N 3 15W
**Wellington** NZ 73 P2 41 17S 174 46E
**Wells** England 19 E3 51 13N 2 39W
**Wells-next-the-Sea** England 17 D3 52 58N 0 51E
**Wels** Austria 52 F4 48 10N 14 02E
**Welshpool** Wales 14 C2 52 40N 3 09W
**Welton** England 13 F2 54 47N 3 00W
**Welwyn Garden City** England 16 C2 51 48N 0 13W
**Wem** England 14 D2 52 51N 2 44W
**Wemyss Bay** Scotland 10 E2 55 55N 4 53W
**Wendover** England 16 C2 51 46N 0 46W
**Wensum** *r.* England 17 D3 52 45N 1 10E
**Wenzhou** China 69 M3 28 02N 120 40E
**West Berkshire** *u.a.* England 16 B2 51 28N 1 00W
**West Bridgford** England 15 E2 52 56N 1 08W
**West Bromwich** England 15 E2 52 31N 1 59W
**Westbury** England 16 A2 51 16N 2 11W
**West Dunbartonshire** *u.a.* Scotland 11 E3 56 02N 4 39W
**Westerham** England 16 D2 51 16N 0 05E
**Western Ghats** *mts.* India 66 C3/D2 15 30N 74 00E
**Western Isles** *u.a.* Scotland 8 B2/B3 57 45N 7 30W
**WESTERN SAHARA** 80 B6/B7
**Western Sayan** *mts.* Russia 63 M7 52 30N 92 30E
**West Felton** England 14 D2 52 49N 2 58W
**West Haddon** England 15 E2 52 20N 1 04W
**Westhill** Scotland 9 G2 57 11N 2 16W
**West Horsley** England 16 C2 51 16N 0 26W
**West Indies** *is.* Caribbean Sea 95 K4/L4 22 00N 69 00W
**West Kilbride** Scotland 10 E2 55 42N 4 51W
**West Kirby** England 12 C2 53 22N 3 10W
**West Loch Tarbert** Western Isles Scotland 8 B2/C2 57 55N 6 52W
**West Loch Tarbert** Argyll and Bute Scotland 10 D2 55 48N 5 31W
**West Lothian** *u.a.* Scotland 11 F2 55 50N 3 36W
**Westmeath** *co.* RoI 20 D3 53 30N 7 30W
**West Mersea** England 17 D2 51 47N 0 55E
**West Midlands** *co.* England 15 D2 52 35N 2 00W
**Weston-super-Mare** England 19 E3 51 21N 2 59W
**Westport** RoI 20 B3 53 48N 9 32W
**Westport Bay** RoI 20 B3 53 49N 9 37W
**Westray** *i.* Scotland 7 A2 59 18N 3 00W
**West Sussex** *co.* England 16 C1 51 00N 0 25W
**West Thurrock** England 17 D2 51 28N 0 20E
**West Virginia** *state* USA 91 K4 39 00N 81 00W
**Westward Ho!** England 18 C3 51 02N 4 15W
**West Wittering** England 16 C1 50 47N 0 54W
**West Yorkshire** *co.* England 13 E2 53 50N 1 30W
**Wetherby** England 13 E2 53 56N 1 23W
**Wetzlar** Germany 52 D5 50 33N 8 30E
**Wexford** *co.* RoI 21 E2 52 25N 6 35W
**Wexford** RoI 21 E2 52 20N 6 27W
**Wexford Bay** RoI 21 E2 52 25N 6 10W
**Wexford Harbour** RoI 21 E2 52 20N 6 25W
**Wey** *r.* England 16 C2 51 18N 0 30W
**Weybridge** England 16 C2 51 22N 0 28W
**Weyburn** Canada 89 Q3 49 39N 103 51W
**Weymouth** England 19 E2 50 37N 2 25W
**Whaley Bridge** England 12 E2 53 20N 1 59W
**Whalley** England 12 D2 53 50N 2 24W
**Wharfe** *r.* England 12 D3 54 10N 2 05W
**Wharfedale** *v.* England 12 D3 54 05N 2 00W
**Wheatley** England 16 B2 53 22N 0 52W
**Whernside** *sum.* England 12 D3 54 14N 2 23W
**Whickham** England 13 E3 54 57N 1 40W
**Whitburn** England 13 E3 54 57N 1 21W
**Whitburn** Scotland 11 F2 55 52N 3 42W
**Whitby** England 13 F3 54 29N 0 37W
**Whitchurch** Shropshire England 15 D2 52 58N 2 41W
**Whitchurch** Buckinghamshire England 16 C2 51 53N 0 51W
**Whitchurch** Hampshire England 16 B2 51 14N 1 20W
**Whitefield** England 12 D2 53 34N 2 18W
**Whitegate** RoI 21 C1 51 50N 8 14W
**Whitehaven** England 12 C3 54 33N 3 35W
**Whitehead** NI 10 D1 54 45N 5 43W
**Whitehills** Scotland 9 G2 57 40N 2 35W
**White Horse, Vale of** England 16 B2 51 35N 1 30W
**Whiten Head** Scotland 9 E3 58 34N 4 32W
**Whiteparish** England 16 B2 51 01N 1 39W
**Whitfield** England 17 E2 51 09N 1 17E
**Whithorn** Scotland 11 E1 54 44N 4 25W
**Whitland** Wales 14 B1 51 50N 4 37W
**Whitley Bay** *tn.* England 13 E4 55 03N 1 25W
**Whitney, Mount** USA 90 C4 36 35N 118 17W
**Whitsand Bay** England 18 C2 50 20N 4 25W
**Whitstable** England 17 E2 51 22N 1 02E
**Whittington** England 14 C2 52 52N 3 00W
**Whittlesey** England 16 C3 52 34N 0 08W
**Whitton** England 13 F2 53 43N 0 38W
**Whitworth** England 12 D2 53 40N 2 10W
**Wichita** USA 91 G4 37 43N 97 20W
**Wick** Scotland 9 F3 58 26N 3 06W
**Wick** *r.* Scotland 9 F3 58 26N 3 10W
**Wickford** England 17 D2 51 38N 0 31E
**Wickham** England 16 B1 50 54N 1 10W
**Wickham Market** England 17 E3 52 09N 1 22E
**Wicklow** RoI 21 E2 52 59N 6 03W
**Wicklow** *co.* RoI 21 E2 52 55N 6 25W
**Wicklow Mountains** RoI 21 E2/E3 53 00N 6 20W
**Widnes** England 12 D2 53 22N 2 44W
**Wiener Neustadt** Austria 53 H3 47 49N 16 15E
**Wierden** Neths 48 E5 52 21N 6 35E
**Wiesbaden** Germany 52 D5 50 05N 8 15E
**Wigan** England 12 D2 53 33N 2 38W
**Wigston** England 15 E2 52 36N 1 05W
**Wigton** England 12 C3 54 49N 3 09W
**Wigtown** Scotland 11 E1 54 52N 4 26W
**Wigtown Bay** Scotland 11 E1 54 45N 4 20W
**Wil** Switzerland 52 D3 47 28N 9 03E
**Wilberfoss** England 13 F2 53 57N 0 53W
**Wilhelmshaven** Germany 52 D6 53 32N 8 07E
**Willington** England 13 E3 54 43N 1 41W
**Williton** England 19 D3 51 10N 3 20W
**Wilmslow** England 12 D2 53 20N 2 15W
**Wilton** England 16 B2 51 05N 1 52W
**Wiltshire** *co.* England 16 A2/B2 51 30N 2 00W
**Wimborne Minster** England 19 F2 50 48N 1 59W
**Wincanton** England 19 E3 51 04N 2 25W
**Winchelsea** England 17 D1 50 55N 0 42E
**Winchester** England 16 B2 51 04N 1 19W
**Windermere** *tn* England 12 D3 54 23N 2 54W
**Windermere** *l.* England 12 D3 54 20N 3 10W
**Windhoek** Namibia 82 C3 22 34S 17 06E
**Windsor** Canada 89 U2 42 18N 83 00W
**Windsor** England 16 C2 51 29N 0 38W
**Windsor and Maidenhead** *u.a.* England 16 C1 50 25N 0 50W
**Windward Islands** Lesser Antilles 95 M2 12 30N 62 00W
**Wingate** England 13 E3 54 55N 1 23W
**Wingerworth** England 13 E2 53 13N 1 28W
**Winkleigh** England 18 D2 50 51N 3 56W
**Winnipeg** Canada 89 R3 49 53N 97 10W
**Winnipeg, Lake** Canada 89 R4 52 30N 97 30W
**Winnipegosis, Lake** Canada 89 Q4 52 10N 100 00W
**Winschoten** Neths 48 F6 53 07N 7 02E
**Winscombe** England 19 E3 51 28N 2 52W
**Winsford** England 12 D2 53 11N 2 31W
**Winslow** England 16 C2 51 57N 0 54W
**Winterbourne** England 15 D1 51 30N 2 31W
**Winterthur** Switz. 52 D3 47 30N 8 45E
**Winterton-on-Sea** England 17 E3 52 43N 1 42E
**Wirksworth** England 13 E2 53 05N 1 34W
**Wisbech** England 17 D3 52 40N 0 10E
**Wisconsin** *state* USA 91 H6/J6 45 00N 90 00W
**Wishaw** Scotland 11 F2 55 47N 3 56W
**Wisła** *r.* Poland 53 J6 53 00N 19 00E
**Wismar** Germany 52 E6 53 54N 11 28E
**Wissembourg** France 50 H5 49 02N 7 57E
**Witham** *r.* England 13 F2 53 05N 0 40W
**Witham** England 17 D2 51 48N 0 38E
**Witheridge** England 18 D2 50 55N 3 42W
**Withernsea** England 13 G2 53 44N 0 02E
**Witley** England 16 C2 51 08N 0 39W
**Witney** England 16 B2 51 48N 1 29W
**Wittenberg** Germany 52 F5 51 53N 12 39E
**Wittenberge** Germany 52 E6 52 59N 11 45E
**Wittlich** Germany 52 C4 49 59N 6 54E
**Wiveliscombe** England 19 D3 51 03N 3 19W
**Wivenhoe** England 17 D2 51 52N 0 58E
**Woburn Sands** *tn.* England 16 C3 52 01N 0 39W
**Woking** England 16 C2 51 20N 0 34W
**Wokingham** England 16 C2 51 25N 0 51W
**Wokingham** *u.a.* England 16 C1 50 25N 0 55W
**Wolfsberg** Austria 52 G3 46 50N 14 50E
**Wolfsburg** Germany 52 E6 52 27N 10 49E
**Wollaston** England 15 F2 52 16N 0 41W
**Wollongong** Australia 73 K4 34 25S 150 52E
**Wolsingham** England 13 E3 54 44N 1 52W
**Wolverhampton** England 15 D2 52 36N 2 08W
**Wolverton** England 16 C3 52 04N 0 50W
**Wombourne** England 15 D2 52 32N 2 11W
**Wombwell** England 13 E2 53 32N 1 24W
**Woodbridge** England 17 E3 52 06N 1 19E
**Woodford Halse** England 15 E2 52 10N 1 13W
**Woodhall Spa** England 13 F2 53 09N 0 14W
**Woodlands** *tn.* Canada 92 A4 50 12N 97 40W
**Woodstock** England 16 B2 51 52N 1 21W
**Wool** England 19 E2 50 41N 2 14W
**Woolacombe** England 18 C3 51 10N 4 13W
**Wooler** England 11 G2 55 33N 2 01W
**Wootton Bassett** England 16 B2 51 33N 1 54W
**Worcester** England 15 D2 52 11N 2 13W
**Worcester** USA 91 M5 42 17N 71 48W
**Worcestershire** *co.* England 15 D2 52 15N 2 15W
**Workington** England 12 C3 54 39N 3 33W
**Worksop** England 13 E2 53 18N 1 07W
**Worms** Germany 52 D4 49 38N 8 23E
**Worms Head** Wales 14 B1 51 34N 4 20W
**Worsbrough** England 13 E2 53 33N 1 29W
**Worthing** England 16 C1 50 48N 0 23W
**Wotton-under-Edge** England 15 D1 51 39N 2 21W
**Wrangle** England 13 G2 53 02N 0 07E
**Wrath, Cape** Scotland 9 D3/E3 58 37N 5 01W
**Wrexham** *tn* Wales 14 C3 53 03N 3 00W
**Wrexham** *u.a.* Wales 14 C3/D3 53 00N 3 00W
**Writtle** England 17 D2 51 44N 0 26E
**Wrocław** Poland 53 H5 51 05N 17 00E
**Wrotham** England 17 D2 51 19N 0 19E
**Wroughton** England 16 B2 51 31N 1 48W
**Wroxham** England 17 E3 52 42N 1 24E
**Wuhan** China 69 K4 30 35N 114 19E
**Wuppertal** Germany 52 C5 51 15N 7 10E
**Würzburg** Germany 52 D4 49 48N 9 57E
**Wuxi** China 69 M4 31 35N 120 19E
**Wye** England 17 D2 51 11N 0 56E
**Wylye** England 16 B2 51 08N 2 01W
**Wylye** *r.* England 16 A2 51 08N 2 13W
**Wymondham** England 17 E3 52 34N 1 07E
**Wyoming** *state* USA 90 E5 43 00N 108 00W

## X

**Xi'an** China 69 J4 34 16N 108 54E

## Y

**Yakutsk** Russia 63 Q9 62 10N 129 50E
**Yamoussoukro** Côte d'Ivoire 80 C3 6 50N 5 20W
**Yangon** Myanmar 71 B7 16 47N 96 10E
**Yaoundé** Cameroon 80 G2 3 51N 11 31E
**Yarmouth** England 16 B1 50 42N 1 29W
**Yaroslavl'** Russia 62 F8 57 34N 39 52E
**Yate** England 15 D1 51 32N 2 25W
**Yateley** England 16 C2 51 20N 0 51W
**Yatsushiro** Japan 70 B1 32 32N 130 35E
**Yatton** England 19 E3 51 24N 2 49W
**Yaxley** England 16 C3 52 31N 0 16W
**Yeadon** England 13 E2 53 52N 1 41W
**Yealmpton** England 18 D2 50 21N 3 59W
**Yekaterinburg** Russia 62 J8 56 52N 60 35E
**Yell** *i.* Scotland 7 C3 60 35N 1 10W
**Yellowknife** Canada 88 N6 62 30N 114 29W
**Yelverton** England 18 C2 50 30N 4 05W
**YEMEN REPUBLIC** 64/65 D2/E2
**Yeovil** England 19 E2 50 57N 2 39W
**Yerevan** Armenia 57 F3 40 10N 44 31E
**Yokohama** Japan 70 C2 35 27N 139 38E
**York** England 13 E2 53 58N 1 05W
**York** *u.a.* England 13 E2 53 56N 1 02W
**York, Cape** Australia 73 H8 10 42S 142 32E
**Yorkshire Wolds** England 13 F3 54 00N 0 45W
**York, Vale of** England 13 E3 54 10N 1 20W
**Youghal** RoI 21 D1 51 51N 7 50W
**Youghal Bay** RoI 21 D1 51 55N 7 48W
**Yoxford** England 17 E3 52 16N 1 30E
**Ystalyfera** Wales 14 C1 51 47N 3 47W
**Ystrad Aeron** Wales 14 B2 52 10N 4 10W
**Ystradgynlais** Wales 14 C1 51 47N 3 45W
**Ystwyth** *r.* Wales 14 C2 52 20N 3 55W
**Ytterhogdal** Sweden 49 D3 62 10N 14 55E
**Yucatan** *p.* Mexico 94 G3 19 00N 89 00W
**Yuci** China 69 K5 37 40N 112 44E
**Yukon** *r.* Canada/USA 88 D6 63 30N 159 40W
**Yukon Territory** *territory* Canada 88 H6/J6 64 15N 135 00W
**Yumen** China 68 G5 39 54N 97 43E
**Yverdon** Switz. 52 C3 46 47N 6 38E
**Yvetot** France 50 E5 49 37N 0 45E

## Z

**Zaanstad** Neths 48 C5 52 27N 4 49E
**Zadar** Croatia 54 F6 44 07N 15 14E
**Zafra** Spain 51 B3 38 25N 6 25W
**Żagań** Poland 52 G5 51 37N 15 20E
**Zagreb** Croatia 54 F6 45 48N 15 58E
**Zagros Mountains** Iran 65 E5/F5 32 45N 48 50E
**Zakopane** Poland 53 J4 49 17N 19 54E
**Zákynthos** *i.* Greece 55 J2 37 45N 20 50E
**Zambezi** *r.* Zambia/Zimbabwe 82 D4 16 00S 23 00E
**ZAMBIA** 82 D5/E5
**Zamora** Spain 51 C4 41 30N 5 45W
**Zanzibar** Tanzania 83 F1 6 10S 39 12E
**Zaragoza** Spain 51 E4 41 39N 0 54W
**Zaria** Nigeria 80 F4 11 01N 7 44E
**Zeebrugge** Belgium 48 B4 51 20N 3 13E
**Zell-am-See** *tn.* Austria 52 F3 47 19N 12 47E
**Zenica** Bosnia-Herzegovina 55 G6 44 11N 17 53E
**Zermatt** Switz. 52 C3 46 01N 7 45E
**Zhengzhou** China 69 K4 34 45N 113 38E
**Zhmerynka** Ukraine 53 P4 49 00N 28 02E
**Zhytomyr** Ukraine 53 P5 50 18N 28 40E
**Zibo** China 69 L5 36 51N 118 01E
**Zielona Góra** Poland 53 G5 51 57N 15 30E
**ZIMBABWE** 82 E4/F4
**Zinder** Niger 80 F4 13 46N 8 58E
**Zlín** Czech Rep. 53 H4 49 14N 17 40E
**Zug** Switz. 52 D3 47 10N 8 31E
**Zugspitze** *mt.* Austria 52 E3 47 25N 11 00E
**Zürich** Switz. 52 D3 47 23N 8 33E
**Zvolen** Slovakia 53 J4 48 34N 19 08E
**Zwettl** Austria 52 G4 48 37N 15 11E
**Zwickau** Germany 52 F5 50 43N 12 30E
**Zwolle** Neths 48 E5 52 31N 6 06E